DIVERSE TOPICS IN
THEORETICAL
AND
MATHEMATICAL PHYSICS

DIVERSE TOPICS IN

THEORETICAL

AND

MATHEMATICAL PHYSICS

Roman Jackiw

Massachusetts Institute of Technology
Cambridge, USA

World Scientific
Singapore • New Jersey • London • Hong Kong

Published by

World Scientific Publishing Co. Pte. Ltd.
5 Toh Tuck Link, Singapore 596224
USA office: 27 Warren Street, Suite 401-402, Hackensack, NJ 07601
UK office: 57 Shelton Street, Covent Garden, London WC2H 9HE

Library of Congress Cataloging-in-Publication Data
Jackiw, Roman W.
    Diverse topics in theoretical and mathematical physics/ Roman Jackiw
      p. cm.
    Includes bibliographical references.
    ISBN-13 978-981-02-1696-2 -- ISBN-10 981-02-1696-3
    ISBN-13 978-981-02-1697-9 (pbk) -- ISBN-10 981-02-1697-1 (pbk)
    1. Mathematical physics. I. Title
    QC20.J215 1995
    530.1--dc20
                                     95-16549
                                         CIP

British Library Cataloguing-in-Publication Data
A catalogue record for this book is available from the British Library.

The editor and publisher would like to thank the following publishers of the various journals and books
for their assistance and permissions to include the selected reprints found in this volume:

Springer-Verlag
Cambridge University Press
The MIT Press
Plenum Publishing Corporation
Birkhäuser Verlag AG
European Physical Society
American Institute of Physics

Gordon and Breach Science Publishers
Academic Press London
Akademiya Nauk SSR
Kyoto University
Elsevier Science Publishers B. V.

Printed in Singapore

Preface

Publishing one's collected/selected research papers always struck me as a vanity that is truly in vain. Such addition to today's book glut serves no purpose: research papers are readily available in archival journals, where they are more legible than in their book – collected reprint format. Moreover, the initial report on a research topic hardly ever tells the story properly and completely. Original papers may have historial interest, but rare is the physicist who can fill an entire book with historically significant works.

For these reasons, a few years ago, I declined a call by the Pied Piper from World Scientific to publish a selection of my research articles, with the excuse that the project was premature: I hoped that my best work still lay ahead. But in the end, I could not resist his tune. However, to remain true to my opinion, I decided to collect review lectures, rather than reports of original research — lectures delivered at meetings, schools, fests and memorials. Descriptions of such events are hard to find: libraries only haphazardly collect their proceedings, and research journals shy from their informal tone. And yet lectures are invaluable in that their text follows more closely the development of ideas and reveals the personal viewpoint of the author.

I chose for this book descriptions of my favorite work in mathematical physics, and I thank Dr. K. K. Phua for giving me the occasion to recollect this material. Also I took this opportunity to publicize once again topics that fascinate me but have not yet resonated strongly with the public: for example, classical sources in Yang-Mills theory and mean field descriptions of non-equilibrium processes. Texts were edited only lightly: blatant missteps were corrected, but there remains plenty evidence of my wondering in the dark.

While preparing the collection, I was reminded of the many pleasant settings where I offered these lectures. I thank the organizers of the various events for inviting and hosting me and giving me opportunities to describe my work.

R. Jackiw
June 1995
Cambridge, MA

Contents

Section I

ANOMALIES AND FRACTIONAL CHARGE

The subject of anomalous or quantum mechanical symmetry breaking has broadened enormously from its obscure beginnings in a description of the neutral pion's decay into two photons; but it has not deepened significantly: we still do not understand which fact about Nature is responsible for the anomalies in our mathematical description. At the same time, we can see that Nature knows about them: chiral anomalies determine low-energy pseudoscalar meson processes, scaling anomalies control high-energy behavior in particle physics and critical behavior in statistical physics, while the charges of fundamental fermions (leptons and quarks) in the "standard" particle physics model are such that gauge anomalies are avoided. Moreover, fractional charge — a surprise first seen in the physics of polyacetylene and now forming an important ingredient in descriptions of the quantum Hall effect — can also be viewed as a manifestation of the anomaly phenomenon.

Paper I.1

Anomaly phenomena were publicly discussed for the first time at the ICTP conference on renormalization theory (Trieste, August 1969). My presentation focused on anomalous commutators — an approach to the subject that re-emerged in 1984.

NON-CANONICAL BEHAVIOR IN
CANONICAL THEORIES

Lecture at the Trieste Conference on Renormalization Theory, Trieste, Italy, August 1969. CERN Report TH-1065

A. I shall summarize for you some work that has been done, in the years 1968–1969, which clarifies the nature of commutators in model field theories. The motivation for detailed investigations of commutators is the success that current algebra has had in two broad categories of application: low energy theorems and high energy sum rules. The purpose of examining field theoretical models is that the current commutators, which form the basis for current algebra, were originally abstracted from wide classes of field theories in which, it was thought, these objects are independent of detailed dynamics. Thus, in spite of the impossibility of unravelling the dynamics, *i.e.*, of solving the theory, it was hoped that certain results could be derived by clever exploitation of these model independent relations. These "true" results are of practical importance when they make experimental predictions; they are of mathematical interest for the structure of the theory when they make theoretical statements about convergence of higher order corrections, and the like.

The force of the present investigation is that many such formal results are not true in explicit calculations. By this I mean, in the first instance, that when a specific current algebraic prediction is compared to the explicitly calculated result, *the two disagree*. Examples that I mention here are the following three:

1) the Sutherland-Veltman theorem predicts that the effective coupling constant for $\pi^0 \rightarrow 2\gamma$ should vanish for zero pion mass in any theory which exhibits current algebra, PCAC, and electromagnetism minimally and gauge invariantly coupled.[1a] However, in the σ model this object does not vanish[1b];

2) the Callan-Gross sum rule predicts that in a large class of quark type models the longitudinal cross-section, for total electroproduction off protons, vanishes rapidly in a certain high energy limit.[2a] Explicit calculation of this cross-section yields a much more slowly vanishing result[2b];*

3) the Preparata-Weisberger argument that radiative corrections to beta decay are finite in a fermion triplet model where the fermions carry charges

---

*Added Note: the reader of this volume should be alerted to the fact that calculations reported here were performed before the full import of the renormalization group and of asymptotic freedom in gauge theories was appreciated.

1, 0, 0[3a] is confounded by a calculation: the radiative correction in fact diverges.[3b]

It must be emphasized that the results, which are quoted above as evidence for the conflict between formal reasoning and explicit calculation, are in no way ambiguous. They are well-defined consequences of the dynamics of the theory, and in the first two examples, they are finite numbers. Evidently, the formal properties are not maintained.

Faced with this incontrovertible conflict between the solutions of the theory and its formal properties, the trend has been to abandon these formal properties. A variety of equivalent descriptions for this anomalous and non-canonical behaviour has been presented in the literature. Thus one speaks of the impossibility of maintaining gauge invariance and PCAC for the vector, vector, axial-vector triangle graph in the conventional theory[4]; modifications of equations of motion for the axial current due to singularities in the definition of the axial current[5]; modifications of Ward identities for three-point functions[6]; non-cancellation of Schwinger terms and sea-gull terms,[7] *i.e.*, violations of Feynman's conjecture[8]; violations of the Bjorken-Johnson-Low theorem[9,10]; and finally, and this point of view I shall adopt today, modifications of commutators from their canonical value.[11]

Since the one common tool, used in obtaining the formal results, is the commutator algebra that various operators of the theory satisfy, a unified approach to these anomalies can be developed by adopting the view-point that the commutators, when calculated by the same techniques as the solutions of the theory, are anomalous, *i.e.*, disagree with their canonical value. This point of view, which however does have disadvantages — I shall mention them later — will now be explained.

B. The task therefore is to calculate equal time commutators. The first difficulty that is encountered is that in perturbation theory, at least, the equal time commutator is ambiguous, and the result one gets will depend on what rules one adopts towards the handling of ambiguous expressions. A related, further problem is that in calculating commutators of operators, which themselves are products of other operators, ambiguities and infinities arise in forming these products; I have in mind the construction of currents which are bilinear in fermion fields. There have appeared in the literature many calculations of commutators which yield different results; this variety is traceable to the various ways one can choose to handle the attendant ambiguities.

Fortunately, since we must maintain close connection with the method of solution of the theory, *i.e.*, with renormalized perturbation theory, some of the arbitrariness is removed. In the first place, the problem of defining products of operators need not be separately solved here, since renormalized perturbation already gives finite matrix elements for currents. Furthermore, we recall that in renormalized perturbation, one calculates T products of operators or more precisely, covariant T^* products. These are in general finite. They *may* be ambiguous, reflecting the ambiguities of how the renormalization procedure is carried out, *i.e.*, at which point the subtraction is made to render a particular Feynman diagram finite. However, such ambiguities are *always* polynomials in the external momenta, and the T product, which is defined as the T^* product with all polynomial terms removed, is unaffected by these ambiguities. Thus the solution of the theory provides unambiguous and finite T products, and it behooves us to define the commutator from these T products.

We therefore, choose as the *definition* of the commutator the Bjorken-Johnson-Low formula[9]

$$i \int d^3x e^{-i\mathbf{q}\cdot\mathbf{x}} \langle\alpha|[A(\mathbf{x},0), B(0)]|\beta\rangle$$

$$= \lim_{q_0 \to \infty} q_0 \int d^4x e^{iqx} \langle\alpha|T A(x) B(0)|\beta\rangle. \tag{1}$$

This definition has several advantages. These are: (a) In nonsingular situations (1) coincides with other definitions for the commutator. Indeed Johnson and Low[8] showed that the commutator defined by (1) is equivalent to

$$\langle\alpha|[A(\mathbf{x},0), B(0)]|\beta\rangle = \lim_{\eta \to 0}\{\langle\alpha|T A(\mathbf{x},\eta)B(0)|\beta\rangle - \langle\alpha|T A(\mathbf{x},-\eta)B(0)|\beta\rangle\}. \tag{2}$$

(b) When the Low equation is unsubstracted (1) is expressible as a sum over spectral functions[9]; such sums are relevant for high energy sum rules. (c) By construction, this commutator will be the commutator relevant for high energy sum rules, even in singular situations.

The program is therefore simple. To calculate the matrix element of a commutator of two operators (it is of course, assumed that these operators exist in renormalized perturbation theory) calculate, by Feynman rules, the Green's function, *i.e.*, the T^* product, drop all polynomials in the external momentum q, *i.e.*, the sea-gulls, and form the limit (1). If the limit is diverging, we interpret this as the statement that the commutator has diverging matrix elements.

C. As an example, let us calculate in quantum electrodynamics the vacuum expectation value of the $[j^0 j^i]$ equal time commutator. We begin by considering the vacuum polarization tensor (here and below j^μ is the fermion bilinear $\bar{\psi}\gamma^\mu\psi$)

$$\Pi^{\mu\nu}(q) = \int d^4 x e^{iqx} \langle 0|T^* j^\mu(x) j^\nu(0)|0\rangle . \tag{3}$$

Apart from irrelevant sea-gulls, one obtains (in lowest order perturbation theory)

$$\Pi^{0i}(q) = \frac{q_0 q^i}{(2\pi)^2} \int_0^1 dx x(1-x) \log \left| \frac{m^2 - x(1-x)q^2}{x(1-x)m^2} \right|, \tag{4a}$$

$$q_0 \Pi^{0i}(q) = \frac{q^i}{(2\pi)^2} \left[\frac{q_0^2}{6} \log \frac{q_0^2}{m^2} + m^2 - \frac{1}{6}\mathbf{q}^2 \right] + O\left(\frac{1}{q_0^2}\right). \tag{4b}$$

As $q_0 \to \infty$, a quadratically diverging term proportional to q^i is encountered; also a finite term proportional to $\mathbf{q}^2 q^i$ is present. Thus in position space, the commutator gives a quadratically divergent term proportional to the first derivative of the δ-function, and another finite term involving *three* derivatives of a δ-function. The quadratically divergent object is the usual Schwinger term, an unexpected term appears — a triple derivative of the δ-function.[12]

D. In the Table, we list all the commutators that have been so far calculated by these techniques by various people. In the first column, we list the naive canonical result. In the second column appears the formal "prediction" that has been obtained from this canonical result, with an X indicating the fact that it violates explicit calculation. The third column gives the commutator as calculated by the method explained above. The offending graphs which lead to the anomalous behavior are drawn in column four. Remarks are contained in column five. Finally, references to the literature comprise column six.

E. Let me now explain in some detail the precise mechanism which allows non-canonical results to appear in this method. Consider the diagrams of Fig. 1, which represent the lowest order contribution to

$$\int d^4 x e^{iqx} \langle \psi|T j^\mu(x) j^\nu(0)|\psi\rangle.$$

(The states ψ and the straight lines are fermions.)

The graphs are equal to

$$\gamma^\mu S(p+q)\gamma^\nu + \gamma^\nu S(p-q)\gamma^\mu . \tag{5a}$$

The $1/q_0$ term, which gives the commutator is got by replacing $S(p \pm q)$ by $\pm i\gamma^0/q_0$, its asymptotic form for large q_0. Thus (5a) becomes asymptotically

$$\frac{i}{q_0}[\gamma^\mu\gamma^0\gamma^\nu - \gamma^\nu\gamma^0\gamma^\mu].\qquad (5b)$$

This is indeed the lowest order contribution to the commutator (times i/q_0).

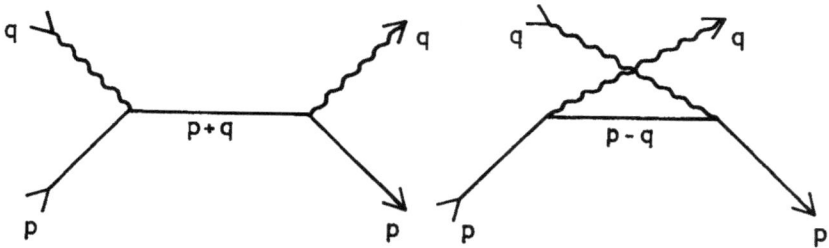

Fig. 1.

Next, consider two radiative correction diagrams of Fig. 2, where the dotted line denotes an [unspecified] exchange that effects the radiative correction.

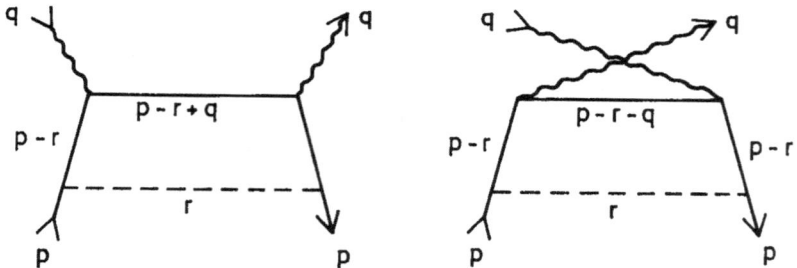

Fig. 2.

Note that these diagrams yield perfectly finite and unambiguous results. If for purposes of calculating the i/q_0 term in the expression represented by Fig. 2 (*i.e.*, the commutator) one could replace $S(p - r \pm q)$ again by $\pm i\gamma^0/q_0$, one would get i/q_0 times the diagram of Fig. 3. This is indeed the relevant radiative correction to the matrix element of the commutator. However, such a replacement ignores the fact that the integration variable r also gets large and the region $r > q_0$ may contribute significantly. Indeed in the present example, the expression of Fig. 3 is (superficially) divergent if the exchanged (dotted) propagator falls as $1/r^2$ for large r — indicating that the asymptotic form of Fig. 2 *cannot be obtained by this formal replacement.* Thus the i/q_0 term in the asymptotic expression for Fig. 2 may differ from Fig. 3 and one says that the commutator is anomalous since it does not agree with its canonical value. Indeed explicit calculation shows that the $1/q_0$ asymptote indeed differs from the *formal* result in Fig. 3. This then is the origin of the anomalous term in the commutators.

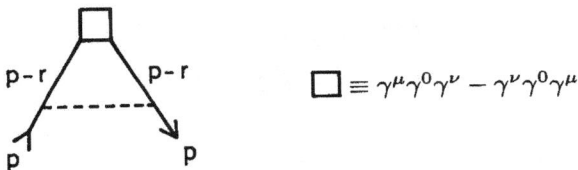

$$\square \equiv \gamma^\mu \gamma^0 \gamma^\nu - \gamma^\nu \gamma^0 \gamma^\mu$$

Fig. 3.

I wish to emphasize here that the occurrence of a divergence is a signal for non-canonical behavior. However, this criterion must be used advisedly — it is the *superficial* divergence that is relevant to the discussion. The fact that any of the conventional methods of integration (symmetric integration, gauge invariant integration, *etc.*) may render the expression finite is not relevant. Indeed, most of the anomalies given in the Table are finite, but they are finite only by convention — their integral representation is superficially divergent.

F. As was stated above, the commutators given in the Table will by definition reproduce the correct high energy behavior of Green's functions. The conflict between formal high energy theorems and explicit high energy behavior has thus been removed. How should low energy theorems, such as the Sutherland-Veltman, be modified so as to remove the conflict there? To analyze this problem, we recall that all low energy theorems are gotten with the help of Ward identities, and we now must inquire how the Ward identities are to be modified in view of all the anomalies in the commutators.

This modification is effected as follows. To derive a Ward identity, one must first construct a covariant T^* product, since in general the T product is not covariant. If the Schwinger term in the $[A^\mu, B^\nu]$ equal-time commutator is known, as it is in our applications, it is easy to construct a sea-gull which covariantizes the T product. (Here A^μ and B^ν are any two operators.) The procedure is[13]: write the equal-time commutator in an explicitly covariant fashion with the help of a unit timelike vector n

$$\delta(x \cdot n)[A^\mu(x),\, B^\nu(0)] = S^{\mu\nu}(n)\delta^4(x) + S^{\mu\nu/\alpha}(n)P_{\alpha\beta}\partial^\beta\delta^4(x)$$
$$P_{\alpha\beta} \equiv g_{\alpha\beta} - n_\alpha n_\beta \,.$$
$$\tag{6}$$

Then the sea-gull, which covariantizes

$$TA^\mu(x)B^\nu(0) \equiv \theta(x \cdot n)A^\mu(x)B^\nu(0) + \theta(-x \cdot n)B^\nu(0)A^\mu(x)$$

is given by

$$C^{\mu\nu}(x, n) = C^{\mu\nu}(n)\delta^4(x)\,, \tag{7a}$$

$$C^{\mu\nu}(n) = \int dn'_\alpha S^{\mu\nu/\alpha}(n')\,, \tag{7b}$$

i.e.,

$$T^* A^\mu(x)B^\nu(0) = TA^\mu(x)B^\nu(0) + \delta^4(x)\int dn'_\alpha S^{\mu\nu/\alpha}(n')$$

is independent of n.[14] Of course the sea-gull is not unique; any additional term of the form

$$C_0^{\mu\nu}\delta^4(x)$$

not depending on n, will also maintain Lorentz covariance.[15] The sea-gull is made unique by requiring the cancellation of Schwinger terms and divergences of sea-gulls. It is easy to show that imposing this requirement on the μ Ward identity, requires

$$n_\mu C^{\mu\nu}(n) = 0\,. \tag{8a}$$

Similarly, the requirement on the index ν forces

$$n_\nu C^{\mu\nu}(n) = 0\,. \tag{8b}$$

By adding covariant terms to

$$\int dn'_\alpha S^{\mu\nu/\alpha}(n')$$

it is *always* possible to satisfy either (8a) or (8b).[13] However, *both* requirements (8a) and (8b) overdetermine the sea-gull and it may not be possible to satisfy them. The *necessary* and *sufficient* condition for the possibility of satisfying *both* requirements is the absence of Schwinger term in the $[A^0,\, B^0]$ equal-time commutator, *i.e.*, $n_\mu n_\nu S^{\mu\nu/\alpha} = 0.$[13] It is seen that the commutators relevant to the Sutherland-Veltman theorem possess such a Schwinger term, and there is no cancellation between all the Schwinger terms and sea-gulls. This lack of cancellation modifies the low energy theorem — exactly in the desired fashion. This also shows that Feynman's conjecture is wrong, in the general case.

To see explicitly how this occurs for the $j^\mu j_5^\nu$ Ward identity, we note from the Table that the Schwinger term is

$$S^{\mu\nu/\alpha}(n) = c[\,{}^*F^{\mu\gamma}(g_\gamma^\alpha n^\nu + g^{\alpha\nu} n_\gamma) + {}^*F^{\nu\gamma}(g_\gamma^\alpha n^\mu + g^{\alpha\mu} n_\gamma)] \qquad (9a)$$

where c is a non-vanishing constant, and $^*F$ is dual to the electromagnetic field tensor F : $^*F^{\mu\nu} \equiv \frac{1}{2}\varepsilon^{\mu\nu\alpha\beta} F_{\alpha\beta}$. Therefore the sea-gull which covariantizes the T product is $C^{\mu\nu}(n)\delta^4(x)$

$$C^{\mu\nu}(n) = c[\,{}^*F^{\mu\gamma} n_\gamma n^\nu + {}^*F^{\nu\gamma} n_\gamma n^\mu] + C_0^{\mu\nu}. \qquad (9b)$$

Imposing (8a) requires

$$n_\mu C^{\mu\nu}(n) = 0 = c\,{}^*F^{\nu\gamma} n_\gamma + n_\mu C_0^{\mu\nu} \qquad (10a)$$

while (8b) forces

$$n_\nu C^{\mu\nu}(n) = 0 = c\,{}^*F^{\mu\gamma} n_\gamma + C_0^{\mu\nu} n_\nu. \qquad (10b)$$

From (10a) we then have

$$C_0^{\mu\nu} = c\,{}^*F^{\mu\nu} \qquad (11a)$$

while from (10b) we get

$$C_0^{\mu\nu} = -c\,{}^*F^{\mu\nu}. \qquad (11b)$$

Thus it is impossible to choose $C_0^{\mu\nu}$ so that both Ward identities are satisfied; though it is possible to satisfy any one.

G. We have eliminated the contradictions between formal reasoning and explicit calculation by modifying the commutators in the fashion indicated in the Table. However, these definitions for the commutators do have certain disadvantages which prevent one from using them in other applications. These disadvantages are the following: (a) the initial development of the theory rests on *canonical* commutators. For example, in deriving the equation for the vertex function involving some kernel K

$$\Gamma^\mu = \gamma^\mu + K\Gamma^\mu, \qquad (12)$$

the fact that the lowest order term γ^μ occurs with the factor 1, rather than some other value, follows from the commutator

$$[\psi(\mathbf{x}, t),\ j^\mu(\mathbf{y}, t)] = \gamma^0 \gamma^\mu \psi(x)\delta(\mathbf{x} - \mathbf{y})\,. \tag{13}$$

On the other hand, we have abandoned this commutator in order to reproduce the correct asymptotic behavior for Γ^μ. Thus there appears to be a conflict between our anomalous commutators and the structure of the canonical theory. (b) A second disadvantage is that these commutators no longer retain algebraic properties, for example, the Jacobi identity is not in general satisfied.[16]

One may show by a rather different method how the commutators become affected by the interaction.[17] Let us calculate $[\psi(\mathbf{x}, t),\ j^i(\mathbf{y}, t)]$. We do not wish to take the canonical definition for $j^i = \bar\psi \gamma^i \psi$, nor to use canonical commutators since such procedures are suspect, in our present context. We will use a commutator which has proven itself to be reliable

$$[\psi(\mathbf{x}, t),\ j^0(\mathbf{y}, t)] = \psi(x)\delta(\mathbf{x} - \mathbf{y})\,. \tag{14}$$

To calculate $[\psi(\mathbf{x}, t),\ j^i(\mathbf{y}, t)]$ we first observe that no gradients of delta functions have been encountered in this term. Thus we may just as well integrate over y without loss of generality. By current conservation it then follows that

$$\begin{aligned}
\int d^3y[\psi(\mathbf{x}, t),\ j^i(\mathbf{y}, t)] &= -\int d^3y\, y^i[\psi(\mathbf{x}, t),\ \partial_j j^j(\mathbf{y}, t)] \\
&= \int d^3y\, y^i \frac{\partial}{\partial t}[\psi(\mathbf{x}, t),\ j^0(\mathbf{y}, t)] - \int d^3y\, y^i[\dot\psi(\mathbf{x}, t),\ j^0(\mathbf{x}, t)]\,.
\end{aligned} \tag{15}$$

Use of the commutator (14), and the equation for ψ, which by definition is taken to be

$$(i\slashed\partial - m)\psi(y) = g\eta(y)\,, \tag{16}$$

casts (15) into the form

$$\begin{aligned}
\int d^3y[\psi(\mathbf{x}, t), j^i(\mathbf{y}, t)] &= \gamma^0 \gamma^i \psi(x) \\
&+ i\gamma^0 g \int d^3y\, y^i \big([\eta(\mathbf{x}, t),\ j^0(\mathbf{y}, t)] - \eta(x)\delta(\mathbf{x} - \mathbf{y})\big)\,.
\end{aligned} \tag{17}$$

In the naive calculation one sets

$$[\eta(\mathbf{x}, t),\ j^0(\mathbf{y}, t)] = \eta(x)\delta(\mathbf{x} - \mathbf{y})\,, \tag{18}$$

since $\eta(x)$ is thought to be of the form $O(x)\psi(x)$ and O is some operator which commutes with j^0. However, a product such as $O(x)\psi(x)$ is ill-defined, and one cannot use the naive theory. Thus, in addition to the interaction-independent term in (17), there may very well be additional, interaction-dependent terms that arise from the singularities in the product. Comparison of (17) with the desired result in the Table

$$[\psi(\mathbf{x},t),\, j^i(\mathbf{y},t)] = \left(1 - \frac{g^2}{8\pi^2}\right)\gamma^0\gamma^i\psi(x)\delta(\mathbf{x}-\mathbf{y})\,, \tag{19}$$

indicates that (19) may be regained from (17) if (18) is changed to[18]

$$[\eta(\mathbf{x},t),\, j^0(\mathbf{y},t)] = \eta(\mathbf{x},t)\delta(\mathbf{x}-\mathbf{y}) + \left(\frac{i}{8\pi^2}\right)g\gamma^i\psi(x)\partial_i\delta(\mathbf{x}-\mathbf{y})\,. \tag{20}$$

H.   All our calculations have been performed in lowest order perturbation theory. Therefore, the question naturally arises: Are these anomalies peculiar to perturbation theory, or are they an essential feature of the complete theory?

It is easy to show *non-perturbatively* the necessary presence of anomalies; that is, I shall now demonstrate that the assertion that the complete theory is canonical and all formal manipulations are allowed is self-contradictory. Consider the $[j^0,\, j^i]$ commutator. Canonically it vanishes. Therefore, already the necessary existence of a single derivative of a delta function indicates that the theory cannot be canonical. Furthermore, one can also demonstrate non-perturbatively that either there exists a triple derivative of a δ-function in the commutator, or else the fermion field anti-commutator must vanish; *i.e.*, anomalies other than the ones dictated by positivity and Lorentz covariance (like the usual Schwinger term) must be present. To show this, consider the j^i to be defined as

$$j_\varepsilon^i(y) \equiv \bar{\psi}(y + \varepsilon/2)\gamma^i\psi(y - \varepsilon/2),$$

where ε is spacelike, to be taken to zero at the end of the calculation. [One cannot quarrel with this definition for the current, since we *assume* the theory to be entirely canonical, therefore insensitive to any intermediate methods of regularization, and we then demonstrate the self-contradictory nature of the assumption.] The vacuum expectation value of the commutator is now proportional to

$$\lim_{\varepsilon \to 0} \operatorname{Tr} \gamma^i G(\varepsilon)[\delta(\mathbf{x}-\mathbf{y}-\varepsilon/2) - \delta(\mathbf{x}-\mathbf{y}+\varepsilon/2)]\,. \tag{21a}$$

Here G is the fermion propagator in position space. Expanding the delta functions in power of ε replaces (21a) by

$$- \lim_{\varepsilon \to 0} \mathrm{Tr}\, \gamma^i G(\varepsilon) \left[\varepsilon^k \partial_k \delta(\mathbf{x} - \mathbf{y}) + \frac{1}{24} \varepsilon^k \varepsilon^l \varepsilon^m \partial_k \partial_l \partial_m \delta(\mathbf{x} - \mathbf{y}) + O(\varepsilon^5) \right]. \quad (21b)$$

Therefore, the third derivative of the δ-function will survive if $G(\varepsilon)$ behaves as $1/\varepsilon^3$ for small ε. Moreover, formal canonical reasoning yields the result that the canonical equal-time commutation relations for the fermion fields imply that the complete propagator goes as $1/\not{p}$ for large p in momentum space; thus in position space $G(\varepsilon)$ *does* go as $1/\varepsilon^3$. The conclusion is that canonical results must be abandoned either for the $[j^0,\ j^i]$ commutator or for the $[\psi,\ \bar{\psi}]_+$ anticommutator, (or both).[19] Similar arguments can be constructed for all the anomalies given in the Table; and it must be concluded that the theory cannot be entirely canonical, although the present argument does not specify where the anomalies are present. It is also evident that to *all* orders in perturbation theory the anomalies must be present. Consider for example the calculation of the effective $\pi^0 \to 2\gamma$ decay constant for zero pion mass in the σ model. Since the lowest order calculation is non-vanishing, higher order effects which of course are non-divergent cannot remove the constant contribution for arbitrary coupling constant. Therefore, the commutators *must* be modified, *i.e.*, the high energy behavior of Green's functions will be non-canonical.[20]

I.   Can one somehow overcome these anomalies; that is can a theory be set up without the contradictions between formal manipulations and explicit results? Only limited investigation has been devoted to this question. The analysis presented above in connection with the relevance of dynamics to commutators, Eqs. (14)–(20), offers a hint how one may modify the theory to effect agreement between formal and explicit results. Evidently, one must modify the dynamics somewhat, by adding additional terms to the Lagrangian. For example, in the discussion of Eqs. (14)–(20) one wants to modify the naive source of the fermion field, $\eta(x) = O(x)\psi(x)$, by an additional term, so that the commutator between the source and j^0 no longer has the gradient term, as in (20), but coincides with the naive result (18). An explicit example of such modification can be given in reference to the breakdown of the Sutherland-Veltman theorem. By adding a contact term to the σ-model Lagrangian $\varphi^* F^{\mu\nu} F_{\mu\nu}$, where φ is the pion field, the non-zero decay constant can be cancelled away yielding zero, as is required by the formal argument. An alternative modification to the theory is to prescribe non-standard regularization techniques, which eliminate the non-zero coupling constant.[21] These modifications however, probably lead to non-renormalizable perturbation series.

Ultimately, the question whether our anomalies are present or absent in Nature, rather than in formal systems, must be answered experimentally. The anomalies in neutral axial vector resolve the problem of the Sutherland-Veltman theorem for pion-decay. The electroproduction data are still uncertain, so one does not know whether σ_L/σ_T decreases as rapidly as Callan and Gross predicted from formal arguments; or whether it decreases more slowly, as the present results indicate. In any case, this question will be soon settled.

If it turns out that these anomalies are of physical importance, it would be of great value to develop techniques for calculating them which do not rest so heavily on explicit evaluation in perturbation theory.

Finally, it is interesting to point out that the current activity in this field has antecedents from the earliest days of quantum field theory: P. Jordan seems to have been the first to realize that the quantization procedure can distort canonical structure.[22]

REFERENCES

(for the Text)

1.
  a. D. G. Sutherland, *Nucl. Phys.* **B2**, 433 (1967); M. Veltman, *Proc. Roy. Soc.* **A301**, 107 (1967).
  b. J. S. Bell and R. Jackiw, *Nuovo Cim.* **60**, 47 (1969). This contradiction between formal reasoning and explicit calculation in closely related contexts has also been discussed by J. Steinberger, *Phys. Rev.* **76**, 1180 (1949); H. Fukuda and Y. Miyamoto, *Prog. Theor. Phys.* **4**, 347 (1949); J. Schwinger, *Phys. Rev.* **82**, 664 (1951); K. Johnson, *Phys. Lett.* **5**, 253 (1963); S. L. Adler, *Phys. Rev.* **177**, 2426 (1969).

2.
  a. C. Callan and D. J. Gross, *Phys. Rev. Lett.* **22**, 156 (1969);
  b. R. Jackiw and G. Preparata, *Phys. Rev. Lett.* **22**, 975, (E)1162 (1969); *Phys. Rev.* **185**, 1748 (1969); S. L. Adler and W.-K. Tung, *Phys. Rev. Lett.* **22**, 978 (1969), *Phys. Rev.* **D 1**, 2846 (1970).

3.
  a. G. Preparata and W. Weisberger, *Phys. Rev.* **175**, 1965 (1968);
  b. S. L. Adler and W.-K. Tung, *Phys. Rev. Lett.* **22**, 978 (1969).

4. J. Steinberger, *Phys. Rev.* **76**, 1180 (1949); J. S. Bell and R. Jackiw, *Nuovo Cim.* **60**, 47 (1969); S. L. Adler, *Phys. Rev.* **177**, 2426 (1969).

5. C. Hagen, *Phys. Rev.* **177**, 2622 (1969); R. Jackiw and K. Johnson, *Phys. Rev.* **182**, 1459 (1969); K. G. Wilson, *Phys. Rev.* **179**, 1499 (1969); B. Zumino, in *Proc. of Topical Conference on Weak Interactions*, CERN, Geneva, p. 361 (1969).

6. K. G. Wilson, *Phys. Rev.* **181**, 1909 (1969); I. Gerstein and R. Jackiw, *Phys. Rev.* **181**, 1955 (1969); W. Bardeen, *Phys. Rev.* **184**, 1849 (1969); D. Amati, C. Bouchiat and J. L. Gervais, *Nuovo Cim.* **65A**, 55 (1970).

7. R. Jackiw and K. Johnson, *Phys. Rev.* **182**, 1459 (1969); S. L. Adler and D. Boulware, *Phys. Rev.* **184**, 1740 (1969); D. J. Gross and R. Jackiw, *Nucl. Phys.* **B14**, 269 (1969).

8. R. Feynman, unpublished, conjectured the cancellation between Schwinger terms and gradients of sea-gulls — a conjecture that has been of great importance in deriving Ward-Takahashi identities; see also J. S. Bell, *Nuovo Cim.* **50**, 129 (1967), who shows that this desired cancellation is a consequence of the gauge properties of the theory.

9. J. D. Bjorken, *Phys. Rev.* **148**, 1467 (1966) as well as K. Johnson and F. Low, *Prog. Theor. Phys.* (Kyoto), Suppl. No. **37–38**, 74 (1966), proved that the high-energy behavior of a T product is governed by a commutator. This mathematically true statement becomes a *theorem*, which can be violated, when it is assumed that the relevant commutator is canonical. Such an assumption has always been made in practical applications.

10. A. I. Vainshtein and B. L. Ioffe, *Zh. Eksp. Teor. Fiz., Pis'ma Redakt.* **6**, 917 (1967) [English translation: *Sov. Phys. JETP Lett.* **6**, 341 (1967)]; R. Jackiw and G. Preparata, *Phys. Rev. Lett.* **22**, 975 (E)1162 (1969), *Phys. Rev.* **185**, 1748 (1969); S. L. Adler and W.-K. Tung, *Phys. Rev. Lett.* **22**, 978 (1969); *Phys. Rev. D* **1**, 2846 (1970).

11. T. Goto and I. Imamura, *Prog. Theor. Phys.* (Kyoto) **14**, 396 (1955); J. Schwinger, *Phys. Rev. Lett.* **3**, 296 (1959); K. Johnson and F. Low, *Prog. Theor. Phys.* (Kyoto), Suppl. No. **37–38**, 74 (1966); J. S. Bell, *Nuovo Cim.* **47A**, 616 (1967); R. Jackiw and K. Johnson, *Phys. Rev.* **182**, 1459 (1969); S. L. Alder and D. Boulware, *Phys. Rev.* **184**, 1740 (1969); R. Jackiw and G. Preparata, *Phys. Rev.* **185**, 1748 (1969); D. Boulware and R. Jackiw, *Phys. Rev.* **186**, 1442 (1969);

12. This calculation is discussed in detail by D. Boulware and R. Jackiw, *Phys. Rev.* **186**, 1442 (1969), where it is also explained how a *third* derivative of the delta function appears, in spite of formal proofs that the commutator contains only *one* derivative of delta function. For these proofs see T. Goto and I. Imamura, *Prog. Theor. Phys.* (Kyoto) **14**, 396 (1955) and D. J. Gross and R. Jackiw, *Phys. Rev.* **163**, 1688 (1967).

13. This technique was developed by D. J. Gross and R. Jackiw, *Nucl. Phys.* **B14**, 269 (1969).

14. It can be shown that in any Lorentz covariant theory, the Schwinger terms are such that the line integral $\int dn_\alpha S^{\mu\nu/\alpha}$ is path-independent, see Ref. 13.

15. It is also possible to add Lorentz invariant sea-gulls which involve gradients of delta functions. In instances of interest for us here, this possibility does not arise. For the general discussion, see Ref. 13.

16. K. Johnson and F. Low, *Prog. Theor. Phys.* (Kyoto), Suppl. No. **37–38**, 74 (1966).

17. The development here follows R. Jackiw and G. Preparata, *Phys. Rev.* **185**, 1929 (1969).

18. Note that gauge covariance is preserved by the anomalous commutator (20)

$$\int d^3y [\eta(\mathbf{x}, t), j^0(\mathbf{y}, t)] = \eta(x).$$

19. Incidentally, we note that this argument shows that canonical reasoning implies that either the wave function renormalization constant Z_1, or the charge renormalization Z_3 vanishes. For if Z_1 is finite, then the propagator *does* go as $1/\varepsilon^3$ in position space and the ordinary Schwinger term is quadratically divergent. That means that the spectral representation for the Schwinger term, $\int_0^\infty \rho(a^2)da^2$, diverges and $\rho(a^2) \to$ constant for large a^2. But Z_3^{-1} is given by $\int_0^\infty [\rho(a^2)/a^2]da^2$ which evidently is logarithmically divergent.

20. However, any argument based solely on perturbation theory can be circumvented if the perturbative expansion diverges. Also one may hope that for a particular value of the coupling constant, the anomaly disappears, although it is unlikely that *all* anomalies disappear for *one* value of the coupling constant. The existence of non-perturbative arguments, given in the text, shows such optimistic speculations to be unrealistic.

21. J. S. Bell and R. Jackiw, *Nuovo Cim.* **60**, 47 (1969).

22. P. Jordan, *Z. Phys.* **93**, 464 (1935).

REMARKS

(for the Table)

a) The solid line is a fermion.

b) The wavy line is a vector boson in the Landau gauge (so that $Z_1 = Z_2$ is finite) coupled with strength g to $\bar\psi\gamma^\mu\psi$.

c) $G(p)$ is the unrenormalized fermion propagator.

d) $\Gamma^\mu(p, q)$ is the unrenormalized vertex function.

e) $\times$ in the diagram represents the vector current.

f) The state $|\psi\rangle$ is a fermion state with momentum $\mathbf{p}$, normalized so that $\langle 0|\psi|\psi\rangle = 1$.

g) Crossed diagrams must also be included.

h) Schwinger[7] showed that positivity and Lorentz covariance forces the commutator to be non-zero. Previously, Goto and Imamura[8] derived a representation for this object, which involved *one* derivative of the δ-function. Their result is not verified by calculation.

i) $\tilde{\times}$ in the diagram is the axial vector current.

j) $|\gamma\rangle$ is a one-photon state.

k) $^*F^{\mu\nu}$ is the dual electromagnetic tensor.

l) c is a constant.

m) For purposes of comparison with the text, the commutator has been written in explicitly covariant notation, with the help of a unit timelike vector n, and $P_{\alpha\beta} \equiv g_{\alpha\beta} - n_\alpha n_\beta$.

n) α, β are bosons.

o) φ is a scalar or pseudoscalar field.

p) Some derivations of Weinberger's first sum rule assume a c number Schwinger term.[14] In spite of the presence of q number Schwinger terms, this theorem remains true.

REFERENCES

(for the Table)

1. H. Lehmann, *Nuovo Cim.* **11**, 342 (1954).
2. R. Jackiw and G. Preparata, *Phys. Rev.* **185**, 1929 (1969).
3. G. Preparata and W. Weisberger, *Phys. Rev.* **175**, 1965 (1968).
4. C. Callan and D. J. Gross, *Phys. Rev. Lett.* **22**, 156 (1969).
5. R. Jackiw and G. Preparata, *Phys. Rev. Lett.* **22**, 975, (E)1162 (1969).
6. S. L. Adler and W.-K. Tung, *Phys. Rev. Lett.* **22**, 978 (1969).
7. J. Schwinger, *Phys. Rev. Lett.* **3**, 296 (1959).
8. T. Goto and I. Imamura, *Prog. Theor. Phys.* (Kyoto) **14**, 396 (1955).
9. D. Boulware and R. Jackiw, *Phys. Rev.* **186**, 1442 (1969).
10. D. Sutherland, *Nucl. Phys.* **B2**, 433 (1967); M. Veltman, *Proc. Roy. Soc.* **A301**, 107 (1967).
11. K. Johnson and F. Low, *Prog. Theor. Phys.* (Kyoto), Suppl. No. **37–38**, 74 (1966).
12. R. Jackiw and K. Johnson, *Phys. Rev.* **182**, 1459 (1969).
13. S. L. Alder and D. Boulware, *Phys. Rev.* **184**, 1740 (1969).
14. S. Weinberg, *Phys. Rev. Lett.* **18**, 507 (1967).
15. T. Nagylaki, *Phys. Rev.* **158**, 1534 (1967).

Table

Canonical commutator	False theorem	Effective commutator	Offending diagrams	Remarks	References
$\langle 0|[\psi(\mathbf{x},t),\bar\psi(\mathbf{y},t)]_+|0\rangle = \gamma^0\delta(\mathbf{x}-\mathbf{y})$	$G(p)\xrightarrow[p\to\infty]{} i/\not p$ (Lehmann) X	$\left(1-\dfrac{3g^2}{32\pi^2}\right)\gamma^0\delta(\mathbf{x}-\mathbf{y})$		a, b, c	1, 2
$\langle 0|[\psi(\mathbf{x},t),j^0(\mathbf{y},t)]|\psi\rangle = \delta(\mathbf{x}-\mathbf{y})$ $\langle 0|[\psi(\mathbf{x},t),j^i(\mathbf{y},t)]|\psi\rangle = \gamma^0\gamma^i\delta(\mathbf{x}-\mathbf{y})$	$\left.\begin{array}{c}\\\end{array}\right\}\lim_{q_0\to\infty}\Gamma^\mu(p,q)\to\gamma^\mu$ X	$\delta(\mathbf{x}-\mathbf{y})$ $\left(1-\dfrac{g^2}{8\pi^2}\right)\gamma^0\gamma^i\delta(\mathbf{x}-\mathbf{y})$		a, b, d, e, f	2
$\langle\psi|[j^i(\mathbf{x},t),j^j(\mathbf{y},t)]|\psi\rangle =$ $\delta(\mathbf{x}-\mathbf{y})i\varepsilon^{ijk}\langle\psi|j_k^5(x)|\psi\rangle$	Preparata-Weisberger theorem about finite radiative corrections X	$\delta(\mathbf{x}-\mathbf{y})i\varepsilon^{ijk}\left(1-\dfrac{3g^2}{16\pi^2}\right)$ $\times\langle\psi|j_k^5(x)|\psi\rangle$		a, b, e, f, g	3, 4, 5, 6
$\displaystyle\int d^3x\langle\psi|[j^i(\mathbf{x},t),j^j(\mathbf{y},t)]|\psi\rangle =$ $(\delta^{ij}p^2-p^ip^j)A+\delta^{ij}B$ X	Callan-Gross sum rule for electroproduction	$(\delta^{ij}p^2-p^ip^j)A+\delta_{ij}B$ $+\delta^{ij}p^2B'$			
$\langle 0|[j^0(\mathbf{x},t),j^i(\mathbf{y},t)]|0\rangle = 0$	See remark h	$S\partial^i\delta(\mathbf{x}-\mathbf{y})$ $-\dfrac{1}{24\pi^2}\nabla^2\partial^i\delta(\mathbf{x}-\mathbf{y})$		a, e, h	7, 8, 9

Table (*Continued*)

Canonical commutator	False theorem	Effective commutator	Offending diagrams	Remarks	References
$\langle 0\|[j^0(x,t), j_5^\mu(y,t)]\|\gamma\rangle = 0$ $\langle 0\|[j^\mu(x,t), j_5^0(y,t)]\|\gamma\rangle = 0$	Sutherland-Veltman theorem about $\pi^0 \to 2\gamma$ decay X	$\delta([x-y]\cdot n)$ $\times \langle 0\|[j^\mu(x,t), j_5^\nu(y,t)]\|\gamma\rangle$ $= -cn^\nu P_{\alpha\beta}$ $\times \delta^4(x-y)\partial^\alpha \langle 0\|{}^*F^{\mu\beta}(x)\|\gamma\rangle$ $+c(g_\delta^\mu[g_\gamma^\alpha n^\nu + g^{\alpha\nu} n_\gamma]$ $+g_\delta^\nu[g_\gamma^\alpha n^\mu + g^{\alpha\mu} n_\gamma])$ $\times P_{\alpha\beta}\partial^\beta \delta^4(x-y)$ $\times \langle 0\|{}^*F^{\delta\gamma}(y)\|\gamma\rangle$		a, e, g, i, j, k,l,m	10, 11, 12, 13
$\langle \alpha\|[j^0(x,t), j^i(y,t)]\|\beta\rangle = 0$	See remark p	= 0 if α and β are vector or pseudovector particles; $= c(\alpha\|\varphi^2(y)\|\beta)\partial^i\delta(x-y)$ if α and β are scalar or pseudovector particles		a, e, l, n, o, p	9, 14, 15

In addition to the entries of this Table, there exist determinations of a large number of commutators from the triangle graph by Johnsonand Low[11]

Paper I.2

Niels Bohr also speculated on quantal symmetry breaking. For his centennial symposium (Copenhagen, May 1985) I attempt a "physical" description of anomalies.

QUANTUM MECHANICAL SYMMETRY BREAKING

Recent Developments in Quantum Field Theory,
J. Ambjørn, B. Durhus and J. Petersen, eds. (North-Holland, Amsterdam, 1985)

1. INTRODUCTION

My talk will be concerned with anomalies, and I shall suggest various research topics that arise from some contemporary ideas about them. But first let me take note of the occasion being marked by this Conference and place the subject in a historical setting which, as it happens, includes Niels Bohr, the originator of much of our quantum mechanical thinking.

When physicists build theories to explain fundamental interactions, we want a high degree of symmetry characterizing the correct model. This is because, first, it is aesthetically satisfying to suppose that the ultimate laws of Nature are very symmetric, and second, as a practical matter, it is important that theories with symmetry are less afflicted with divergences than those without. However, observed natural phenomena do not exhibit a large amount of symmetry, and this gap between theory and experiment must be bridged.

Explicit symmetry breaking — the oldest and most straightforward way of reducing a symmetry — need not concern us, since obviously it is of no fundamental significance. A better idea is due to Heisenberg. Inspired by examples drawn from many-body physics, he suggested that energetic stability considerations can lead to *spontaneous symmetry breaking*, also in particle physics, so that the lowest energy, equilibrium state — the vacuum — need not possess the full symmetries of the dynamical equation.[1] His idea proved to be most successful, and came to fruition within particle physics in the modern theory of pseudoscalar mesons, in the unification of electromagnetism with weak interactions, and is believed to be an ingredient for the unified theory which we are seeking. Still so far, we have not understood the dynamical reason why a symmetric state is energetically unstable.

There is, however, another more subtle mechanism for removing from the solutions of a quantized theory a symmetry which is present on the Lagrangian level. This is the *anomaly phenomenon*, which is now widely appreciated in physics, but remains poorly understood. Here, the symmetry is not violated by explicit terms in the Lagrangian, nor do energetic or stability considerations select a non-symmetric equilibrium state. Rather, the very process of quantizing, specifically second quantizing, the theory destroys the symmetry. There is no evidence for this in the formal, classical equations of the model, and the mechanism was a surprising discovery; hence, it has been called *anomalous*, but *quantum mechanical symmetry breaking* is a better name. Anomalous or quantum mechanical symmetry breaking afflicts symmetries associated with masslessness: both scale and conformal invariance as well as chiral invariance of massless fermions are broken by quantum effects. Although the former is important for the description of phase transitions in condensed matter physics and of high-energy processes in particle physics, I shall not dwell on it here. My emphasis is on chiral anomalies.

That effects of quantization may interfere with classical symmetries and conservation laws is an idea from the beginnings of modern physics, when Bohr, resisting Einstein's light quantum, proposed that energy is not conserved in electromagnetic processes. Presumably, time-translation invariance would be absent also. Later, before existence of the neutrino became established, Bohr reiterated and extended his suggestion to include momentum non-conservation in β-decay. Of course, experiment has shown that this position is untenable; space-time translations *are* symmetries of quantal Nature so that energy and momentum *are* conserved.[2] Nevertheless, we may say that our present point of view towards scale and conformal symmetry breaking was prefigured by Bohr's intuition concerning effects of quantization on space-time symmetries.

Closer to my emphasis today on anomalously broken chiral symmetries is the story of gauge invariance in quantum electrodynamics. Before correct computational methods were fixed, it was not clear that the quantized theory maintains gauge invariance. Indeed, Wentzed claimed that the theory predicts a photon mass and presumably a corresponding violation of charge conservation for Dirac fermions.[3] Again, subsequent developments established gauge invariance. Therefore, the problem was removed, but its descendant forms our present understanding that massless Weyl fermions cannot in general possess gauge invariant interactions, nor can their charge be conserved, owing to quantum mechanical violation of symmetry. Moreover, for massless Dirac fermions, the charges in each of the two separate chiralities cannot be conserved, although their sum is conserved.

Nature requires that ungauged chiral symmetry of massless fermions be broken. This is deduced from details in the spectrum of pseudoscalar mesons: chiral symmetry for the constituent quarks would suppress meson decay modes and produce mass degeneracies, neither of which are seen experimentally. Since current models contain no mechanism for breaking all the relevant symmetries, the occurrence of anomalies is a phenomenologically welcome and useful result. On the other hand, models with anomalously broken gauge symmetries are inconsistent, as far as we know. I shall speak further about these so-called *anomalous gauge theories*, which thus far have no role in physics.

2. ORIGIN OF CHIRAL ANOMALIES

Let us recall the various ways we have of establishing that a chiral anomaly

occurs when massless fermions interact with a gauge field. First, there are
the original perturbative calculations of Feynman diagrams which show that
chiral Ward identities cannot be maintained. These calculations can be performed in momentum space, where the effect arises from momentum rout-
ing ambiguities,[4] or in position space , where singularities in the product of
fermion bilinears provide the operative mechanism.[5] A second derivation, also
known from the earliest investigations observes that the algebra of chiral generators, when calculated in perturbation theory, fails to close.[6] Later, I shall say
more about modern developments in this approach. Third, in the functional
integral formulation, the anomaly arises because the fermion measure is not
chirally invariant.[7] Fourth, we now have mathematical, specifically topological
and cohomological, reasons for understanding that the functional determinant
of non-Abelian massless Weyl [chiral] fermion in interaction with gauge fields
cannot be defined gauge invariantly.[8]

While all arguments are convincing and striking in the variety of routes
they offer to the same goal, they lack direct physical immediacy; one is still
left with the central physical puzzle of the chiral anomaly. This puzzle may be
stated in the following way. Consider massless Dirac fermion field ψ, in even-
dimensional space-time, interacting with an external electromagnetic gauge
field A_μ. Dynamics is governed by the Lagrangian,

$$\mathcal{L} = \bar{\psi}(i\not{\partial} - e\not{A})\psi \tag{1a}$$

which may be decomposed into left and right Weyl fermion pieces.

$$\begin{aligned} \psi &= \psi_L + \psi_R \\ \mathcal{L} &= \bar{\psi}_L(i\not{\partial} - e\not{A})\psi_L + \bar{\psi}_R(i\not{\partial} - e\not{A})\psi_R = \mathcal{L}_L + \mathcal{L}_R \,. \end{aligned} \tag{1b}$$

Why is it in the second quantized theory that the separate left and right charges
are not conserved, even though there is no apparent interaction between the
left and right worlds, so that in the first quantized theory the left and right
probability currents are conserved?

A detailed analysis of the second quantized theory gives the answer, but
again in terms of a formal rather than physical concept: the Dirac negative
energy sea cannot be defined in a gauge invariant way, separately for the
left and right portions of the model.[9] This is a consequence of gauge-field
configurations that give rise to zero-eigenvalue modes in the Dirac equation
in two dimensions lower.[10] For the four-dimensional theory, we observe that
the two–dimensional Dirac operator in a constant magnetic field possesses zero

nodes. When the full four-dimensional background gauge field in (1) includes uch components, the energy spectrum in the first quantized theory cannot e divided into positive [particle] and negative [anti-particle] states in a gauge nvariant manner, separately for the left and right components. It is the insistence on gauge invariance in the second quantized Dirac theory that produces a quantum mechanical coupling between the left and right worlds.

More specifically, the Lagrangian (1) in four dimensional space-time leads o a three-dimensional Hamiltonian problem for determining the modes to be second quantized

$$Hu_E = [\alpha \cdot (\mathbf{p} - e\mathbf{A}) + eA^0]u_E = E(A)u_E. \qquad (2)$$

With a background gauge potential chosen so that A_x and A_y produce the constant magnetic field B in the z-direction, A_z constant, and vanishing A_0, the two-dimensional zero modes give rise to an energy-momentum dispersion aw in (2) of

$$E(A) = p_z - eA_z \qquad (3a)$$

for the right-handed fermions, and

$$E(A) = -p_z + eA_z \qquad (3b)$$

for the left-handed ones. The zero of each branch cannot be defined in a gauge invariant manner. Moreover, when both branches are included, the ground state [at fixed A_z] of the second quantized theory is defined by "filling" the negative energy levels [at fixed A_z], and leaving the positive levels [at fixed A_z] "empty". However, if A_z is varied adiabatically, empty levels move to positive energies and filled levels move to negative energy [or *vice versa*]. This creates or destroys an amount of charge proportional to $B\delta A_z$ for each chirality, but leaves total charge conserved.

Similarly, in a six-dimensional model, the four dimensional instanton produces the zero mode. The two-dimensional [Schwinger] model realizes this anomaly-producing mechanism trivially, since in two dimensions fewer there is nothing there, and the eigenvalue is obviously zero, while the Hamiltonian problem with vanishing A^0 and constant A^1 obviously possesses eigenvalues of the form (3).[11]

In two dimensions, moreover, we can present the coupling between the left and right worlds very explicitly, because the model is solved.[12] Owing to the two-dimensional identity $i\gamma^\mu\gamma_5 = \varepsilon^{\mu\nu}\gamma_\nu$, one verifies that only one light-cone component of A_μ couples to the right fermions, and the other to the left

$$\mathcal{L}_{(2)} = \bar{\psi}_L\gamma^-(i\partial_- - eA_-)\psi_L + \bar{\psi}_R\gamma^+(i\partial_+ - eA_+)\psi_R. \qquad (4)$$

The "plus" and "minus" components are constructed from the space and time components by the rule $\pm \equiv (0 \pm 1)/\sqrt{2}$. The gauge invariant, effective quantum action is known[12]

$$
\begin{aligned}
-i \ln \det(i\slashed{\partial} - e\slashed{A}) &= \frac{e^2}{2\pi} \int A_\mu \left(g^{\mu\nu} - \frac{\partial^\mu \partial^\nu}{\Box} \right) A_\nu \\
&= -\frac{e^2}{4\pi} \int A_- \frac{\partial_+}{\partial_-} A_- - \frac{e^2}{4\pi} \int A_+ \frac{\partial_-}{\partial_+} A_+ + \frac{e^2}{2\pi} \int A_+ A_- .
\end{aligned}
\tag{5}
$$

The last contribution, a contact term unambiguously dictated by gauge invariance, puts into evidence the quantum mechanical left-right coupling. Note also that properly gauge invariant determinant of Dirac fermions is not merely the product of left- and right-handed determinants — gauge invariance can force contact terms that spoil the factorization.

The two-dimensional model may also be viewed as providing the essence for the higher-dimensional anomaly. We begin in $2d$ dimensional space-time, and argue as follows: 2 dimensions lowers, *i.e.*, in $2d - 2$ dimensions, a zero mode may be established with the help of the index theorem. Existence of the zero mode is assured, provided there is a $2d - 2$-dimensional anomaly, which in turn requires a zero mode in $2d - 4$ dimensions, established by the $2d - 4$-dimensional index theorem and anomaly, *etc.* Thus, in a very precise way, the two-dimensional Abelian anomaly[13] is at the center of the entire anomaly phenomenon.

The above is the most "physical" description of chiral anomalies known to me; but still it uses the unphysical, formal construct of a Dirac sea, and negative energy states. However, it should be recalled that charge fractionization, another unexpected effect of second quantized fermions, is also understood in terms of distortions in the negative energy Dirac sea.[14]

We have learned much from mathematicians about the topological and cohomological necessity of anomalies,[8] but perhaps physics can, in its turn, advance mathematical concepts by insisting on the fact that the essence of the anomaly lies beyond present topological/cohomological ideas. The latter involve integrated, global quantities, like the Chern-Pontryagin number, yet the anomaly is local. Moreover, anomalies are present even in the absence of obstructions, like in Abelian [U(1)] theories, as in the discussed example which, being two-dimensional, hardly possesses any structure, save the anomaly. The U(1) anomaly, on the other hand, appears to be the heart of the matter, not only for the non-Abelian anomalies, but also for the non-perturbative ones.[15]

Thus, it seems to me that we are not yet at the end of the physics nor of the mathematics that can emerge from understanding anomalies. I expect that in this framework, we shall find answers to questions about the precise nature of the vacuum or ground state.

3. ANOMALOUS GAUGE THEORIES

Let us now consider an anomalous theory: right-handed Weyl fermions interacting with a gauge field. Apparently, gauge invariance cannot be maintained owing to the anomaly; second quantization of the coupled gauge field-matter system is problematical. In terms of our earlier discussion, the negative energy chiral anti-fermions cannot be separated gauge invariantly from the positive energy chiral fermions.

The nature of the problem has recently been couched in mathematical terms. We consider first the fermion sector and view the gauge potential, with $A^0 = 0$, as an externally prescribed field — to be quantized later, if possible. Within the fermionic theory, we may construct the unitary operator $U(g) = \exp G_\theta$ that implements the [topologically trivial] gauge transformation $g = e^\theta$. The infinitesimal generator G_θ is $\int d\mathbf{r}\, \theta_a(\mathbf{r})[\delta_a(\mathbf{r}) - i\rho_a(\mathbf{r})]$, where $\delta_a(\mathbf{r}) \equiv -\mathcal{D}_{ab} \cdot [\delta/\delta A_b(\mathbf{r})]$ effects an infinitesimal gauge transformation on $\mathbf{A}_a$ and ρ_a, the fermion charge density, performs the same job on the fermion degrees of freedom; $\mathcal{D}$ is the covariant derivative. The occurrence of anomalies in the generator's commutator algebra[6] has led to the suggestion that the operators U give a projective representation for the gauge group

$$g_1 g_2 = g_{12}$$
$$U(g_1)U(g_2) = e^{i2\pi\omega_2(g_1,g_2)}U(g_{12}), \tag{6}$$

where the 2-cocycle, $2\pi\omega_2$, can be determined by *a priori* arguments when it is cohomological non-trivial.[16]

While such projective representations of transformation groups are familiar in quantum mechanics, this is the first time they make an appearance in quantized gauge theory. Moreover, since gauge invariance is conventionally enforced as the first-class constraint that $U(g)$ leave physical states invariant [for topological trivially gauge transformations], we see that owing to the projective composition law the constraint cannot be satisfied, and the anomalous gauge theory loses gauge invariance. Note, however, that the cocycle occurs even in the cohomologically trivial Abelian theory, for which the original calculation[6] of anomalous commutators was performed. Yet no *a priori* mathematical argument is available to establish existence of the cocycle in the cohomologically trivial case. Here again, we have further evidence that the anomaly goes beyond present mathematical ideas.

What to do with an anomalous theory, other than simply abandoning it? The conventional remedy is to adjust fermion content so that the anomaly is absent — *i.e.*, the 2-cocycle vanishes. This principle leads to a physical prediction, which is thus far satisfied: the number of leptons must match the number of quarks.[17] Moreover, the same principle has focused attention on the most recent version of the string.[18]

Nevertheless, one may still inquire whether an anomalous theory can yield consistent physics. One suggestion is implicit in Faddeev's[16] work: perhaps gauge invariance in the quantum theory should not be implemented as a first-class constraint, but as a higher order one. Thus far, nothing has come from this idea, and it seems to me that further problems abound: the constraints do not properly commute with the electric field, nor do they commute with the Hamiltonian. [However, since an anomalous current divergence can be expressed as a derivative of a gauge non-covariant quantity, one *can* construct a conserved, gauge non-invariant object — but it is unclear that this may be used as a symmetry generator in a Lorentz-invariant theory.]

An alternative position, recently taken by Rajaraman and me, is to give up the constraints altogether, *i.e.*, abandon gauge invariance and view anomalies as a mechanism for gauge symmetry breaking.[19] The original arguments[17] for cancelling anomalies called attention to two possible pitfalls in our adventurous approach: (1) renormalizability may be lost, (2) unitarity may be lost. Rajaraman and I have nothing to say about renormalizability; indeed, we study a finite, two-dimensional model — the chiral Schwinger model with gauge coupling of strength e — which is anomalous. Owing to the model's simplicity, its spectrum may be analyzed. We find that unitarity is preserved; the vector particle acquires a mass, which, however, is not calculable but lies between $e/\sqrt{\pi}$ and infinity; also, there are massless excitations, which appear to be deconfined fermion. Of course, this two-dimensional model is far from reality, and accepting anomalies runs counter to current practice, which has led to the determination of a unique pair of string models. Nevertheless, I find it attractive to speculate that weak interactions, with their deconfined fermions and massive vector mesons, are a remnant of an anomalous chiral gauge theory.

4. 3-COCYCLES

I have mentioned 2-cocycles — they arise in a projective representation of a transformation group and introduce a phase in the composition law. Familiar also are 1-cocycles: they are phases that represent the action on vectors. For example, the action of Galilean transformations on quantum mechanical wave functions includes a phase, as does the action of gauge transformations on quantum field theoretical wave functional in gauge theories with topological Chern-Simons terms.[20]

How about 3-cocycles? These have not arisen in the mathematical discussions of representation theory, but several physicists[21] have pointed out that a non-vanishing 3-cocycle, $2\pi\omega_3$, is a measure of non-associativity in the operator algebra of a representation

$$\Big(U(g_1)U(g_2)\Big)U(g_3) = e^{i2\pi\omega_3(g_1,g_2,g_3)}U(g_1)\Big(U(g_2)U(g_3)\Big). \tag{7}$$

An example can be found in the quantum mechanics of a point particle with charge e, moving at $\mathbf{r}$ in an external magnetic field $\mathbf{B}(\mathbf{r})$, which is not divergence free: $\nabla \cdot \mathbf{B} \neq 0$. The Hamiltonian for this dynamics does not see the magnetic field, because the magnetic field does no work on the charged particle

$$H = \frac{1}{2}v^2. \tag{8}$$

[We set the mass of the particle to unity.] In order that the Lorentz force law be obtained by commutation with the Hamiltonian, we require

$$[r^i, \ r^j] = 0, \qquad [r^i, \ v^j] = i\delta^{ij}, \qquad [v^i, \ v^j] = ie\varepsilon^{ijk}B^k \tag{9}$$

and the Lorentz law follows.

$$\begin{aligned} \dot{r}^i &= i[H, \ r^i] = v^i \\ \dot{v}_i &= i[H, \ v^i] = e\varepsilon^{ijk}v^j B^k. \end{aligned} \tag{10}$$

Finite translations of $\mathbf{r}$ are represented by

$$U(\mathbf{a}) = e^{i\mathbf{a}\cdot\mathbf{v}} \tag{11}$$

since

$$U(\mathbf{a})\mathbf{r}U^{-1}(\mathbf{a}) = \mathbf{r} + \mathbf{a}. \tag{12}$$

However, these do not represent the translation group faithfully since one finds from (9)

$$U(\mathbf{a}_1)U(\mathbf{a}_2) = e^{-ie\Phi}U(\mathbf{a}_1 + \mathbf{a}_2), \tag{13}$$

where Φ is the flux through the triangle at $\mathbf{r}$ formed from $\mathbf{a}_1$ and $\mathbf{a}_2$; see Figure 1.

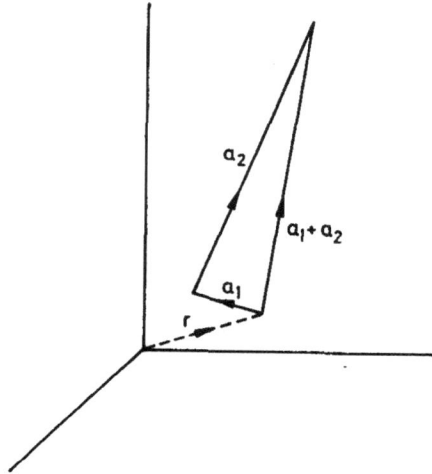

Fig. 1. The triangle at $\mathbf{r}$ through which the flux Φ is calculated.

Moreover, by considering the triple product $U(\mathbf{a}_1)U(\mathbf{a}_2)U(\mathbf{a}_3)$, associated in the two different ways as in (7), one finds a 3-cocycle, given by $-e$ times the total flux out of the tetrahedron formed at $\mathbf{r}$ from $\mathbf{a}_1$, $\mathbf{a}_2$ and $\mathbf{a}_3$; see Figure 2.

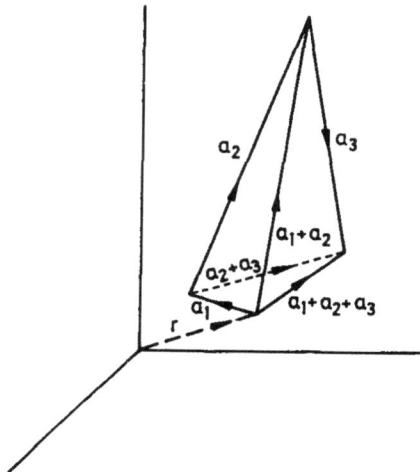

Fig. 2. The tetrahedron at $\mathbf{r}$ through which the flux determining the 3-cocycle is calculated.

Of course, when $\nabla \cdot \mathbf{B}$ vanishes, so does the total flux through any closed surface; there is no 3-cocycle and $\mathbf{v}$ may be realized by linear operators on a Hibert space: $\mathbf{v} = -i\nabla - e\mathbf{A}(r)$, $\mathbf{B}(r) = \nabla \times \mathbf{A}(r)$.

When there are magnetic sources, the flux is non-zero, but we may still achieve associativity provided ω_3 is an integer, since then $e^{i2\pi\omega_3} = 1$. This requirement forces: (1) $\nabla \cdot \mathbf{B}$ to consist of localized delta functions, so that the total flux not vary continuously when the $\mathbf{a}_i$'s change; *i.e.*, the sources must be point-monopoles; (2) since a monopole of strength g produces the cocycle $-4\pi eg$, eg must satisfy the Dirac quantization condition. In this way, removal of the 3-cocycle, which is necessary for conventional quantum mechanics with associative operators on Hibert space, limits magnetic sources to quantized Dirac monopoles. Other magnetic sources lead to a non-associative algebra.

Finally, note that for infinitesimal generators, a non-vanishing 3-cocycles, *i.e.*, non-associativity, implies failure of the Jacobi identity. Indeed, from (9) one deduces that[22]

$$\left[v^1, \left[v^2, v^3\right]\right] + \left[v^2, \left[v^3, v^1\right]\right] + \left[v^3, \left[v^1, v^2\right]\right] = e\nabla \cdot \mathbf{B}. \qquad (14)$$

[The triple products are associated in the obvious way.]

5. EXTERNAL U(1) CONNECTIONS IN FIELD THEORY

Is there anything similar in field theory? Violations of the Jacobi identity had been found years ago, but not in gauge group generators, rather in the old U(6)×U(6) current algebra.[23] While it may be interesting to frame this into a coherent mathematical picture, it seems far removed from current interest. However, you may be surprised to hear that an "external magnetic field" of particle mechanics has a sensible analog in gauge field theory.

The aspect of particle dynamics in an external magnetic field that is of interest for the field theoretic generalization, is that the magnetic field is invisible in the Hamiltonian, but reappears in the velocity commutator, as in (8) and (9) *i.e.*, the canonical momentum does not coincide with the velocity.

We have grown accustomed to topological terms contributing to gauge field actions — the θ parameter multiplying the Chern-Pontryagin density in even dimensions, the Chern-Simons term with quantized coefficient in odd dimensions.[20] Since both are topological world scalars, not involving the metric tensor, they do not contribute to the energy-momentum tensor, and the Hamiltonian retains its conventional form

$$H = \int dr \left(\frac{1}{2} E_a^i E_a^i + \frac{1}{4} F_a^{ij} F_a^{ij}\right). \qquad (15)$$

However, the canonical field momenta differ from the field "velocities" — $\mathbf{E}_a$, and the difference may be ascribed to an external U(1) field connection. The U(1) field curvature — the analog of the external magnetic field — is determined by the equal-time commutator of the field velocities.

For theories in even-dimensional space-time with a Chern-Pontryagin density, the commutator vanishes — the external field curvature is zero, the connection is flat. This is to be expected, since the topological term does not affect equations of motion, and the connection is a pure functional gauge. It can be gauged away, and this is the familiar procedure, which shifts the vacuum angle from the Lagrangian to the state.[20] In this way, the situation is similar to point vortices on a plane.

The Chern-Simons term in odd-dimensional space-time gives rise to a non-vanishing external field curvature, since it does modify equations of motion. In the three-dimensional theory for example, the equal-time commutator [which lives in two-space] reads

$$i[E_a^i(\mathbf{r}), E_b^j(\mathbf{r}')] = \frac{m}{4\pi}\delta_{ab}\varepsilon^{ij}\delta(\mathbf{r}-\mathbf{r}') \qquad m = 0, \pm1, \ldots . \qquad (16)$$

The curvature may be described as a constant, functional external U(1) "magnetic" field. That the coefficient is quantized brings out the analogy, once again, between the quantization of Chern-Simons parameters in field theory and the point Dirac monopole.

While this viewpoint towards topological contributions to the gauge field action[24] does not produce new results, it suggests a direction for further investigation. Are there other forms of external U(1) field connections and curvatures that may be added to conventional gauge dynamics?

REFERENCES

1. H. Dürr, W. Heisenberg, H. Mitter, S. Schlieder and K. Yamazaki, *Z. Natur-forsch.* **14a**, 441 (1959).

2. N. Bohr, *Nature* **138**, 25 (1936).

3. G. Wentzel, *Phys. Rev.* **74**, 1070 (1948).

4. H. Fukuda and K. Miyamoto, *Prog. Theor. Phys.* (Kyoto) **4**, 347 (1949); J. Steinberger, *Phys. Rev.* **76**, 1180 (1949); J. S. Bell and R. Jackiw, *Nuovo Cim.* **60A**, 47 (1969); S. L. Adler, *Phys. Rev.* **177**, 2426 (1969). Bell and I were directly motivated by Steinberger's calculation, who described it to us over coffee on the CERN patio. Although both the earlier and later papers deal with neutral pion decay, the notion that a symmetry — the chiral symmetry of pion

physics — can be broken by quantum effects appears only in the work from the late 1960's, since that symmetry was unknown in the earlier period.

5. J. Schwinger, *Phys. Rev.* **82**, 664 (1951); C. Hagen, *Phys. Rev.* **177**, 2622 (1969); R. Jackiw and K. Johnson, *Phys. Rev.* **182**, 1459 (1969).

6. Jackiw and Johnson, Ref. 5; S. L. Adler and D. Boulware, *Phys. Rev.* **184**, 1740 (1969). These anomalous commutators are sometimes called "Schwinger terms", although this is confusing since Schwinger examined non-canonical contributions to the commutator between time and space components of the current, while here the commutator between time components is under discussion. Anomalies in time-space component commutators were established earlier by T. Goto and T. Imamura, *Prog. Theor. Phys.* (Kyoto) **14**, 396 (1955) and J. Schwinger, *Phys. Rev. Lett.* **3**, 296 (1959).

7. K. Fujikawa, *Phys. Rev. Lett.* **42**, 1195 (1979).

8. The Abelian anomaly is related to the Atiyah-Singer index theorem; see A. Schwarz. *Phys. Lett.* **67B**, 172 (1977); L. Brown, R. Carlitz and C. Lee, *Phys. Rev. D* **16**, 417 (1977); R. Jackiw and C. Rebbi, *Phys. Rev. D* **16**, 1052 (1977). The non-Abelian anomaly is related to the index theorem in two dimensions higher; see M. Atiyah and I. Singer, *Proc. Natl. Acad. Sci.* (USA) **81**, 2597 (1984); L Alvarez-Gaumé and P. Ginsparg, *Nucl. Phys.* **B243**, 449 (1984).

9. This argument was developed by many people; I follow the presentation of R. Feynman.

10. Note that the topological arguments in Ref. 8, which establish gauge non-invariance of the chiral fermion determinant, use zero modes in two dimensions higher.

11. For a Hamiltonian analysis of two-dimensional models, see N. Manton, *Ann. Phys.* (NY) **159**, 220 (1985).

12. J. Schwinger, *Phys. Rev.* **128**, 2425 (1962).

13. K. Johnson, *Phys. Lett.* **5**, 253 (1963).

14. R. Jackiw and C. Rebbi, *Phys. Rev. D* **13**, 3398 (1976); W. P. Su, J. R. Schrieffer and A. Heeger, *Phys. Rev. Lett.* **42**, 1698 (1979) and *Phys. Rev. B* **22**, 2099 (1980). For a review see "Fermion Fractionization", reprinted in this volume on p. 79.

15. E. Witten, *Phys. Lett.* **117B**, 324 (1982). The relation between non-perturbative anomalies and the U(1) anomaly was found by J. Goldstone, and is explained by R. Jackiw in *Relativity Groups and Topology II*, B. DeWitt and R. Stora, eds. (North-Holland, Amsterdam, 1984) and in S. Treiman, R. Jackiw, B. Zumino and E. Witten, *Current Algebra and Anomalies* (Princeton University Press, Princeton NJ/World Scientific, Singapore, 1983).

16. L. Faddeev, *Phys. Lett.* **145B**, 81 (1984); L. Faddeev and S. Shatashvili, *Theor. Mat. Fiz.* **60**, 206 (1984) [English translation: *Theor. Math. Phys.* **60**, 770 (1984)]; J. Mickelsson, *Comm. Math. Phys.* **97**, 361 (1985); S. Y. Jo, *Phys. Lett.* **163B**, 353 (1985).

17. D. J. Gross and R. Jackiw, *Phys. Rev.* D **6**, 477 (1972); C. Bouchiat, J. Iliopolous and Ph. Meyer, *Phys. Lett.* **38B**, 519 (1972).
18. M. Green and J. Schwarz. *Phys. Lett.* **149B**, 117 (1984).
19. R. Jackiw and R. Rajaraman, *Phys. Rev. Lett.* **54**, 1219 (1985); see "Update on Anomalous Theories", reprinted in this volume on p. 55.
20. For a review, see Jackiw, Ref. 15.
21. R. Jackiw, *Phys. Rev. Lett.* **54**, 159 (1985); B. Grossman, *Phys. Lett.* **152B**, 92 (1985); Y. S. Wu and A. Zee, *Phys. Lett.* **152B**, 98 (1985).
22. H. Lipkin, W. Weisberger and M. Peshkin, *Ann. Phys.* (NY) **53**, 203 (1969). When the three-fold Jacobi identity fails and an algebra is non-associative, one may impose a four-fold identity, the so-called Malcev identity, which requires that $\nabla \cdot \mathbf{B}$ be constant. When this fails, one can impose a five-fold identity, *etc.*
23. K. Johnson and F. Low, *Prog. Theor. Phys.* (Kyoto) Suppl. **37–38**, 74 (1966).
24. R. Jackiw, in *Quantum Field Theory and Quantum Statistics*, I. Batalin, C. Isham and G. Vilkovisky, eds. (A. Hilger, Bristol, UK, 1985); M. Asorey and P. Mitter, *Phys. Lett.* **153B**, 147 (1985).

Paper I.3

In *Memoriam* M. A. B. Bég, who was interested in the quantum mechanics of δ-function potentials, I show how the two-dimensional case provides the simplest example of anomalous symmetry breaking; specifically scale symmetry is lost after quantization.

DELTA FUNCTION POTENTIALS
IN TWO- AND THREE-DIMENSIONAL
QUANTUM MECHANICS

M. A. B. Bég Memorial Volume, A. Ali and P. Hoodbhoy, eds. (*World Scientific, Singapore, 1991*)

I. INTRODUCTION

Two- and three-dimensional δ-function interactions in Schrödinger theory are the formal non-relativistic limits for the scalar field ϕ^4 self-interactions of relativistic quantum field theory in $(2+1)$- and $(3+1)$-dimensional space-time, respectively. The quantum mechanical problems possess non-trivial dynamics if infinite renormalization or self-adjoint extension of the Hamiltonians is performed. The field theory is known to exist for the lower dimensionality, but for the higher dimensionality it is conjectured to be trivial. Evidently, the non-relativistic limits, supplemented by renormalization or self-adjoint extension, do not show this variety. Also the planar δ-function interaction formally admits an $SO(2,1)$ dynamical symmetry, but quantization necessarily spoils the invariance, putting into evidence the simplest example of quantum mechanical symmetry breaking. In this pedagogical essay, dedicated to the memory of M. A. B. Bég, work initiated by him is elaborated.

Baqi Bég was an eminent particle theorist who never lost sight of the *physical* goals of our profession — so much more difficult to attain than the purely mathematical. Nevertheless, he used mathematical tools with ease, and at various times during the development of modern fundamental physics Bég illuminated crucial phenomenological/experimental issues. The importance of Baqi's work and the esteem in which it is held are well exemplified by "Bég's Theorem" of nuclear physics, which "surprised" Peierls,[1] "Bég's Sum Rule" of current algebra, discussed in Adler's and Dashen's definitive book on that subject[2]; his many contributions to the quark model, through the framework of higher symmetries, a selection of which is collected in Dyson's reprint volume[3]; and also his attempts to complete the "standard model" by urging a dynamical mechanism for its spontaneous symmetry breaking, as documented in the collection of sources for these ideas.[4]

Spontaneous symmetry breaking in the standard model concerned Bég in the final period of his research. The tension between the model's unquestioned phenomenological success and the theoretical inadequacy of the scalar field (Higgs) mechanism for its spontaneous symmetry breaking informed his activity. Taking account of the conjectured non-existence of the scalar field self-interaction in four-dimensional space-time,[5] Baqi on the one hand attempted to do away with the Higgs sector of the model, replacing its function by a dynamical mechanism[4]; on the other hand, he tried characteristically to extract phenomenologically useful and experimentally verifiable information from the conjectured triviality of Higgs field dynamics.[6]

In order to understand better the nature of the scalar field ϕ^4 self-interaction Bég and Furlong[7] had the good idea to consider the non-relativistic limit of the model. When that limit is taken formally, particle number is conserved, the interaction between particles becomes the zero-range δ-function and dynamics is governed by tractable quantum mechanics. But even though one is dealing with quantum mechanics, there arise ultraviolet divergences, reminiscent of quantum field theory and renormalization is required. Bég and Furlong then showed that the δ-function interaction in three-dimensional space gives rise to a trivial S-matrix, when the bare/unrenormalized coupling constant is finite, a result that does not disagree with triviality of the relativistic theory, but of course, neither does it establish triviality relativistically. The same conclusion was later obtained by K. Huang[8] in an independent investigation.

Owing to the distress inflicted on the standard model by the absence of the Higgs interaction, various proposals have been made for defining a non-trivial relativistic ϕ^4 theory. While one cannot deem these attempts convincingly successful, it is natural to inquire whether one can evade the Bég-Furlong-Huang result in the non-relativistic theory.

In this essay, which is inspired by Baqi's work and is dedicated to his memory, I point out that indeed procedures are available for defining a non-trivial δ-function interaction in three dimensions. Moreover, the same methods work in the same way in two dimensions; indeed they *must* be used if triviality is to be avoided there, because also the planar δ-function interaction needs renormalization, which extinguishes a finite bare/unrenormalized coupling.

While the positive non-relativistic results are pleasing, they do not of course illuminate the situation in the relativistic quantum field theories, though it is reassuring that in three-dimensional space-time the relativistic ϕ^4 interaction is known to be alive and well, just like the "improved" non-relativistic δ-function interaction.

The procedures for defining two- and three-dimensional δ-function interactions are two fold. One may simply perform *infinite* renormalization, arriving at amplitudes parametrized by a finite (by definition) *renormalized* coupling, which in the case of attraction may be alternatively expressed in terms of an uncalculable bound state energy. More satisfactory, especially within a mathematical frame, is the view that the δ-function interaction is merely a self-adjoint extension to a formally Hermitian, non-interacting Hamiltonian on a space with one point removed. The parameter in the extension is finite and plays the role of renormalized coupling strength.

These approaches to the δ-function interaction are as old as the subject.

The first analyses, in three dimensions, were performed in physicists' terms by Bethe, Peierls and Fermi.[9] Mathematically, rigorous treatments begin with the work of Berezin and Faddeev[10] and now, there is even a monograph on the subject.[11] These days, the two-dimensional δ-function interaction and the equivalent self-adjoint extension have arisen in discussions of point particle dynamics in $(2+1)$-dimensional gravity[12] and in Chern-Simons gauge theory (Aharonov-Bohm/Ehrenberg-Siday interaction).[13]

Section II is devoted to qualitative remarks about quantum mechanical interactions and their symmetries. The δ-function Hamiltonians that we shall consider are introduced. The two-dimensional model is especially noteworthy, because on the classical/formal level it possesses a dilation symmetry, which is then *necessarily* destroyed by quantization — an effect seen in quantum field theory as the anomaly phenomenon,[14] but not previously identified in quantum mechanics.

In Section III, the Bég-Furlong-Huang calculation is reconsidered but with infinite renormalization yielding non-trivial dynamics, and this is repeated in two dimensions. The same results are regained by the method of self-adjoint extension, whose symmetry breaking properties in two dimensions are highlighted.

In Section IV, the Aharonov-Bohm/Ehrenberg-Siday effect in the Dirac equation is shown to lead in an equivalent Schrödinger equation to a δ-function interaction, which for consistency *must* be interpreted as a self-adjoint extension. In this way, the identification of the δ-function interaction with the self-adjoint extension is made complete.

The final Section V comprises concluding remarks.

II. DISCUSSION

The typical quantum mechanical Hamiltonian operator consists of a kinetic term involving spatial derivatives and a potential function of position. Of course, the expression appears Hermitian. However, appearances can be misleading, and experience shows that when the short-distance behavior of the potential function is the same or more singular than that of the kinetic operator pathologies can mar the eigenvalue problem. These arise because the Hamiltonian operator though formally Hermitian ("symmetric" in mathematical terminology) is not self-adjoint (the domain of definition of the operator does not coincide with the domain of the adjoint).

Some familiar examples: The non-relativistic r^{-2} potential shares an r^{-2} singularity with the Laplacian kinetic operator; when the potential is too strongly attractive (the strength depends on the dimension of space) the bound state spectrum is not discrete. For the Dirac-Coulomb problem, both the kinetic term $\boldsymbol{\alpha} \cdot \frac{1}{i} \boldsymbol{\nabla}$ and the r^{-1} Coulomb potential behave as an inverse length as short distances, and the bound state energies become complex for sufficiently strong attraction.

Less familiar, but similar examples arise with vector potentials in Dirac theory: the Hamiltonian $\boldsymbol{\alpha} \cdot (\frac{1}{i} \boldsymbol{\nabla} - \mathbf{A})$, is not self-adjoint when $\mathbf{A} \propto r^{-1}$ at small r, as for a point monopole in three dimensions[15] or a point vortex in two dimensions.[13]

The above information points to the following conclusions about Hamiltonians with δ-function potentials in non-relativistic Schrödinger equations for various dimensions

$$H = \frac{1}{2}p^2 + v\delta(\mathbf{r}) = -\frac{1}{2}\nabla^2 + v\delta(\mathbf{r}) \,. \tag{2.1}$$

(Mass and $\hbar$ are set to unity.) In two and three dimensions, where a δ-function scales as r^{-n}, $n = 2$ and 3, the short-distance singularity of the potential is respectively comparable to, and more singular than the kinetic term. Consequently, we anticipate difficulties with the eigenvalue equation

$$H\psi = E\psi \,. \tag{2.2}$$

Only the one-dimensional $\delta(x)$ potential presents a simple problem — one that is found in most quantum mechanical texts — but a $\delta'(x)$ potential exhibits pathologies similar to the above higher-dimensional cases.[11]

The two-dimensional δ-function [and the one-dimensional δ'-function] are additionally interesting in that the Hamiltonian does not contain dimensional parameters: v in (2.1) is dimensionless. This property, shared by the r^{-2} potential in any number of dimensions, renders the theory scale invariant, at least formally.

Specifically, what is meant here is that there exists a dilation operator D

$$D = tH - \frac{1}{4}(\mathbf{r} \cdot \mathbf{p} + \mathbf{p} \cdot \mathbf{r}) \tag{2.3}$$

that implements the dilation transformation on the dynamical variable $\mathbf{r}$,

$$\delta_D \mathbf{r} = i[D, \, \mathbf{r}] = t \cdot \dot{\mathbf{r}} - \frac{1}{2}\mathbf{r} \tag{2.4}$$

showing that $\mathbf{r}$ has the scale dimension $-1/2$. Moreover, commuting with the Hamiltonian gives

$$i[D, \, H] = H \tag{2.5}$$

i.e., H has unit scale dimension. Therefore, D is a constant of motion

$$\frac{dD}{dt} = i[H, D] + \frac{\partial D}{\partial t} = 0 \,. \tag{2.6}$$

Establishing (2.5) requires the identity

$$i[\mathbf{r} \cdot \mathbf{p}, \; \delta(\mathbf{r})] = \mathbf{r} \cdot \boldsymbol{\nabla}\delta(\mathbf{r}) = -2\delta(\mathbf{r}) \,. \tag{2.7}$$

[In one dimension, what is needed is $i[xp, \delta'(x)] = x\delta''(x) = -2\delta'(x)$.] This insures that the interaction scales with $\mathbf{r}$ as r^{-2}, so that its scale dimension is unity.

One consequence of scale invariance is that quantum scattering phase shifts must be energy-independent (there is no scale to give an energy dependence!) — a fact which is explicitly verified by the energy-independent phase shifts of the r^{-2} potential.[16]

Another consequence, which follows from (2.3) in those simple models where $\mathbf{r} \cdot \mathbf{p} + \mathbf{p} \cdot \mathbf{r} = dr^2/dt$, is that D may be written as a total time derivative

$$D = \frac{d}{dt}\left(\frac{t^2}{2}H - \frac{1}{4}r^2\right) \,.$$

This reveals the presence of one more constant of motion

$$K = -t^2 H + 2tD + \frac{1}{2}r^2 \,, \tag{2.8}$$

$$\frac{dK}{dt} = 0 \tag{2.9}$$

which is the conformal generator that transforms $\mathbf{r}$ according to

$$\delta_K \mathbf{r} = i[K, \; \mathbf{r}] = t^2 \dot{\mathbf{r}} - t\mathbf{r} \,. \tag{2.10}$$

Here we have another example of how scale invariance sometimes (but not always!) implies conformal invariance.

The operators H, K, D close on commutation: in addition to (2.5) it is true that

$$i[K, \; D] = K \,, \tag{2.11}$$

$$i[H, \; K] = -2D \,. \tag{2.12}$$

These commutators not only verify (2.9), but also show that the invariance algebra is $SO(2, 1)$.

To conclude: $SO(2, 1)$ is a symmetry group of the r^{-2} potential — a well-known fact that can be maintained quantum mechanically, provided there is not too much attraction[17] — and formally appears also to be a symmetry of the $\delta(\mathbf{r})$ potential in two dimensions. [$SO(2, 1)$ is also known to be a quantum mechanical symmetry of the non-relativistic point magnetic monopole[18] and point vortex.[19]]

Our reason for entering upon this discussion of $SO(2, 1)$ dynamical symmetry is that we shall soon establish the remarkable result that any quantum mechanical definition of the apparently $SO(2, 1)$ invariant, two-dimensional $\delta(\mathbf{r})$ potential, necessarily violates $SO(2, 1)$ invariance. Doubtlessly, this is the most elementary manifestation of quantum mechanical symmetry breaking.

III. SOLVING THE δ-FUNCTION SCHRÖDINGER EQUATION

A. Renormalization

We seek scattering solutions to (2.1) and (2.2) for the spatial dimensionalities n, $n = 2, 3$, which we treat simultaneously.

In terms of momentum-space wave functions

$$\phi(\mathbf{p}) = \int d^n r e^{i\mathbf{p}\cdot\mathbf{r}} \psi(\mathbf{r}) \tag{3.1}$$

equations that we solve are

$$\frac{1}{2}(p^2 - k^2)\phi(\mathbf{p}) = -v\psi(0)$$

$$\frac{k^2}{2} = E. \tag{3.2}$$

The scattering solutions are

$$\phi(\mathbf{p}) = (2\pi)^n \delta(\mathbf{p} - \mathbf{k}) - \frac{2v}{p^2 - k^2 - i\varepsilon}\psi(0). \tag{3.3}$$

Evidently, the scattering amplitudes are proportional to $v\psi(0)$, which is self-consistently determined from (3.3) by

$$\psi(0) = \int \frac{d^n p}{(2\pi)^n} \varphi(\mathbf{p}) = 1 - 2v I_n(-k^2 - i\varepsilon)\psi(0)$$

$$v\psi(0) = \left(\frac{1}{v} + 2I_n(-k^2 - i\varepsilon)\right)^{-1} \tag{3.4}$$

where I_n is the integral

$$I_n(z) = \int \frac{d^n \mathbf{p}}{(2\pi)^n} \frac{1}{p^2 + z} \tag{3.5}$$

which diverges in the ultraviolet for $n \geq 2$.

To make progress, we regulate, for example by limiting $|\mathbf{p}|$ at Λ. It follows for large Λ that

$$I_2^\Lambda(z) = \frac{1}{4\pi} \ln \frac{\Lambda^2}{z}, \tag{3.6}$$

$$I_3^\Lambda(z) = \frac{1}{2\pi^2}\Lambda - \frac{1}{4\pi}\sqrt{z}. \tag{3.7}$$

Alternatively, one may use dimensional regularization and find

$$I_2^\varepsilon(z) = \frac{1}{4\pi} \ln \frac{4\pi \, e^{-\gamma + 1/\varepsilon}}{z}, \tag{3.8}$$

$$I_3^\varepsilon(z) = -\frac{1}{4\pi}\sqrt{z}. \tag{3.9}$$

In (3.8) we calculate with $n = 2 - 2\varepsilon$ and take the limit $\varepsilon \to 0$, obtaining a result identical to (3.6). In (3.9) there is no dependence on ε, the departure from three dimensions; a finite answer is obtained as is characteristic for dimensionally regulated, odd-dimensional integrals. From (3.4) $v\psi(0)$ is determined for the two cases as

$$n = 2: \quad v\psi(0) = \left(\frac{1}{v} + \frac{1}{\pi} \ln \frac{\Lambda}{\mu} - \frac{1}{\pi} \ln \frac{k}{\mu} + \frac{i}{2} \right)^{-1}, \tag{3.10}$$

$$n = 3: \quad v\psi(0) = \left(\frac{1}{v} + \frac{1}{\pi^2}\Lambda + \frac{ik}{2\pi} \right)^{-1} \tag{3.11}$$

except that with dimensional regularization $\frac{1}{\pi^2}\Lambda$ is absent from (3.11). In (3.10) μ is a convenient normalization point.

As Λ is removed to infinity, $v\psi(0)$ and therefore the scattering amplitudes vanish, both for $n = 2$ and $n = 3$, *provided v is finite*. At $n = 3$, the result of Bég-Furlong and Huang is thus regained, but note the curiosity that with dimensional regularization the three-dimensional answer is cut-off independent and finite. Moreover, the two-dimensional scattering amplitude vanishes for both regularization procedures.

In the spirit of quantum field theory, it is very plausible to take the bare coupling to be cut-off dependent and to introduce a renormalized coupling constant g, in terms of which (3.10) and (3.11) read

$$n = 2: \quad v\psi(0) = \left(\frac{1}{g} - \frac{1}{\pi}\ln\frac{k}{\mu} + \frac{i}{2}\right)^{-1}, \tag{3.12}$$

$$n = 3: \quad v\psi(0) = \left(\frac{1}{g} + \frac{ik}{2\pi}\right)^{-1}. \tag{3.13}$$

These are finite and well-defined, provided $1/v$ absorbs by definition the cut-off dependence

$$n = 2: \quad \frac{1}{g} = \frac{1}{v} + \frac{1}{\pi}\ln\frac{\Lambda}{\mu}, \tag{3.14}$$

$$n = 3: \quad \frac{1}{g} = \frac{1}{v} + \frac{1}{\pi^2}\Lambda. \tag{3.15}$$

To obtain the scattering amplitude, we present (3.3) in position space

$$\psi(\mathbf{r}) = e^{i\mathbf{k}\cdot\mathbf{r}} - 2vG_k(r)\psi(0), \tag{3.16}$$

where $G_k(r)$ are the Green's functions appropriate to the two dimensionalities

$$(-\nabla^2 - k^2)G_k(r) = \delta(\mathbf{r}), \tag{3.17}$$

$$n = 2: \quad G_k(r) = \frac{i}{4}H_0^{(1)}(kr) \xrightarrow[r\to\infty]{} \frac{1}{2\sqrt{2\pi kr}}e^{i(kr+\pi/4)}, \tag{3.18}$$

$$n = 3: \quad G_k(r) = \frac{1}{4\pi}\frac{e^{ikr}}{r}. \tag{3.19}$$

Upon identifying the scattering amplitude from the asymptotic behavior of the scattering wave function,

$$n = 2: \quad \psi(\mathbf{r}) \to e^{i\mathbf{k}\cdot\mathbf{r}} + \frac{1}{\sqrt{r}}f(\theta)e^{i(kr+\pi/4)}, \tag{3.20}$$

$$n = 3: \quad \psi(\mathbf{r}) \to e^{i\mathbf{k}\cdot\mathbf{r}} + \frac{1}{r}f(\theta)e^{ikr} \tag{3.21}$$

we obtain

$$n = 2: \quad f(\theta) = -\frac{1}{\sqrt{2\pi k}}v\psi(0) = -\frac{1}{\sqrt{2\pi k}}\left(\frac{1}{g} - \frac{1}{\pi}\ln\frac{k}{\mu} + \frac{i}{2}\right)^{-1}, \quad (3.22)$$

$$n = 3: \quad f(\theta) = -\frac{1}{2\pi}v\psi(0) = -\left(\frac{2\pi}{g} + ik\right)^{-1}. \quad (3.23)$$

Only s-wave scattering takes place, whose phase shifts may be read off from standard formulas

$$n = 2: \quad f(\theta) = \frac{1}{i\sqrt{2\pi k}}\sum_{m=-\infty}^{\infty}(e^{2i\delta_m} - 1)e^{im\theta}, \quad (3.24)$$

$$n = 3: \quad f(\theta) = \frac{1}{2ik}\sum_{l=0}^{\infty}(e^{2i\delta_l} - 1)P_l(\cos\theta). \quad (3.25)$$

Comparison with (3.22) and (3.23) yields

$$n = 2: \quad \text{ctn}\,\delta_0 = \frac{1}{\pi}\ln\frac{k^2}{\mu^2} - \frac{2}{g}, \quad (3.26)$$

$$n = 3: \quad \tan\delta_0 = -\frac{gk}{2\pi}. \quad (3.27)$$

Note that the two-dimensional scattering phase shift has acquired an $\ln k$ dependence, in clear violation of scale invariance, except of course at $g = 0$, which corresponds to no interaction.

With attractive δ-functions there exist bound states, provided v is renormalized. Then in the scattering amplitudes the renormalized coupling constant may be replaced by the bound state energy, which in two dimensions is an example of dimensional transmutation within quantum mechanics.[20]

The bound state, momentum space wave functions $\phi_B(\mathbf{p})$ satisfy (3.2) with $E = -B$, and the solution are

$$\phi_B(\mathbf{p}) = -\frac{2v\psi_B(0)}{p^2 + 2B}. \quad (3.28)$$

This implies

$$\psi_B(0) = -2v\int\frac{d^n\mathbf{p}}{(2\pi)^n}\frac{1}{p^2 + 2B}\psi_B(0) \quad (3.29a)$$

or

$$-\frac{1}{2v} = I_n(2B) \qquad (3.29b)$$

which should be used to determine B in terms of v, but of course (3.29) exhibits divergences that shall be discussed presently. First, let us express everything in terms of the bound state energies B.

The wave function normalizations fix $2v\psi_B(\mathbf{0})$

$$n = 2: \quad \phi_B(\mathbf{p}) = \sqrt{8\pi B}\,\frac{1}{p^2 + 2B}\,, \qquad (3.30)$$

$$n = 3: \quad \phi_B(\mathbf{p}) = (128\pi^2 B)^{1/4}\frac{1}{p^2 + 2B}\,. \qquad (3.31)$$

Equation (3.29b) may be used to eliminate $1/v$ in the scattering solutions. In (3.4) $v\psi(\mathbf{0})$ becomes

$$v\psi(\mathbf{0}) = \frac{1}{2}[I_n(-k^2 - i\varepsilon) - I_n(2B)]^{-1}\,. \qquad (3.32)$$

Since a single subtraction renders I_n finite for $n = 2$ and 3, the scattering amplitudes can be expressed in terms of "physical" quantities

$$n = 2: \quad f(\theta) = -\frac{1}{\sqrt{2\pi k}}\left(\frac{1}{2\pi}\ln\frac{2B}{k^2} + \frac{i}{2}\right)^{-1}\,, \qquad (3.33)$$

$$n = 3: \quad f(\theta) = -\left(\sqrt{2B} + ik\right)^{-1}\,. \qquad (3.34)$$

Because of the divergences, bound state energies are not calculable: from (3.6), (3.7) [or (3.8), (3.9)] and (3.29b) we get

$$n = 2: \quad -\frac{1}{v} = \frac{1}{2\pi}\ln\frac{\Lambda^2}{2B}\,, \qquad (3.35)$$

$$n = 3: \quad -\frac{1}{v} = \frac{1}{\pi^2}\Lambda - \frac{1}{2\pi}\sqrt{2B}\,. \qquad (3.36)$$

The Λ dependence may be combined with $1/v$ and hidden in the renormalized coupling $1/g$ as in (3.14) and (3.15). Then B is related to g by

$$n = 2: \quad \sqrt{2B} = \mu e^{\pi/g}\,, \qquad (3.37)$$

$$n = 3: \quad \sqrt{2B} = \frac{2\pi}{g}\,. \qquad (3.38)$$

This is also seen in (3.22), (3.23): the scattering amplitudes possess poles on the positive imaginary k axis corresponding to the above bound state energies. Note that (3.38) requires g to be positive, but no sign restriction on g need be made in (3.37).

Formulas (3.35) and (3.36) may be used to put into evidence a physical effect. A regulated expression for the δ-function potential is posited

$$V_n(\mathbf{r}) = \frac{c_n v}{2\pi R^{n-1}} \delta(r - R)$$

$$c_2 = 1, \qquad c_3 = \frac{1}{2} \tag{3.39}$$

that effectively reproduces the potential in (2.1) when $R \to 0$. V_n supports s-wave bound states. The binding energies B are obtained by matching the discontinuities at $r = R$ in the logarithmic derivatives of the wave functions against the coefficients of the δ-function (3.39). This leads to the equations

$$n = 2: \quad -\frac{1}{v} = \frac{1}{\pi} I_0(\sqrt{2B}R) K_0(\sqrt{2B}R) \approx \frac{1}{2\pi} \ln \frac{4e^{-2\gamma}/R^2}{2B}, \tag{3.40}$$

$$n = 3: \quad -\frac{1}{v} = \frac{1}{2\pi R} \frac{1 - e^{\sqrt{8B}R}}{\sqrt{8B}R} \approx \frac{1}{2\pi R} - \frac{1}{2\pi}\sqrt{2B}. \tag{3.41}$$

The second approximate equalities are valid as the regulator is removed and R is small; they are seen to reproduce (3.35) and (3.36) with $\Lambda \propto R^{-1}$. Similar results are obtained by a square well or lattice regularization of the δ-function.[21]

B. Self-Adjoint Extension

Although regularization and infinite renormalization of the δ-function interaction strength produces physically sensible (unitary) and non-trivial scattering amplitudes, and also the possibility of bound states, it would seen preferable to arrive at the results without introducing the mathematically awkward "infinite" quantity $\Lambda \propto R^{-1}$. This can be achieved through the method of self-adjoint extension.[11]

Consider the free Schrödinger operator

$$H^0 = \frac{1}{2}p^2 = -\frac{1}{2}\nabla^2. \tag{3.42}$$

This is Hermitian and self-adjoint when acting on functions that are finite. However, for two and three dimensions H^0 remains self-adjoint even when the finiteness requirement is relaxed: functions are permitted to diverge at isolated points, provided they remain square integrable; and also a boundary condition, consistent with self-adjointness of H^0, needs to be specified. We take a single point of divergence, the origin, and the required boundary condition is imposed on s-waves; it involves an arbitrary parameter, the self-adjoint extension parameter λ[11]

$$n = 2: \quad \lim_{r \downarrow 0} \frac{\psi(r)}{\ln r} = \frac{\lambda}{\pi} \lim_{r \downarrow 0} \left[\psi(r) - \lim_{r' \downarrow 0} \frac{\psi(r')}{\ln r'} \ln r \right], \qquad (3.43)$$

$$n = 3: \quad \lim_{r \downarrow 0} r\psi(r) = -\frac{\lambda}{2\pi} \lim_{r \downarrow 0} [\psi(r) + r\psi'(r)]. \qquad (3.44)$$

This defines the extended Hamiltonians H^λ; $\lambda = 0$ corresponds to the conventional free Hamiltonian with regular wave functions.

It is obvious that with (3.43) and (3.44) one can find s-wave eigenfunctions of H^λ that differ from the regular non-interacting ones $J_0(kr)$ (two dimensions) and $\sin kr/r$ (three dimensions) because the irregular solution is now acceptable. For positive energy $E = k^2/2$ we have

$$n = 2: \quad \psi(r) = J_0(kr) - \tan \delta Y_0(kr), \qquad (3.45)$$

$$n = 3: \quad \psi(r) = \frac{1}{r}(\sin kr + \tan \delta \cos kr). \qquad (3.46)$$

From (3.43) one determines that δ in (3.45) coincides with the phase shift δ_0 of (3.26) when $1/g$ is identifies with $1/\lambda - \gamma/\pi - (1/\pi) \ln \mu/2$, while with (3.44) δ in (3.46) is the same as δ_0 (3.27), with $\lambda = g$. Also there are bound states, $E = -B$

$$n = 2: \quad \psi_B(r) = \sqrt{\frac{2B}{\pi}} K_0(\sqrt{2B}r), \qquad (3.47)$$

$$n = 3: \quad \psi_B(r) = \left(\frac{B}{2\pi^2} \right)^{1/4} \frac{e^{-\sqrt{2B}r}}{r}. \qquad (3.48)$$

The binding energies, fixed by satisfying (3.43) and (3.44), agree with (3.37) and (3.38), once the above identifications between g and λ are made. As before the sign of $\lambda(g)$ is immaterial for the two-dimensional bound state, while for the three-dimensional bound state, it must be positive. Finally, it is readily verified that (3.47), (3.48) are the Fourier transforms of (3.30), (3.31).

In conclusion, we see that the method of self-adjoint extension provides a description of renormalized δ-function potentials in two and three spatial dimensions for the following three reasons:

1) It is *a priori* plausible to describe a Hamiltonian with a δ-function potential as a free Hamiltonian on a space with one point deleted plus a boundary condition specifying what happens at that point
2) The boundary conditions (3.43) and (3.44) permit a $\ln r$ and a r^{-1} singularity in the two- and three-dimensional wave functions., as is seen in (3.45), (3.47) and (3.46), (3.48). The effect of the Laplacian on these is indeed a δ-function.
3) Most convincing is the fact that the scattering data and bound state spectra arising from the renormalized δ-function interactions are reproduced by the self-adjoint extensions.

In the next Section, another reason for viewing a δ-function potential as a self-adjoint extension is given.

The boundary condition (3.43) implied by the self-adjoint extension in two dimensions shows also why dilation symmetry is broken quantum mechanically. Observe first that the logarithms occurring in that equation introduce a scale for r, which is not dilation invariant. More formally, we can demonstrate that D is not defined on our space. Consider any s-wave energy eigenfunction $\psi_E(r)$. From (2.3) it follows that

$$D\psi_E(r) = tE\psi_E(r) + \frac{i}{2}(r\partial_r + 1)c\psi_E(r) \,. \tag{3.49}$$

The boundary condition (3.43) requires that at small r, $\psi_E(r)$ is proportional to $1 + \frac{\lambda}{\pi} \ln r$. But then the last term in (3.49) is proportional to $1 + \frac{\lambda}{\pi} + \frac{\lambda}{\pi} \ln r$, which means that $D\psi_E$ satisfies (3.43) and exists in the Hilbert space only for $\lambda = 0$.[22]

IV. AHARONOV-BOHM/EHRENBERG-SIDAY INTERACTION

The two-dimensional δ-function potential arises also in a problem involving a point magnetic vortex. The interpretation as a self-adjoint extension serves to explain a puzzle that occurs in this context.

Let us first describe the problem and the puzzle. Consider the Hamiltonian for a planar Dirac particle interacting with a magnetic field B, described by the vector potential $\mathbf{A}$, $B = \varepsilon^{ij}\partial_i A^j$

$$H = \boldsymbol{\alpha} \cdot (\mathbf{p} - \mathbf{A}) + \beta m \,. \tag{4.1}$$

In (2+1) space-time dimensions, Dirac matrices are 2×2 and may be chosen to be the Pauli matrices: $\alpha^i = \sigma^i$, $i = 1, 2$; $\beta = \sigma^3$.

The Dirac eigenvalue equation for the two-component spinor $\chi = \begin{pmatrix} \chi_+ \\ \chi_- \end{pmatrix}$

$$\left(\boldsymbol{\alpha} \cdot (\mathbf{p} - \mathbf{A}) + \beta m \right) \chi = \varepsilon \chi \qquad (4.2)$$

may be iterated and decoupled

$$\left((\mathbf{p} - \mathbf{A})^2 - \beta B \right) \chi = (\varepsilon^2 - m^2) \chi . \qquad (4.3\text{a})$$

Thus we arrive at two Schrödinger equations for the two components $\chi_\pm$

$$\left[\frac{1}{2} \left(\frac{1}{i} \boldsymbol{\nabla} - \mathbf{A} \right)^2 \mp \frac{1}{2} B \right] \chi_\pm = E \chi_\pm$$

$$E = \frac{1}{2} (\varepsilon^2 - m^2) . \qquad (4.3\text{b})$$

Now for the puzzle: we consider a point vortex, as in an idealized description of the Aharonov-Bohm/Ehrenberg-Siday effect

$$B = \Phi \delta(\mathbf{r})$$

$$A^i = -\frac{\Phi}{2\pi} \varepsilon^{ij} \frac{r^j}{r^2} = \frac{\Phi}{2\pi} \partial_i \theta . \qquad (4.4)$$

Since the magnetic field vanishes almost everywhere, the vector potential is a pure gauge almost everywhere — it is expressed in (4.4) as a gradient of the angle θ, $\tan \theta = \frac{y}{x}$, $r = (x, y)$. [Consistency requires the amusing formula $\varepsilon^{ij} \partial_i \partial_j \theta = 2\pi \delta(\mathbf{r})$.] Since $\mathbf{A}$ is a pure gauge it may be removed from the equations by defining

$$\chi = e^{i\nu\theta} \chi^0 \qquad (4.5)$$

where $\nu = \Phi/2\pi$. However, since χ is single-valued,

$$\chi|_{\theta=2\pi} = \chi|_{\theta=0} \qquad (4.6\text{a})$$

it follows that χ^0 satisfies

$$\chi^0|_{\theta=2\pi} = e^{-2\pi\nu i} \chi^0|_{\theta=0} . \qquad (4.6\text{b})$$

So χ^0 is "multivalued."

When the change of variables (4.5) is made in the Dirac equation (4.2) we find that χ^0 satisfies the free Dirac equation

$$(\boldsymbol{\alpha} \cdot \mathbf{p} + \beta m)\chi^0 = \varepsilon \chi^0 \qquad (4.7)$$

and the interaction is entirely hidden in the angular boundary condition (4.6b). On the other hand, changing to χ^0 in the Schrödinger equation (4.3), removes the vector potential but leaves the magnetic field, which is here a δ-function,

$$\left(-\frac{1}{2}\nabla^2 \mp \pi\nu\delta(\mathbf{r})\right)\chi_\pm^0 = E\chi_\pm^0 \qquad (4.8)$$

while the boundary condition, on $\chi_\pm^0$ remain as in (4.6b). The question now presents itself: does χ^0 satisfy a free equation as in (4.7) or does it experience an additional δ-function interaction as in (4.8)? It appears puzzling that both equations are true.

The answer to the question and the resolution to the puzzle resides in the self-adjoint extension that has to be performed on the Dirac equation. No matter whether the equation is taken with its interaction as in (4.2), or without an interaction as in (4.7), but with the angular boundary condition (4.6b), it is impossible to satisfy it for $\nu \neq 0$ with wave functions that are everywhere finite. One must admit an infinite but normalizable solution and once this is allowed, a further boundary condition must be specified. In other words, neither Dirac operator is self-adjoint and a self-adjoint extension is required.[13] On the other hand, in view of what was explained in Section III, the Schrödinger equation (4.8) with a δ-function should be viewed as the free equation with self-adjoint extension. Therefore, regardless whether one works with (4.7) or (4.8), the Hamiltonian is non-interacting, there is an angular boundary condition (4.6b) that recalls the presence of the vortex, and there is further radial boundary condition specifying the self-adjoint extension.

It must be emphasized that an important difference exists between the self-adjoint extensions of the free Schrödinger Hamiltonian $H_S^0 \equiv -\nabla^2/2$ with conventional angular boundary conditions and of the free Dirac Hamiltonian $H_D^0 \equiv \boldsymbol{\alpha} \cdot \boldsymbol{\nabla}/i + \beta m$ with vortex angular boundary conditions. In the former, no extension is needed; H_S^0 has finite eigenfunctions and an extension represents an additional interaction — the δ-function. On the other hand, the free vortex Dirac Hamiltonian H_D^0 does not possess finite eigenfunctions; an extension is required and it represents further information (beyond total flux) that must be specified when describing the physical attributes of the already posited vortex.

In contrast to the Schrödinger case, the extension in the Dirac equation is not a matter of choice and does not reflect *additional* interactions.

V. CONCLUSION

There can be no doubt that a δ-function interaction in two- and three-dimensional Schrödinger theory can be defined, and the method of self-adjoint extension allows dispensing with infinite renormalization. Of course, the relation of these non-relativistic theories to relativistic field theories in $(2+1)$- and $(3+1)$-dimensional space-time is purely formal. Thus, one cannot draw any definite conclusions about the field theory models.

In $(1+1)$-dimensional space-time, the ϕ^4 relativistic interaction rigorously goes over to the non-relativistic Schrödinger theory of a one-dimensional δ-function.[23] The possibility of carrying out the proof relies on the mildness of that field theory's ultraviolet divergences. It should also be feasible to carry out an analysis of the non-relativistic limit for the super-renormalizable $(2+1)$-dimensional ϕ^4 theory. While this model is known to exist, the presumed Schrödinger theory δ-function limit shows some unexpected features: the need for infinite renormalization, which is not necessary in the field theory; the existence of a bound state, regardless of the sign of the renormalized coupling — it is as if only an attractive non-relativistic theory exists.

In conclusion, while the status of relativistic, $(3+1)$-dimensional ϕ^4 field theory remains unsettled, the non-relativistic theory is not necessarily trivial. Indeed, a non-trivial scattering amplitude exists, but its construction is a subtle task. It remains to be seen whether a subtle construction of the field theory is possible.

ACKNOWLEDGEMENT

I was introduced to this subject by Baqi Bég at Rockefeller University, where he hosted me many times. I shall miss his hospitality and our physics discussions. I am grateful for informative conversations with E. D'Hoker, J. Dimock, A. Jaffe and B. Simon.

REFERENCES

1. R. Peierls, *Surprises in Theoretical Physics* (Princeton University Press, Princeton, NJ, 1979).
2. S. L. Adler and R. Dashen, *Current Algebras* (W. A. Benjamin, New York, NY, 1968).
3. F. Dyson, *Symmetry Groups in Nuclear and Particle Physics* (W. A. Benjamin, New York, NY, 1966).
4. E. Farhi and R. Jackiw, *Dynamical Gauge Symmetry Breaking* (World Scientific, Singapore, 1982).
5. K. G. Wilson, *Phys. Rev.* B **4**, 3184 (1971).
6. M. A. B. Bég, "Triviality and Higgs Mass Bounds: A Status Report," invited plenary talk at the Higgs Particle/Particles Workshop, Erice, Italy, 15–26 July 1989; "Triviality", invited lecture at the 1989 Summer School on High Energy Physics and Cosmology, Trieste, Italy, 26 June – 18 August 1989.
7. M. A. B. Bég and R. Furlong, *Phys. Rev.* D **31**, 1370 (1985).
8. K. Huang, *Int'l. Jnl. Mod. Phys.* **A4**, 1037 (1989).
9. H. Bethe and R. Peierls, *Proc. Roy. Soc.* (London) **148A**, 146 (1935); E. Fermi, *Ricerca Scientifica* **7**, 13 (1936) [English translation: E. Fermi, *Collected Papers* Vol. 1, Italy 1921–1938 (University of Chicago Press, Chicago, IL, 1962)].
10. F. Berezin and L. Faddeev, *Dokl. Akad. Nauk* (USSR) **137**, 1011 (1961) [English translation: *Sov. Math. Dokl.* **2**, 372 (1961)].
11. S. Albeverio, F. Gesztesy, R. Høegh-Krohn and H. Holden, *Solvable Models in Quantum Mechanics* (Springer, Berlin, 1988). This contains extensive references to the research literature.
12. P. Gerbert and R. Jackiw *Comm. Math. Phys.* **124**, 229 (1989).
13. P. Gerbert, *Phys. Rev.* D **40**, 1346 (1989).
14. For a review see S. Treiman, R. Jackiw, B. Zumino and E. Witten, *Current Algebra and Anomalies* (Princeton University Press/World Scientific, Princeton, NJ/Singapore, 1985).
15. A. Goldhaber, *Phys. Rev.* D **16**, 1815 (1977); C. Callias, *Phys. Rev.* D **16**, 3068 (1977).
16. R. Jackiw, *Phys. Today* **25**, No. 1, 23 (1972).
17. V. de Alfaro, S. Fubini and G. Furlan, *Nuovo Cim.* **A34**, 569 (1976). Although dilation symmetry is maintained quantum mechanically, composite operators carry anomalous scale dimension; see G. Parisi and F. Zirilli, *J. Math. Phys.* **14**, 243 (1973).
18. R. Jackiw, *Ann. Phys.* (NY) **129**, 183 (1980).
19. R. Jackiw, *Ann. Phys.* (NY) **201**, 83 (1990).
20. C. Thorn, *Phys. Rev.* D **19**, 639 (1979); K. Huang, *Quarks, Leptons and Gauge Fields* (World Scientific, Singapore, 1982).

21. S. J. Dong and C. N. Yang, *Rev. Math. Phys.* 1, 139 (1989).
22. The $SO(2, 1)$ symmetry breaking features of self-adjoint extensions were previously discussed by E. D'Hoker and L. Vinet, *Comm. Math. Phys.* **97**, 391 (1985).
23. J. Dimock, *Comm. Math. Phys.* **57**, 51 (1977).

Paper I.4

Existence of anomalous gauge theories removes one motivation for the re-emergence of higher dimensional super strings. Physical realizations of anomalous theories are provided by edge states in the quantum Hall effect. My talk in Santiago, Chile (December 1985) was on the occasion of the re-establishment of humane institutions in that country.

UPDATE ON ANOMALOUS THEORIES

Quantum Mechanics of Fundamental Systems I,
C. Teitelbom, ed. (Plenum, New York, NY, 1988)

1. WHAT IS AN ANOMALOUS THEORY?

It is known that the quantization procedure can spoil classical symmetries. The problem afflicts continuous chiral symmetries and gravitational symmetries of massless (Weyl) fermions, the former in any even-dimensional space-time, the latter in space-times with dimensionality $4k + 2$, $k = 0, 1, \ldots$. [Similar quantum breaking afflicts discrete symmetries (P, T) in odd dimensions, and scale/conformal symmetries in any dimension; we shall not be concerned with these.] As a consequence, the symmetry current, whose classical conservation is assured by Noether's theorem, ceases to be conserved after quantization. We call such a current *anomalous;* it possesses an *anomalous divergence,* and the coupling of gauge fields to this current becomes problematical.[1]

The difficulty with gauge field couplings is seen from the equations of motion for the gauge field A_μ (taken to be an anti-Hermitian matrix in the Lie algebra of the internal symmetry group):

$$F_{\mu\nu} \equiv \partial_\mu A_\nu - \partial_\nu A_\mu + [A_\mu, A_\nu], \tag{1}$$

$$D_\mu F^{\mu\nu} \equiv \partial_\mu F^{\mu\nu} + [A_\mu, F^{\mu\nu}] = J^\nu. \tag{2}$$

Since $D_\mu D_\nu F^{\mu\nu} = 0$, Eq. (2) requires $D_\mu J^\mu = 0$. If, on the other hand, the current J^μ acquires an anomalous divergence

$$D_\mu J^\mu = \text{anomaly} \neq 0, \tag{3}$$

it appears that the theory can be consistent only on the subspace where the anomaly vanishes. Moreover, aside from issues of dynamical consistency, questions about gauge invariance and unitarity arise. Finally, renormalizability of the quantum theory must be re-examined, since that desirable property is frequently linked with gauge invariance.

Another viewpoint is gotten from a functional integral formulation. Upon integrating over the chiral fermions that couple to the gauge field, one is left with an effective action:

$$W(A) = -i \ln \mathcal{D}(A), \tag{4}$$

$$\mathcal{D}(A) \equiv \det (\not{\partial} + \not{A}). \tag{5}$$

Under a gauge transformation by the group element g,

$$A_\mu \rightarrow A_\mu^g \equiv g^{-1} A_\mu g + g^{-1} \partial_\mu g \tag{6}$$

which can be taken infinitesimally,

$$g = I + \theta + \cdots, \tag{7}$$

$$\delta A_\mu = D_\mu \theta. \tag{8}$$

$\mathcal{D}(A)$ changes as

$$\delta D \equiv \mathcal{D}(A + D\theta) - \mathcal{D}(A) = \int \text{tr}\, (D_\mu \theta) \frac{\delta}{\delta A_\mu} \mathcal{D}(A)\,. \qquad (9)$$

The variation is the vacuum matrix element of the current in the presence of an external gauge field:

$$\langle J^\mu \rangle_A = -\frac{\delta}{\delta A_\mu} \mathcal{D}(A)\,. \qquad (10)$$

Hence

$$\delta \mathcal{D} = \int \text{tr}\, \theta \mathcal{D}_\mu \langle J^\mu \rangle_A \qquad (11)$$

and the effective action for chiral fermions is not gauge invariant when the fermion current is not conserved (in its matrix elements). Thus the total action, gauge field action $I(A)$ plus effective fermion action $W(A)$, loses gauge invariance and it is unclear whether the gauge invariant $I(A)$ can be consistently combined with the gauge noninvariant $W(A)$.

Models that possess the above-described gauge pathology are called *anomalous theories*.[2]

An approach to anomalous theories that has been advocated in the past is to eliminate the anomaly, *i.e.*, the (chiral) fermion content is adjusted so that the anomalous divergence vanishes.[1] More specifically, in four dimensions, the anomalous divergence reads[1]

$$(D_\mu J^\mu)_a \propto \frac{1}{24\pi^2} \text{tr}\, T^a \partial_\mu \epsilon^{\mu\alpha\beta\gamma} \left(A_\alpha \partial_\beta A_\gamma + \frac{1}{2} A_\alpha A_\beta A_\gamma \right)\,. \qquad (12)$$

Here, the T^a's comprise anti-Hermitian matrices that provide a basis for the Lie algebra, in the representation of the fermions:

$$A_\alpha = A_\alpha^a T^a\,. \qquad (13)$$

The proportionality constant is fixed by the number and chirality of fermions that interact with A_μ. The most important term in the divergence is the first. Because of cyclicity of the trace, it involves

$$\frac{i}{2} d_{abc} \equiv \text{tr}\, T^a \{T^b,\, T^c\} \qquad (14)$$

where the "b" and "c" indices refer to the gauge fields that couple to the current "a", as represented by the triangle diagram of Fig. 1.

The trace is unaffected by group transformations, hence an invariant statement for the absence of anomalies in four dimensions is that fermions belong to representations for which d_{abc} vanishes. (One can show that the rest of the anomaly also vanishes once d_{abc} does.)

This requirement of anomaly cancellation has been widely accepted and has two notable successes: First, when applied to the electroweak unified $SU(2) \times U(1)$ theory, which is potentially anomalous, it predicts that the number of quarks balances against the number of leptons, thus providing us with the only theoretical explanation of this apparently true experimental fact. Second, when applied to the construction of superstring models, the rank of the internal symmetry group is fixed uniquely at 16 and the group is essentially predicted. All the other imponderables in the string program make this one fixed fact very important indeed.

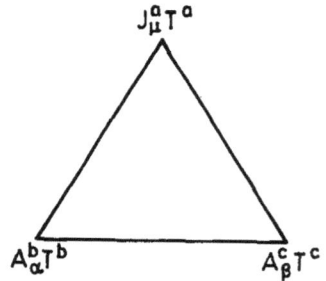

$$J_\mu^a T^a$$

$$A_\alpha^b T^b \qquad\qquad A_\beta^c T^c$$

Fig. 1. Triangle graph that spoils gauge invariance in a four-dimensional chiral gauge theory.

Precisely because of the central role that anomaly cancellation plays in modern theoretical particle physics, it is worthwhile looking more deeply into the matter, and inquiring whether a gauge theory that does possess anomalies is in fact meaningless, or whether some sense can be made of it. The question is especially appropriate at the present time, when as it happens a consistent mathematical framework for discussing anomalies has emerged.

The conclusion of several investigations is that unitarity and positivity need not be destroyed by the anomaly. Renormalizability of an anomalous theory remains unsettled.

DISCUSSION

C. TEITELBOIM: I would like to ask you something very naive: Is it possible to have anomalies in quantum mechanics, as opposed to quantum field theory?

R. JACKIW: Yes, several examples are given in the exercises to my Les Houches lectures, reprinted in Ref. 1. Also the apparently scale-invariant 2-dimensional

δ-function potential leads to a quantum mechanics with anomalously broken scale invariance.

TEITELBOIM: So there can be a case where something is anomalously broken, without infinities.

JACKIW: Quantum mechanics does not usually suffer from field-theoretic ultraviolet infinities. However, there are infinities associated with "infinite-dimensional matrices," *etc.* Indeed, in the examples mentioned by me in Les Houches, one is evaluating determinants of differential operators. Moreover, δ-function interactions in quantum mechanics give rise to short-distance singularities, which are similar to field-theoretic infinities.

2. MATHEMATICALLY COHERENT FRAME FOR ANOMALIES

When anomalous nonconservation of currents was first encountered, this unexpected phenomenon was associated with the ultraviolet divergences of perturbation theory.[1] Later, a Hamiltonian analysis of the effect related it to symmetry-breaking aspects of the filled "negative energy sea" — again a quantity that cannot be described in finite terms.[1]

However, more recent discussion of the anomaly make use of various well-defined mathematical characteristic classes: Chern-Pontryagin, Chern-Simons, *etc.*, and attention is drawn to the fact that various quantities of interest like the anomalous divergence of a chiral current, the anomalous response to gauge transformation of a fermion determinant, *etc.* are expressed in terms of these.[1] The mathematical connection has come to a sharper focus in the characterization of an anomalous gauge theory by the fact that commutators of gauge transformation generators are anomalous, and do not follow the Lie algebra of the gauge group.[3]

It was appreciated already in the earliest days of research on anomalies that fermion charge densities $\rho_a \equiv J_a^0$ satisfy anomalous commutators.[1] Since ρ_a generates gauge transformations only on the fermion degrees of freedom, it is natural, in a mathematical/algebraic framework for the gauge theory, to consider the complete generator, G_a, given by ρ_a supplemented by $-(DE)_a$. [E_α^i is the non-Abelian electric field F_a^{i0} — it is the negative of the momentum conjugate to the canonical coordinate A_a^i.] G_a generates the complete gauge transformation, which modifies the gauge field degrees of freedom as well. It was conjectured[3] and then verified[4] that the anomaly in the commutator of the G_a's is also given by a mathematically determined quantity:

$$[G_a(\mathbf{r}),\ G_b(\mathbf{r}')] = i f_{abc} G_c(\mathbf{r})\delta(\mathbf{r}-\mathbf{r}')$$
$$\pm \frac{i}{24\pi^2}\varepsilon^{ijk}\mathrm{tr}\{T^a,\ T^b\}\partial_i A_j \partial_k \delta(\mathbf{r}-\mathbf{r}')\,. \tag{15}$$

The sign is determined by the fermion chirality, and we observe that d_{abc} controls the anomaly in the commutator as well. (The third representation matrix making up d_{abc} resides in the gauge field: $A_j \equiv A_j^c T^c$.)

A finite gauge transformation with $g = e^{\theta^a T^a}$ is implemented in the quantum theory by the unitary operator

$$U(g) = \exp\left(i\int \theta^a G_a\right)\,. \tag{16}$$

The fact (15) that the Lie algebra of the generators does not follow that of the group, but possesses an extension, means that the composition law for the operators U differs from the group composition law:

$$g_1 g_2 = g_{12}\,, \tag{17}$$
$$U(g_1)U(g_2) = e^{-i2\pi\omega_2(A;g_1,g_2)}U(g_{12})\,. \tag{18}$$

The additional phase, called a 2-cocycle, when expanded for g_1 and g_2 near the identity reproduces the extension in (15).

Thus, the anomaly phenomenon is just the statement that in the quantized field theory, the gauge group is represented projectively, as in (18).

Before proceeding with the discussion of anomalous theories, let us pause for a mathematical message about cocycles.

3. MATHEMATICAL ASIDE ON COCYCLES

The mathematical theory for representing transformation groups by unitary operators on a linear vector or Hilbert space meets its objective correlative — its specific application — in quantum mechanics, which naturally is concerned with operators acting on physical states. When one is dealing with an abstract group of transformations, g, which obey the composition law (17) and which transform variables q according to a definite rule,

$$g \xrightarrow{g} q^g \tag{19}$$

then the simplest representation of this on functions of q, $\Psi(q)$, is obtained by associating with g an operator $U(g)$, which implements (19) as,

$$U(g)qU^{-1}(g) = q^g \tag{20}$$

acts on $\Psi(q)$ according to,

$$U(g)\Psi(q) = \Psi(q^g) \tag{21}$$

and satisfies a composition law that parallels (17):

$$U(g_1)U(g_2) = U(g_{12}). \tag{22}$$

However, it is possible to elaborate this simplest realization of the transformation group by introducing phases into formulas (21) and (22). Such phases are called *cocycles* and the various conditions that they must satisfy have been elaborated in mathematics.

It has been known for some time that cocycles are present in the quantum mechanics of point particles. As mentioned above, modern gauge field theory makes use of them as well, to give a coherent mathematical framework for the anomaly phenomena.

3.1. 1-Cocycle

In the simplest generalization of (21) and (22), a phase factor is inserted in the action of the representation:

$$U(g)\Psi(q) = e^{-i2\pi\omega_1(q;g)}\Psi(q^g). \tag{23}$$

Consistency with (22) requires that ω_1 satisfy

$$\omega_1(q^{g_1}; g_2) - \omega_1(q; g_{12}) + \omega_1(q; g_1) = 0. \qquad \text{(mod integer)} \tag{24}$$

A quantity $\omega_1(q; g)$ depending on one member of the transformation group, g, and possibly on the variable acted upon, q, is called a *1-cocycle* if it obeys (24).

Representing Galilean boosts in quantum mechanics makes use of a 1-cocycle. The Abelian group of Galileo transformations of the position vector $\mathbf{r}$ (q in the general discussion) is defined to act as $\mathbf{r} \rightarrow \mathbf{r} + \mathbf{v}t$. The transformation is labeled by $\mathbf{v}$ (corresponding to g in the general discussion) and is implemented by the operator

$$U(\mathbf{v}) = e^{i\mathbf{v}\cdot(\mathbf{p}t-m\mathbf{r})}. \tag{25}$$

[The second term in the exponential involving $m\mathbf{r}$ is present so that in the Galilean invariant theory of a nonrelativistic free particle with mass m, the exponent is the appropriate constant of motion: $(d/dt)(\mathbf{p}t - m\mathbf{r}) = \dot{\mathbf{p}}t + \mathbf{p} - m\dot{\mathbf{r}} = 0$ when $\mathbf{p} = m\dot{\mathbf{r}}$, $\dot{\mathbf{p}} = 0$.] One verifies with the help of the Heisenberg algebra

$$[r^i, r^j] = 0 = [p^i, p^j],$$
$$[r^i, p^j] = i\,\delta^{ij} \tag{26}$$

that

$$U(\mathbf{v})\mathbf{r}\,U^{-1}(\mathbf{v}) = \mathbf{r} + \mathbf{v}t,$$
$$U(\mathbf{v})\mathbf{p}\,U^{-1}(\mathbf{v}) = \mathbf{p} + m\mathbf{v} \tag{27}$$

and that (22) is satisfied, but $U(\mathbf{v})$ acts on wave functions $\Psi(\mathbf{v})$ as in (22) with

$$2\pi\omega_1(\mathbf{r}; \mathbf{v}) = m\mathbf{v} \cdot \mathbf{r} + \frac{1}{2}mv^2 t \tag{28}$$

which satisfies (24).

3.2. 2-Cocycle

In the next generalization, a phase is introduced into the composition law (22):

$$U(g_1)U(g_2) = e^{-i2\pi\omega_2(q;g_1,g_2)}U(g_{12}). \tag{29}$$

A consistency condition on ω_2 follows from the assumed associativity of the composition law. If

$$[U(g_1)U(g_2)]U(g_3) = U(g_1)[U(g_2)U(g_3)] \tag{30}$$

then

$$\omega_2(q^{g_1}; g_2, g_3) - \omega_2(q; g_{12}, g_3) + \omega_2(q; g_1, g_{23}) - \omega_2(q; g_1, g_2) = 0.$$
$$\text{(mod integer)} \tag{31}$$

When a quantity depends on two group elements, g_1 and g_2, and possibly on q, and also satisfies (31), it is called a *2-cocycle*. Representations that make use of 2-cocycles are called *projective* or *ray* representations and they occur frequently in quantum mechanics.

Indeed, the Heisenberg commutator algebra (26) indicates that translations on phase space: $\mathbf{r} \to \mathbf{r} + \mathbf{a}$, $\mathbf{p} \to \mathbf{p} + \mathbf{b}$ are represented by the operator

$$U(\mathbf{a}, \mathbf{b}) = e^{i(\mathbf{a}\cdot\mathbf{p} - \mathbf{b}\cdot\mathbf{r})} \tag{32}$$

i.e.,

$$U(\mathbf{a}, \mathbf{b})\mathbf{r}\, U^{-1}(\mathbf{a}, \mathbf{b}) = \mathbf{r} + \mathbf{a}$$
$$U(\mathbf{a}, \mathbf{b})\mathbf{p}\, U^{-1}(\mathbf{a}, \mathbf{b}) = \mathbf{p} + \mathbf{b}. \tag{33}$$

$U(\mathbf{a}, \mathbf{b})$ composes according to (29) with

$$2\pi\omega_2(\mathbf{r}; \mathbf{a}_1\mathbf{b}_1, \mathbf{a}_2\mathbf{b}_2) = \frac{1}{2}(\mathbf{a}_1 \cdot \mathbf{b}_2 - \mathbf{a}_2 \cdot \mathbf{b}_1). \tag{34}$$

The U's associated and ω_2 obeys the consistency condition (31). (Coordinate and momentum shifts form a symmetry operation for a free particle: Galilean transformations supplemented by spatial translations are realized in this way.) So, following Weyl and Bargmann, one may accurately say that the essence of quantum mechanics is the 2-cocycle, leading to ray or projective representation of the Abelian translation group.

When dealing with a continuous or Lie group of transformations, the discussion may be carried out in infinitesimal terms. Corresponding to the finite group element g, there is the infinitesimal quantity θ, $g = e^\theta$, and the composition law (17) is reflected in a Lie algebra:

$$[\theta_1, \, \theta_2] = \theta_{12}. \tag{35}$$

Suppose further that we are dealing with dynamical variables q, governed by Lagrangian L. The infinitesimal action of the transformation on q is given by an infinitesimal version of (19)

$$q \rightarrow q + \delta_\theta q. \tag{36}$$

When no cocycles occur, the representative of θ, that is, the infinitesimal generator G, where $U(g) = e^{iG}$, is given by $(\delta_\theta q)\partial L/\partial \dot{q} = (\delta_\theta q)p$, so that its action on wave functions is realized by $(\delta_\theta q)(1/i)\partial/\partial q$. Moreover, the generators satisfy the Lie algebra of the group (35),

$$[iG_1, \, iG_2] = iG_{12} \tag{37}$$

which is a consequence of (17) for infinitesimal quantities.

In the presence of a 1-cocycle, the generator becomes

$$G = (\delta_\theta q)p - \delta\omega_1 \tag{38}$$

where $\delta\omega_1$ is (proportional to) the infinitesimal part of the 1-cocycle; compare with (25) and (28). The 1-cocycle condition (24), in infinitesimal form, ensures that the modified generators (38) continue to satisfy the Lie algebra of the group (35).

A 2-cocycle indicates that the generators' Lie algebra acquires an extension:

$$[iG_1, \ iG_2] = iG_{12} - \delta\omega_2 \ . \tag{39}$$

Here $\delta\omega_2$ is [proportional to] the infinitesimal portion of the 2-cocycle, whose consistency condition (31) ensures that (39) does not contradict the Jacobi identity,

$$(G_1, \ [G_2, \ G_3]) + (G_2, \ [G_3, \ G_1]) + (G_3, \ [G_1, \ G_2]) = 0 \ . \tag{40}$$

This is the infinitesimal version of (30). The nonvanishing Heisenberg commutator (26) is an example of an extended Lie algebra (39).

3.3. 3-Cocycle

Not unexpectedly, the 3-cocycle involves abandoning associativity (30), and for infinitesimal generators, the Jacobi identity (40) fails. It is important to appreciate that nonassociating quantities cannot be represented by well-defined linear operators acting on a vector or Hilbert space since, by definition, operations on vectors necessarily associate. Hence, the quantities discussed below are abstract, algebraic objects obeying formal relations.

To encounter a 3-cocycle, we replace (30) by a formula that includes a phase:

$$[U(g_1)U(g_2)]U(g_3) = e^{-i2\pi\omega_3(q;g_1,g_2,g_3)}U(g_1)[U(g_2)U(g_3)] \ . \tag{41}$$

By considering fourfold products, and associating in different ways, it is established that consistency requires that ω_3 satisfy the 3-cocycle condition:

$$\omega_3(q^{g_1}; g_2, g_3, g_4) - \omega_3(q; g_{12}, g_3, g_4) + \omega_3(q; g_1, g_{23}, g_4)$$
$$- \omega_3(q; g_1, g_2, g_{34}) + \omega_3(q; g_1, g_2, g_3) = 0 \ . \quad \text{(mod integer)} \tag{42}$$

This arcane structure in fact arises in elementary Hamiltonian dynamics. Consider a particle with mass m and charge e, moving at $\mathbf{r}$ in an external magnetic field $\mathbf{B}(\mathbf{r})$. The dynamical equations express the Lorentz force law:

$$m\dot{\mathbf{v}} = e\mathbf{v} \times \mathbf{B} \ ,$$
$$\dot{\mathbf{r}} = \mathbf{v} \ . \tag{43}$$

The velocity of light is scaled to unity. We do not assume that **B** is necessarily divergence free, *i.e.*, there may be magnetic sources. Since a magnetic field does no work, the energy is purely kinetic, and the Hamiltonian does not see **B**:

$$H = \frac{1}{2}mv^2 . \tag{44}$$

In order that the dynamical equations (43) emerge as canonical Hamiltonian equations, we postulate the following commutation relations for **r** and **v** (in the classical theory, these would be Poisson brackets):

$$[r^i, \, r^j] = 0 , \tag{45a}$$

$$[r^i, \, mv^j] = i\delta^{ij} , \tag{45b}$$

$$[mv^i, \, mv^j] = ie\varepsilon^{ijk}B^k . \tag{45c}$$

Thus, the magnetic field, which is invisible in (44), reappears in (45c) so that (43) may be regained as

$$\dot{r} = i[H, \, \mathbf{r}] = \dot{\mathbf{v}} , $$
$$m\dot{\mathbf{v}} = i[H, \, m\mathbf{v}] = e\mathbf{v} \times \mathbf{B} . \tag{46}$$

If now (45) is subjected to the test of the Jacobi identity, we find that triple velocity commutators sum to

$$[mv^1, \, [mv^2, \, mv^3]] + [mv^2, \, [mv^3, \, mv^1]] + [mv^3, \, [mv^1, \, mv^2]] = e\boldsymbol{\nabla} \cdot \mathbf{B} . \tag{47}$$

When $\boldsymbol{\nabla} \cdot \mathbf{B}$ is zero, we are dealing with an associative Lie algebra, and $m\mathbf{v}$ may be represented by $(1/i)\boldsymbol{\nabla} - e\mathbf{A}$, $\mathbf{B} = \boldsymbol{\nabla} \times \mathbf{A}$. However, for $\boldsymbol{\nabla} \cdot \mathbf{B} \neq 0$ a 3-cocycle is present.

To understand the nonassociativity, let us consider the finite quantities

$$U(\mathbf{a}) = e^{i\mathbf{a}\cdot m\mathbf{v}} . \tag{48}$$

These represent translations of **r**, in that $U(\mathbf{a})\mathbf{r}U^{-1}(\mathbf{a}) = \mathbf{r} + \mathbf{a}$. However, the representation is not faithful, because from (45c) it follows that

$$U(\mathbf{a}_1)U(\mathbf{a}_2) = e^{-ie\Phi}U(\mathbf{a}_1 + \mathbf{a}_2) , \tag{49}$$

where Φ is the flux through the triangle at **r** formed from $\mathbf{a}_1$ and $\mathbf{a}_2$; see Fig. 2. Moreover, by considering the triple product $U(\mathbf{a}_i)U(\mathbf{a}_2)U(\mathbf{a}_3)$, associated in the two different ways, one finds

$$[U(\mathbf{a}_1)U(\mathbf{a}_2)]U(\mathbf{a}_3) = e^{-i2\pi\omega_3}U(\mathbf{a}_1)[U(\mathbf{a}_2)U(\mathbf{a}_3)] \tag{50}$$

where the 3-cocycle $2\pi\omega_3$ is e times the flux out of the tetrahedron at $\mathbf{r}$ formed from $\mathbf{a}_1$, $\mathbf{a}_2$, and $\mathbf{a}_3$; see Fig. 3.

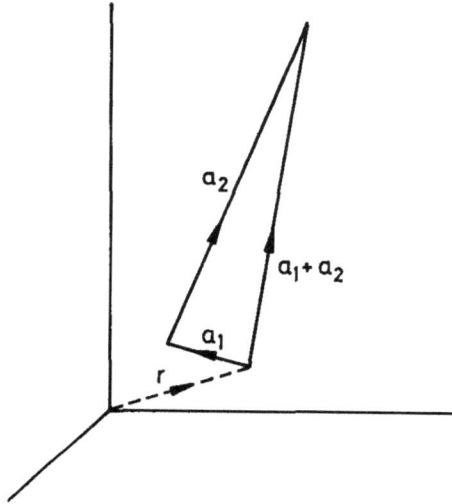

Fig. 2. Triangle at $\mathbf{r}$, defined by two translations $\mathbf{a}_1$ and $\mathbf{a}_2$, through which the flux Φ is calculated.

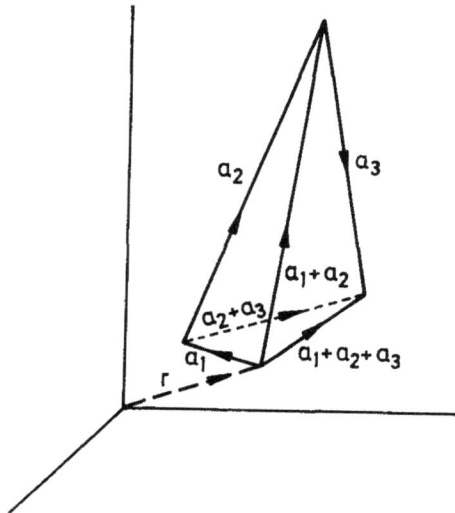

Fig. 3. Tetrahedron at point $\mathbf{r}$, defined by three translations $\mathbf{a}_1$, $\mathbf{a}_2$, and $\mathbf{a}_3$. The 3-cocycle is proportional to the flux out of the tetrahedron.

When $\nabla \cdot \mathbf{B} = 0$, no flux emanates from a closed surface; the 3-cocycle vanishes and associativity is regained. When there are sources, $\nabla \cdot \mathbf{B} \neq 0$, the flux is nonzero, but associativity will prevail if ω_3 is an integer, since then $e^{-i2\pi\omega_3} = 1$. This requirement forces (1) $\nabla \cdot \mathbf{B}$ to consist of delta functions so that the total flux not vary continuously when the $\mathbf{a}_i$ change, *i.e.*, sources must be point-monopoles; (2) since a monopole of strength g gives rise to flux $4\pi g$, hence, produces the cocycle $2eg$, eg must satisfy Dirac's quantization condition, *i.e.*, it must be an integral multiple of $1/2$. In this way, the removal of the 3-cocycle, which is necessary for conventional quantum mechanics with associative operators on Hilbert space, limits magnetic sources to quantized Dirac monopoles. Other magnetic sources lead to a nonassociative algebra. Quantum mechanics with associative operators on Hilbert space, limits magnetic sources to quantized Dirac monopoles. Other magnetic sources lead to a nonassociative algebra.

Finally, note that even for quantized monopoles, where the finite 3-cocycle is invisible since $e^{-i2\pi\omega_3} = 1$, there is a remnant of nonassociativity in the nonvanishing of the right-hand side of (47), which for a monopole at $\mathbf{r}_0$, reads $4\pi eg\delta(\mathbf{r} - \mathbf{r}_0) = 2\pi n\delta(\mathbf{r} - \mathbf{r}_0)$. This does not interfere with an operator formulation on a space with one point — the location of the monopole — excluded.

Further generalization is possible; one may define 4- and higher cocycles, but thus far no role has been found for these in physics.

3.4. Beyond Quantum Mechanics

The above concerns quantum mechanics. In quantum field theory, cocycles arise in the description of anomalies, and we see in (15) the role that a 2-cocycle plays. The 1-cocycle also is relevant: it measures the gauge noninvariance of the fermionic determinant (5). Define a quantity $\omega_1(A; g)$ by

$$\mathcal{D}(A^g) = e^{-2\pi i\omega_1(A;g)}\mathcal{D}(A). \tag{51}$$

From the gauge group composition law, $g_1 g_2 = g_{12}$, it follows that $\mathcal{D}((A^{g_1})^{g_2}) = \mathcal{D}(A^{g_{12}})$. Hence $\omega_1(A; g)$ must satisfy (24), *i.e.*, it is a 1-cocycle. We further recognize that the haphazard-looking formula (12) must coincide with the infinitesimal, first-order in θ, contribution to $\omega_1(A; g)$, because according to (5) and (11) the anomalous divergence is an infinitesimal 1-cocycle. Moreover, from the 1-cocycle condition we see that the quantity $2\pi\omega_1(A; h)$ transforms under a gauge transformation $A \rightarrow A^g$, $h \rightarrow g^{-1}h$ exactly the same way as $W(A)$ does in (4):

$$W(A^g) = W(A) - 2\pi\omega_1(A; g)$$
$$2\pi\omega_1(A^g; g^{-1}h) = 2\pi\omega_1(A; h) - 2\pi\omega_1(A; g). \qquad (52)$$

Therefore, one may view the 1-cocycle as an effective action that transforms under gauge transformations in the same anomalous way as the effective action of chiral fermions. Necessarily, there is present a field other than the gauge field A, *viz.* the "chiral" field h. When the 1-cocycle is used in this way it is called the *Wess-Zumino term*.[1]

Please note that in (51) and (52) the cocycle functions in a way somewhat different from the representation theory examples discussed earlier: unlike in (23), ω_1 goes not arise from the action of an operator on a vector.

What about field theoretic 3-cocycles? There is evidence of Jacobi identity violation for commutators of spatial components of fermionic currents even in a noninteracting theory.[1,5] However, the structure of the commutators is ambiguous and fraught with divergences. Moreover, spatial components are involved, but they are not generators of symmetry transformations. Hence, no coherent algebraic framework has been given for this pathology.

Another commutator anomaly, hinting at an anomaly-related 3-cocycle, has been encountered in the course of verifying (15).[6] When the commutator of two electric fields is computed in perturbation theory, one finds a nonvanishing result:

$$[E_a^i(\mathbf{r}), E_b^j(\mathbf{r}')] = \frac{i}{48\pi^2}\varepsilon^{ijk}d_{abc}A_c^k\delta(\mathbf{r} - \mathbf{r}'). \qquad (53)$$

If we view $\mathbf{E}$ as a "velocity" in field space — an interpretation that is appropriate since $\mathbf{E} = -\dot{\mathbf{A}}$ in the Weyl, $A^0 = 0$, gauge, then the right-hand side of (53) defines a $U(1)$ curvature in field space — a kind of functional "magnetic field"; compare (45c). (Commutators with $\mathbf{A}$ — a "coordinate" in field space — show no anomalies.) Moreover, the triple $\mathbf{E}$ commutator fails to satisfy the Jacobi identity.

Finally, we note that in string theory, violations of the Jacobi identity — potential 3-cocycles — have been found.[7]

However, none of these fascinating structures have been coherently described by a nonassociative algebra. It is interesting to recall that already many years ago in a well-known body of work, P. Jordan investigated nonassociative algebras for quantum mechanics. Also, P. Dirac predicted a physical role for nonassociativity (1969 Rouse Ball lecture in Cambridge University, as communicated to me by H. Gottlieb).

DISCUSSION

I. SCHMIDT: when you introduce a 3-cocycle, as in (50), do you still assume that a 2-cocycle arises in the composition, as in (29)?

R. JACKIW: In general, nothing is said about the composition law (29). But it is easy to show that if a phase is present there and also in (50), then (31), which is a consequence of associativity, is modified by the occurrence of the 3-cocycle on the right-hand side. Hence, in that case, the phase in composition law should not be called a cocycle.

M. HENNEAUX: Are the velocities which fail to satisfy the Jacobi identity linear operators?

JACKIW: No, they are not; linear operators acting on vectors necessarily associate by definition.

D. GROSS: Is it that $m\mathbf{v}$ cannot be $\mathbf{p} - e\mathbf{A}$ because $\mathbf{A}$ is not well defined?

JACKIW: $\mathbf{A}$ has not been introduced. I am giving a gauge-invariant discussion, without $\mathbf{A}$. Indeed, we cannot have a vector potential in the region that $\nabla \cdot \mathbf{B} \neq 0$.

GROSS: What goes wrong if you don't have associativity?

JACKIW: It is not clear. Certainly, you cannot have conventional quantum mechanics with linear operators on a Hilbert space. Perhaps you can do something with density matrices, but I don't know how far you can go.

HENNEAUX: But even when the Dirac quantization condition for a monopole holds, you cannot realize the velocities as linear operators.

JACKIW: Since the failure of the Jacobi identity for a monopole is located at one point — the position of the monopole — one can realize $m\mathbf{v}$ as $\mathbf{p} - e\mathbf{A}$ away from that point. However, as is well known, $\mathbf{A}$ is not globally defined.

P. VAN NIEUWENHUIZEN: If your formulas are not about linear operators, to what do they refer?

JACKIW: They are algebraic relations between quantities that I merely write on the blackboard. Whether they have some realization (representation) remains to be determined.

4. APPROACHES TO QUANTIZING AN ANOMALOUS THEORY

Since the gauge transformation generator $\mathbf{G}_a$ coincides with the time component of $-D_\mu F^{\mu\nu} + J^\nu$, the field equation motion (2) requires that $\mathbf{G}_a$ vanish — this is Gauss's law. Moreover, if the spatial components of the field equation vanish, so must the time component, as a consequence of Lorentz invariance.

In a Hamiltonian, classical canonical or quantum mechanical, context $\mathbf{G}_a = \rho_a - (\mathbf{D} \cdot \mathbf{E})_a$ involves only canonical variables and no time derivatives ($-\mathbf{E}_a$ is the canonical momentum in the Weyl gauge), hence its vanishing must be viewed as a constraint rather than an equation of motion.

Constraints are conveniently divided into first class, when they close under Poisson bracketing, and second class otherwise. For first-class constraints, the quantization procedures is standard, except that a further condition is imposed: physical states must be annihilated by the constraint. This is how a nonanomalous gauge theory is quantized, and the requirement $G_a|\Psi\rangle = 0$ ensures that sates are invariant against homotopically trivial gauge transformations.

Second-class constraints cannot annihilate states, because their commutators do not close and a modification of the quantization rules is required to implement them.

Therefore, it is natural to suggest that an anomalous gauge theory can be consistently quantized, provided that Gauss's law is recognized as a second-class constraint and that the quantization procedure is appropriately altered.[8] Moreover, it has been conjectured that this procedure produces an effective action containing the Wess-Zumino term (1-cocycle), in addition to the gauge field and chiral fermion contributions.[8] In this way, the total action would regain gauge invariance: the anomalous response to gauge transformations of the chiral fermions would be cancelled by the explicit gauge noninvariance of the Wess-Zumino term. The chiral field of the Wess-Zumino term signals the emergence of new degrees of freedom, which are seen in this scenario as a consequence of the anomaly.

This program of modified canonical quantization has thus far not been realized. The principal difficulty is that the second-class nature of the constraint algebra emerges as a quantization anomaly only after the theory has already been quantized by conventional methods.

Nevertheless, various indicators are encouraging.

4.1. Decoupling Anomaly-Cancelling Fermions

Consider a potentially anomalous theory with spontaneous gauge symmetry breaking, effected by the Higgs mechanism, and fermion content properly adjusted to cancel the anomaly. By setting various parameters in the Higgs sector to infinity one may decouple some of the chiral fermions, and for the remaining ones the anomaly no longer cancels. It is found that the decoupling fermions leave behind the Wess-Zumino term, so that gauge invariance is retained. (The chiral field here is the Higgs field.[9])

The above may be viewed as a method for quantizing an anomalous gauge theory. Indeed, there is precedent for effecting difficult canonical quantization by first enlarging the theory in such a way that quantization is straightforward, and then removing the addition by sending parameters to infinity.[10] Note, however, that here the chiral field (Higgs field) retains a kinetic term, which is not needed for gauge invariance and may upset renormalizability.

4.2. Integrating over All Gauge Potentials

It has been observed that an effective action for chiral fermions that includes the Wess-Zumino term can be obtained from the functional integral, supplemented by the Faddeev-Popov trick. One begins with a formal functional integral over *all* configurations of the gauge potential, including gauge copies[11]:

$$Z = \int \mathcal{D}A \exp i[I(A) + W(A)]. \tag{54}$$

[The chiral fermion fields have been integrated to yield $W(A)$.] Next unity is resolved in the Faddeev-Popov way:

$$I = \Delta_f(A) \int \mathcal{D}g \delta[f(A^{g^{-1}})]. \tag{55}$$

Here $f(A) = 0$ is a gauge fixing, and $\Delta_f(A)$ is the gauge invariant Faddeev-Popov determinant which renders (55) true. Upon multiplying (54) by unity, as presented in (55), changing orders of functional integration, and changing the integration variable from A to A^g, one is left with the announced results:

$$Z = \int \mathcal{D}A \mathcal{D}g \Delta_f(A) \delta[f(A)] \exp i\,[I(A^g) + W(A^g)]$$
$$= \int \mathcal{D}A \mathcal{D}g \Delta_f(A) \delta[f(A)] \exp i\,[I(A) + W(A) - 2\pi\omega_1(A^g)]. \tag{56}$$

However, one cannot view this as a true derivation from quantum mechanical first principles. The Faddeev-Popov trick is just that — a trick that short-circuits a lengthy canonical analysis of constraints. It is known to be a valid trick in conventional gauge theories, but misses the mark in more subtly constrained situations. In the present context, it is not at all clear why one should start, as in (54), with an integral over *all* gauge potentials. One may worry that integrating over all gauge configurations will destroy positivity and/or unitarity, owing to the indefinite metric. Indeed, a soluble example, which we now discuss, shows that unitarity can be lost in this way.

5. CHIRAL SCHWINGER MODEL

In view of the uncertainties mentioned above with the general program, it is very satisfying that in some simple models one can analyze the situation completely.[12] Especially useful is the chiral Schwinger model: two-dimensional electrodynamics with massless Weyl fermions of one chirality interacting with $U(1)$ gauge field. The Lagrangian is

$$\mathcal{L} = -\frac{1}{4}F^{\mu\nu}F_{\mu\nu} + \bar\psi\gamma^\mu\left[i\partial_\mu + eA_\mu\frac{(1+i\gamma_5)}{2}\right]\psi\,, \quad \gamma_5 = i\gamma^0\gamma^1\,. \tag{57}$$

Owing to the well-known simplifications that hold in two dimensions, the chiral fermion determinant may be evaluated. The effective action is

$$I_{\text{eff}} = I(A) + \frac{e^2}{8\pi}\int A_\mu[ag^{\mu\nu} - (g^{\mu\alpha} + \varepsilon^{\mu\alpha})\frac{\partial_\alpha\partial_\beta}{\Box}(g^{\beta\nu} - \varepsilon^{\beta\nu})]A_\nu\,, \tag{58}$$

where a is a constant, left undetermined in the evaluation of the determinant. For no value of a is the answer gauge invariant — that is why the theory is anomalous. Since a cannot be determined, the strength of the contact term $\int A^2$ is arbitrary. In a gauge invariant theory, contact terms are fixed by the requirement of gauge invariance; here we have no such principle.

An equivalent description is in terms of a bosonized action

$$I(A,\phi) = I(A) + \frac{1}{2}\int d^2x\left[\partial_\mu\phi\partial^\mu\phi + \frac{e}{\sqrt\pi}(g^{\mu\nu} - \varepsilon^{\mu\nu})\partial_\mu\phi A_\nu + \frac{ae^2}{4\pi}A_\mu A^\mu\right]\,, \tag{59}$$

where the a term now arises from ambiguities in the bosonization procedure, Functionally integrating over ϕ in (59) yields (58).

The theory described by (58) or (59) may be solved. One finds for $a \geq 1$ a unitary, positive definite model. The vector meson acquires mass m:

$$m^2 = \frac{e^2}{4\pi}\frac{a^2}{a-1}\,. \tag{60}$$

This lies anywhere between e^2/π and infinity, in contrast to the conventional Schwinger model, where gauge invariance fixes the corresponding constant a and the gauge boson mass.

It is noteworthy that a consistent, but not completely determined theory has emerged. (The one-parameter ambiguity at first caused confusion,[13] but now has been rederived by many different methods.[14]) In a sense, one can view the above calculation, with its loss of gauge invariance, as a calculation within the gauge invariant chiral action with Wess-Zumino term (56), in the "unitary" gauge $g = 1$, where the Wess-Zumino term vanishes. Alternatively, one can regard the result as coming from an anomaly-free theory, where additional anomaly-canceling fermions have been decoupled, by sending parameters to infinity. However, neither of these alternative viewpoints explains why only a range of the undetermined parameter $[a \geq 1]$ gives a unitary theory, while unitarity is lost outside this range $[a < 1]$.

The bosonized action explicitly incorporates the anomaly, and canonically quantizing (59) does indeed involve second-class constraints. Carrying out the details of this quantization reproduces the results obtained by the action integral.[15]

In all of these derivations, one fact remains: a consistent, Lorentz invariant[16] theory of a massive vector meson emerges from an anomalous gauge theory. Similar results can be established for the non-Abelian generalization, though the analysis is more formal and less explicit, since the model cannot be solved completely.[17]

Unfortunately, nothing is learned from the two-dimensional exercises about renormalizability; these two-dimensional theories are finite. Thus it is not known whether four-dimensional anomalous theories are renormalizable, but one may expect that anomalies do not spoil unitarity and positivity. This expectation is borne out by a formal canonical analysis of a bosonized action that describes the low-energy dynamics of an anomalous four-dimensional gauge theory.[18] One finds that gauge symmetry is broken, and the gauge field is massive.

DISCUSSION

D. GROSS: How can you see that the constraint is second class before you quantize?

R. JACKIW: As I mentioned, that is precisely the obstacle to carrying out the constrained quantization program with anomalies — they arise only after the theory has been quantized, but then they change the way that the theory should be quantized.

C. TEITELBOIM: To get started, you have to put in $\hbar$ classically.

JACKIW: Something like that. You can write an effective (bosonized) action, as in (59), which reproduces the anomalies by explicit (classical) reasoning. Then you can quantize, and you find you are dealing with second-class constraints.

6. GRAVITATIONAL THEORIES

In 2 mod 4 dimensions, chiral fermions induce gravitational anomalies.[1] This is to be expected since gravity theory may be formulated as a gauge theory, invariant against local Lorentz transformations. The current which now is the symmetric energy-momentum tensor $\theta_{\mu\nu}$, is not (covariantly) conserved in the presence of chiral fermions. Just as for gauge anomalies, which arise in 2 mod 2 dimensions, the two-dimensional case is the most important one, but gravity in two dimensions has its own pecularities, which have nothing to do with anomalies.

The usual equation governing gravity, involving the Einstein tensor, $G_{\mu\nu}$,

$$G_{\mu\nu} - \Lambda g_{\mu\nu} = G\theta_{\mu\nu}, \qquad G_{\mu\nu} \equiv R_{\mu\nu} - \frac{1}{2}g_{\mu\nu}g^{\alpha\beta}R_{\alpha\beta} \qquad (61)$$

where $R_{\mu\nu}$ is the Ricci curvature, Λ a cosmological constant, $g_{\mu\nu}$ the metric, and G the gravitational constant, requires conservation of $\theta_{\mu\nu}$, since the left-hand side possesses this property. However, in two dimensions (61) cannot be postulated, since $G_{\mu\nu}$ vanishes identically. Correspondingly, the quantity $\int d^2x\sqrt{-g}R$, $g \equiv \det g_{\mu\nu}$, $R \equiv g^{\mu\nu}R_{\mu\nu}$, whose variation gives $G_{\mu\nu}$ in dimension greater than 2, is a surface term in two dimensions, with vanishing variation.

Therefore, two-dimensional gravity must be based on an equation different from (61), and the anomaly problem for two-dimensional gravity becomes more subtle than a mere mismatch between the conservation properties of two sides of Einstein's equation, as in (61).

Nevertheless, the essence of the problem remains the same: the chiral fermion determinant (which may be computed exactly[19]) loses local Lorentz invariance (general coordinate invariance can always be maintained). Also, it possesses an arbitrariness, analogous to that parametrized by "a" in the gauge theory case.

The noninvariant chiral fermion effective action may be combined with an invariant action for two-dimensional gravity whose form has been derived from the three-dimensional Einstein-Hilbert action by dimensional reduction[20]:

$$I(g) = \int d^2x N\sqrt{-g}(R + 2\Lambda). \qquad (62)$$

Here N is a scalar Lagrange multiplier. The analysis of the complete theory cannot be carried out completely, since the model is nontrivial. But again, formal indications support unitarity with massless excitations for a range of the arbitrary parameter.[21]

7. CONCLUSIONS AND OUTLOOK

It should not be constructed that my present pursuit of anomalous theories repudiates the anomaly cancellation principle that D. Gross and I enunciated in response to the Weinberg-Salam model in the 1970s. In its original formulation, that model is anomalous, yet a conventional quantization was proposed; this is inconsistent. The remedy we put forward cancels the anomaly with a quark-lepton balance, which is verified but not completely established experimentally: the top quark has not been discovered and the τ neutrino need not be different from the e and μ neutrinos. Of course, absence of the sixth quark can be accommodated at any given time if its mass is beyond experimental range. However, should ν_τ prove not to be a new particle, anomalies cannot cancel, and new ideas — perhaps those presented here — will be necessary.

Of course, renormalizability of a four-dimensional anomalous theory remains an open problem, which is under consideration.[22] The hope is that infinities arising from the anomaly cancel against those from the Wess-Zumino term. Verification of this is complicated in the absence of a kinetic term for the Wess-Zumino chiral field.[23] Should a positive result hold, it could very well be coupled with the emergence of a mass for the gauge fields, as is seen in all the examples studied thus far. Gauge symmetry breaking in electroweak theories would then appear in a new light: the weak forces are mediated by massive vector mesons whose mass breaks the gauge symmetry owing to chiral anomalies; these are absent in the chirally symmetric electromagnetic channel with its massless photon.

While uncertainties remain about four-dimensional anomalous theories, the complete success in two dimensions counts as an important result, in view of the many physical applications and the central role for the string program that two-dimensional models enjoy.

REFERENCES AND NOTES

1. For a review and references to the original literature, see S. Treiman, R. Jackiw, B. Zumino and E. Witten, *Current Algebra and Anomalies* (Princeton University Press, Princeton, NJ/World Scientific, Singapore, 1985); *Anomalies, Geometry, Topology,* W. Bardeen and A. White, eds. (World Scientific, Singapore, 1985).

2. For earlier review see R. Jackiw, in *Proceedings of the Oregon Meeting,* R. Hwa, ed. (World Scientific, Singapore, 1986, p. 772); C. Viallet, in *Super Field Theories,* H. C. Lee, V. Elias, G. Kunstatter, R. B. Mann and K. S. Viswanathan, eds. (Plenum, New York, NY, 1987, p. 399).

3. L. Faddeev, *Phys. Lett.* **145B**, 81 (1984); J. Mickelsson, *Comm. Math. Phys.* **97**, 361 (1985); I. Singer, *Asterisque*, 323 (1985).

4. S.-G. Jo, *Phys. Lett.* **163B**, 353 (1985); M. Kobayashi, K. Seo and A. Sugamoto, *Nucl. Phys.* **B273**, 607 (1986). Let me note here that other purported verifications, using regulated fixed-time procedures for calculating the

anomalous commutator, in contrast to the BJL method of the above-cited investigations, are in fact incomplete: L. Faddeev and S. Shatashvili's paper, *Phys. Lett.* **167B**, 225 (1986), contains errors, while I. Frenkel and I. Singer did not complete their calculations. A non-BJL analysis, which apparently also confirms the conjectured form of the commutator, is by A. Niemi and G. Semenoff, *Phys. Rev. Lett.* **56**, 1019 (1986). This interesting paper relates the anomaly phenomenon to Berry's adiabatic phase; see also G. Semenoff, in *Super Field Theories*, H. C. Lee, V. Elias, G. Kunstatter, R. B. Mann and K. S. Viswanathan, eds. (Plenum, New York, NY, 1987, p. 407).

5. Recent investigations with references to the older literature are D. Levy, *Nucl. Phys.* **B282**, 367 (1987); Y.-Z. Zhang, *Phys. Lett.* **189B**, 149 (1987).
6. Jo, Niemi and Semenoff, Ref. 4.
7. G. Horowitz and A. Strominger, *Phys. Lett.* **185B**, 45 (1987); A. Strominger, *ibid.* **187B**, 149 (1987).
8. L. Faddeev, in *Supersymmetry and Its Applications*, G. Gibbons, S. Hawking and P. Townsend, eds. (Cambridge University Press, Cambridge, UK, 1985, p. 41); Faddeev and Shatashvili, Ref. 4.
9. E. D'Hoker and E. Farhi, *Nucl. Phys.* **B248**, 59, 77 (1984).
10. J. Reiff and M. Veltman, *Nucl. Phys.* **B13**, 545 (1969).
11. O. Babelon, F. Schaposnik and C. Viallet, *Phys. Lett.* **177B**, 385 (1986); K. Harada and I. Tsutsui, *Phys. Lett.* **183B**, 311 (1987); N. Falck and G. Kramer, *Ann. Phys.* (NY) **176**, 330 (1987).
12. R. Jackiw and R. Rajaraman, *Phys. Rev. Lett.* **54**, 1219, 2060(E), **55**, 2224(C) (1985).
13. C. Hagen, *Phys. Rev. Lett.* **55**, 2223(C) (1985); A. Das, *Phys. Rev. Lett.* **55**, 2126 (1985).
14. R. Banerjee, *Phys. Rev. Lett.* **56**, 1889 (1986); J. Webb, *Z. Phys.* **C31**, 301 (1986); M. Chanowitz, *Phys. Lett.* **171B**, 280 (1986); K. Harada, T. Kubota and I. Tsutsui, *Phys. Lett.* **173B**, 77 (1986); I. Halliday, E. Rabinovici, A. Schwimmer and M. Chanowitz, *Nucl. Phys.* **B268**, 413 (1986); H. Girotti, H. Rothe and K. Rothe, *Phys. Rev. D* **34**, 592 (1986); F. Schaposnik and J. Webb, *Z. Phys.* **C34**, 367 (1987); R. Ball, *Phys. Lett.* **183B**, 315 (1987); K. Harada and I. Tsutsui, *Prog. Theor. Phys.* (Kyoto) **78**, 878 (1987).
15. R. Rajaraman, *Phys. Lett.* **154B**, 305 (1985); J. Lott and R. Rajaraman, *Phys. Lett.* **165B**, 321 (1985); H. Girotti, H. Rothe and K. Rothe, *Phys. Rev. D* **33**, 514 (1986).
16. Lorentz noninvariant results have also been obtained by Halliday *et al.* and Chanowitz, Ref. 14; A. Niemi and G. Semenoff, *Phys. Lett.* **175B**, 439 (1986). But these authors use a Lorentz noninvariant gauge, which cannot be justified in a gauge noninvariant theory.
17. R. Rajaraman, *Phys. Lett.* **162B**, 148 (1985).

18. R. Rajaraman, *Phys. Lett.* **184B**, 369 (1987).
19. H. Leutwyler, *Phys. Lett.* **153B**, 65 (1985).
20. R. Jackiw, in *Quantum Theory of Gravity*, S. Christensen, ed. (A. Hilger, Bristol, UK, 1984, p. 403); *Nucl. Phys.* **B252**, 343 (1985); C. Teitelboim, in *Quantum Theory of Gravity*, S. Christensen, ed. (A. Hilger, Bristol, UK, 1984, p. 327).
21. K.-K. Li, *Phys. Rev.* D **34**, 2292 (1986).
22. L. Faddeev (private communication).
23. E. D'Hoker and E. Farhi, in unpublished research, have convinced themselves that their model (Ref. 9), with a kinetic term for the Wess-Zumino chiral field, is not renormalizable.

Paper I.5

Participants at the Nuffield workshop on quantum gravity (London, August 1981) remarked that the word "polyacetylene" had never previously been heard in their discussions of general relativity. Nevertheless, the topological effects, first realized in this mundane linear polymer, may very well have application to theories of space and time.

FERMION FRACTIONIZATION

Quantum Structure of Space and Time, M. Duff and C. Isham, eds. *(Cambridge University Press, Cambridge, UK, 1982)*

I. INTRODUCTION

I shall discuss a novel quantum mechanical/topological phenomenon that has become a focus for contemporary research in theoretical and experimental physics as well as in mathematics. The effect is as easy to describe as it is puzzling to comprehend: when a fermion is introduced to a soliton, the fermion can split up — it can fractionize — and its quantum numbers are shared by the resulting states. In this way, fractional charge can emerge in a theory where all basic constituents carry integral charges.

This unexpected behavior was first noted in purely mathematical, but otherwise unmotivated, studies of soliton-fermion interactions.[1] The current interest arises from the circumstance that a one-dimensional fermion-soliton system can be physically realized in condensed matter situations, as a description of an electron [fermion] in the presence of a domain wall [soliton]. In polyacetylene, for example, the peculiar experimental effects that have been observed[2] are explained by the peculiarities of fermions in the presence of solitons.[3] What makes the theory especially beautiful is that its predictions depend little on details of any particular model. Rather, very general topological and quantum mechanical considerations suffice to establish the occurrence of charge fractionization. But of course the practical relevance of the result depends on the specific physical setting in which it is encountered.

There are many topics to discuss: the original calculations in relativistic quantum field theory and the subsequent ones in condensed matter field theory which established fermionic solitons with charge 1/2, per degree of freedom[1,3,4]; the physical situation of polyacetylene and similar quasi-one dimensional organic polymers, which provide a theoretical and experimental laboratory for soliton physics[2,3]; the recently posited generalizations which give rise to other fractional charges in condensed matter systems, as well as in intriguing relativistic field theories[5,6]; and finally, the mathematical concepts which relate this phenomenon to the non-trivial topological structures that are present.

All this cannot be covered here; therefore I shall select for discussion only the relativistic field theory calculations, with an emphasis on the topological properties which give rise to charge 1/2 and, in a generalization, to arbitrary fractions. However, since the polyacetylene story is so important for establishing the physical actuality of the phenomenon, I shall review it briefly and pictorially.

II. THE POLYACETYLENE STORY

Polyacetylene is a material consisting of parallel chains of carbon atoms, with electrons moving primarily along the chains, while hopping between chains is strongly suppressed. Consequently, the system is effectively one-dimensional, and can be represented as in Fig. 1a. The distance between carbon atoms is about 1 Å.

If the atoms are considered to be completely stationary, *i.e.*, rigidly attached to their equilibrium lattice sites, electron hopping along the chain is a structureless phenomenon, described by a probability amplitude, which to be sure depends on the distance between the atoms, but since that quantity never changes neither does the hopping amplitude.

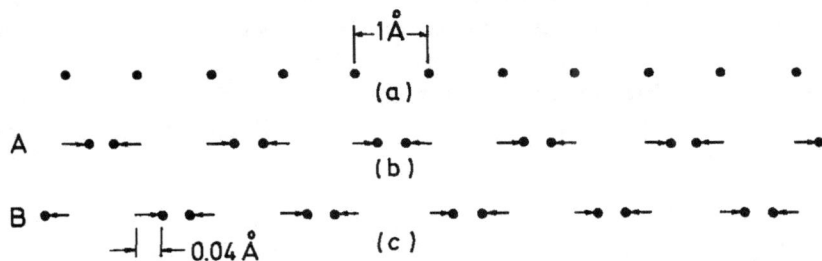

Fig. 1(a). The rigid lattice of polyacetylene; the carbon atoms are 1 Å apart. **(b), (c).** The effect of Peierls' instability is to shift the carbon atoms .04 Å to the right **(A)** or to the left **(B)**, thus giving rise to a double degeneracy.

However, the atoms can be displaced from their rigid lattice positions, for a variety of reasons, like zero-point motion, thermal excitation, *etc.* It might be thought that these effects merely give rise to small oscillations about the rigid-lattice sites, and produce only a slight fuzzing of the undistorted-lattice situation.

In fact this is not correct; something more dramatic takes place. Rather than oscillating about the rigid-lattice site, the atoms first shift a distance of about .04 Å and then proceed to oscillate around the new, slightly distorted location. That this should happen was predicted by Peierls, and is called the Peierls instability.[7] Due to reflection symmetry, there is no difference between a shift to the right or a shift to the left; the material chooses one or the other, thus breaking spontaneously the symmetry, and giving rise to doubly

degenerate vacua, called A and B, as is illustrated in Figs. 1b and 1c. The
chemical bonding patterns are illustrated in Figs. 2a and 2b, where the double
bond connects atoms that are closer together, and the single bond those that
are further apart than in the [unphysical] rigid lattice.

If the displacement is described by a field ϕ which depends on the position
x along the lattice, the so-called phonon field, then Peierls instability, as well
as detailed dynamical calculations[3] indicate that the energy density $V(\phi)$, as
a function of constant ϕ, has the double-well shape, depicted in Fig. 3a. The
symmetric point $\phi = 0$ is unstable; the system in its ground state must choose
one of the two equivalent ground states $\phi = \phi_0 = \pm.04$ Å. In the ground states,
the phonon field has uniform values, independent of x; as is shown in Fig. 3b.

By now, it is widely appreciated that whenever the ground state is degenerate, there frequently exist additional stable states of the system, for which
the phonon field is non-constant. Rather, as a function of x, it interpolates,
when x passes from negative to positive infinity, between the allowed ground
states. These are the famous solitons, or kinks. For polyacetylene, they are
also depicted in Fig. 3b, and they correspond to domain walls which separate
regions with vacuum A from those with vacuum B [solitons], and vice versa
[anti-solitons].

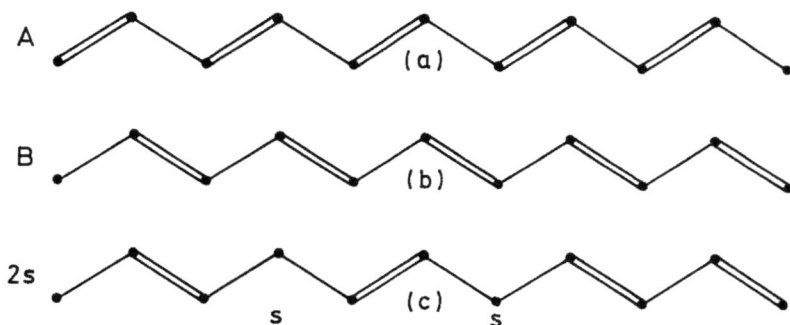

Fig. 2(a), (b). Pattern of chemical bonds in vacua A and B. **(c).** Two solitons inserted
into vacuum B.

Consider now a polyacetylene sample in the B vacuum, but with two
solitons along the chain, as depicted in Fig. 2c. Let us count the number
of links in the sample without solitons (Fig. 2b) and compare with the
number of links where two solitons are present (Fig. 2c). It suffices to
examine the two chains only in the region where they differ, *i.e.*, between

the two solitons. Vacuum B exhibits 5 links, while the addition of two solitons decreases the number of links to 4. *The two soliton state exhibits a deficit of one link.* If now we imagine separating the two solitons a great distance, so that they act independently of one another, *then each soliton carries a deficit of half a link, and the quantum numbers of the link, for example the charge, are split between the two states.* This is the essence of fermion fractionization.

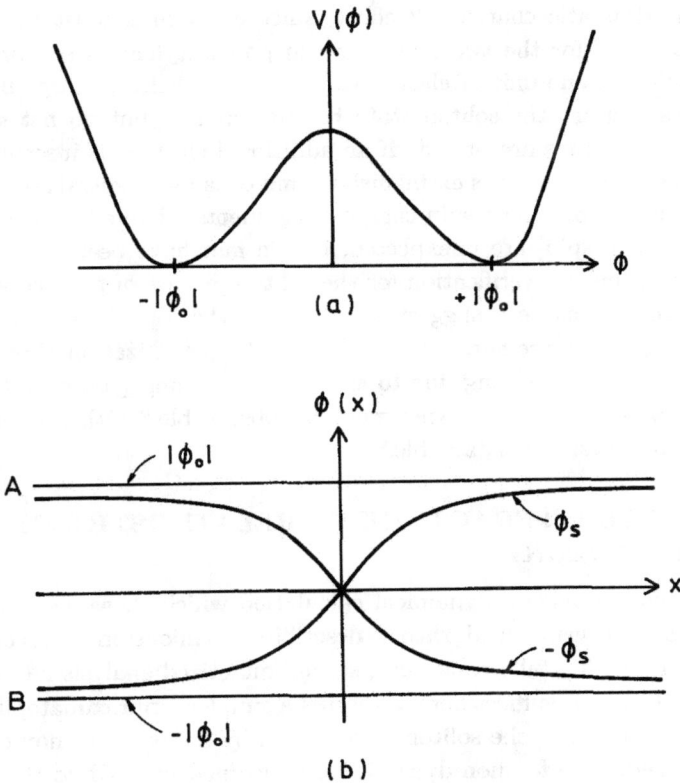

Fig. 3(a). Energy density $V(\phi)$, as a function of a constant phonon field ϕ. The symmetric stationary point, $\phi = 0$, is unstable. Stable vacua are at $\phi = +|\phi_0|$, (A) and $\phi = -|\phi_0|$, (B). **(b)** Phonon fields corresponding to stable states. The two constant fields, $\pm|\phi_0|$, correspond to the two vacua (A and B). The two kink fields, $\pm\phi_s$, interpolate between the vacua and represent domain walls.

It should be emphasized that we are not here describing the familiar situation of an electron moving around a two-center molecule, spending "half" the time with one nucleus and "half" with the other. Then one might say that the electron is split in half, on the average; however fluctuations in any quantity are large since the one-nucleus state is not an eigenstate of any physical observable. In our soliton example, the fractionization is without fluctuations; in the limit of infinite separation one achieves an eigenstate with fractional eigenvalues.

We must however remember that the link in fact corresponds to *two* states: an electron with spin up and another with spin down. This doubling obscures the dramatic charge $1/2$ effect, since everything must be multiplied by 2 to account for the two states. So in polyacetylene, a soliton carries a charge deficit of one unit of electric charge. Nevertheless, charge fractionization leaves a spur: the soliton state has net charge, but no net spin, since all the electron spins are paired. If an additional electron is inserted into the sample, the charge deficit is extinguished, and obtains a neutral state, but now there is a net spin. These spin-charge assignments [charged — without spin, neutral — with spin] are unexpected, but in fact have been observed,[2] and provide experimental verification for the soliton picture of polyacetylene.

Moreover, it has been suggested that in another one-dimensional system, TTF-TCNQ under pressure, the fundamental fractionization should be $1/3$ per state, and the doubling due to spin would no longer obscure the charge fractions, which should be experimentally observable.[5] Other fractions also appear to be physically attainable.[6]

III. CHARGE $1/2$ FRACTIONS IN RELATIVISTIC QUANTUM FIELD THEORY

I shall now provide a dynamical calculation which shows how charge $1/2$ arises in a relativistic field theory describing fermions in interaction with solitons. Although a fully consistent, second quantized analysis has developed for this problem,[8] it suffices here to discuss a simpler, approximate, first quantized approach, where the soliton is described by an external, non-dynamical c-number field. The fermion dynamics are governed by a Dirac Hamiltonian, $H(\phi)$, which also depends on a background field ϕ, with which the fermions interact. In the vacuum sector, ϕ takes on a constant value ϕ_0, appropriate to the vacuum. When a soliton is present, ϕ becomes the appropriate, static soliton profile ϕ_s. We need not be any more specific; in particular, the discussion is *not* limited to one spatial dimension. Thus in one dimension the soliton can be the kink of Fig. 3b; in two, it is an Abrikosov [Nielsen-Olesen] vortex[9]; in three, the 't Hooft-Polyakov monopole.[10] We need not even insist on the explicit soliton profiles that solve the corresponding non-linear field equations; all that we require is that the topology [*i.e.*, the large distance behavior] of the profiles be non-trivial. The one-dimensional model is relevant for polyacetylene; the Dirac Hamiltonian arises not because the electrons are relativistic; but rather it emerges in a certain well-formulated approximation [which ignores electron spin][11] to the microscopic theory.[3]

To analyze the system we need the eigenmodes, both in the vacuum and soliton sectors

$$H(\phi_0)\psi_E^0 = E^0\psi_E^0 \,, \tag{3.1}$$

$$H(\phi_s)\psi_E^s = E^s\psi_E^s \,. \tag{3.2}$$

As is familiar, there will be in general negative energy solutions and positive energy solutions. [For polyacetylene, the negative energy solutions correspond to the states in the valence band; the positive energy ones, to the conduction band.] In the ground state, all the negative energy levels are filled, and the ground state charge is the integral over all space of the charge density $\rho(x)$; which in turn is constructed from all the negative energy solutions

$$\rho(x) = \int_{-\infty}^0 dE\rho_E(x) \tag{3.3}$$

$$\rho_E(x) = \psi_E^*(x)\psi_E(x) \,.$$

Of course, integrating (3.3) over x will produce an infinity; to renormalize, we measure all charges relative to the ground state in the vacuum sector. Thus the soliton charge is

$$Q = \int dx \int_{-\infty}^0 dE\{\rho_E^s(x) - \rho_E^0(x)\} \,. \tag{3.4}$$

Equation (3.4) may be completely evaluated without explicitly specifying the soliton profile, nor actually solving for the negative energy modes, provided H possesses a further property. We assume that there exists a conjugation symmetry which takes positive energy solutions of (3.1) and (3.2) into negative energy solutions. That is, we assume that there exists a unitary matrix M, such that

$$M\psi_E = \psi_{-E} \,. \tag{3.5}$$

An immediate consequence, crucial to the rest of the argument, is that the charge density at E is an even function of E

$$\rho_E(x) = \rho_{-E}(x)\,. \tag{3.6}$$

[Clearly, one may allow a phase factor, or even complex conjugation, in the right-hand side of (3.5), without affecting (3.6)]. Charge conjugation is an example of the required symmetry; it holds in the polyacetylene theory.

Whenever one solves a conjugation symmetric Dirac equation, with a topologically interesting background field, like a soliton, there always are, in addition to the positive and negative energy solutions related to each other by conjugation, self-conjugate, normalizable zero-energy solutions. That this is indeed true can be seen by explicit calculation, and will be presently demonstrated when we exemplify the general discussion. However, the occurrence of the zero mode is also predicted by very general mathematical theorems about differential equations. These so-called "index theorems" count the zero eigenvalues, and insure that the number is non-vanishing whenever the topology of the background is non-trivial.[12] We shall assume that there is just one zero mode, described by a normalizable wave function ψ_0. This minimal situation can be achieved by taking a minimal, non-trivial background topology.[1]

To evaluate (3.4). we first recall that the wave functions are complete both in the soliton sector and in the vacuum sector

$$\int_{-\infty}^{\infty} dE \psi_E^*(x)\psi_E(y) = \delta(x-y)\,. \tag{3.7}$$

As a consequence, it follows that

$$\int_{-\infty}^{\infty} dE[\,\rho_E^s(x) - \rho_E^0(x)] = 0\,. \tag{3.8}$$

In the above completeness integral over all energies, we record separately the negative energy contributions, the positive energy contributions, and for the soliton, the zero-energy contribution. Since the positive energy charge density is equal to the negative energy one, by virtue of (3.6), we conclude that (3.8) may be equivalently written as an integral over negative E

$$\int_{-\infty}^{0} dE[2\rho_E^s(x) + \psi_0^*(x)\psi_0(x) - 2\rho_E^0(x)] = 0\,. \tag{3.9}$$

After a rearrangement of terms, Eq. (3.9) shows that

$$Q = -\frac{1}{2}\int dx\,\psi_0^*(x)\psi_0(x) = -\frac{1}{2}\,. \tag{3.10}$$

This is the final result: the soliton's charge is $-1/2$; a fact that follows from completeness [Eq. (3.7)] and conjugation symmetry [Eq. (3.6)]. It is seen in (3.10) that the zero-energy mode is essential to the conclusion. The existence of the zero modes in the conjugation symmetric case is assured by the non-trivial topology of the background field. The result is otherwise completely general, and holds in any number of dimensions.

IV. ARBITRARY FRACTIONAL CHARGE IN RELATIVISTIC QUANTUM FIELD THEORY

Since charge $1/2$ follows from completeness of the wave functions in the presence of a conjugation symmetry, generalizing to arbitrary fractions requires abandoning the conjugation symmetry. The original suggestion for a generalization came from condensed matter physics,[5] and now has been realized in relativistic quantum field theory.[6] However, as we shall see, the field theoretical version is not yet completely satisfactory: the fractionization is arbitrary, parameterized by the arbitrary amount of conjugation symmetry breaking; there is no understanding how this arbitrariness is fixed. [In the condensed matter application,[5] physical considerations determine all parameters, leading to definite fractions.]

I shall present a simple one-dimensional model, for which it will be possible to show that the soliton's charge is a transcendental function of the parameters. The calculation will demonstrate that the effect again depends little on the details, and requires almost no explicit computation; though of course, more is needed than in the very general charge $1/2$ examples. It is important to know that the generalization is not solely a one-dimensional effect; similar results have been established in three dimensions.[8]

Consider the following Dirac Hamiltonian $H(\phi)$, depending on a background field ϕ

$$H(\phi) = \sigma^2 p + \sigma^1 \phi + \sigma^3 \varepsilon$$
$$p = \frac{1}{i}\frac{d}{dx}.$$

(4.1)

Since we are in one spatial dimension, the Dirac algebra is realized by 2×2 matrices; we take them to be the Pauli matrices: Dirac's α matrix is σ^2, β is σ^1, and σ^3 corresponds to a pseudoscalar coupling. When $\varepsilon = 0$, there exists a conjugation symmetry since σ^3 anticommutes with $H|_{\varepsilon=0}$. However, with non-vanishing ε, which we take to be positive, the conjugation symmetry is lost.

To evaluate the charge (3.4), we need the negative energy modes, both in the vacuum sector $\phi = \phi_0$, and in the soliton sector $\phi = \phi_s$. All we need to know about the soliton profile is that it interpolates between the vacua

$$\varphi_s(x)_{x \to \pm\infty} = \pm|\phi_0|\,. \tag{4.2}$$

The vacuum sector is trivial: the wave functions are plane waves and the spectrum is continuous, beginning at $\pm\sqrt{\phi_0^2 + \varepsilon^2}$; $E^0(k) = \pm\sqrt{k^2 + \phi_0^2 + \varepsilon^2}$.

In the soliton sector, let us note first the existence of a discrete bound state at $E^s = \varepsilon$. The wave function is proportional to

$$\begin{pmatrix} e^{-\int^x dx' \phi_s(x')} \\ 0 \end{pmatrix}\,.$$

This is normalizable precisely because ϕ_s tends to opposite limits at opposite ends of the real line. Observe that in the conjugation symmetric limit, $\varepsilon = 0$, this is the zero-energy bound state mentioned before. It is present here as well, but shifted from the symmetric value. Its occurrence will play an important role in our analysis.

To proceed, we need more explicit information about the eigenmodes. Upon writing ψ_E as $\begin{pmatrix} u \\ v \end{pmatrix}$, and working out the Pauli matrices, we see that v is determined by u

$$v = \frac{1}{E+\varepsilon}(\partial_x + \phi)u\,. \tag{4.3}$$

The function u satisfies a Schrödinger-like equation

$$(-\partial_x^2 + \phi^2 - \phi')u = (E^2 - \varepsilon^2)u$$
$$\phi' = \frac{d}{dx}\phi \tag{4.4}$$

with a "potential" $\phi^2 - \phi'$, which is the constant ϕ_0^2 in the vacuum sector, and tends to ϕ_0^2 at infinite x in the soliton sector.

The soliton sector Schrödinger equation has a bound state,

$$u^s = e^{-\int^x dx' \phi_s(x')};$$

this is just the Dirac state at $E = \varepsilon$, previously identified. We shall assume that the soliton profile is sufficiently weak, so that it supports no other normalizable solutions to (4.4) in the soliton sector. [It is obvious how to modify the analysis when additional bound states are present.] The remaining eigenmodes of (4.4) lie in the continuum, which begins at $E^2 - \varepsilon^2 = \phi_0^2$.

It is now straightforward to construct the negative energy solutions of the Dirac equation, from the continuum solutions u_k of the Schrödinger equation

$$\psi_k = \begin{pmatrix} \sqrt{\dfrac{E+\varepsilon}{2E}} u_k \\[2mm] \dfrac{-1}{\sqrt{2E(E+\varepsilon)}}(\partial_x + \phi)u_k \end{pmatrix}$$

$$H(\phi)\psi_k = E\psi_k$$

$$E = -\sqrt{k^2 + \phi_0^2 + \varepsilon^2} .$$

(4.5)

Proper normalization of ψ_k is assured, when u_k is properly normalized.

The charge density, at a given [negative] energy E, is given by $\psi_k^*(x)\psi_k(x)$, which according to (4.5) is

$$\rho_k(x) = \frac{E+\varepsilon}{2E}|u_k(x)|^2 + \frac{1}{2E(E+\varepsilon)}|(\partial_x + \phi)u_k|^2$$

$$= |u_k(x)|^2 + \frac{1}{4E(E+\varepsilon)}\partial_x^2|u_k(x)|^2$$

$$+ \frac{1}{2E(E+\varepsilon)}\partial_x\big[|u_k(x)|^2\phi\big] .$$

(4.6)

The second equality follows from the first by virtue of the Schrödinger equation. The charge is the integral of the above over all x and k in the soliton sector, minus a similar integral in the vacuum sector. However, in the vacuum sector $|u_k|^2$ is a constant, as is ϕ, so the last two terms in (4.6) vanish. This leaves

$$Q = \int_{-\infty}^{\infty} dx \int_{-\infty}^{\infty} \frac{dk}{(2\pi)}\Big[|u_k^s(x)|^2 - |u_k^0(x)|^2\Big]$$

$$+ \int_{-\infty}^{\infty} \frac{dk}{(2\pi)}\frac{1}{4E(E+\varepsilon)}\big\{\partial_x|u_k^s(x)|^2 + 2|u_k^s(x)|\phi_s(x)\big\} \Big|_{x=-\infty}^{x=\infty} .$$

(4.7)

The double integral on the right hand side can be evaluated by completeness. The u_k^0 represent all the Schrödinger modes in the vacuum sector, while the u_k^s are one short of a complete set in the soliton sector, since the normalizable bound state is not included among them. [It corresponds to positive energy Dirac mode; the negative energy modes are constructed from the continuum Schrödinger solutions.] Hence, the first term on the right-hand side of Eq. (4.7) contributes -1 to Q. To evaluate the remaining terms, we need to know the Schrödinger modes, in the presence of the soliton, but only at $x = \pm\infty$. These may be expressed in terms of transmission and reflection coefficients

$$u_k(x) \xrightarrow[x\to]{} Te^{ikx}$$

$$u_k(x) \xrightarrow[x\to]{} e^{ikx} + Re^{-ikx} . \qquad (4.8)$$

We thus get

$$Q = -1 + \int_{-\infty}^{\infty} \frac{dk}{(2\pi)} \frac{|\phi_0|}{2E(E+\varepsilon)} \{|T|^2 + (|R|^2 + 1)\} , \qquad (4.9)$$

where oscillatory terms have been dropped and the plus sign between the term from $x = +\infty$ and the terms from $x = -\infty$ arises because of the sign reversal in $\phi_s(x)$. Unitarity

$$|T|^2 + |R|^2 = 1 \qquad (4.10)$$

permits the final evaluation

$$Q = -\frac{1}{\pi} \tan^{-1} \left| \frac{\phi_0}{\varepsilon} \right| . \qquad (4.11)$$

In the conjugation symmetric limit, $\varepsilon \to 0$, and the previous result, $Q = -1/2$, is regained.

V. DISCUSSION

The results presented here demonstrate a novel and fascinating quantum mechanical phenomenon, which was previously unsuspected. The emergence of fractional quantum numbers is related to other examples of odd quanta in topologically non-trivial settings, like the occurrence of spin 1/2 in a bosonic theory,[13] or the presence of hidden variables [vacuum angles] in the description of physical states.[14]

Charge fractionization possesses an elegant mathematical description. Observe that the charge (3.4) may also be written as an integral of a charge density

$$Q = \int_{-\infty}^{\infty} dx \rho(x) , \qquad (5.1a)$$

$$\rho(x) = -\frac{1}{2\pi} \frac{d}{dx} \tan^{-1}\left(\frac{\phi_s(x)}{\varepsilon}\right) . \qquad (5.1b)$$

It has been suggested that for weakly varying background fields the above formula holds even when ε depends on x.[6] It has been further suggested that for weakly time-varying background fields, (5.1b) is the time-component of a conserved topological current[6]

$$
\begin{aligned}
j^\mu &= -\frac{1}{2\pi}\varepsilon^{\mu\nu}\partial_\nu\,\tan^{-1}\left(\frac{\phi_s}{\varepsilon}\right) \\
&= \frac{1}{2\pi}\varepsilon^{\mu\nu}\varepsilon_{ab}\hat\phi_a\partial_\nu\hat\phi_b
\end{aligned}
\tag{5.2}
$$
$$
\hat\phi_1 = \phi_s/\sqrt{\phi_s^2 + \varepsilon^2}
$$
$$
\hat\phi_2 = \varepsilon/\sqrt{\phi_s^2 + \varepsilon^2}\,.
$$

I have already mentioned that similar effects are found in other dimensions. For example, in three spatial dimensions, where the monopole is the topological soliton, the analogue to (5.2) is[6]

$$
\begin{aligned}
j^\mu =\;&\frac{1}{12\pi^2}\varepsilon^{\mu\alpha\beta\gamma}\varepsilon_{abcd} \\
&\times\left\{\hat\phi_a(\mathcal{D}_\alpha\hat\phi)_b(\mathcal{D}_\beta\hat\phi)_c(\mathcal{D}_\gamma\hat\phi)_d + \frac{3}{4}eF_{\alpha\beta,ab}\hat\phi_c(\mathcal{D}_\gamma\hat\phi)_d\right\}\,.
\end{aligned}
\tag{5.3a}
$$

Here the $SU(2)$ monopole is considered within a $SU(2) \times SU(2)$ gauge theory. Taking the fourth component of the scalar field to be ε, and the fourth components of the gauge field to be zero, while setting the remaining components to the monopole profile, one finds the same result as in one dimension

$$
Q = -\frac{1}{\pi}\tan^{-1}\left|\frac{\phi_0}{\varepsilon}\right|\,.
\tag{5.3b}
$$

There is another mathematical connection worth describing. In the formula for the charge (3.4), assume for a moment that the states are normalizable, and the negative energy integral is replaced by a sum. It is then clear that Q measures the difference between the number of negative eigenmodes of the Dirac equation in the presence of the soliton, relative to the situation without the soliton. A more continuous description is achieved by parametrizing the background field by a parameter α such that ϕ varies continuously from ϕ_0 to ϕ_s, as α passes from $-\infty$ to ∞; for example,

$$
\phi(x,\alpha) = [\theta(-\alpha)\phi_0 + \theta(\alpha)\phi_s(x)]\tanh|\alpha|\,.
\tag{5.4}
$$

The energy eigenvalues E of the Dirac equation will now depend on α, $E(\alpha)$, and in the course of α's variation some number will flow across $E = 0$. The charge is a measure of this spectral flow, which is topologically determined.[15] Of course, if the system were finite so that the levels are truly discrete and the wave functions normalizable, that number would be an integer; however, on the infinite space, with continuum wave functions, a non-integral value is possible.

What is the physical import of fermion fractionization? It is clearly relevant to condensed matter phenomena, especially in one dimension. Moreover, particle physicists are intrigued by the quark-like charge assignments, which emerge dynamically, rather than being postulated *a priori*. However, it is not at all apparent how one might replace the successful hadron phenomenology based on quarks, by one based on fractionally charged solitons. Nevertheless, let me record one more amusing formula. In the models considered, the mass of the fermion is $M_F = \sqrt{\phi_0^2 + \varepsilon^2}$. We may eliminate ε in favor of the charge, with the help of (4.11), which is also true in three dimensions, see (5.3b). There results a mass formula,

$$M_F = \frac{|\phi_0|}{|\sin \pi Q|} \qquad (5.5)$$

which has the intriguing property that it takes the same value, both for $Q = \frac{1}{3}$ and $Q = \frac{2}{3}$.

In gravity theory it should be possible to establish similar effects. One needs to develop the concept of a static, soliton-like background geometry. Then the solutions of the Dirac equation on the solitonic space should have zero modes, which would be interpreted as states with fractional fermionic quantum numbers. However, none of this has been worked out as yet.

It is very satisfying to contemplate a physical idea, with range of application from chemistry, to condensed matter physics, to quantum field theory and gravity theory, and finally to mathematics and topology. This is certainly striking evidence for the unity of all these disciplines, and provides another example of the power of mathematics to uncover unexpected physical phenomena.

REFERENCES

1. R. Jackiw and C. Rebbi, *Phys. Rev. D* **13**, 3398 (1976).
2. For a review of the experiments, see A. Heeger, *Comments Solid State Physics* **10**, 53 (1981).
3. W. P. Su, J. R. Schrieffer and A. Heeger, *Phys. Rev. Lett.* **42**, 1698 (1979) and *Phys. Rev. B* **22**, 2099 (1980); see also M. Rice, *Phys. Lett.* **71A**, 152 (1979).
4. R. Jackiw and J. R. Schrieffer, *Nucl. Phys.* **B190** [FS3], 253 (1981).
5. W. P. Su and J. R. Schrieffer, *Phys. Rev. Lett.* **46**, 738 (1981).

6. J. Goldstone and F. Wilczek, *Phys. Rev. Lett.* **47**, 968 (1981).

7. R. Peierls, *Quantum Theory of Solids* (Clarendon Press, Oxford, UK, 1955).

8. The approach to fermionic problems, based on J. Goldstone and R. Jackiw, *Phys. Rev. D* **11**, 1486 (1975), is given in Ref. 1. For a summary, see R. Jackiw, Rev. Mod. Phys. **49**, 681 (1977).

9. A. Abrikosov, *Zh. Eksp. Teor. Fiz.* **32**, 1442 (1957) [English translation: JETP **5**, 1174 (1957)]; H. Nielsen and P. Olesen, *Nucl. Phys.* **B61**, 45 (1973).

10. G. 't Hooft, *Nucl. Phys.* **B79**, 276 (1974); A. Polyakov, *Zh. Eksp. Teor. Fiz., Pis'ma Redakt.* **20**, 430 (1974) [English translation: *JETP Lett.* **20**, 194 (1974)].

11. H. Takayama, Y. Lin-Liu and K. Maki, *Phys. Rev. B* **21**, 2388 (1980).

12. For a summary of the index theorems, see for example T. Eguchi, P. Gilkey and A. Hanson, *Phys. Rep.* **66**, 213 (1980).

13. R. Jackiw and C. Rebbi, *Phys. Rev. Lett.* **36**, 1116 (1976); P. Hasenfratz and G. 't Hooft, *Phys. Rev. Lett.* **36**, 1119 (1976); J. Friedman and R. Sorkin, *Phys. Rev. Lett.* **44**, 1100 (1980).

14. S. Coleman, R. Jackiw and L. Susskind, *Ann. Phys.* (NY) **93**, 267 (1975); R. Jackiw and C. Rebbi, *Phys. Rev. Lett.* **37**, 172 (1976); C. Callan, R. Dashen and D. Gross, *Phys. Lett.* **B63**, 334 (1976).

15. M. F. Atiyah, V. K. Patodi and I. M. Singer, *Math. Proc. Camb. Phil. Soc.* **79**, 71 (1976).

Paper I.6

Remembering John Bell: How a discrepancy in the current algebra formulation of the decay of the neutral pion into two photons was resolved in terms of a quantum mechanical mechanism for symmetry breaking that has become an essential ingredient of models in particle physics.

THE CHIRAL ANOMALY

Europhysics News **22**, 76 (1991)

John Bell and I met and became acquainted in 1967, when I went to CERN for a year-long research visit, soon after finishing my doctoral studies at Cornell. At that time, particle physics theory was dominated, as it happens from time-to-time, by a single idea: there was broad agreement among theorists what the important problems were and how they should be solved, although today one can hardly remember the details of that program. But attaching my scientific activity to a consensus was not my ambition; I had much admired the independent attitude of one of my research supervisors at Cornell, Ken Wilson. So I looked among the staff at CERN for someone who pursued interesting issues that were neither "central" nor "important", and I was delighted to find such a scientist in John Bell. Moreover, he was generous in giving his time: he tolerated my coming to his office and appeared willing to discuss without limit. I appreciated the magnitude of his generosity only years later when I too became installed in an office and people began coming in and taking my time to talk about things.

The Pion Decay Puzzle

There began for us a period of wide-ranging conversations — and not only about physics — which acquainted me with the many issues that concerned John. But nothing was then said about his work on quantum mechanics — he did not at that time describe it to me and I did not know of it. Current algebra interested John very much. Within its framework, one can understand the low energy behaviour of elementary particles, without making a commitment to a dynamical model, unknown in the 1960's. (Today's "standard model" gives a dynamical justification, but still resists solution in the low energy domain; so current algebra remains vital.) The approach seemed successful, complete and exhausted by the late 1960's, yet there remained discrepancies between theoretical predictions and experimental verification.

John was particularly impressed with an analysis by his good friend Martinus Veltman and also D. Sutherland, to the effect that the neutral pion could not decay into two photons if the charge-neutral and gauge invariant axial vector (chiral) current is conserved, as it was then taken to be in current algebra applications. Because the decay does, in fact, occur in Nature, while the Sutherland/Veltman argument appeared incontrovertible, John stressed that the subject of current algebra must not be closed until this puzzle is resolved and urged a study of the chiral current.

This was the second time I received such advice: in my final student days, Wilson suggested a critical examination of the apparent conservation of the axial vector current in the Baker-Johnson-Willey theory of massless electrodynamics, with which he had his own disagreements.

I was therefore willing to research this topic, but since the existing discussions were straightforward and the conclusions immediate, it was hard to see how a useful probe could be launched. I asked fellow theorists for suggestions but the subject did not spark interest. I do recall two mathematically oriented colleagues, Henri Epstein and Raymond Stora, offering a diagnosis that in retrospect proved prescient: in their opinion one could not rely on current algebra analyses because physicists treat cavalierly singular products of distributions. But their prognosis that a cure will be found if one uses rigorous rather than heuristic mathematics did not appeal to me. In fact, the decisive suggestion did not come from a theorist but from an experimentalist.

Reformulating the σ-model

One of CERN's civilized activities, to which John frequently invited me, consists of taking an afternoon drink in the cafeteria, where we would continue our conversations together with people who joined us. On one occasion, Jack Steinberger — John's friend and collaborator on a CP formalism — was at the table and asked about our current interests. When we described to him the $\pi^0 \to 2\gamma$ puzzle, he expressed amazement that theorists should still be pursuing a process that he, an experimentalist, calculated almost twenty years earlier. He had found excellent agreement with experiment, while also noting a discrepancy between results obtained when the pion coupled to nucleons by pseudovector and pseudoscalar interactions (pions, nucleons and photons were the only particles in Steinberger's model, and it was believed that equivalent results emerge for pseudovector and pseudoscalar pion-nucleon coupling).

There, at that table, came to us the realization that Steinberger's calculation would be identical to the one performed in the dynamical framework of the σ-model, which was constructed to realize current algebra explicitly. We reasoned that within the σ-model, we could satisfy the current algebraic assumptions of Sutherland/Veltman and also obtain good experimental agreement in view of Steinberger's result, thereby resolving the $\pi^0 \to 2\gamma$ puzzle.

Guided by Steinberger's paper (at that time, we were not familiar with the work of his contemporaries, H. Fukuda and Y. Miyamoto, and only dimly aware of subsequent contributions by J. Schwinger), we quickly established that the amplitude describing correlations between the three currents appearing in the problem — two vector currents to which the two photons couple and one axial vector current to which the pion couples — is given in lowest order (one

loop) perturbation theory by the now famous triangle graph depicted in Fig. 1.
The amplitude is determined by Feynman rules only up to an overall ambiguity,
owing to ultraviolet divergences, even though the amplitude is finite. Moreover,
while the ambiguity may be resolved by enforcing current conservation, it is
impossible to maintain conservation of all *three* currents, as was assumed in
the current algebra calculation.

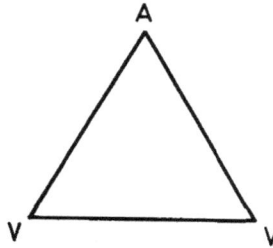

Fig. 1. — Triangle Feynman diagram for the three-current correlation amplitude that
leads to quantum mechanical symmetry breaking. Two vector currents V and one axial
vector current A form the three corners, at which virtual fermions are created/annihilated.
Propagation of the virtual fermions, indicated by solid lines, prevents the currents from being
conserved. A similar effect arises with three axial vector currents.

Thus, we found that the σ-model's symmetries, which underlie current
algebra and which should guarantee the conservation of the respective currents,
cannot be maintained when the model is quantized. In the absence of these
symmetries, pion decay is no longer forbidden.

Anomalous Symmetry Breaking

Our work resolved the $\pi^0 \to 2\gamma$ problem, by exposing a purely quantum
mechanical mechanism for symmetry breaking, which is the modern interpretation of Steinberger's discrepancy, and these days is described as "anomalous
breaking of a symmetry", although once the surprise has worn off, it is better
named "quantum mechanical symmetry breaking".[1]

The "anomaly" was substantiated by S. Adler, who working independently
came to a similar conclusion about the (absence of) symmetries in massless
electrodynamics, and building on our work established with W. Bardeen, the
important fact that higher perturbative orders do not modify the one-loop
calculation of pion decay. Further confirmation came from Wilson, who used

our theory as a case study for his non-Lagrangian models of current algebra, based on his operator product expansion. The early period of research on this subject culminated in a phenomenological description of quantum mechanical symmetry breaking in terms of an effective Lagrangian, constructed by J. Wess and B. Zumimo who apparently were not aware of our result.[2]

In time, this work, which arose from clearing up a corner of current algebra, grew to affect much of particle physics. It became an important ingredient of model building, both for speculative string theories and for the conventional "standard model", where, among other things, it enforces colour triality and explains the numerical equality of quark and lepton degrees of freedom, thus predicting the existence of the elusive top quark.[2]

John maintained an amused interest as our calculation became transformed in various contexts, and was shown to be a consequence of diverse physical and mathematical considerations[2]: symmetry breaking aspects of the Dirac sea (R. Feynman), anomalous transformation properties of the functional integral measure (K. Fujikawa), the necessary effect of high-energy modes on low-energy physics (V. Gribov), quantum field theoretic manifestation of Berry's phase, local version of the Atiyah-Singer index, and cohomological properties of gauge groups (L. Faddeev). The last two mathematical connections seeded a remarkable collaboration between mathematics and physics, which is still flourishing. On the other hand, the physical world itself became threatened by the anomaly because, as Gerard 't Hooft showed,[3] it catalyzes baryon decay, but fortunately at a sufficiently slow rate to cause no immediate concern.

In spite of these wide-ranging generalizations, John preferred the simple triangle graph calculation. He always stressed the element of choice that exists in resolving the calculational ambiguity, thus putting different faces on the nature of the anomaly — a freedom that is obscured in the more abstract and high-powered approaches.

Indeed John was rather diffident about the entire matter. In this he showed one of his striking qualities: modesty about his own work, praise for the work of others, but skepticism in the face of inflated claims, even if they were extolling his own contributions. When he eventually described to me his famous analysis of the foundations of quantum mechanics, he called that research "a hobby".

Fractional Charge Quantum Numbers

After I left CERN in 1968, we had many occasions to meet and talk about interesting topics, but our discussions never again resulted in a joint publication. The closest we came to this happened when I described to John the phenomenon of fractional charge quantum number,[4] which had become physically relevant.[5,6] He found this interesting but characteristically was at first skeptical that a fractional value could be an eigenvalue. Upon elaborating the precise circumstances in which a sharp observable arises, he published with R. Rajaraman[7] an analysis that contributed to the understanding and acceptance of this fascinating idea that today has also gained wide currency among physicists.

In all my contacts with John, I was always made aware of his overwhelming intellectual precision and honesty. These are the qualities that made him such an incisive critic and therefore a wonderful colleague. Precisely, this attitude lay behind his scientific achievements, which are informed by clarity of observation about previously murky subjects.

The same attitude characterized his approaches outside science, for example to social and political questions. Many physicists profess humane and liberal values, but often these become obscured by personal emotion and prejudice. In the last quarter century, issues of Vietnam, Ireland and Palestine offer a dramatic opportunity for displaying social conscience in search of justice. John recognized and spoke on these matters clearly. Already in 1967/68, I heard him analyze America's role in Vietnam in terms that did not gain acceptance until years later; his opinions on the two other tragedies remain in the minority even today, but one hopes that here too his ideas are merely ahead of their time.

I liked John very much and together with many colleagues, I shall miss him. He was an outstanding scientist and helped us do good science, which is one reason why we become physicists. Moreover, many enter our field not only for the opportunity of exploring Nature in its most fundamental workings, but also for what we perceive as the purity and honesty of the profession. These qualities sometimes get submerged by pressure of personal ambition, struggling for achievement and recognition, but John Bell never lost them, and in this way, he reminded us of the ethical motivation for becoming a physicist.

ADDITIONAL REFERENCES

1. See "Quantum Mechanical Symmetry Breaking", reprinted in this volume on p. 21.
2. For detailed discussion and a guide to the literature, see S. Treiman, R. Jackiw, B. Zumino and E. Witten, *Current Algebra and Anomalies* (Princeton University Press/World Scientific, Princeton, NJ/Singapore, 1985).
3. G.'t Hooft, *Phys. Rev. Lett.* **37**, 8 (1976).
4. R. Jackiw and C. Rebbi, *Phys. Rev. D* **13**, 3398 (1976).

5. W.-P. Su, J. R. Schrieffer and A. Heeger, *Phys. Rev. Lett.* **42**, 1968 (1979); R. Jackiw and J. R. Schrieffer, *Nucl. Phys.* **B190**, 253 (1981).

6. See "Fermion Fractionization", reprinted in this volume on p. 79.

7. R. Rajaraman and J. S. Bell, *Phys. Lett.* **116B**, 151 (1982); J. S. Bell and R. Rajaraman, *Nucl. Phys.* **B220** [FS8] 1 (1983).

Section II

GAUGE THEORIES AND GRAVITY

Positing a mathematical principle to guide model building has rarely resulted in physically relevant and phenomenologically interesting theories. The apparent exception provided by the gauge principle actually proves the rule, since that idea emerged from the physical theories of Maxwell electrodynamics and Einstein gravity, though it was not identified until the development of quantum mechanics, where upon the non-Abelian Yang-Mills theory was invented by analogy to the older theories. These days, the gauge principle continues guiding construction of mathematically interesting models (Chern-Simons terms, lower-dimensional gravity), which rapidly acquire physical importance (high-temperature field theory, planar condensed matter physics, cosmic strings).

Paper II.1

Specifying a gauge field theory obviously requires selecting a gauge group. In four dimensional space-time, specifying the quantized theory needs fixing a further parameter — : the vacuum angle. This intriguing and still un-understood effect has analogies in condensed matter physics and reflects the topological non-trivial properties of gauge groups, whose significance was appreciated in the mid-nineteen-seventies, thanks to the earlier discoveries of anomalies in the nineteen-sixties. At the Spring Meeting of the American Physical Society, these results were described in a joint mathematics/physics presentation, which marked the beginning of a fruitful and unexpected collaboration between physics and mathematics that is still flourishing, and which was eulogized by the well-known mathematician, I. Singer.

YANG-MILLS VACUUM AS A BLOCH WAVE

Lecture at the Spring Meeting, American Physical Society Washington DC, April 1977

In the last two-and-a-half years, non-perturbative, semi-classical approximation methods for extracting physically interesting content from field theories have been developed. Whenever we encounter a model which cannot be solved, we must resort to approximation techniques in order to describe the physics contained in that model, and the simplest, most widely used approximation is the Born perturbative series. However, from experience with quantum mechanics of point particles, we have learned that quantum systems will also exhibit non-perturbative effects, and we have succeeded in uncovering such non-perturbative phenomena in quantum field theory, which after all is also a quantum system, albeit a very complicated one.

We have now seen a wide variety of non-perturbative behavior; here I shall describe one such effect, which is present in non-Abelian gauge theories and which is the field theoretic analogue of quantum tunneling — a very familiar non-perturbative property that has not been previously discussed in the context of a quantum field theory.

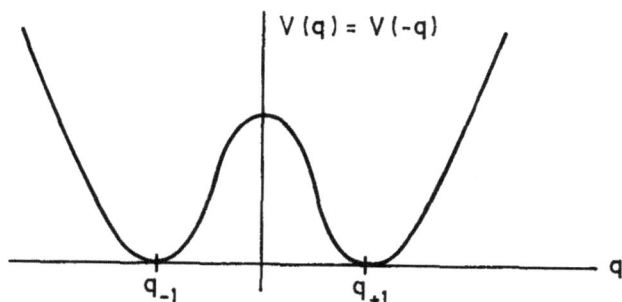

Fig. 1. Example of a potential profile $V(q) = \frac{\lambda^2}{2}(q^2 - q_0^2)^2$.

In order to introduce the type of analysis that I wish to apply to a field theory, let me begin first with a very simple example from point particle mechanics. Let us consider the one-dimensional problem of a particle moving in a symmetric potential of the form in Fig. 1 and let us set ourselves the task of giving an accurate, but approximate description of the ground state wave function, that is of the lowest energy solution to the Schrödinger equation

$$\left[-\frac{1}{2}\frac{d^2}{dq^2} + V(q) \right] \Psi(q) = E_0 \Psi(q)$$

$$E_0 \approx 0 .$$

A semi-classical, approximate, but accurate description of the quantal ground state begins by identifying the *classical* zero-energy configurations of the system, that is by solving the classical zero-energy equations of motion

$$0 = \frac{1}{2}\dot{q}^2(t) + V(q) \ .$$

We expect that the quantum wave function will localize the particle at the classical zero-energy position. Due to quantum fluctuations, the localization is not infinitely sharp — the wave function is not a delta-function — but is well-approximated by a Gaussian $\psi(q; q_c)$ centered as the classical position q_c. Correspondingly, the energy eigenvalue E_0 is not zero, but has some small quantal magnitude.

Let us observe that in this problem, there are *two* classical zero-energy configurations: $q = q_{-1}$, $q = q_{+1}$ — the classical ground state is degenerate. Consequently, we can form *two* Gaussians, one centered at q_{-1}, the other at q_{+1}. However, neither Gaussian respects the symmetry of the problem $q \rightarrow -q$, and therefore is not a good candidate for the ground state wave function. Rather we take linear combinations which respect the symmetry

$$\Psi_{\pm}(q) = \psi_+(q) \pm \psi_-(q) \ ,$$
$$\psi_{\pm}(q) = \psi(q; q_{\pm 1}) \ .$$

At this stage, we have arrived at two candidates for approximations to the ground state wave function with eigenvalues $E_{\pm}$, and the next question is whether the quantal ground state is doubly degenerate as is the classical ground state, $[E_+ = E_-]$, or whether quantum effects lift the degeneracy $[E_+ \neq E_-]$. Of course, we know that the degeneracy is lifted by quantum tunneling; that is by the possibility of quantum mechanical motion in the classically forbidden region, $q_{+1} \geq q \geq q_{-1}$. Is there any hint of this in the classical equations? Certainly, there is no real zero-energy motion in the forbidden region — however, it is possible to find a classical signal for quantum tunneling by the following trick. In the classical zero-energy equations, let us replace the time variable t by imaginary time x_4: $t \rightarrow -ix_4$

$$0 = -\frac{1}{2}\dot{q}^2(x_4) + V(q) \ .$$

Now the classical equation has a solution [for the example with Fig. 1 it is $q(x_4) = q_0 \tanh q_0 \lambda x_4$], which has the property that as the imaginary time

x_4 passes from $-\infty$ to $+\infty$, the solution passes from one of the classical zero-energy configurations $[q_{-1}]$ to the other $[q_{+1}]$. We interpret the existence of this imaginary time solution as a signal in classical physics for quantum mechanical tunneling, and we conclude that there is no quantum mechanical degeneracy; the energy levels are split.

I have given you rather much detail on this familiar example, so that I can now abstract a procedure for determining approximately, but accurately, the properties of the ground state in a quantum theory. The procedure involves four steps. First, enumerate the classical zero-energy configurations; second, construct Gaussians at each zero-energy configuration; third, form linear superpositions to respect symmetries; fourth, test for tunneling by solving zero-energy, imaginary-time equations.

Before applying this procedure to the problem of interest — a non-Abelian gauge theory — let me illustrate it once more, and briefly, in yet another simple quantum mechanical system which, as we shall see shortly, is analagous to the gauge theory. I consider one dimensional motion in a periodic potential of the following form depicted in Fig. 2.

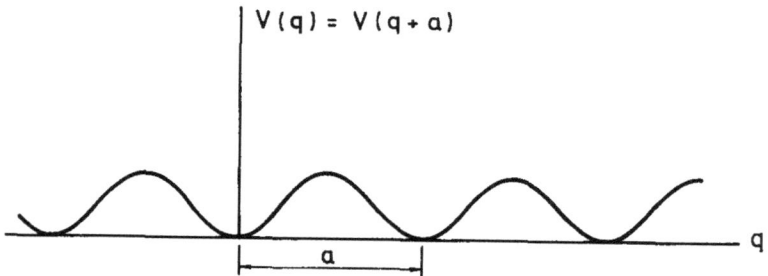

Fig. 2. Periodic quantum mechanical potential, which provides an analogy for the potential energy (density) in $(3 + 1)$-dimensional non-Abelian Yang-Mills theory.

The zero-energy configurations are $q = an$, $n = 0, \pm1, \ldots$; but the Gaussians formed around any one of them, $\psi_n(q) = \psi(q; an)$, do not respect the symmetry of the problem $q \to q + a$. Hence we form linear superpositions

$$\Psi_\theta(q) = \sum_n e^{in\theta} \psi_n(q) .$$

The translation symmetry is realized by

$$\mathcal{G}\Psi_\theta(q) = \Psi_\theta(q + a) = e^{i\theta}\Psi_\theta(q) ,$$

where $\mathcal{G}$ is the operator that translates by a. We thus arrive at a family of wave functions which are candidates for the ground state. The last question that must be answered is whether the states are degenerate in energy or whether their energy eigenvalues depend on θ. So we change time to imaginary-time and look for zero-energy solutions which interpolate, as imaginary time passes from $-\infty$ to $+\infty$, between the classical, real-time zero-energy configurations. Such solutions do exist; we conclude that tunneling lifts the degeneracy and we see that there exists an energy band of levels, $E(\theta)$. Thus with the semi-classical analysis, we arrive at an entirely accurate description of a Bloch wave.

Let me now, at last, turn to the physically interesting case of an $SU(2)$ Yang-Mills theory, described by the Lagrangian density

$$\mathcal{L} = -\frac{1}{4}F_a^{\mu\nu}F_{a\mu\nu}$$
$$F_a^{\mu\nu} = \partial^\mu A_a^\nu - \partial^\nu A_a^\mu + g\,\varepsilon_{abc}A_b^\mu A_c^\nu$$
$$a = 1, 2, 3 \; .$$

The Hamiltonian, that is the total energy, takes the form

$$H = \int d^3x\, \frac{1}{2}[E_a^2 + B_a^2]$$
$$E_a = F_a^{i0}$$
$$B_a = -\frac{1}{2}\varepsilon^{ijk}F_a^{jk}$$

and the quantum theory is developed in the gauge $A_a^0 = 0$. We can then pass to a Schrödinger picture, and we seek a solution of the functional Schrödinger equation

$$\int d^3x\, \frac{1}{2}\left[-\frac{\delta^2}{\delta A_a(x)\delta A_a(x)} + B_a(x)B_a(x)\right]\Psi(\mathbf{A}) = E\Psi(\mathbf{A}) \; .$$

For the vacuum functional we need the lowest energy solution to the equation. Let us analyze this problem according to the procedure previously outlined.[1] The first step is to enumerate the classical zero-energy configurations. Since $H = \int d^3x\frac{1}{2}[E_a^2 + B_a^2]$, we must have $\mathbf{E}_a = \mathbf{B}_a = 0$. Certainly this is achieved when $\mathbf{A}_a = 0$; however note that any $\mathbf{A}_a$ which is a pure gauge leads to vanishing energy. A pure gauge field is conveniently described by a matrix notation. Define $(1/g)\mathbf{A} = \mathbf{A}_a(\sigma_a/2i)$ where the σ_a are Pauli matrices. Then a pure gauge potential is given by $\mathbf{A} = -U^{-1}\boldsymbol{\nabla}U$ where U is any unitary

2×2 matrix. Of course, this potential is just a gauge transform of the vanishing potential, $U = I$, and we must decide whether any physical significance is to be attached to such gauge copies. Before making this decision, let me enumerate the kinds of gauge matrices U that occur.

To effect the enumeration, I begin with a restriction: We demand that $U \xrightarrow[x \to \infty]{} I$; at infinity at least we always have the trivial configurations. The categories of U are now the following. Category *zero* comprises $U = I$, as well as all those U's which are continuously deformable to $U = I$; that is U's for which we can construct a $U(\alpha)$ such that $U(1) = U$, $U(0) = I$. Category *one* comprises those U's that are not continuously deformable to I, but are continuously deformable to some canonical, representative matrix, say

$$U_1 = \frac{x^2 - \lambda^2}{x^2 + \lambda^2} - \frac{2i\lambda\boldsymbol{\sigma} \cdot \mathbf{x}}{x^2 + \lambda^2} \ .$$

In category *two*, we put matrices that are not deformable to I or U_1, and so forth. One can continue with this enumeration; indeed one can prove the following theorem: There are as many categories as there are integers n, $n = 0, \pm 1, \ldots$, and the representative matrix is $U_n = (U_1)^n$. With each category of gauge matrices, there is an associated gauge potential, which is a pure gauge and carries zero energy

$$\mathbf{A}_{(n)} = -U_n \boldsymbol{\nabla} U_n$$

and we see that the gauge transformations of the theory fall into two classes. There are those that gauge transform the gauge potentials and remain within a given category; we may call them "little" gauge transformations — they are deformable to the identity. Also there are the so-called "big" gauge transformations, which take the gauge potential from one category to another one, and are not deformable to the identity.

Now we recall that the whole procedure of fixing a gauge in a gauge theory is set up to remove the gauge freedom associated with infinitesimal gauge transformations — *i.e.*, gauge transformations, which are deformable to the identity, and the new suggestion, which we are here making, is that the big gauge transformations, which are not continuously deformable to the identity, are physically relevant; and gauge potentials should be counted as physically distinct if they are related only by big gauge transformations.

Having accepted this suggestion, we may straightforwardly complete our program of investigating the ground state. We form functional Gaussians peaked around each $\mathbf{A}_n$, $\psi_n(\mathbf{A}) = \psi(\mathbf{A}; \mathbf{A}_n)$; but the theory is gauge invariant, even under big gauge transformations, hence to respect this symmetry we must form linear combinations

$$\Psi_\theta(\mathbf{A}) = \sum_n e^{in\theta} \psi_n(\mathbf{A}),$$

$$\mathcal{G}\Psi_\theta(\mathbf{A}) = e^{-i\theta}\Psi_\theta(\mathbf{A}),$$

where $\mathcal{G}$ is the operator that effects the big gauge transformation. In this way, we arrive at a family of states parametrized by an angle θ, and to finish the investigation we must decide whether or not all these states are degenerate in energy or whether tunneling will lift the classical degeneracy. We need to know therefore whether there exist imaginary-time, zero-energy solutions which interpolate between the allowed real-time zero-energy field configurations. When $t \to -ix^4$, the Hamiltonian becomes

$$H \underset{t \to -ix^4}{=} \frac{1}{2}\int d^3x\,[-E_a^2 + B_a^2]$$

and vanishes for $\mathbf{E}_a = \pm\mathbf{B}_a$. This equation was first solved by a soviet of physicists[2] and the solution does behave in just the right way for us to conclude that there is tunneling: The lowest energy states form a band $E(\theta)$; and it is truly remarkable that an angle which was not initially present in the Lagrangian, now appears. Let me assure any of you who may be harboring the suspicion that θ is some foolish gauge artifact, that it is possible to demonstrate the gauge invariance of this picture. Thus we conclude that the Yang-Mills vacuum is a Bloch wave.

The physical picture here presented changes radically when the system is enlarged to include coupling to massless fermions. In that case, the theory acquires a chiral $U(1)$ symmetry. However, the gauge invariant axial vector current J_5^μ is not conserved due to the triangle anomaly

$$\partial_\mu J_5^\mu = \frac{1}{8\pi^2}\,\mathrm{tr}\,{}^*F^{\mu\nu}F_{\mu\nu}$$

$$^*F_{\mu\nu} = \frac{1}{2}\varepsilon_{\mu\nu\alpha\beta}F^{\alpha\beta}\,.$$

Since $\mathrm{tr}\,{}^*F^{\mu\nu}F_{\mu\nu}$ is a total divergence of a gauge variant object

$$\mathrm{tr}\,{}^*F^{\mu\nu}F_{\mu\nu} = \partial^\mu\left\{4\varepsilon_{\mu\nu\alpha\beta}\,\mathrm{tr}\left(\frac{1}{2}A^\nu\partial^\alpha A^\beta + \frac{1}{3}A^\nu A^\alpha A^\beta\right)\right\}$$

a gauge variant, conserved current can be defined. The corresponding time-independent constant of motion, Q_5, which commutes with the Hamiltonian, turns out to be gauge invariant under the little gauge transformations, but not under the big ones

$$\mathcal{G} Q_5 \mathcal{G}^{-1} = Q_5 + 2$$
$$i\,[H,\, Q_5] = 0$$
$$i\,[H,\, \mathcal{G}] = 0\ .$$

Since our vacuum states diagonalize $\mathcal{G}$, they are not annihilated by Q_5; indeed the exponential of Q_5 acts as a shifting operator,

$$e^{-i(\theta'/2)Q_5}\,\Psi_\theta = \Psi_{\theta+\theta'}$$

and the commutativity of Q_5 with the Hamiltonian indicates that in the presence of fermions, the energy ceases to depend on θ; tunneling is suppressed and the vacuum becomes degenerate, with the degeneracy breaking spontaneously the $U(1)$ chiral symmetry, without any emergence of massless Goldstone bosons.

What is the physical import of all this? Let me first tell you that detailed calculations[3] validate the qualitative picture that I have here described: all effects are weak and non-perturbative, as befits tunneling phenomena. In the absence of a better understanding of the physical content in a non-Abelian gauge theory, I cannot be very definite about detailed, physically meaningful consequences. Certainly the ability to break spontaneously the chiral $U(1)$ symmetry without introducing Goldstone bosons offers the possibility for the resolution of the long standing $U(1)$ problem in quark models. Also the emergence of an angle in the theory, which can be shown to be CP violating, is a tantalizing phenomenon. More generally, I find it extremely interesting that tunneling can happen and can lead to weak interactions which violate all kinds of symmetries of the Lagrangian.

In conclusion, let me observe that the possibility of obtaining information about a quantum theory by semi-classical methods requires a complete understanding of the classical theory. We have therefore been engaged in many calculations for classical Yang-Mills theory. We have sought other solutions to the field equations, studied their stability and infinitesimal deformations. We have exposed various unexpected symmetries and examined the coupling of other fields to the Yang-Mills theory. It was soon evident to us that the many results which we were obtaining were reflecting an underlying, general order that we wanted to put into evidence. It then transpired that we became acquainted with the work of colleagues in mathematics who during the last decade have developed just exactly the necessary framework for understanding these varied results. It is rare in contemporary times that the concerns of modern mathematicians and modern theoretical physicists should coincide; therefore we are especially pleased that such a conjuntion of interests has emerged. I shall now yield to my colleague from MIT; Professor I. Singer who, together with Professor M. Atiyah of Oxford, is responsible for many of the relevant results. Professor Singer has kindly agreed to describe his work but here, I shall only record his poetic response to our joint physics/mathematics efforts.

> In this day and age
> The physicist sage
> Writes page after page
> On the current rage
> The gauge
>
> Mathematicians so blind
> Follow slowly behind
> With their clever minds
> A theorem they'll find
> Duly written and signed
>
> But gauges have flaws
> God hems and haws
> As the curtain He draws
> O'er His physical laws
> It may be a lost cause.　　　　　I. Singer

REFERENCES

1. R. Jackiw and C. Rebbi, *Phys. Rev. Lett.* **37**, 172 (1976); C. Callan, R. Dashen and D. J. Gross, *Phys. Lett.* **63B**, 334 (1976).
2. A. Belavin, A. Polyakov, A. Schwartz and Y. Tyupkin, *Phys. Lett.* **59B**, 85 (1975).
3. G.'t Hooft *Phys. Rev. Lett.* **37**, 8 (1976) and *Phys. Rev. D* **14**, 3432 (1976).

Paper II.2

Classical Yang-Mills equations with static sources comprise an intricate dynamical system. In contrast to the analogous and simpler problem in Maxwell theory, physical relevance of such classical Yang-Mills fields has never been established. Nevertheless, the mathematical issues are fascinating and were described at the Montréal Meeting on Geometrical and Topological Methods in Gauge Theories (September 1979).

BIFURCATION AND STABILITY
IN YANG-MILLS THEORY
WITH SOURCES

Lecture Notes in Physics: **129**, *59 (Springer, Berlin 1980)*

1. INTRODUCTION

In my lecture, I shall discuss some recent work on solutions to classical Yang-Mills theory. The investigations that I shall summarize study the field equations with static, external [non-dynamical] sources. The physical, quantum-mechanical significance of such solutions has not thus far been as profound as that of solitons, where sources are dynamical *i.e.*, monopoles with Higgs-field sources; nor as that of instantons, where sources are absent but the equations are continued to imaginary time. Nonetheless, the new results are interesting in their differences from the Abelian counterpart and should suggest intuition about the physical content of non-Abelian gauge-quantum field theory. Moreover, structurally the equations are sufficiently intricate to provide a most interesting example of analysis in mathematical physics.

There is now available a variety of solutions for review; but only recently did a pattern emerge which allows for a comprehensive description. A summary of this is presented in Section 2. Section 3 is devoted to an account of stability properties. I begin by recalling the theory of stability — a subject widely studied by physicists in former times, but now, in its general form, largely forgotten. The general theory does not rely on minima of the energy and is found to be applicable to the Yang-Mills model. It comes as no surprise that the non-Abelian structure lets the gauge theory share with a top the phenomenon of stable configurations that do not minimize the energy. The Section concludes with an assessment of the stability of various solutions. Finally, a list of open questions and problems for further research comprise the concluding Section 4.

2. SOLUTIONS TO YANG-MILLS THEORY WITH SOURCES

2.1 Preliminaries

The field equations with which we are concerned are

$$\mathcal{D}_\mu F^{\mu\nu} = \delta^{\nu 0}\rho\,,\tag{2.1a}$$

$$F_{\mu\nu} = \partial_\mu A_\nu - \partial_\nu A_\mu + [A_\mu,\, A_\nu]\,,\tag{2.1b}$$

$$\mathcal{D}_\mu = \partial_\mu + A_\mu\,.\tag{2.1c}$$

We study the $SU(2)$ theory with coupling strength scaled to unity and use interchangeably component notation and anti-Hermitian matrix notation; *eg.*, ρ_a, $a = 1, 2, 3$; $\rho = \rho_a \sigma^a / 2i$, σ^a = Pauli matrices. The source ρ is taken to be a given, time-independent function, $\partial_t \rho = 0$. Equation (2.1a) carries with it an integrability condition; the right-hand side must be covariantly conserved.

In the present circumstance, that requirement reduces to

$$[A^0, \rho] = 0.$$ (2.2)

The energy of the system is given by a positive, gauge-invariant formula

$$\mathcal{E} = \frac{1}{2} \int d^3\mathbf{r} \left\{ \mathbf{E}_a^2 + \mathbf{B}_a^2 \right\},$$

$$E_a^i = F_a^{i0}, \qquad B_a^i = -\frac{1}{2}\varepsilon^{ijk} F_a^{jk}.$$ (2.3)

The class of solutions which I shall be describing here is delimited by the requirement of finite energy. This means that sources must also be well-behaved; a condition which will not be spelled out in detail, beyond noting that point sources are excluded; ρ is an extended function.

When it comes to discuss stability, we shall want a Hamiltonian formulation for the field equations (2.1); since they are locally gauge-invariant this is not straightforward. In order to overcome the familiar difficulty, we do not pick a gauge; rather we take the variations used for obtaining the equations of motion to be constrained by Gauss' law, which is the $\nu = 0$ component of (2.1a). Specifically, we take the Hamiltonian to coincide with $\mathcal{E}$, viewed as a functional of independent variables $\mathbf{E}$ and $\mathbf{A}$, while $\mathbf{B}$ is constructed in the usual way from $\mathbf{A}$

$$\mathbf{B}_a = \boldsymbol{\nabla} \times \mathbf{A}_a - \frac{1}{2}\varepsilon_{abc}\mathbf{A}_b \times \mathbf{A}_c.$$ (2.4)

$-\mathbf{E}$ is identified with the canonical momentum conjugate to $\mathbf{A}$, and the constraint of Gauss' law is imposed with the help of a Lagrange multiplier, here called A_a^0. Hence unrestricted variation can be performed on

$$\bar{\mathcal{E}} = \mathcal{E} - \int d^3\mathbf{r} A_a^0 (\boldsymbol{\nabla} \cdot \mathbf{E}_a - \varepsilon_{abc}\mathbf{A}_b \cdot \mathbf{E}_c - \rho_a).$$ (2.5)

In this way, the Yang-Mills equations are obtained

$$0 = -\frac{\delta\bar{\mathcal{E}}}{\delta A_a^0} = \boldsymbol{\nabla} \cdot \mathbf{E}_a - \varepsilon_{abc}\mathbf{A}_b \cdot \mathbf{E}_c - \rho_a, \qquad \text{(Gauss' law constraint)} \quad (2.6a)$$

$$\partial_t \mathbf{E}_a = \frac{\delta\bar{\mathcal{E}}}{\delta\mathbf{A}_a} = \boldsymbol{\nabla} \times \mathbf{B}_a - \varepsilon_{abc}\mathbf{A}_b \times \mathbf{B}_c - \varepsilon_{abc}A_b^0 \mathbf{E}_c, \qquad \text{(Ampère's law)} \quad (2.6b)$$

$$\partial_t \mathbf{A}_a = -\frac{\delta\bar{\mathcal{E}}}{\delta\mathbf{E}_a} = -\mathbf{E}_a - \boldsymbol{\nabla} A_a^0 + \varepsilon_{abc}\mathbf{A}_b A_c^0. \qquad \text{(Definition of $\mathbf{E}_a$)} \quad (2.6c)$$

Note that static solutions [all time-derivatives vanish] are critical points of the energy, subject to the constraint of Gauss' law.[1,2]

Presentation of solutions is complicated by the gauge covariance of (2.1): if A_μ solves the equations with source ρ, then the equations with a gauge-rotated source ρ'

$$\rho' = U^{-1}\rho U \qquad (2.7a)$$

are solved by gauge transforming the previous

$$A'_\mu = U^{-1}A_\mu U + U^{-1}\partial_\mu U . \qquad (2.7b)$$

[Here U is an $SU(2)$ matrix.] Two solutions related as above describe the same physical situation and we shall view them as the same solution but presented in different "gauge frames". Frequently, we shall speak of an "Abelian gauge frame" — one in which the source points in the third direction

$$\rho_a = \delta_{a3}q . \qquad (2.8)$$

Of course, results for gauge-invariant quantities like the energy, are frame independent.

In addition to the above gauge covariance, there is present also a gauge invariance with respect to gauge transformations which leave the source unchanged. From (2.2), we see that gauge transformations with U constructed from A^0 are of this type. Thus it is always possible to pass to the temporal gauge where A^0 vanishes, without changing the gauge frame.

Solutions naturally fall into two classes: those that exist for arbitrary sources and those that require a critical, finite source strength. We list these in turn.

2.2 Arbitrary Sources

Four different types of solutions will be discussed in this sub-Section, two static, two time-dependent. The latter provide a well-defined generalization of the former.

The most obvious Yang-Mills solution is the static Coulomb one which is readily presented in the Abelian gauge frame, where it is given by the regular solution to Poisson's equation[3]

$$A_a^0 = \delta_{a3}\,\phi , \qquad (2.9a)$$

$$\mathbf{A}_a = 0 , \qquad (2.9b)$$

$$\phi = -\frac{1}{\nabla^2}q . \qquad (2.9c)$$

An alternate description, still in the Abelian gauge frame, is gotten by passing to the temporal gauge

$$A_a^0 = 0, \tag{2.10a}$$

$$\mathbf{A}_a = \delta_{a3}\boldsymbol{\nabla}\phi\, t. \tag{2.10b}$$

The energy of this, according to (2.3), is the familiar Coulomb expression

$$\mathcal{E}_c = \frac{1}{2}\int q\,\frac{-1}{\nabla^2}\,q = \frac{1}{8\pi}\int d^3\mathbf{r}\,d^3\mathbf{r}'\,\frac{q(\mathbf{r})\,q(\mathbf{r}')}{|\mathbf{r}-\mathbf{r}'|}. \tag{2.11}$$

Note that in the Abelian frame, the solution vanishes with the source.

The next solution is a time-dependent generalization of the above. It shares with the Coulomb solution the [gauge-invariant] property that the magnetic field vanishes. From Ampère's law it follows that, in the temporal gauge, vanishing **B** implies a static electric field. Equations (2.4) and (2.6c) require the electric field to be [gauge equivalent to] a gradient of a scalar [matrix] function Φ which further must satisfy

$$[\boldsymbol{\nabla}\Phi, \boldsymbol{\nabla}\Phi] = 0 \tag{2.12}$$

a condition which is easily fulfilled; see Ref. 4. Thus we have, for ϕ's satisfying (2.12),

$$A'^0 = 0, \tag{2.13a}$$

$$\mathbf{A}' = \boldsymbol{\nabla}\Phi\, t. \tag{2.13b}$$

[A primed notation is used as reminder that the solution is being presented in a gauge frame other than the Abelian one.] The source which gives rise to such a field is determined by Gauss' law

$$\nabla^2\Phi = -\rho'. \tag{2.13c}$$

There are as many configurations in the above category as there are functions Φ consistent with (2.12). However, our interest is only in those for which ρ' is gauge equivalent to the Abelian frame formula (2.8); only then are we dealing with solutions to the same problem as the Coulomb one

$$\rho' = U\rho U^{-1} = U\frac{\sigma^3}{2i}U^{-1}q. \tag{2.14}$$

When (2.14) holds, we can express the solution in the Abelian frame, in the temporal gauge

$$A^0 = 0,$$ (2.15a)

$$\mathbf{A} = -\mathbf{E}t - U^{-1}\boldsymbol{\nabla} U,$$ (2.15b)

$$\mathbf{E} = -U^{-1}\boldsymbol{\nabla}\Phi U.$$ (2.15c)

The energy of the above is given by a Coulomb-type formula

$$\mathcal{E} = \frac{1}{2}\int \rho_a'\frac{-1}{\nabla^2}\rho_a' = \frac{1}{8\pi}\int d^3\mathbf{r}\, d^3\mathbf{r}'\, \frac{\rho_a'(\mathbf{r})\rho_a'(\mathbf{r}')}{|\mathbf{r}-\mathbf{r}'|}.$$ (2.16)

To recapitulate, the solution for a given source (2.8) is constructed by choosing a gauge function U; computing ρ' from (2.14); Φ from (2.13c); and finally, the potentials from (2.13a) and (2.13b) or (2.15). When (2.12) is met, one has solutions which in general are essentially time-dependent — a time-translation cannot be compensated by a gauge-transformation. The only member of the family with gauge-artifactual time-dependence is the Coulomb one where $U = I$. By continuously deforming I to U, one passes continuously from the static Coulomb solution to its time-dependent generalization.[5] In Section 3 it will be demonstrated that the energy (2.16) can be lowered by an arbitrary amount below the Coulombic value $\mathcal{E}_c$.

Our third solution is again static, but it differs from the Coulomb one by the property that in the Abelian frame it does not vanish with the source; rather it becomes a pure gauge

$$A_\mu\Big|_{\substack{\text{zero}\\\text{source}}} = U^{-1}\partial_\mu U.$$ (2.17)

[In the absence of sources, finite-energy solutions are necessarily trivial[6]; thus the potentials either vanish or are pure gauges.] A closed expression for this solution has not been given; only a formula perturbative in the source is available. So that we can speak of orders of perturbation, we shall take the source to be $O(Q)$ where Q is a convenient scale of magnitude for the source. [For example, Q can be an overall factor.] This solution is most economically presented by first transforming out of the Abelian frame with the gauge function U, occurring in (2.17)

$$\rho' = U\rho U^{-1} = U\frac{\sigma^3}{2i}U^{-1}q.$$ (2.18a)

In the new frame, the vector potentials vanish with the source. Perturbative formulas for them are[1]

$$A'^0 = \Phi + O(Q^3),\qquad(2.18\text{b})$$

$$\mathbf{A}' = \frac{1}{\nabla^2}[\Phi, \boldsymbol{\nabla}\Phi] + O(Q^4),\qquad(2.18\text{c})$$

$$\Phi = -\frac{1}{\nabla^2}\rho'.\qquad(2.18\text{d})$$

Primes remind that the quantities are displayed in a non-Abelian frame. The gauge function U is not arbitrary but must be chosen so that the consistency condition (2.2) is satisfied. It is a consequence of that equation and of (2.18b) that we must have

$$[\Phi, \nabla^2\Phi] = 0.\qquad(2.19)$$

The following is the temporal gauge equivalent to (2.18)

$$A'^0 = 0,\qquad(2.20\text{a})$$

$$\mathbf{A}' = \boldsymbol{\nabla}\Phi\, t + \left(\frac{1}{2}t^2 + \frac{1}{\nabla^2}\right)[\Phi, \boldsymbol{\nabla}\Phi] + O(Q^3).\qquad(2.20\text{b})$$

The electric field is $O(Q)$; the magnetic field, $O(Q^2)$

$$\mathbf{E}' = -\boldsymbol{\nabla}\Phi - t[\Phi, \boldsymbol{\nabla}\Phi] + O(Q^3),\qquad(2.20\text{c})$$

$$\mathbf{B}' = \boldsymbol{\nabla}^{-1} \times [\Phi, \boldsymbol{\nabla}\Phi] + O(Q^3),\qquad(2.20\text{d})$$

$$\boldsymbol{\nabla}^{-1} = \boldsymbol{\nabla}/\nabla^2.$$

The time dependence in (2.20) is of course a consequence of the gauge choice, as comparison with (2.18) shows.] In the primed frame, the solution appears similar to the Coulomb one (2.9) or (2.10), save that the non-vanishing commutator $[\Phi, \boldsymbol{\nabla}\Phi]$ prevents the expressions from closing. Hence, we call the above a "non-Abelian Coulomb" solution to contrast it with the "Abelian Coulomb" discussed at the outset. The energy further exhibits similarities with the Abelian Coulomb case. The formula to lowest order in Q follows from (2.3), (2.18d) and (2.20c)

$$\mathcal{E} = \frac{1}{2}\int \rho'_a \frac{-1}{\nabla^2}\rho'_a + O(Q^4) = \frac{1}{8\pi}\int d^3\mathbf{r}\, d^3\mathbf{r}'\, \frac{\rho'_a(\mathbf{r})\rho'_a(\mathbf{r}')}{|\mathbf{r} - \mathbf{r}'|} + O(Q^4).\qquad(2.21)$$

A specific example of a non-Abelian Coulomb solution is given when the source in the Abelian frame is spherically symmetric

$$\rho_a = \delta_{a3}\, q(r).\qquad(2.22\text{a})$$

One then verifies that (2.19) is satisfied with the charge density in the radial frame;

$$\rho'_a = \hat{r}^a q(r) \tag{2.22b}$$

i.e., U is the gauge transformation which rotates the third axis into the radial axis. A further interesting feature is that the present solution carries less energy than the corresponding Coulomb one[7]

$$\mathcal{E} = \frac{1}{8\pi} \int d^3r d^3\mathbf{r}' \frac{q(r)q(r')}{|\mathbf{r} - \mathbf{r}'|} \hat{r} \cdot \hat{r}' + O(Q^4) < \mathcal{E}_c . \tag{2.22c}$$

The fourth and last solution, that I mention in this sub-Section, generalizes the static non-Abelian Coulomb in a time-dependent fashion, quite similarly to the way that the second solution, Eqs. (2.13), generalizes the Abelian Coulomb. It is constructed by choosing an arbitrary gauge transformation U and transforming the source once again

$$\rho'' = U\rho'U^{-1} . \tag{2.23}$$

We use double primes to distinguish this source from ρ — the source in the Abelian frame — and from ρ' — the source in the gauge transformed frame where the non-Abelian Coulomb solution has a simple perturbative expansion, see (2.18) or (2.20). Next we take the regular solution of Poisson's equation

$$\Phi = -\frac{1}{\nabla^2}\rho'' \tag{2.24}$$

and build the time-dependent solution perturbatively in Φ, in the temporal gauge[5]

$$A''^0 = 0 , \tag{2.25a}$$

$$\mathbf{A}'' = \boldsymbol{\nabla}\Phi\, t + \left(\frac{t^2}{2} + \frac{1}{\nabla^2}\right)([\Phi, \boldsymbol{\nabla}\Phi] - \boldsymbol{\nabla}^{-1}[\Phi, \nabla^2\Phi]) + O(Q^3) . \tag{2.25b}$$

One readily computes the electric field, which is $O(Q)$

$$\mathbf{E}'' = -\boldsymbol{\nabla}\Phi - t([\Phi, \boldsymbol{\nabla}\Phi] - \boldsymbol{\nabla}^{-1}[\Phi, \nabla^2\Phi]) + O(Q^3) . \tag{2.25c}$$

The $O(Q^2)$ magnetic field

$$\mathbf{B}'' = \boldsymbol{\nabla}^{-1} \times [\Phi, \boldsymbol{\nabla}\Phi] + O(Q^3) \tag{2.25d}$$

is of the same form as in the static non-Abelian Coulomb solution, (2.20d), which is here included when $U = I$. Just as in the Abelian situation, by continuously changing I to U, we obtain a continuous deformation of the static non-Abelian Coulomb into its time-dependent generalization. In both cases, the magnetic field retains the same form during the deformation. Once again, the $O(Q^2)$ energy is given by a Coulombic formula, as follows from (2.3), (2.24) and (2.25c)

$$\mathcal{E} = \frac{1}{2} \int \rho''_a \frac{-1}{\nabla^2} \rho''_a + O(Q^4) = \frac{1}{8\pi} \int d^3r\, d^3r' \frac{\rho''_a(\mathbf{r})\rho''_a(\mathbf{r}')}{|\mathbf{r}-\mathbf{r}'|} + O(Q^4). \quad (2.26)$$

In Section 3, we show that also the above energy can lie below the corresponding energy of the static solution (2.22c) by an arbitrary amount. The similarities between the four solutions should be apparent. Indeed, if for different gauge frames a common [unprimed] notation is used, a master formula which presents all four may be given.[5] Define first the vector $\mathbf{C}$

$$\mathbf{C} = [\Phi, \nabla\Phi] \qquad \Phi = \frac{-1}{\nabla^2}\rho. \quad (2.27)$$

Then in the temporal gauge, set

$$A^0 = 0$$

$$\mathbf{A} = \frac{-1}{\nabla^2}\left[\nabla\rho\, t + \left(\frac{1}{2}t^2 + \frac{1}{\nabla^2}\right)\nabla\times\nabla\times\mathbf{C}\right]. \quad (2.28)$$

The static, Abelian Coulomb has vanishing $\mathbf{C}$, while vanishing $\nabla\times\mathbf{C}$ leads to its time-dependent generalization as is seen from (2.12) and (2.13). Furthermore, according to (2.19) and (2.20), vanishing $\nabla\cdot\mathbf{C}$ corresponds to the static, non-Abelian Coulomb solution while its time-dependent generalization (2.25) has no restrictions on $\mathbf{C}$. For the first two solutions (2.28) is exact; for the last two it is accurate up to $O(Q^2)$. The $O(Q^2)$ formula

$$\mathcal{E} = \frac{1}{2}\int \rho_a \frac{-1}{\nabla^2}\rho_a + O(Q^4) \quad (2.29)$$

gives the complete energy for the exact solutions, and the $O(Q^2)$ contribution to the perturbative ones. For the two static solutions, the quantity in (2.29) is stationary against variations of ρ_a which preserve its length [gauge transformations].[7] The Coulomb solution is seen to maximize (2.29).

2.3 Sources with Critical Strength

When the source strength Q increases, the previous solutions continue to be present. For the Abelian Coulomb, and its time-dependent generalization, the closed expressions given above hold for arbitrary Q. For the non-Abelian

Coulomb with its time-dependent generalization, one must calculate perturbatively terms higher order in Q; a tedious procedure with unknown convergence properties. Alternatively one can do numerical computations.

Furthermore as Q increases, solutions appear which require a critical, minimal source strength to support them. Very little is known about these, and the numerical method is presently the only effective means of investigation. We review one such example.[1,8]

When the source is radially symmetric, as in (2.22a), we have the spherically symmetric Abelian Coulomb solution. Also by passing to the radial frame (2.22b) we can exhibit the perturbative non-Abelian Coulomb solution. By iterating Eqs. (2.18) a few orders in Q, it is found that the form of the potentials remains within the following *Ansatz*

$$A^0 = \frac{\hat{r} \cdot \boldsymbol{\sigma}}{2i} \frac{1}{r} f(r/r_0) \,, \tag{2.30a}$$

$$\mathbf{A} = \frac{\hat{r} \times \boldsymbol{\sigma}}{2i} \frac{1}{r} [a(r/r_0) - 1] \,. \tag{2.30b}$$

Here r_0 is a length scale. In this sub-Section, we shall always remain in the radial frame,

$$\rho_a = \frac{\hat{r}^a}{r_0^3} q(r/r_0) \tag{2.31}$$

hence, primes on the potentials are dropped. The above *Ansatz* is postulated for the complete static solution and the mode functions satisfy the following non-linear differential equations, which are all that remain of Gauss' and Ampère's laws

$$- f'' + \frac{2a^2}{x^2} f = xq \,, \tag{2.32a}$$

$$- a'' + \frac{a^2 - 1 - f^2}{x^2} a = 0 \,. \tag{2.32b}$$

All functions depend only on $x = r/r_0$, and the dash indicates differentiation with respect to that variable. More general radially summetric *Ansätze* can be given, but it has been proven that static, radial solutions necessarily fall into the above restrictions.[1] [We emphasize that the Abelian Coulomb solution does not lie within the *Ansatz* (2.30), and cannot be found in the solutions to (2.32); in the radial frame, the Abelian Coulomb solution is not radially symmetric.] Requiring finiteness of the energy

$$\mathcal{E} = \frac{4\pi}{r_0} \int_0^\infty dx \left[(a')^2 + \frac{1}{2x^2}(a^2 - 1)^2 + \frac{1}{2}(f')^2 + \frac{1}{x^2} f^2 a^2 \right] \tag{2.33}$$

— the above is the form that (2.3) takes within the *Ansatz* (2.30) — imposes boundary conditions at the origin and at infinity. At the origin the potentials must vanish rapidly: $f(0) = 0$, $a(0) = 1$, $A^0(0) = 0$, $\mathbf{A}(0) = 0$. At infinity two types of behavior are allowed: type I, where the potentials vanish as in the origin; type II, where the vector potential tends to a nontrivial pure gauge, $a(\infty) = -1$,

$$\mathbf{A} \xrightarrow[r\to\infty]{} i\,\frac{\hat{r} \times \boldsymbol{\sigma}}{r} = -(i\boldsymbol{\sigma}\cdot r)\boldsymbol{\nabla}(-i\boldsymbol{\sigma}\cdot\hat{r}).$$

The type I solution is the previously perturbatively encountered non-Abelian Coulomb $[a = 1 + O(Q^2)$, hence a never equals $-1]$. The type II is a new, non-perturbative solution.

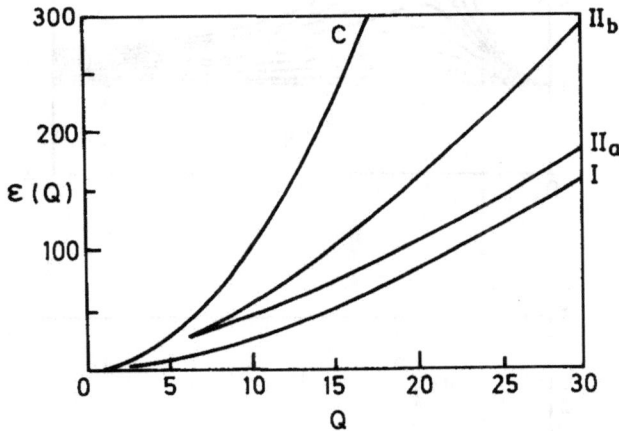

Fig. 1. Energy, in units of $2\pi/r_0$ as a function of Q for a delta-shell source of strength Q. The curve C is the Abelian Coulomb parabola. The curve I is the non-Abelian Coulomb solution. Curves IIa and IIb are the two branches of the bifurcating solution. The bifurcation point occurs at $Q = 5.835$.

Numerical computation confirms the above, with the further surprise that type II comes in two branches, once Q exceeds a critical magnitude.[1] Hence, we call this the "bifurcating" solution. Figure 1 shows a plot of the energy versus source strength for solutions with a delta-shell charge density

$$\rho_a = \frac{\hat{r}^a}{r_0^2}Q\delta(r - r_0). \tag{2.34}$$

The Coulomb parabola [which does not lie within the *Ansatz*] is also exhibited for comparison. Note that the non-Abelian Coulomb [type I] carries lower energy than the Abelian Coulomb for all Q, even outside the perturbative regime. The bifurcation point where the two type II solutions first occur is found numerically to be $Q = 5.835$. In Figs. 2, 3 and 4 the mode functions f and a are displayed for the various solutions. All figures are taken from Ref. 1.

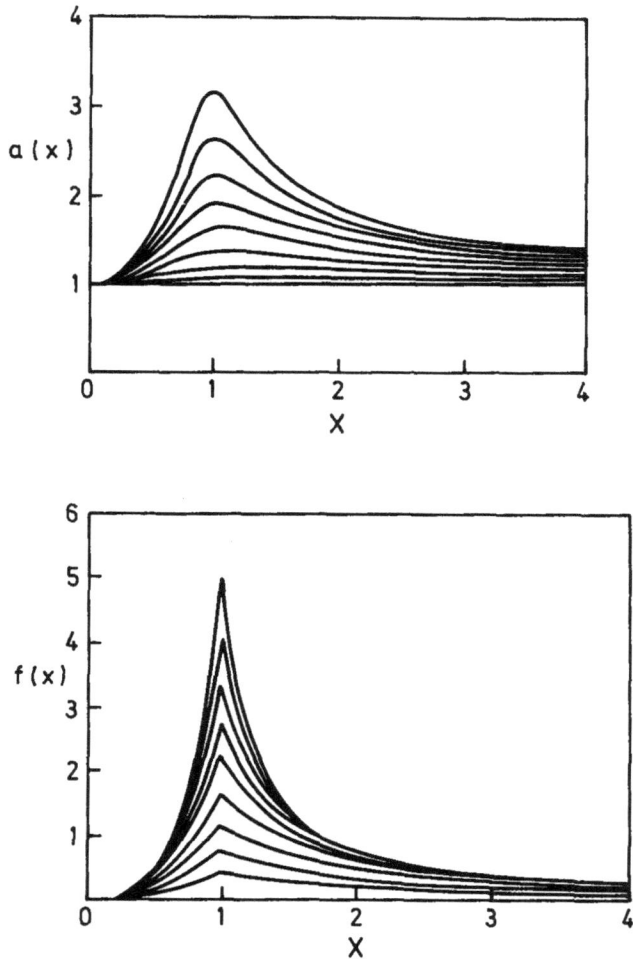

Fig. 2. Profiles of the functions a and f for the type I solution. Starting from the lowest curves, the values of Q in a delta-shell source are 1.41, 2.53, 4.04, 6.43, 10.05, 14.18, 19.93, 23.38 and 41.72.

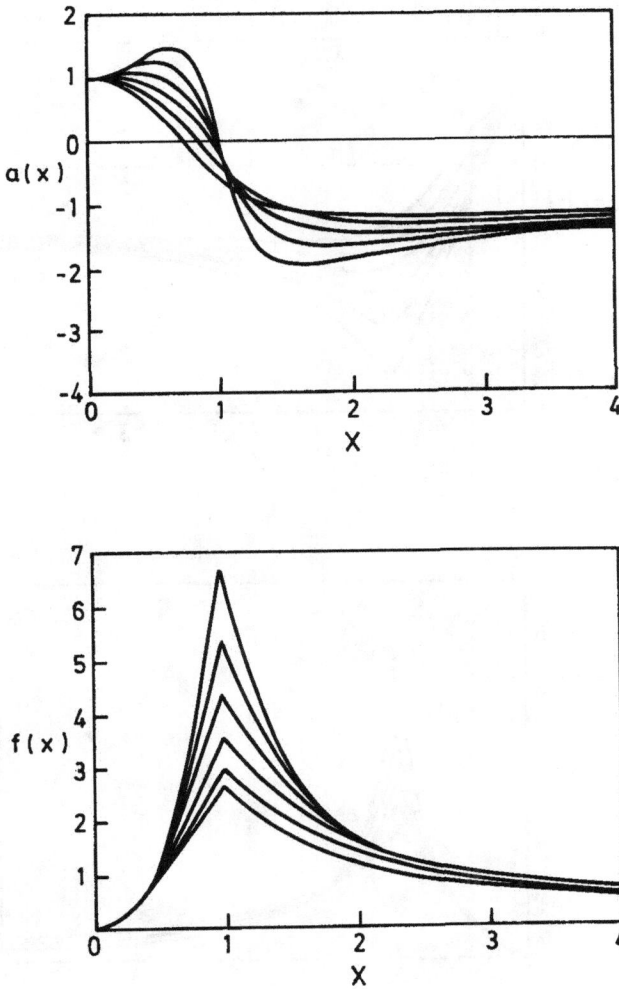

Fig. 3. Profiles of the functions a and f for the upper branch of the type II bifurcating solution with a delta-shell source. $Q=5.86$, 6.44, 8.09, 11.05, 15.71 and 23.19. Correspondence
between the individual curves and these values of Q is established by the fact that, as Q
increases, so do $a''(0)$ and $f(1)$.

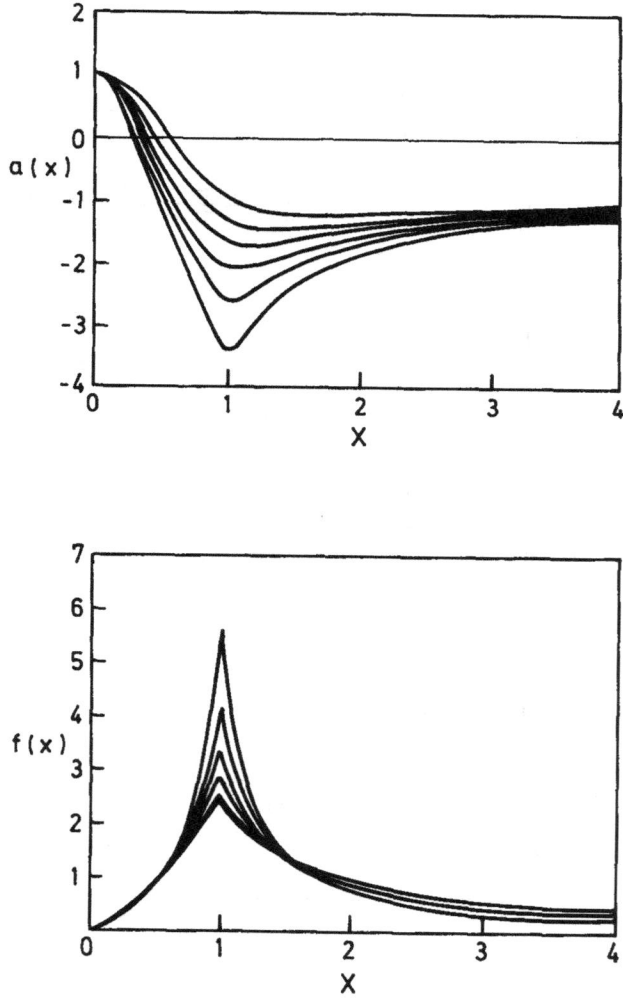

Fig. 4. Profiles of the functions a and f for the lower branch of the type II bifurcating solution with a delta-shell source. $Q=6.53$, 8.61, 12.10, 17.85, 28.06 and 49.16. For these curves, as Q increases $a''(0)$ decreases and $f(1)$ increases.

3. STABILITY IN DYNAMICAL SYSTEMS

3.1 Review of the General Theory

We consider a time-translation invariant system whose equations of motion for the $2N$ dynamical variables P_n and Q_n, $n = 1, \cdots, N$, can be obtained from a Hamiltonian $H(P, Q)$, which is also the conserved energy $\mathcal{E}$

$$\dot{P}_n = -\frac{\partial H(P, Q)}{\partial Q_n}, \qquad \dot{Q}_n = \frac{\partial H(P, Q)}{\partial P_n}. \qquad (3.1)$$

A static solution, one for which $\dot{P}$ and $\dot{Q}$ vanish, is a critical point of H, and *vice versa*, stationary points of the energy define static solutions. [An over-dot means differentiation with respect to time.]

We wish to ascertain whether a static solution $\{P^{(S)}, Q^{(S)}\}$ is stable. "Stable" by definition will mean the following: Take a configuration of the form $\{P^{(S)} + \delta P, Q^{(S)} + \delta Q\}$, substitute in (3.1) and linearize about $\{P^{(S)}, Q^{(S)}\}$ to obtain linear equations for the fluctuating quantities $\{\delta P, \delta Q\}$. When the linear equations produce exponential growth in time for the fluctuations, the solution is unstable; otherwise, it is stable. In other words, for stable motion the small quantities $\{\delta P, \delta Q\}$ fluctuate harmonically in time with real frequency, while complex frequencies signal instability.

The above criterion for stability is also in accord with quantum-mechanical ideas. The first quantum correction to the energy of a state involves the fluctuation frequencies. That quantity must be real for the state to be quantum-mechanically stable.

Note that growth in time of the fluctuations smaller than exponential, say polynomial, is not a sign of instability. In such a circumstance, the eigenfrequencies are degenerate, but still real and the quantal energy remains real.

An intuitively appealing idea is that stability should be connected with minimizing the energy: the static solution which is a stationary point should also be a [local] minimal point. More precisely, the minimality condition is the requirement that only non-negative eigenvalues characterize the quadratic Hamiltonian matrix, $\mathcal{H}$, defined by expanding $H(P, Q)$ about $\{P^{(S)}, Q^{(S)}\}$ and retaining quadratic terms in $\{\delta P, \delta Q\}$ [Linear terms vanish since the expansion is about a critical point.]

$$H(P, Q) = H(P^{(S)}, Q^{(S)}) + \frac{1}{2}\delta P_n T_{nm} \delta P_m + \delta P_n G_{nm} \delta Q_m$$

$$+ \frac{1}{2}\delta Q_n V_{nm} \delta Q_m + \cdots = H(P^{(S)}, Q^{(S)}) + \frac{1}{2}\tilde{X}\mathcal{H}X + \cdots \qquad (3.2)$$

$$\mathcal{H} = \begin{pmatrix} T & G \\ \tilde{G} & V \end{pmatrix} \qquad X = \begin{pmatrix} \delta P \\ \delta Q \end{pmatrix}. \tag{3.3}$$

[The tilde indicates transposition.] The minimality condition demands

$$\det(\mathcal{H} - \lambda I) = 0 \to \lambda \ge 0. \tag{3.4}$$

In fact, minimality is a sufficient condition for stability — a result, known as Dirichelet's theorem, which will become apparent below — but by no means is it a necessity.[9] There are indeed familiar physical systems [tops, gyroscopes, planetary configurations] which are stable, even though their energy is not locally minimal. To derive a more general criterion we expand (3.1) around the static solution and find

$$\mathcal{H}X = i\eta \dot{X}, \tag{3.5}$$

$$\eta = \begin{pmatrix} 0 & -iI \\ iI & 0 \end{pmatrix}. \tag{3.6}$$

By making a monochromatic *Ansatz* for X,

$$X = e^{-i\omega t} x. \tag{3.7}$$

We recognize the [constant] x as symplectic eigenvectors of $\mathcal{H}$ with symplectic eigenvalue ω

$$\mathcal{H}x = \omega \eta x. \tag{3.8}$$

It is clear that our definition of stability requires the ω's be real; this is known as Liapunov's theorem[9]

$$\det(\mathcal{H} - \omega \eta) = 0 \to \omega \text{ real}. \tag{3.9}$$

The point is that (3.9) is in general different from (3.4) and can be satisfied when (3.4) fails.

If (3.8) is premultiplied by $x^\dagger$, where the dagger indicates transposition and complex conjugation

$$x^\dagger \mathcal{H}x = \omega x^\dagger \eta x, \tag{3.10}$$

we see that the left-hand side is real, $\mathcal{H}$ being real symmetric, hence Hermitian. Also $x^\dagger \eta x$ is real since η is Hermitian, and we conclude that ω can fail to be real only when $x^\dagger \mathcal{H}x$ and $x^\dagger \eta x$ vanish. So when $\mathcal{H}$ is positive definite, ω is real and Dirichelet's theorem is established: minimality implies stability. But of course ω may be real without $\mathcal{H}$ being positive definite.

One may consider η as a metric in the vector space of the x's. Then (3.8) is the condition that $x^\dagger \mathcal{H}x$ be stationary against variations of x which preserve the symplectic length $x^\dagger \eta x$. Instability can occur only when there are zero-length symplectic eigenvectors of $\mathcal{H}$. The eigenvalue equation in (3.9) is relevant to the program of diagonalizing $\mathcal{H}$ by symplectic matrices, just as the corresponding equation in (3.4) arises when diagonalizing with orthogonal matrices. [A matrix M is symplectic when $\tilde{M}\eta M = \eta$.]

The conditions (3.4) and (3.9) are clearly different, and no simple relationship exists between the two in the general case. In practice, we can specialize somewhat. Firstly, the kinetic energy matrix T in (3.2) and (3.3) is taken to be positive definite: with an appropriate definition of coordinates, we may choose it to be the identity. Secondly, the off-diagonal matrix G arising from mixed $p - q$ terms in the Hamiltonian, which are frequently called gyroscopic or Coriolis terms, is always anti-symmetric, when the theory is derivable from a Lagrangian. The reason is that any symmetric piece in such velocity dependent forces corresponds to a total time-derivative in the Lagrangian and may be dropped. Thus we are led to a simpler form for $\mathcal{H}$

$$\mathcal{H} = \begin{pmatrix} I & G \\ -G & V \end{pmatrix}$$

$$\tilde{G} = -G, \qquad \tilde{V} = V.$$
(3.11)

With this $\mathcal{H}$, the symplectic eigenvalue problem (3.9) reduces to

$$[(i\omega + G)(i\omega + G) + V]\delta Q = 0 \tag{3.12a}$$

and the stability condition becomes

$$\det\left(2i\omega G + G^2 + V - \omega^2 I\right) = 0 \to \omega \text{ real}. \tag{3.12b}$$

To compare the above with the minimality condition (3.4), we impose that requirement not on $\mathcal{H}$, but on the equivalent matrix $\tilde{M}\mathcal{H}M$ which has the same positivity properties as $\mathcal{H}$. Here

$$M = \begin{pmatrix} I & -G \\ 0 & I \end{pmatrix}, \qquad \tilde{M}\mathcal{H}M = \begin{pmatrix} I & 0 \\ 0 & G^2 + V \end{pmatrix}.$$

Then (3.4) becomes equivalent to

$$\det\left(G^2 + V - \lambda I\right) = 0 \to \lambda \geq 0. \tag{3.13}$$

This is analogous to (3.12b), but an obvious difference exists when G is present. When the gyroscopic forces are absent, the two conditions coincide and ω^2 may be identified with λ. In that case instability occurs only for imaginary ω. In the presence of gyroscopic terms, there exist stable static solutions which do not minimize the energy, while instability can exist with complex ω. When $\mathcal{H}$ is as in (3.11), the condition for instability, $x^\dagger \eta x = 0$, is equivalent to

$$\mathrm{Re}\,\omega = i\frac{\delta Q_n^* G_{nm} \delta Q_m}{\delta Q_n^* \delta Q_n}\,. \tag{3.14}$$

We shall use the phrase "gyroscopic stability" when we wish to distinguish this form of stability from the more familiar "energetic stability". A hint for gyroscopic stability occurs when we can find arbitrarily close to a static solution oscillatory fluctuations that lower the energy. As we shall show, such configurations exist in the Yang-Mills theory. Instability would be indicated when there are, arbitrarily close to the static solution, time-dependent solutions which decrease the energy and grow exponentially in time.

To conclude this review of stability theory, let us remark that although we discussed Dirichelet's sufficient condition in terms of the energy constant of motion, a similar criterion can be formulated by reference to other constants of motion. This generalization is useful when analyzing solutions invariant with respect to the symmetry transformation which is associated with the constant in question.[9]

3.2 Stability Analysis for Yang-Mills Theory

We turn now to the stability analysis of static solutions for the Yang-Mills equations. But before we use the ideas sketched above, we must recognize that there are two ways in which the Yang-Mills field theory differs from the simple Hamiltonian. Firstly, rather than $2N$ degrees of freedom, there is an infinite number. This causes matrices to be replaced by differential operators, summations by integrations, *etc.*, thus raising questions of convergence and uniformity. We shall not concern ourselves with this complication, even though there will be occasion to refer to it in the course of our development. Secondly, the Hamiltonian formulation now has constraints. This has already been dealt with in Section 2. Here we observe that the small fluctuation equations, which follow from (2.6) by linearizing around a static solution A_a^μ, are

$$0 = \boldsymbol{\mathcal{D}}_{ab} \cdot \delta \mathbf{E}_b + \varepsilon_{abc} \mathbf{E}_b \cdot \delta \mathbf{A}_c\,, \tag{3.15a}$$

$$\mathcal{D}_{ab}^0 \delta \mathbf{E}_b = \boldsymbol{\mathcal{D}}_{ab} \times \delta \mathbf{B}_b + \varepsilon_{abc} \mathbf{E}_b \delta A_c^0 - \varepsilon_{abc} \mathbf{B}_b \times \delta \mathbf{A}_c\,, \tag{3.15b}$$

$$\delta \mathbf{E}_a = -\mathcal{D}_{ab}^0 \delta \mathbf{A}_b - \boldsymbol{\mathcal{D}}_{ab} \delta A_b^0\,. \tag{3.15c}$$

$\delta \mathbf{B}_a$ is short-hand for $\boldsymbol{\mathcal{D}}_{ab} \times \delta \mathbf{A}_b$. The quadratic energy, $\mathcal{E}^{(2)}$, obtained by expanding (2.3) around the static solution and by using Gauss' law constraint for the fluctuations, is precisely of the form (3.11)

$$\mathcal{E}^{(2)} = \frac{1}{2} \int d^3\mathbf{r} \, \{ (\delta\mathbf{E}_a)^2 + 2\delta E_a^i (\varepsilon_{acb} A_c^0) \delta A_b^i$$

$$+ (\delta\mathbf{B}_a)^2 - \delta A_a^i (\varepsilon^{ikj} \varepsilon_{acb} B_c^k) \delta A_b^j \} \,. \tag{3.16}$$

In particular, an anti-symmetric gyroscopic term is present

$$G = \delta^{ij} \delta(\mathbf{r} - \mathbf{r}') \varepsilon_{abc} A_c^0 \,. \tag{3.17}$$

Equations (3.15) can also be obtained by taking (3.16) to be the quadratic Hamiltonian, and varying it subject to the constraint (3.15a) which is implemented with the help of a Lagrange multiplier δA_a^0. In other words, unconstrained variations are performed on

$$\bar{\mathcal{E}}^{(2)} = \mathcal{E}^{(2)} - \int d^3\mathbf{r} \, \delta A_a^0 (\mathcal{D}_{ab} \cdot \delta\mathbf{E}_b + \varepsilon_{abc} \mathbf{E}_b \cdot \delta\mathbf{A}_c) \tag{3.18}$$

and Eqs. (3.15a), (3.15b), and (3.15c) emerge respectively as

$$0 = -\frac{\delta\bar{\mathcal{E}}^{(2)}}{\delta(\delta A_a^0)} \,, \tag{3.19a}$$

$$\partial_t \delta\mathbf{E}_a = \frac{\delta\bar{\mathcal{E}}^{(2)}}{\delta(\delta\mathbf{A}_a)} \,, \tag{3.19b}$$

$$\partial_t \delta\mathbf{A}_a = -\frac{\delta\bar{\mathcal{E}}^{(2)}}{\delta(\delta\mathbf{E}_a)} \,. \tag{3.19c}$$

With a monochromatic *Ansatz* for the time-dependence, the above take on the form of a symplectic eigenvalue problem, equivalent to stationarizing. $\mathcal{E}^{(2)}$, subject to the constraint that the symplectic length of $\begin{pmatrix} -\delta\mathbf{E}_a \\ \delta\mathbf{A}_a \end{pmatrix}$ be fixed, and subject to the constraint of Gauss' law for the fluctuations. In short, the Yang-Mills model is seen to fit the general theory quite nicely.

Let us comment on some of the properties of the small fluctuation equations (3.15). An integrability condition follows from (3.15b). By taking the covariant divergence, one finds that the infinitesimal version of (2.2) must be satisfied

$$\varepsilon_{abc} \delta A_b^0 \rho_c = 0 \,. \tag{3.20}$$

Also *vice versa;* (3.20) and the integrability condition on (3.15b) imply, together with (3.15c), that the covariant time-derivative of the right-hand side in (3.15a) vanishes.

Equations (3.15) possess a local gauge invariance

$$\delta\mathbf{E}_a \rightarrow \delta\mathbf{E}_a - \varepsilon_{abc}\mathbf{E}_b\theta_c \,, \tag{3.21a}$$

$$\delta A_a^0 \rightarrow \delta A_a^0 - \partial_t\theta_a - \varepsilon_{abc}A_b^0\theta_c \,, \tag{3.21b}$$

$$\delta\mathbf{A}_a \rightarrow \delta\mathbf{A}_a + \boldsymbol{\nabla}\theta_a - \varepsilon_{abc}\mathbf{A}_b\theta_c \,. \tag{3.21c}$$

Here, θ_a is a local function which must be parallel to the source

$$\varepsilon_{abc}\theta_b\rho_c = 0 \,. \tag{3.22}$$

[There is also a gauge covariance: a gauge transformation on the background fields is compensated by a homogeneous gauge transformation on the fluctuating quantities. We shall not make use of this property.]

It is clear from (3.20) and (3.22) that the external charge density defines a direction in group space which we can call the "electromagnetic" direction, while the orthogonal directions can be termed "charged". Thus A_a^0, δA_a^0, θ_a and ρ_a all lie in the electromagnetic direction and vanish in the charged direction. This reduces the allowed gauge transformations, in that the last term in (3.21b) must vanish. Observe also that the gyroscopic term (3.17) affects only the charged direction; the electromagnetic fluctuations are free of gyroscopic terms.

It is possible to derive a gauge invariant fluctuation equation in the following way. The quantity

$$\begin{aligned}\mathbf{e}_a &= \delta\mathbf{E}_a + \varepsilon_{abc}A_a^0\delta\mathbf{A}_c \\ &= -\partial_t\delta\mathbf{A}_a - \boldsymbol{\mathcal{D}}_{ab}\delta A_b^0 \end{aligned} \tag{3.23}$$

is gauge invariant with respect to the gauge transformations (3.21). By taking a covariant time-derivative of (3.15b) we arrive after some steps at

$$\mathcal{D}_{ab}^0\mathcal{D}_{bc}^0\mathbf{e}_c + \boldsymbol{\mathcal{D}}_{ab} \times \boldsymbol{\mathcal{D}}_{bc} \times \mathbf{e}_c - \varepsilon_{abc}\mathbf{B}_b \times \mathbf{e}_c = 0 \,. \tag{3.24a}$$

[This is most readily obtained in the $\delta A^0 = 0$ gauge, which can always be achieved with the transformation (3.21b).] With a monochromatic *Ansatz*

$$\mathbf{e}_a = e^{-i\omega t}\chi_a \tag{3.24b}$$

(3.24a) may be written as

$$[i\omega\delta_{ab} + \varepsilon_{abm}A_m^0]\,[i\omega\delta_{bc} + \varepsilon_{bcn}A_n^0]\chi_c^i + V_{ac}^{ij}\chi_c^j = 0 \,, \tag{3.24c}$$

$$V_{ac}^{ij} = \varepsilon^{ikm}\mathcal{D}_{ab}^m\mathcal{D}_{bc}^n\varepsilon^{nkj} - \varepsilon^{ikj}\varepsilon_{abc}B_b^k \,. \tag{3.24d}$$

We see that (3.24c) is precisely of the form (3.12a), again showing that the Yang-Mills theory follows the analysis described in sub-Section 3.1.

Equations (3.24) are gauge invariant and involve the unconstrained $\mathbf{e}_a$. It is remarkable that such equations can be derived; the possibility to do so is intimately linked with the existence of an external charge density which defines a direction with respect to which the small fluctuations are constrained by (3.20).[5]

3.2.1 Abelian Coulomb Solution

For the Abelian Coulomb solution, the general stability theory is easily applied. The small oscillation equations are best presented by introducing complex quantities in the charged directions 1 and 2

$$\delta\mathbf{E} = \frac{1}{\sqrt{2}}(\mathbf{E}_1 + i\mathbf{E}_2),$$

$$\delta\mathbf{A} = \frac{1}{\sqrt{2}}(\mathbf{A}_1 + i\mathbf{A}_2) \tag{3.25}$$

$$\mathbf{e} = \frac{1}{\sqrt{2}}(\mathbf{e}_1 + i\mathbf{e}_2).$$

Equation (3.24a) in the electromagnetic direction, 3, decouples completely

$$\partial_t^2 \mathbf{e}_3 + \boldsymbol{\nabla} \times \boldsymbol{\nabla} \times \mathbf{e}_3 = 0 \tag{3.26}$$

while in the charged direction we have simply

$$(\partial_t + i\phi)^2 \mathbf{e} + \boldsymbol{\nabla} \times \boldsymbol{\nabla} \times \mathbf{e} = 0$$
$$\nabla^2 \phi = -q. \tag{3.27}$$

The electromagnetic fluctuations are free; the charged ones describe the motion of charged vector mesons in an external electric field with a potential ϕ.[10]

Detailed analysis of the equations can be performed in frequency space. Note that the electromagnetic equation involves ω^2 as an eigenvalue of a Hermitian operator, hence it is real. Only the issue remains whether ω^2 is positive or negative. In the charged equation, there appears $(\omega - \phi)^2$ and ω can be complex; it is not related to the eigenvalue of a Hermitian operator. This difference reflects the fact previously remarked upon: in the electromagnetic direction there are no gyroscopic terms, hence stability is equivalent to minimality. In the charged direction, gyroscopic terms are present; they are responsible for the more complicated equation.

The electromagnetic fluctuations are obviously stable. Those in the charged directions are stable in the absence of the external potential and by continuity they remain stable for a sufficiently small external potential.[3] As the external charge density increases in strength, an instability is expected to appear. This is not the instability of the Klein-Gordon equation in a $1/r$ [Coulomb] potential, which has previously been remarked upon, and which is a consequence of the [presumably unphysical] singularity at the origin.[11] In our examples, the potentials are non-singular. Instead it is the instability of the Klein-Gordon equation in a strong external field.[12]

In spite of stability for weak sources, we expect as a consequence of the gyroscopic terms to find modes which, though oscillatory, lower the energy. These can be readily exhibited, without passing to frequency space. We remain with the first-order equations (3.15), and seek a solution with $\delta \mathbf{B}_a = 0$. In that case, the charged portions of (3.15) with an Abelian Coulomb solution as the background field, reduce to

$$0 = \boldsymbol{\nabla} \cdot \delta \mathbf{E} - i \boldsymbol{\nabla} \phi \cdot \delta \mathbf{A} \,, \tag{3.28a}$$

$$0 = (\partial_t + i\phi) \delta \mathbf{E} \,, \tag{3.28b}$$

$$\delta \mathbf{E} = -(\partial_t + i\phi) \delta \mathbf{A} \,, \tag{3.28c}$$

$$\delta \mathbf{B} = \boldsymbol{\nabla} \times \delta \mathbf{A} = 0 \,. \tag{3.28d}$$

The solution of (3.28b) and (3.28c) is

$$\delta \mathbf{A} = [\mathbf{a}_0(\mathbf{r}) + t\mathbf{a}_1(\mathbf{r})] \exp[-it\phi(\mathbf{r})] \,, \tag{3.29a}$$

$$\delta \mathbf{E} = -\mathbf{a}_1(\mathbf{r}) \exp[-it\phi(\mathbf{r})] \,, \tag{3.29b}$$

and to satisfy (3.28a) and (3.28d) we must have

$$\mathbf{a}_0 = \boldsymbol{\nabla} \theta \,, \tag{3.29c}$$

$$\mathbf{a}_1 = -i\theta \boldsymbol{\nabla} \phi - i \boldsymbol{\nabla}^{-1}(\theta q) \,, \tag{3.29d}$$

where θ is an arbitrary function. Finally, there is one more condition: $\boldsymbol{\nabla}^{-1}(\theta q)$ must be parallel to $\boldsymbol{\nabla}\phi$, which can be easily achieved, for example by setting $\theta q = \nabla^2 F(\phi)$, where F is arbitrary. Thus equations (3.28) can be satisfied in terms of one function.

The quadratic energy (3.16) is seen to be negative[13]

$$\mathcal{E}^{(2)} = \int \left[(q\theta) \frac{-1}{\nabla^2} (q\theta^*) - |\theta|^2 (q) \frac{-1}{\nabla^2} (q) \right]$$

$$= \frac{-1}{8\pi} \int d^3\mathbf{r}\, d^3\mathbf{r}' \frac{q(\mathbf{r})q(\mathbf{r}')}{|\mathbf{r}-\mathbf{r}'|} |\theta(\mathbf{r}) - \theta(\mathbf{r}')|^2. \qquad (3.30)$$

This rather peculiar fluctuation gives evidence that the Coulomb solution is gyroscopically stable since the energy decreases below its Coulomb value.[5] Note from (3.29), the linear growth of the fluctuation with time, which however does not produce instability. Because the frequency is position dependent, it is not clear how to locate this mode among superpositions of functions with definite frequency. Nevertheless, we have encountered it already! It is merely the time-dependent generalization of the Abelian Coulomb solution, Eqs. (2.12)–(2.16), when the latter is brought arbitrarily close to the static solution.[14] The energy formula (3.30) can now be recognized as the $O(\theta^2)$ contribution to (2.16), when the source ρ' is a gauge transformation, with gauge function θ, of ρ the source in the standard frame.

3.2.2 Non-Abelian Coulomb Solution

The non-Abelian Coulomb solution, Eqs. (2.18), follows in many respects, at least for weak sources, the behavior of the Abelian Coulomb solution. [Only the weak source regime is amenable to analytic treatment, since our formulas are given by a source strength power series.] The stability equations are now highly coupled, and have not been solved. However, by continuity with the sourceless problem one expects stability for weak sources.[3] Moreover, one can show that this again must be an instance of gyroscopic stability, since the energy is lowered by the time-dependent generalization, presented in Eqs. (2.23)–(2.26), which can be taken arbitrarily close to the non-Abelian Coulomb solution. This is achieved by making ρ'', the source for the time-dependent solution, to be an infinitesimal gauge transformation of ρ', the source in the non-Abelian Coulomb solution. The energy can then be computed from (2.26). The details are the following. Set

$$\rho_a'' = \left(1 - \frac{1}{2}\theta_b\theta_b\right)\rho_a' - \varepsilon_{abc}\theta_b\rho_c'. \qquad (3.31)$$

This assures that ρ_a'' is a gauge transformation of ρ_a' taken through second-order in θ_a, which lies only in the charged direction

$$\theta_a(\mathbf{r})\rho_a'(\mathbf{r}) = 0. \qquad (3.32)$$

It now follows from (2.26) that the $O(Q^2)$ energy is

$$
\mathcal{E} = \frac{1}{8\pi} \int d^3\mathbf{r}\, d^3\mathbf{r}'\, \frac{\rho'_a(\mathbf{r})\rho'_a(\mathbf{r}')}{|\mathbf{r}-\mathbf{r}'|}
$$

$$
- \frac{1}{4\pi} \int d^3\mathbf{r}\, d^3\mathbf{r}'\, \frac{\rho'_a(\mathbf{r})\rho'_b(\mathbf{r}')}{|\mathbf{r}-\mathbf{r}'|}\varepsilon_{abc}\theta_c(\mathbf{r}')
$$

$$
- \frac{1}{16\pi} \int d^3\mathbf{r}\, d^3\mathbf{r}'\, \frac{\rho'_a(\mathbf{r})\rho'_a(\mathbf{r}')}{|\mathbf{r}-\mathbf{r}'|}[\theta_b(\mathbf{r})-\theta_b(\mathbf{r}')]^2 \qquad (3.33)
$$

$$
- \frac{1}{8\pi} \int d^3\mathbf{r}\, d^3\mathbf{r}'\, \frac{\rho'_a(\mathbf{r})\rho'_b(\mathbf{r}')}{|\mathbf{r}-\mathbf{r}'|}\theta_a(\mathbf{r}')\theta_b(\mathbf{r}) + O(Q^4)\,.
$$

The first term is the $O(Q^2)$ non-Abelian Coulomb energy. The second may also be written as $\int d^3\mathbf{r}\, \Phi_a(\mathbf{r})\varepsilon_{abc}\nabla^2\Phi_b(\mathbf{r})\theta_c(\mathbf{r})$, whence it is seen to vanish due to (2.19). The remaining two terms give the energy of the fluctuation. Unlike in (3.30), one cannot determine the sign by inspection, but after some straightforward manipulations, one can show that in the generic spherically symmetric case the terms are negative.[5]

3.2.3 Bifurcating Solutions

The bifurcating solutions, described in Section 2.3, exist only for sufficiently strong sources. Consequently, we have no closed-form expressions to analyze; yet precisely because there is a bifurcation, we can say a considerable amount without explicit computation. Consider first a solution to the static Yang-Mills equations for a definite source ρ — Eqs. (2.6) with the left-hand side of (2.6b) and (2.6c) set to zero. Next, imagine changing the source strength slightly. $\rho \to \rho + \delta\rho$, and looking for a new static solution. If the new solution is regularly related to the old one, the increments in the Yang-Mills fields will satisfy linear equations which are of the same form as the fluctuation equations (3.15), except that $\delta\rho$ occurs in the left-hand side of (3.15a) and time-derivatives are absent in (3.15b) and (3.15c). However, if we are at the bifurcation point, it must be impossible to solve these equations, and this happens if the homogeneous system has a non-trivial solution. In this way, we arrive at the important observation that at the bifurcation point, the stability equations have a zero-eigenvalue mode, and *vice versa:* a zero-eigenvalue mode may signal bifurcation [or generalizations thereof] in the static solutions viewed as functional of the source. It is possible to use general properties of bifurcation phenomena to give a description of the Yang-Mills theory near the bifurcation point. The discussion is complicated by a proliferation of equations and indicies. Therefore, we present here an analysis of a simplified, one-component model, in order to exemplify the general theory. The Yang-Mills problem follows completely the features of the example, and details are presented elsewhere.[5]

Let us consider a non-linear field equation for the field $\Phi(t, \mathbf{r})$ in the presence of an external, static source $\rho(\mathbf{r})$

$$\left(\frac{\partial^2}{\partial t^2} - \nabla^2\right)\Phi + U'(\Phi) = -\rho. \tag{3.34}$$

Here U is a potential for the field, and the dash indicates a differentiation with respect to argument. We suppose that a bifurcation occurs at $\rho = \rho_c$. By hypothesis, ρ_c supports a unique static solution $\phi_c(\mathbf{r})$; with stronger source there exists more than one solution. Correspondingly, there is a real, normalized zero-eigenvalue mode $\psi(\mathbf{r})$ in the small fluctuation equations

$$-\nabla^2\phi_c + U'(\phi_c) = -\rho_c, \tag{3.35a}$$

$$\{-\nabla^2 + U''(\phi_c)\}\psi = 0. \tag{3.35b}$$

Let us now replace ρ by $\rho_c + \varepsilon\delta\rho$, where ε is a small parameter, chosen to be positive, which systematizes the study of the theory around the bifurcation point. It is then appropriate to expand the static field ϕ according to

$$\phi = \phi_c + \varepsilon^{1/2}c\psi + \varepsilon\delta\phi + \cdots \tag{3.36}$$

with c being a numerical factor, which together with $\delta\phi$ is to be determined. Expansion of (3.34) shows that terms independent of ε as well as those of order $\varepsilon^{1/2}$ vanish by virtue of (3.35). The order ε equation leaves

$$\{-\nabla^2 + U''(\phi_c)\}\delta\phi + \frac{1}{2}c\,U'''(\phi_c)\psi^2 = -\delta\rho. \tag{3.37}$$

Equation (3.35b) implies a consistency condition on (3.37)

$$\frac{1}{2}c^2\int d^3\mathbf{r}\,U'''(\phi_c)\psi^3 = -\int d^3\mathbf{r}\,\psi\delta\rho. \tag{3.38}$$

For generic $\delta\rho$, the right-hand side of (3.38) is non-vanishing. We shall further assume that the integral in the left-hand side also is non-vanishing

$$\int d^3\mathbf{r}\,U'''(\phi_c)\psi^3 \neq 0. \tag{3.39}$$

[This assumption is a prerequisite for the subsequent development. Although we have not checked its validity in our Yang-Mills application, the fact that our analysis of the bifurcation is verified in the explicit numerical results, see below, provides an *a posteriori* justification.] Thus (3.38) determines both the magnitude of c and the direction of the bifurcation

$$c^2 = 2 \frac{- \int d^3 \mathbf{r} \psi \delta \rho}{\int d^3 \mathbf{r} \, U'''(\phi_c) \psi^3} . \tag{3.40a}$$

Obviously, the functional sign of $\delta \rho$ must be such that the right-hand is positive. [We seek real solutions, hence c in (3.36) must be real.] Two solutions for c are obtained

$$c = \pm \left| \frac{2 \int d^3 \mathbf{r} \, \psi \delta \rho}{\int d^3 \mathbf{r} \, U'''(\phi_c) \psi^3} \right|^{1/2} \tag{3.40b}$$

and consequently (3.36) shows that ϕ bifurcates around ϕ_c. Once the consistency condition (3.40) is satisfied, (3.37) may be solved for $\delta \phi$.

The energy around the bifurcation point also is readily determined. The energy of a static solution to (3.34) is

$$\mathcal{E} = \int d^3 \mathbf{r} \left[\frac{1}{2} (\boldsymbol{\nabla} \phi)^2 + U(\phi) + \rho \phi \right] . \tag{3.41a}$$

When this is differentiated with respect to ε we find

$$\frac{\partial \mathcal{E}}{\partial \varepsilon} = \int d^3 \mathbf{r} \frac{\delta \mathcal{E}}{\delta \phi(\mathbf{r})} \frac{\partial \phi(\mathbf{r})}{\partial \varepsilon} + \int d^3 \mathbf{r} \phi(\mathbf{r}) \frac{\partial \rho}{\partial \varepsilon} . \tag{3.41b}$$

The first term on the right-hand side vanishes, since a static solution stationarizes the energy, while in the second term we use our assumed form for ϕ and ρ. Thus we find

$$\frac{\partial \mathcal{E}}{\partial \varepsilon} = \int d^3 \mathbf{r} \, (\phi_c + \varepsilon^{1/2} c \psi + \dots) \delta \rho$$

$$\mathcal{E} = \mathcal{E}_c + \varepsilon \int d^3 \mathbf{r} \, \phi_c \delta \rho + \frac{2}{3} c \varepsilon^{3/2} \int d^3 \mathbf{r} \, \psi \delta \rho + \dots . \tag{3.41c}$$

Since c can take on two different signs according to (3.40), the above shows that the energy bifurcates, with an energy difference rising as $\varepsilon^{3/2}$.

Finally, we examine stability of the bifurcating solutions. The oscillatory modes associated with (3.34) satisfy

$$\Phi = \phi + e^{-i\omega_n t} \Psi_n , \tag{3.42a}$$

$$\{-\nabla^2 + U''(\phi)\} \Psi_n = \omega_n^2 \Psi_n . \tag{3.42b}$$

We concentrate on the mode ψ_0, which at the bifurcation point is the zero-eigenvalue mode, $\phi = \phi_c$, $\psi_0 = \psi$, and examine what happens immediately above the bifurcation, where we may set

$$\phi = \phi_c + \varepsilon^{1/2} c\psi + \cdots , \tag{3.43a}$$

$$\Psi_0 = \psi + \delta\Psi_0 . \tag{3.43b}$$

ω_0^2 is taken to be a small quantity, since in lowest order it vanishes. Inserting (3.43) into (3.42b) gives, to first-order in small quantities,

$$\omega_0^2\psi = \{-\nabla^2 + U''(\phi_c)\}\delta\Psi_0 + c\varepsilon^{1/2}U'''(\phi_c)\psi^2 . \tag{3.44a}$$

Equation (3.35) implies a consistency condition on the above; this evaluates ω_0^2

$$\omega_0^2 = c\varepsilon^{1/2} \int d\mathbf{r}\, U'''(\phi_c)\psi^3 . \tag{3.44b}$$

Again, since c can have either sign, we see that for one of the branches ω_0^2 is negative, hence there is an instability. Comparison with (3.40a) and (3.41c) shows that ω_0^2 has the opposite sign from the energy difference. This means that it is the upper branch which is unstable, while the lower branch shares the stability properties of the unique solution at the bifurcation point: if the latter is stable [the zero-eigenvalue mode is the lowest mode] so is the lower branch solution; if there is instability at the bifurcation point, [there exist complex eigenfrequencies] it will persist in the lower mode even beyond the bifurcation point.

When a similar derviation is carried out in the Yang-Mills problem, the conclusions are exactly the same[5]: one establishes that the static mode functions behave as $\pm(Q - Q_c)^{1/2}$ above the critical charge Q_c where bifurcation occurs. The energy difference between the two modes rises as $(Q - Q_c)^{3/2}$, with the upper branch being unstable. Since no analytic information about lower branch is available, we cannot report on its stability, but the general theory insures that it follows the stability properties at the bifurcation point. The fact that the zero-eigenvalue mode is spherically symmetric supports the conjecture that this is the lowest mode and no instability is present. Numerical analysis of the numerically determined results in Figs. 1 to 4 validate the above.

4. Conclusion

Finite-energy solutions to the Yang-Mills equations with arbitrary sources, can be studied perturbatively for weak sources. A rather comprehensive description is available. There exist at least two static solutions, the Abelian and non-Abelian Coulomb, with the latter carrying lower energy. They are accompanied by time-dependent generalizations which are continuous deformations of the static solutions. The time-dependent ones have the important property of lowering the energy, relative to the static configurations. The whole assembly of solutions can be compactly described in terms of the quantity $\mathbf{C} = [\Phi, \nabla\Phi]$. Beyond the perturbative regime, it is difficult to study the problem analytically [save for the Abelian Coulomb case], but numerical investigation does not expose any significant new structure.

Some questions remain. One would like to know how many different non-Abelian Coulomb solutions there are for a fixed source. [Thus far we have found only one.] Also one wonders whether there is a topological distinction between the Abelian and non-Abelian cases; a hint of one arises from the observation that the gauge transformation U, which takes the source from the Abelian frame to the non-Abelian frame in (2.18), is topologically non-trivial.

Solutions which are supported only by sources that exceed a critical strength, are known in isolated examples, but little of a general nature can be said about them at present. Presumably they are always characterized by bifurcations, and one wonders whether the bifurcating solutions are topologically different from the perturbative ones. Again one finds a hint: at large distances, the radial non-Abelian Coulomb solution vanishes rapidly, while the bifurcating one tends to a non-trivial pure gauge. Also one would like to know how to characterize the different bifurcated branches.

Stability for weak sources can be established, but the behavior for stronger sources is thus far unknown, save for the Abelian Coulomb case where an explicit formula allows for computations — the Coulomb solution is unstable beyond a critical source strength. In the bifurcating solutions, the bifurcation point corresponds to a zero-eigenvalue mode in the stability equations, and one of the bifurcated branches, the upper, is unstable. The other branch, the lower one, follows the stability behavior at the bifurcation point.

The most interesting result of the stability analysis for Yang-Mills theory is the observation that both the Abelian and non-Abelian Coulomb solutions, when stable, are gyroscopically stable. Modes which lower the energy, without introducing instability, have been identified. However, it is not clear how they are to be represented by superpositions of conventional monochromatic fluctuations.

The Yang-Mills model shares the physics of a top — stable motion does not minimize the energy. The analogy can be developed. For the top, gyroscopic forces arise from the constraint of conservation of angular momentum. In the Yang-Mills theory, the gyroscopic terms arise by the imposition of the Gauss law constraint

$$-\varepsilon_{abc}\mathbf{A}_b \cdot \mathbf{E}_c + \boldsymbol{\nabla} \cdot \mathbf{E}_a = \rho_a \,.$$

But the left-hand side is the generator of local rotations in group space; it is like a group space angular momentum. $[-\varepsilon_{abc}\mathbf{A}_c \cdot \mathbf{E}_c$ is analogous $\mathbf{q} \times \mathbf{p}.]$ In other words a non-vanishing source establishes at each point in ordinary space a non-vanishing angular momentum in group space, which then stabilizes configurations, which otherwise would be unstable.[15]

While some further computations obviously suggest themselves, especially for strong sources, the most pressing open question concerns the relevance of these mathematical investigations to the quantum physics of Yang-Mills theory.

REFERENCES

1. R. Jackiw, L. Jacobs and C. Rebbi, *Phys. Rev. D* **20**, 474 (1979).
2. P. Sikivie and N. Weiss, *Phys. Rev. D* **20**, 487 (1979).
3. J. Mandula, *Phys. Rev. D* **14**, 3497 (1976).
4. An example of this solution, with a radially symmetric source, $\rho_a = \delta_{a3}\, q(r)$ is constructed by taking $\mathbf{A}_a = -\mathbf{E}_a t + \delta_{a1}\boldsymbol{\nabla}(\alpha r^2 \frac{d\phi}{dr})$,

$$\mathbf{E}_a = \frac{-\hat{r}}{r^2}\frac{1}{\alpha}\left\{ \delta_{a3}\sin\left(\alpha r^2 \frac{d\phi}{dr}\right) + \delta_{a2}\left[1 - \cos\left(\alpha r \frac{d\phi}{dr}\right)\right] \right\}, \quad \phi = -\frac{1}{\nabla^2}q\,.$$

This is of the form (2.15) with $U = \exp\left(\frac{i\sigma^1}{2}\alpha r^2 \frac{d\phi}{dr}\right)$. The energy is

$$\mathcal{E} = 8\pi \int_0^\infty \frac{dr}{\alpha^2 r^2}\sin^2\left(\frac{\alpha}{2}r^2 \frac{d\phi}{dr}\right).$$

Here α is an arbitrary parameter, which when set to zero gives the Abelian Coulomb solution. This configuration is essentially the "total screening solution" found by P. Sikivie and N. Weiss, *Phys. Rev. Lett.* **40**, 1411 (1978), and *Phys. Rev. D* **18**, 3809 (1978); except that they use a discrete parameter rather than our continuously varying α. The feasibility of generalizing the total screening solution was pointed out by P. Pirilä and P. Presnajder, *Nucl. Phys.* **B142**, 229 (1978). The present formulation is given by Jackiw and Rossi, Ref. 5.
5. R. Jackiw and P. Rossi, *Phys. Rev. D* **21**, 426 (1980).
6. S. Coleman, in *New Phenomena in Sub-Nuclear Physics*, A. Zichichi (Plenum, New York, NY, 1977); S. Deser, *Phys. Lett.* **64B**, 463 (1976).
7. The truth of this statement is manifest from (2.22c) for sufficiently small Q, so that the $O(Q^4)$ terms are negligible; and for charge densities $q(r)$ which never

change sign, so that $q(r)q(r')\hat{r} \cdot \hat{r}' < q(r)q(r')$. However, in Ref. 1 it is shown that even for charge densities with varying signs the inequality in (2.22c) is valid, and that numerical computation at large Q confirms the bound; see Fig. 1.

8. Another example of a solution which exists when the source is of sufficient magnitude is Sikivie and Weiss' "magnetic dipole solution"; see Ref. 4. The source which supports this solution is studied by Y. Leroyer and A. Raychaudhuri, *Phys. Rev. D* **20**, 2086 (1979).

9. For an introduction to current, mainly mathematical research on stability of motion, see C. Siegel and J. Moser, *Lectures on Celestical Mechanics* (Springer, Berlin, 1971). A simple discussion, referring to older physics research, is found in H. Jeffreys and B. Jeffreys, *Methods of Mathematical Physics* (Cambridge University Press, Cambridge, UK, 1972).

10. This form for the fluctuation equations in a Coulomb background field was also given by M. Magg, *Phys. Lett.* **74B**, 246 (1978).

11. J. Mandula, *Phys. Lett.* **67B**, 175 (1978); M. Magg, *Phys. Lett.* **74B**, 246 (1978).

12. L. Schiff, H. Snyder and J. Weinberg, *Phys. Rev.* **57**, 315 (1940); K. Johnson, Harvard Ph.D thesis (1954) (unpublished); A. Migdal, *Zh. Eksp, Teor. Fiz.* **61**, 2209 (1972) [English translation: *JETP* **34**, 1184 (1972)]; A. Klein and J. Rafelski, *Phys. Rev. D* **11**, 300 (1975). For the delta-shell source (2.34) the instability sets in at $Q = 1.5$. This value is not related in any transparent way to the magnitude of Q at the bifurcation. It does agree with the strength of a point source at the onset of instability, as determined in Ref. 11. However, this coincidence is a consequence of the scaling properties of the delta-shell source, and is not expected for arbitrary extended sources.

13. The truth of this statement is manifest for charge densities which do not change sign, so that $q(\mathbf{r})q(\mathbf{r}') > 0$. However, for spherically symmetric charge densities the proviso can be removed, see Ref. 14, below.

14. For spherically symmetric charge densities, we may take the solution described in Ref. 4. For small α it becomes an infinitesimal deformation of the Abelian Coulomb with energy $\mathcal{E} = 2\pi \int_0^\infty dr\, r^2 (\phi')^2 - \frac{\alpha^2 \pi}{6} \int_0^\infty dr\, r^6 (\phi')^4$, which is always less than the Coulomb energy, regardless of the sign of the source.

15. That the physics of the top is encountered in the Yang-Mills theory was previously remarked by J. Goldstone and R. Jackiw, *Phys. Lett.* **74B**, 81 (1978). Indeed, it was in the context of the formalism developed in this paper that some of the results summarized here, were first encountered.

Paper II.3

High-temperature field theories are effectively described by a field theory in one lower dimension. This was the motivation in 1992 for introducing the Chern-Simons term — a three-dimensional gauge theoretical structure — into discussions of the physics of four-dimensional Yang-Mills theory. However, the precise sense in which Chern-Simons theory becomes physically relevant at high temperatures was not understood until a decade later. My account was presented on the occasion of S. Fubini's return from CERN to Italy.

TOPOLOGICAL STRUCTURES IN THE STANDARD MODEL AT HIGH T

Fubini Fest, Torino, Italy, February 1994

We meet to wish our dear colleague, Sergio Fubini all the best on his sixty-fifth birthday. Moreover, behind the formality of this calendrically significant instant, there is also our timeless expression of great affection and thanks to Sergio: affection for his sympathetic character and thanks for his activities in our profession, both within the scientific framework and in the broader social context. But for those of us from MIT there is a special feeling of gratitude towards Sergio, and that is because in the late sixties and early seventies, he was with us and helped shape what can now be seen as the most recent golden age of physics, thereby establishing at MIT a tradition that still flourishes today. You have to appreciate the moment: Steven Weinberg was on the faculty completing the standard model, while Garbriele Veneziano and Sergio were inventing what proved to be the physics of the future — string theory. These people have since departed from our University, but Sergio has kept the legacy vital by visiting us — not as frequently as we would have liked — and by encouraging continuing contact with his wonderful and talented Italian compatriot physicists. The people who came from Italy to enrich our department are too numerous to list, but theirs is an ongoing presence, formalized recently by an agreement with the INFN, and the well-spring of all his good fortune is Sergio Fubini, whom we all thank.

Sergio is now gone from MIT, but I am certain he wants to be informed of activity there, so I shall describe one project, with the hope that it meets his criteria of simplicity and symmetry.

These days, as high energy particle colliders become unavailable for testing speculative theoretical ideas, physicists are looking to other environments that may provide extreme conditions where theory confronts physical reality. One such circumstance may arise at high temperature T, which perhaps can be attained in heavy ion collisions or in astrophysical settings. It is natural therefore to examine the high-temperature behavior of the standard model, and here I shall report on recent progress in constructing the high-T limit of QCD.

In studying a field theory at finite temperature, the simplest approach is the so-called imaginary-time formalism. We continue time to the imaginary interval $[0, 1/iT]$ and consider bosonic (fermionic) fields to be periodic (anti-periodic) on that interval. Perturbative calculations are performed by Feynman rules as at zero temperature, except that in the conjugate energy-momentum, Fourier-transformed space, the energy variable p^0 (conjugate to the periodic time variable) becomes discrete — it is $2\pi nT$, (n integer) for bosons. From this one immediately sees that at high temperature — in the limiting case, at infinite temperature — the time-direction disappears, because the temporal interval shrinks to zero. Also only zero-energy processes survive, since "non-vanishing energy" necessarily means high energy owing to the discreteness of the energy variable $p^0 \sim 2\pi nT$, and therefore all modes with $n \neq 0$ decouple at large T. In this way, a Euclidean three-dimensional field theory becomes effective at high temperatures and describes essentially static processes.

While all this is quick and simple, it may be physically inadequate. First of all, frequently one is interested in non-static processes in real time, so complicated analytic continuation from imaginary time needs to be made. Also one may wish to study amplitudes where the real external energy is neither large nor zero, even though virtual internal energies are high.

Large T Feynman graph with external legs carrying limited amounts of energy and internal lines characterized by large momenta because T is large have been dubbed "hard thermal loops." In fact, they are a very important feature of high temperature QCD, because they necessarily arise in a resummed perturbative expansion.[1,2]

Here is the argument. Consider a one-loop amplitude $\Pi_1(p)$

$$\Pi_1(p) \equiv \int dk \; I_1(p, k) \,, \tag{1}$$

given by the graph in Fig. 1.

$$\Pi_1(p) = $$

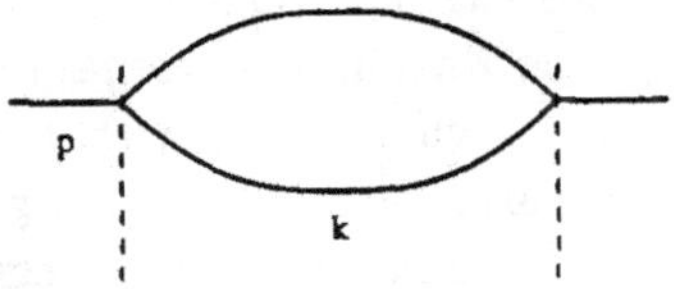

$$\equiv \int dk \; I_1(p, k)$$

Fig. 1. A one-loop amplitude.

Compare this to a two-loop amplitude $\Pi_2(p)$

$$\Pi_2(p) \equiv \int dk \; I_2(p, k) \,, \tag{2}$$

in which Π_1 is an insertion, as in Fig. 2

$$\Pi_2(p) = \quad \overset{\text{(figure)}}{}$$

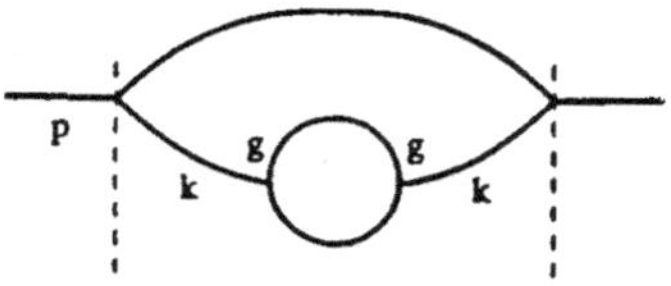

$$\equiv \quad \int dk \; I_2(p,k)$$

Fig. 2. A two-loop amplitude, whose magnitude is comparable to the one-loop amplitude in Fig. 1.

Following Pisarski,[1] I estimate the relative importance of Π_2 to Π_1 by the ratio of their integrands,

$$\frac{\Pi_2}{\Pi_1} \sim \frac{I_2}{I_1} = g^2 \frac{\Pi_1(k)}{k^2} . \tag{3}$$

Here g is the coupling constant, and the k^2 in the denominator reflects that we are considering a massless particle as in QCD. Clearly, the $k^2 \to 0$ limit is relevant to the question whether the higher order graph can be neglected relative to the lower order one. Because one finds that for small k and large T, $\Pi_1(k)$ behaves as T^2, the ratio Π_2/Π_1 is $g^2 T^2/k^2$. Thus when k is $O(gT)$ or smaller, the two-loop amplitude is not negligible compared to the one-loop amplitude. Thus graphs with "soft" external momenta $[O(gT)$ or smaller] have to be included as insertions in higher order calculations.

These so-called "hard thermal loops," *i.e.*, the high-temperature limits of real-time Feynman graphs with finite external momenta, have become the object of much study, which culminated with the discovery[2] of a remarkable simplicity in their structure. Specifically, the generating functional for hard thermal loops with only external gauge field legs, in an $SU(N)$ gauge theory containing N_F fermion species of the fundamental representation is found (i) to be proportional to $(N + \frac{1}{2}N_F)$, (ii) to behave as T^2 at high temperature, and (iii) to be gauge invariant

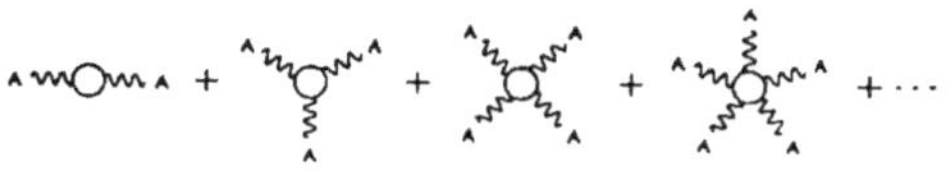

$$\text{(diagrams)} + \cdots = \left(N + \frac{1}{2}N_F\right)\frac{g^2 T^2}{12\pi}\Gamma_{\text{HTL}}(A), \tag{4}$$

$$\Gamma_{\text{HTL}}(U^{-1}AU + U^{-1}dU) = \Gamma_{\text{HTL}}(A). \tag{5}$$

A further kinematical simplification in Γ_{HTL} has also been established. To explain this, we define two light-like four-vectors $Q^{\mu}_{\pm}$ depending on a unit three-vector $\hat{q}$, pointing in an arbitrary direction

$$Q^{\mu}_{\pm} = \frac{1}{\sqrt{2}}(1, \pm\hat{q}) , \tag{6}$$

$$\hat{q} \cdot \hat{q} = 1 , \qquad Q^{\mu}_{\pm}Q_{\pm\mu} = 0 , \qquad Q^{\mu}_{\pm}Q_{\mp\mu} = 1 . \tag{7}$$

Coordinates and potentials are projected onto $Q^{\mu}_{\pm}$

$$x^{\pm} \equiv x_{\mu}Q^{\mu}_{\pm} , \qquad \partial_{\pm} \equiv Q^{\mu}_{\pm}\frac{\partial}{\partial x^{\mu}} , \qquad A_{\pm} \equiv A_{\mu}Q^{\mu}_{\pm} . \tag{8}$$

The additional fact that is now known is that (iv) after separating an ultralocal contribution from Γ_{HTL}, the remainder may be written as an average over the angles of $\hat{q}$ of a functional W that depends only on A_{+}; also this functional is non-local only on the two-dimensional $x_{\pm}$ plane, and is ultralocal in the remaining directions, perpendicular to the $x_{\pm}$ plane. ["Ultralocal" means that any potentially non-local kernel $k(x, y)$ is in fact a δ-function of the difference $k(x, y) = \delta(x - y)$.]

$$\Gamma_{\text{HTL}}(A) = 2\pi \int d^4x \, A_0^a(x)A_0^a(x) + \int d\Omega_{\hat{q}}W(A_+) . \tag{9}$$

These results are established in perturbation theory, and a pertubative expansion of $W(A_+)$, *i.e.*, a power series in A_+, exhibits the above mentioned properties. A natural question is whether one can sum the series, *i.e.*, obtain an expression for $W(A_+)$.

Important progress on this problem was made when it was observed[3] that the gauge-invariance condition can be imposed infinitesimally, whereupon it leads to a functional differential equation for $W(A_+)$, which is best presented as

$$\frac{\partial}{\partial x^+}\frac{\delta}{\delta A_+^a}\left[W(A_+) + \frac{1}{2}\int d^4x A_+^b(x)A_+^b(x)\right]$$
$$-\frac{\partial}{-\partial x^-}[A_+^a] + f^{abc}A_+^b\frac{\delta}{\delta A_+^c}\left[W(A_+) + \frac{1}{2}\int d^4x A_+^d(x)A_+^d(x)\right] = 0 . \tag{10}$$

In other word, we seek a quantity, call it

$$S(A_+) \equiv W(A_+) + \frac{1}{2}\int d^4x A_+^a(x)A_+^a(x) , \tag{11}$$

which is a functional on a two-dimensional manifold $\{x^+, x^-\}$, depends on a single functional variable A_+, and satisfies

$$\partial_1 \frac{\delta}{\delta A_1^a} S - \partial_2 A_1^a + f^{abc} A_1^b \frac{\delta}{\delta A_1^c} S = 0 , \tag{12}$$

$$\text{``1''} \equiv x^+ , \qquad \text{``2''} \equiv -x^- , \qquad A_1^a \equiv A_+^a . \tag{13}$$

Another suggestive version of the above is gotten by defining $A_2^a \equiv \frac{\partial S}{\partial A_1^a}$. Then we need to solve

$$\partial_1 A_2^a - \partial_2 A_1^a + f^{abc} A_1^b A_2^c = 0 . \tag{14}$$

To solve the functional equation and produce an expression for $W(A_+)$, we now turn to a completely different corner of physics, and that is Chern-Simons theory.

The Chern-Simons term is a peculiar gauge theoretic topological structure that can be constructed in odd dimensions, and here we consider it in 3-dimensional space-time

$$I_{\mathrm{CS}} \propto \int d^3x \varepsilon^{\alpha\beta\gamma} \mathrm{Tr} \left(\partial_\alpha A_\beta A_\gamma + \frac{2}{3} A_\alpha A_\beta A_\gamma \right) . \tag{15}$$

This object was introduced into physics over a decade ago, and since that time it has been put to various physical and mathematical uses. Indeed one of our originally stated motivations for studying the Chern-Simons term was its possible relevance to high-temperature gauge theory.[4] Here following Efraty and Nair,[5] we shall employ the Chern-Simons term for a determination of hard thermal loop generating functional, Γ_{HTL}.

Since it is the space-time integral of a density, I_{CS} may be viewed as the action for a quantum field theory in $(2 + 1)$ dimensional space-time, and the corresponding Lagrangian would then by given by a two-dimensional, spatial integral of a Lagrange density

$$I_{\mathrm{CS}} \propto \int dt \, L_{\mathrm{CS}} , \tag{16a}$$

$$I_{\mathrm{CS}} \propto \int d^2x (A_2^a \dot{A}_1^a + A_0^a F_{12}^a) . \tag{16b}$$

I have separated the temporal index (0) from the two spatial ones (1,2) and have indicated time differentiation by an over dot. F_{12}^a is the non-Abelian field strength, defined on a two-dimensional plane

$$F_{12}^a = \partial_1 A_2^a - \partial_2 A_1^a + f^{abc} A_1^b A_2^c . \tag{17}$$

Examining the Lagrangian, we see that it has the form

$$L \sim p\dot{q} - \lambda H(p, q) \,, \tag{18}$$

where A_2^a plays the role of p, A_1^a that of q, F_{12} is like a Hamiltonian and A_0 is like the Lagrange multiplier λ, which forces the Hamiltonian to vanish; here A_0^a enforces the vanishing of F_{12}^a

$$F_{12}^a = 0 \,. \tag{19}$$

The analogy instructs us how the Chern-Simons theory should be quantized.

We postulate equal-time commutation relations, like those between p and q

$$[A_1^a(\mathbf{r}), \ A_2^b(\mathbf{r}')] = i\delta^{ab}\delta(\mathbf{r} - \mathbf{r}') \,. \tag{20}$$

In order to satisfy the condition enforced by the Lagrange multiplier, we demand that F_{12}^a, operating on "allowed" states, annihilate them

$$F_{12}^a| \ \rangle = 0 \,. \tag{21}$$

This equation can be explicitly presented in a Schrödinger-like representation for the Chern-Simons quantum field theory, where the state is a functional of A_1^a. The action of the operators A_1^a and A_2^a is by multiplication and functional differentiation, respectively

$$| \ \rangle \sim \Psi(A_1^a) \,, \tag{22a}$$

$$A_1^a| \ \rangle \sim A_1^a \Psi(A_1^a) \,, \tag{22b}$$

$$A_2^a| \ \rangle \sim \frac{1}{i}\frac{\delta}{\delta A_1^a}\Psi(A_1^a) \,. \tag{22c}$$

This of course, is just the field theoretic analog of the quantum mechanical situation where states are functions of q, the q operator acts by multiplication, and the p operator by differentiation. In the Schrödinger representation, the condition that states be annihilated by F_{12}^a

$$(\partial_1 A_2^a - \partial_2 A_1^a + f_{abc}A_1^b A_2^c)| \ \rangle = 0 \tag{23a}$$

leads to a functional differential equation

$$\left(\partial_1 \frac{1}{i}\frac{\delta}{\delta A_1^a} - \partial_2 A_1^a + f_{abc}A_1^b\frac{1}{i}\frac{\delta}{\delta A_1^c}\right)\Psi(A_1^a) = 0 \,. \tag{23b}$$

If we define S by $\Psi = e^{iS}$ we get equivalently

$$\partial_1 \frac{\delta}{\delta A_1^a} S - \partial_2 A_1^a + f_{abc} A_1^b \frac{\delta}{\delta A_1^c} S = 0 . \tag{23c}$$

This equation comprises the entire content of Chern-Simons quantum field theory. S is the Chern-Simons eikonal, which gives the exact wave functional owing to the simple dynamics of the theory. Also the above eikonal equation is recognized to be precisely the equation for the hard thermal loop generating functional.

The gained advantage is that "acceptable" Chern-Simons states, *i.e.*, solutions to the above functional equations, had been constructed long ago,[6] and one can now take over those results to the hard thermal loop problem. One knows from the Chern-Simons work that Ψ and S are given by a 2-dimensional fermionic determinant, *i.e.*, by the Polyakov-Wiegman expression. While these are not described by very explicit formulas, many properties are understood, and the hope is that one can use these properties to obtain further information about high-temperature QCD processes.

For example one can compute the induced current $j^\mu_{\text{induced}} \sim \frac{\delta \Gamma_{\text{HTL}}}{\delta A_\mu}$, and use this as a source in the Yang-Mills equation, thereby obtaining a non-Abelian generalization of the Kubo equation, which governs the response of the hot gluonic plasma to external disturbances[7]

$$D_\mu F^{\mu\nu} = \frac{m^2}{2} j^\nu_{\text{induced}}$$

$$m = gT\sqrt{\frac{N + N_F/2}{3}} . \tag{24}$$

From the known properties of the fermionic determinant — hard thermal loop generating functional — one can show that j^μ_{induced} is given by

$$j^\mu_{\text{induced}} = \int \frac{d\Omega_{\hat{q}}}{4\pi} \left\{ Q^\mu_+ (a_-(x) - A_-(x)) + Q^\mu_- (a_+(x) - A_+(x)) \right\} , \tag{25}$$

where $a_\pm$ are solutions to the equations

$$\partial_+ a_- - \partial_- A_+ + [A_+, a_-] = 0 , \tag{26a}$$

$$\partial_+ A_- - \partial_- a_+ + [a_+, A_-] = 0 . \tag{26b}$$

Evidently j^μ_{induced}, as determined by the above equations, is a non-local and non-linear functional of the vector potential A_μ.

An alternative, equivalent derivation of the induced current has been given by Blaizot and Iancu,[8] directly from the QCD field equations. Their argument may be succintly put in the language of the composite effective action[9] and makes use of two approximations. The composite action is truncated at the one loop (semi-classical) level — two-particle irreducible graphs are omitted. This comprises the first, dynamical approximation. Then, in the second, kinematical approximation, the stationary conditions on the one-loop action are shown to lead to the gauge invariance equation for Γ_{HTL}.

In the Abelian case, everything commutes and linearizes. One can determine $a_{\pm}$ in terms of $A_{\pm}$

$$a_{\pm} = \frac{\partial_{\pm}}{\partial_{\mp}} A_{\mp} . \tag{27}$$

Incidentally, this formula exemplifies the kinematical simplicity, mentioned above, of hard thermal loops: the nonlocality of $1/\partial_{\pm}$ is entirely in $\{x^+, x^-\}$ plane. With the above form for $a_{\pm}$ inserted into the Kubo equation, the solution can be constructed explicitly. It coincides with the results obtained by Silin long ago, on the basis of the Boltzmann-Vlasov equation,[10] and one sees that ours is the non-Abelian generalization of that physics. In particular, m is recognized as the gauge invariant Debye screening length.

Moreover, it has been possible to extend the Boltzmann-Vlasov transport theory to the non-Abelian case, and to derive the non-Abelian Kubo equation (24)–(26) from purely classical considerations,[11] just as in the Abelian theory.[10] The only non-trivial step involves the realization that the phase space distribution function in the Boltzmann equation must be defined on a phase space, which is enlarged beyond its usual (p, q) variables to include non-Abelian charges as canonical variables.

At the present time, the non-Abelian Kubo equation is under further investigation. It has been possible to find local expressions for the current in the static case[9,12] and in the position-independent case[12]

$$\frac{m^2}{2} j^{\mu}_{\text{induced}} = (-m^2 A_0, \mathbf{0}) \quad \text{(static)} , \tag{28a}$$

$$\frac{m^2}{2} j^{\mu}_{\text{induced}} = \left(0, -\frac{1}{3} m^2 \mathbf{A}\right) \quad \text{(position-independent)} . \tag{28b}$$

The non-Abelian Kubo equation may then by solved, but the physical relevance of the solutions is unclear. In particular, the static solutions are not solitons since their energy is infinite.

A much more interesting result is due to Blaizot and Iancu.[13] They abstract from the Silin solution the plane-wave *Ansatz* $A_{\mu}(x) = A_{\mu}(x \cdot k)$ where k

is an arbitrary constant 4-vector and they determine explicitly the induced current associated with non-Abelian plane waves. In terms of the above, this corresponds to

$$a_{\pm} = \frac{Q_{\pm} \cdot k}{Q_{\mp} \cdot k} A_{\mp} \ . \tag{29}$$

The physics of all these solutions, as well as of other, still undiscovered ones, remains to be eluicidated, and I invite any of you to join in this interesting task.

References

1. R. Pisarski, *Physica* **A158**, 246 (1989).
2. E. Braaten and R. Pisarski, *Nucl. Phys.* **B337**, 569 (1990); J. Frenkel and J. C. Taylor, *Nucl. Phys.* **B334**, 199 (1990).
3. J. C. Taylor and S. M. Wong, *Nucl. Phys.* **B346**, 115 (1990).
4. R. Jackiw, in *Gauge Theories of the Eighties*, R. Ratio and J. Lindfors, eds. *Lecture Notes in Physics* **181**, 157 (Springer, Berlin, 1983).
5. R. Efraty and V. P. Nair, *Phys. Rev. Lett.* **68**, 2891 (1992); *Phys. Rev. D* **47**, 5601 (1993).
6. D. Gonzales and A. Redlich, *Ann. Phys.* (NY) **169**, 104 (1986); G. Dunne, R. Jackiw and C. Trugenberger, *Ann. Phys.* (NY) **149**, 197 (1989).
7. R. Jackiw and V. P. Nair, *Phys. Rev. D* **48**, 4991 (1993).
8. J.-P. Blaizot and E. Iancu, *Phys. Rev. Lett.* **70**, 3376 (1993).
9. R. Jackiw, Q. Liu and C. Lucchesi, *Phys. Rev. D* **49**, 6787 (1994).
10. E. M. Lifshitz and L. P. Pitaevskii, *Physical Kinetics* (Pergamon, Oxford, UK, 1981).
11. P. Kelly, Q. Liu, C. Lucchesi and C. Manuel, *Phys. Rev. Lett.* **72**, 3461 (1994).
12. J.-P. Blaizot and E. Iancu, *Phys. Rev. Lett.* **72**, 3317 (1994).
13. J.-P. Blaizot and E. Iancu, *Phys. Lett.* **B326**, 138 (1994).

Paper II.4

Lower-dimensional gravity theories merit study for pedagogical and practical reasons. The $(2+1)$-dimensional model — planar gravity — was described in Cocoyoc, Mexico (December 1990).

PLANAR GRAVITY

Relativity and Gravitation: Classical and Quantum, J. D'Olivio, E. Nahmad-Achar, M. Rosenbaum, M. Ryan, L. Urrutia and F. Zertuche, eds. (World Scientific, Singapore, 1991)

1. INTRODUCTION

Gravity theory remains rather in the same state it was in Einstein's time: a singularly elegant model from the mathematical point of view that agrees well with the few available experimental facts, which however has resisted incorporation in a modern view on physics: neither quantization of gravity nor its unification with other forces has been possible. Faced with this impasse many physicists, beginning with Einstein's contemporaries,[1] have reacted in the same way: the number of space-time dimensions is increased beyond four and in that (presumably fictitious) world the desired unification and quantization are attempted. Consensus on how the higher-dimensional model should be relevant to our world has not been attained: for some people the extra dimensions are real but small — "curled up" — for others they have no physical reality but are mathematical constructs. On the other hand, a small minority — also from Einstein's time — has sought inspiration in dimensions less than four.[2] To be sure there is no hope of dynamically expanding to four, rather one expects that in the simpler setting, one will acquire information useful for the physical problem. Also there are possible physical applications, if one is phenomenologically describing actual systems that are confined to move in a lower dimension — on a plane, in a line. Finally, an elegant and tractable mathematical structure characterizes lower-dimensional physics and attracts study.

While there are several spaces in which lower-dimensional gravity can be examined — two, one, fewer than one and even in between these — I shall here describe to you work performed by colleagues on two-dimensional models — planar gravity.[3]

The equations for Einstein's theory of gravity — general relativity — can be presented in any space-time with dimension equal to or greater than three: The Einstein tensor

$$G_{\mu\nu} \equiv R_{\mu\nu} - \frac{1}{2}g_{\mu\nu}R \tag{1.1}$$

vanishes in the absence of matter sources,

$$G_{\mu\nu} = 0 \tag{1.2}$$

while in their presence it is proportional to the energy-momentum tensor of matter, $T_{\mu\nu}$

$$G_{\mu\nu} = 2\pi G T_{\mu\nu} . \tag{1.3}$$

Here $R_{\mu\nu}$ and R are traces of the four-index Riemann tensor $R_{\alpha\mu\beta\nu}$ in which all local geometrical information about the space-time is encoded. G is the gravitational coupling constant — the generalization to other dimensions of Newton's constant; in (1.3) G enters with an unconventional normalization that is convenient for the subsequent analysis. The reason that (1.2) and (1.3) cannot be posited in two space-time dimensions is that there $G_{\mu\nu}$ vanishes identically. [However, other geometrical equations have been proposed for gravity on a line.[4]]

It is obvious from (1.3) that when space-time is flat, *i.e.*, when the Riemann tensor vanishes, so also does the Einstein tensor and $T_{\mu\nu}$ must be zero. In general, the converse does not hold: absence of matter implies vanishing Einstein tensor, but the Riemann tensor need not be zero so that empty space-time need not be flat. However, in three dimensions the Riemann tensor is linearly related to the Einstein tensor,

$$R^{\alpha\mu}_{\beta\nu} = \varepsilon^{\alpha\mu\gamma}\varepsilon_{\beta\nu\delta}G^{\delta}_{\gamma} \tag{1.4}$$

so that the vanishing of the latter implies the vanishing of the former: empty space-time is necessarily locally flat.

Several consequences follow immediately: since the vacuum state [empty space-time] is locally flat, there are no gravitational waves in the classical theory, and upon quantization, there are no quantum gravitons. Sources produce curvature, but only locally at the location in space-time of the sources. Forces between sources are not mediated by graviton exchange, since there are no gravitons. Rather interactions arise because the locally flat space-time possesses in the large non-trivial geometrical and topological structure that gives rise to non-trivial motions. It also follows that the non-relativistic limit of Einstein's general relativity in three-dimensional space-time is not planar Newtonian gravity, which involves a conventional force law that decreases with the inverse power of the distance.

It is the purpose of our research program to study in three-dimensional space-time the classical and quantum motions of matter that interacts gravitationally. Since there are no propagating gravitational degrees of freedom, the problem is tractable, and we can learn much about the puzzles that are encountered when a geometrical theory is confronted by quantum mechanics. In four dimensions these puzzles exist as well, and it is my opinion that understanding them is important for understanding quantum gravity; a task quite independent of and perhaps more fundamental than the task of overcoming the unrenormalizable infinities that pollute four-dimensional gravity, but are absent in three dimensions since non-renormalizable graviton exchange does not occur.

To conclude these introductory remarks, I note the following points:

(a) The theory can be elaborated by adding a cosmological constant to the field equations (1.2) and (1.3). The vacuum is then a space of constant curvature, whose sign depends on the sign of the cosmological constant. While some investigations of such models have been performed, I shall not further discuss them here.

(b) Another elaboration of the conventional theory involves adding a topological term, a Chern-Simons term. This addition will be discussed below.

(c) I shall not further discuss the theory without sources, because as far as I can tell it possesses no interesting dynamics, even though some global structure remains: locally flat space-time may still carry non-trivial topology.

2. CLASSICAL SPACE-TIMES

We record several interesting space-times that arise from classical sources. We begin with a single massive but spinless point-particle. Without loss of generality the particle is taken to be at rest at the origin of the coordinate system, *i.e.*, it is described by an energy-momentum tensor all whose components, except the energy density, vanish,

$$\sqrt{\det g_{\mu\nu}}\, T^{00} = M\delta(X)\delta(Y)$$
$$T^{0i} = T^{ij} = 0 \, . \tag{2.1}$$

Here M is the particle mass.

The task is to find the metric or equivalently to give a formula for the line element $(ds)^2$, which is non-trivial only in its spatial components

$$(ds)^2 = (dt)^2 - (d\ell)^2 \, . \tag{2.2}$$

So we need to find expressions for $(d\ell)^2$.

We recognize that we seek a space which is everywhere flat $[T^{\mu\nu}$ vanishes] except at the origin where a δ-function singularity concentrates the curvature. It follows that the desired space is a cone, with the source particle positioned at the apex. It remains to give an analytic description of this obvious geometrical fact.

To solve the Einstein equation (1.3) with sources given by (2.1), it is necessary to choose a coordinate system, and the conical solution looks different in different coordinates. Of course, only the two-dimensional spatial section needs to be considered.

Particularly useful coordinates, which lend themselves to a many-body generalization, are the conformal ones where the metric tensor is a multiple of the flat metric tensor; this can always be locally achieved in two dimensions. The conformally flat spatial metric that solves Einstein's equation then leads to the following spatial interval

$$(d\ell)^2 = \frac{1}{R^{2GM}} \left((dR)^2 + R^2(d\Theta)^2 \right). \tag{2.3}$$

Here the variables range over the conventional circles

$$\begin{aligned} 0 &\leq R \leq \infty \\ -\pi &\leq \Theta \leq \pi. \end{aligned} \tag{2.4}$$

While (2.3) certainly provides the desired solution, it does not seem to produce the cone described earlier. Nor is it manifest that the space is flat except at the origin.

All this can be seen by passing to another coordinate system, attained from (2.3) and (2.4) by a change of variables

$$\begin{aligned} r &= \frac{R^{1-GM}}{1 - GM} \\ \theta &= (1 - GM)\Theta. \end{aligned} \tag{2.5}$$

In terms of r and θ the spatial metric is flat, and the line-element is trivial

$$(d\ell)^2 = (dr)^2 + r^2(d\theta)^2. \tag{2.6}$$

However, the range of the new variables is unconventional — an angular region is excised, since according to (2.4) the range of (r, θ) is

$$\begin{aligned} 0 &\leq r \leq \infty \\ -\pi(1 - GM) &\leq \theta \leq \pi(1 - GM). \end{aligned} \tag{2.7}$$

This describes a cone, with apex determined by GM. [Henceforth, we take $GM \leq 1$. For $GM > 1$, the space changes character and the description becomes more complicated. At $GM = 1$, it is seen from (2.3) that space becomes a cylinder in the variable $r = \ln R$.]

In summary, we see that a point particle of mass M at the origin gives rise to a locally flat space-time, but the global identification of coordinate variables is unconventional and reveals the presence of a massive point-particle: the point

(t, r, θ) is identified with

$$(t, r, \theta) \approx (t, r, \theta + 2\pi(1 - GM)).$$ (2.8a)

In terms of a complex variable description of the space, $z = x + iy$, we identify z with

$$z \approx e^{-2\pi i GM} z.$$ (2.8b)

This is the analog in planar gravity of the Schwarzschild solution.

To find the planar analog of the Kerr solution, we endow our point-particle at the origin with spin S, *i.e.*, now the energy-momentum tensor possess non-trivial energy density and momentum density, the latter giving rise to no momentum but to angular momentum S

$$\sqrt{\det g_{\mu\nu}} T^{00} = M\delta(X)\delta(Y)$$
$$\sqrt{\det g_{\mu\nu}} T^{0i} = \sqrt{\det g_{\mu\nu}} T^{i0} = S\varepsilon^{ij}\partial_j\delta(X)\delta(Y)$$ (2.9)
$$T^{ij} = 0.$$

In the spatially conformal coordinate system, the metric that solves the field equation leads to a space-time interval, which is non-trivial in time as well as space

$$(ds)^2 = (dt + GSd\Theta)^2 - \frac{1}{R^{2GM}}\left((dR)^2 + R^2(d\Theta)^2\right).$$ (2.10)

Once again, by a change of variables one may pass to a locally flat space-time, where the presence of a massive, spinning source is encoded in a non-trivial identification of coordinate variables. Defining new spatial variables as in (2.5) and also a new time variable τ by

$$\tau = t + GS\Theta = t + \frac{GS}{1 - GM}\theta,$$ (2.11)

we see that (2.10) becomes flat,

$$(ds)^2 = (d\tau)^2 - (dr)^2 - r^2(d\theta)^2$$ (2.12)

but the required identification is

$$(\tau, r, \theta) \approx (\tau + 2\pi GS, r, \theta + 2\pi(1 - GM)).$$ (2.13)

Time is helical, space is conical and there are closed time-like contours.

Note that specifying a solution is equivalent to specifying an element of the (2+1)-dimensional Poincaré group that effects the identification (2.13). This remark will be expanded upon later.

The static one-body solution can be generalized to describe N particles located at $\mathbf{R}_i$, with masses M_i and spins $S_i, i = 1, \ldots, N$. One finds in spatially conformal coordinates

$$(ds)^2 = \left(dt + G \sum_{i=1}^N S_i \frac{(\mathbf{R} - \mathbf{R}_i)}{|\mathbf{R} - \mathbf{R}_i|^2} \times d\mathbf{R}\right)^2 - \prod_{i=1}^N \frac{1}{|\mathbf{R} - \mathbf{R}_i|^{2GM_i}} (d\mathbf{R})^2 . \quad (2.14)$$

The passage to locally flat coordinates is effected by defining a new time τ

$$d\tau = dt + G \sum_{i=1}^N S_i \frac{(\mathbf{R} - \mathbf{R}_i)}{|\mathbf{R} - \mathbf{R}_i|^2} \times d\mathbf{R} . \quad (2.15)$$

This hides the spins in complicated identifications on τ. To flatten the spatial interval, it is useful to express it in complex variables $Z = X + iY$, *etc.*

$$(d\ell)^2 = \left(\prod_{i=1}^N \frac{1}{(Z - Z_i)^{GM_i}}\right) dZ \left(\prod_{i=1}^N \frac{1}{(\bar{Z} - \bar{Z}_i)^{GM_i}}\right) d\bar{Z} . \quad (2.16)$$

Thus the definition

$$dz = \left(\prod_{i=1}^N \frac{1}{(Z - Z_i)^{GM_i}}\right) dZ \quad (2.17)$$

gives the flat spatial interval

$$(d\ell)^2 = dz \, d\bar{z} \quad (2.18)$$

but complicated identifications on the complex plane, which generalize (2.8b), reveal the presence of N particles with masses M_i. Unlike in the one-body problem, we cannot express z as a closed form function of Z, but for most purposes the integral expression suffices,

$$z = \int^Z dZ' \prod_{i=1}^N \frac{1}{(Z' - Z_i)^{GM_i}} \quad (2.19)$$

and (2.19) can be explicitly evaluated in special cases.

It is easy to show that the above solution also satisfies self-consistently the geodesic equation. Thus a static N-body configuration exists and is stable in three-dimensional space-time, in contrast to higher dimensions where gravitational attraction would prevent this. This demonstrates vividly the absence of Newtonian attraction in our theory.

With point-particle sources, the two-dimensional space is flat, but curvature is concentrated on a lower-dimensional sub-space: the zero-dimensional collection of points where the particles are located. One may next consider flat space with curvature concentrated on one-dimensional lines; *i.e.*, string sources in the plane, which presumably correspond to domain walls in four-dimensional space-time, just as points on the plane correspond to strings in four-dimensional space-time.

When considering strings, it is natural to allow for tension along the string; otherwise the source is an uninteresting pulverization of the point-particle — a "dust" string.

In the spinless case, the results are simple and startling. There are no open strings, only closed ones. A circular source at $R = a$ is described by an energy-momentum tensor whose non-vanishing components are

$$\sqrt{\det g_{\mu\nu}} T_0^0 = \mu\delta(R - a), \tag{2.20a}$$

$$\sqrt{\det g_{\mu\nu}} T_\theta^\theta = \tau\delta(R - a). \tag{2.20b}$$

Here μ and τ are mass and stress density/per unit length; the total mass is $M = 2\pi a\mu$. The momentum density and the other stress components vanish. With this source, the space-time interval in conformally flat spatial coordinates is

$$(ds)^2 = \begin{cases} \left(1 - 2\pi G a\tau \ln\dfrac{R}{a}\right)^2 (dt)^2 - \left(\dfrac{a}{R}\right)^2 [(dR)^2 + R^2(d\Theta)^2] & R \geq a \\ (dt)^2 - (dR)^2 - r^2(d\Theta)^2 & R \leq a. \end{cases} \tag{2.21}$$

The exterior spatial interval also reads $(da \ln\frac{R}{a})^2 + (ad\Theta)^2$, which is a half-cylinder of radius a extending from infinity to $R = a$, where it is capped by the flat disk of the $R \leq a$ region. Moreover, the total mass $M = 2\pi a\mu$ is given by G^{-1}, so that

$$GM = 1. \tag{2.22}$$

We have seen earlier that for point-particles obeying (2.22) the space is a cylinder; here (2.22) is always obeyed for spinless strings under tension and the space is a capped cylinder.

Although for $\tau > 0$, g_{00} vanishes at a finite distance, this is not a conventional horizon because g_{00} does not change sign, but time does "stand still" there. Clearly, there exist solutions with either sign of τ and unrelated to μ. However, for a relativistic string $\tau = \mu > 0$.

For more discussion on these extended objects and inclusion of spin, please consult the research papers.

3. QUANTUM DYNAMICS

The simplest non-trivial dynamics arises when we consider the interaction of two point-particles with each other. Because the gravitational field is determined by its sources through a formula that is local in time, it is possible to pass to the center-of-mass frame, as in other mechanical problems. The relative coordinate then moves in an effective potential that describes the interaction. The same problem arises without the center-of-mass reduction, but in the limit when one particle's mass becomes much larger than the other.

In view of this, it suffices to consider the problem of a test particle [mass m] moving in the field produced by the source particle [mass M] located at the origin.

The classical motion of a spinless test particle is easy to describe: in flat coordinates there is no deviation from straight-line motion. However, when the unconventional identification (2.13) is performed, we find a classical scattering angle

$$\Delta\theta_{\text{classical}} = \pm\pi GM \qquad (3.1)$$

and a classical time delay,

$$\Delta t_{\text{classical}} = \mp\frac{\pi GS}{1 - GM}, \qquad (3.2)$$

where S is the source particle's spin, and the sign depends on which side the source is passed. The classical trajectories are depicted in Fig. 1 [ignore the dotted lines for the moment]. They depend only on the impact parameter, but not on the energy; the scattering angle does not vary with impact parameter, except in its sign.

Next, we give a quantum mechanical description and to this end, we solve a quantum mechanical equation appropriate to the test particle: Schrödinger or Klein-Gordon for spinless test particles; Driac for spin 1/2 test particles, *etc.* [We do not second quantize the matter degrees of freedom.] The question that still must be considered is what interaction should we use to describe the influence of the source on the test particle.

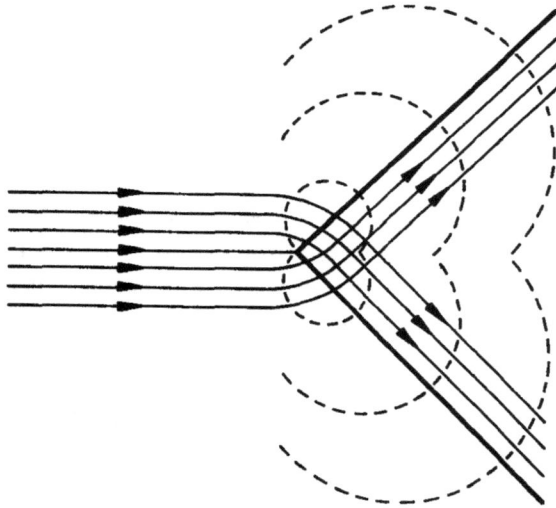

Fig. 1. Qualitative pictorialization for scattering of waves on an obstacle at the origin. The sharp lines are classical trajectories with scattering angle $\pm\pi GM$, see (3.1), the sign depending on which side the trajectory passes the source. The envelope to the right of the source, formed by heavy diagonal lines, is the sharp geometrical shadow. Broken lines represent diffraction on two sharp edges, even though no edge is actually present — the source [conical defect] produced the "edges."

The answer that we propose is that no interaction need be considered; rather we solve the free, non-interacting equation but impose on the solution a coordinate condition that reflects the identification (2.13).

For example, let us consider the simplest case first — a spinless test particle in a spinless source. The equation we propose to solve is the free [square-root] Klein-Gordon,

$$i\frac{\partial}{\partial t}\psi(t;r,\theta) = \sqrt{-\nabla^2 + m^2}\,\psi(t;r,\theta) \tag{3.3}$$

with the requirement that

$$\psi(t;r,\theta) = \psi(t;r,\theta+2\pi\alpha)$$
$$\alpha = 1 - GM. \tag{3.4}$$

[If non-relativistic motion is of interest, the non-local "square root" operator is replaced by $m - \nabla^2/2m$, which leads to the free Schrödinger equation, with boundary conditions (3.4). The mathematical analysis is identical.]

While the above prescription appears eminently plausible, one may wish to derive it from a detailed analysis of the problem. Such a derivation has indeed been given, it relies on the identification of planar Einstein gravity with a Chern-Simons theory based on the (2+1)-dimensional Poincaré group, $ISO(2,1)$. This subject will be discussed in the last Lecture.

The solution of (3.4) proceeds as follows: time is separated in the usual way and the remaining static wave function depending on $\mathbf{r} = (r, \theta)$ is expanded in terms of partial waves $e^{i\ell\theta}$. This too is conventional, except that ℓ, the angular momentum, is non-integral; rather $\alpha\ell$ is an integer. This, of course, is a consequence of the fact that the angular range is $2\pi\alpha$, not 2π. The radial equation involves Bessel functions from which a phase-shift [relative to $\alpha = 1$] may be identified. Then the scattering solution is a superposition of partial waves. The sum can then be performed by using the Schläfli contour representation for the Bessel function. After some shifting of contours, the scattering solution is given by

$$\psi(t; r, \theta) = e^{-iEt} \oint \frac{dz}{2\pi} e^{i\,\mathbf{k}(z)\cdot\mathbf{r}} \frac{1}{1 - e^{iz/\alpha}} \tag{3.5}$$

$$\equiv e^{-iEt} \psi(r, \theta)\,.$$

Here $E = \sqrt{k^2 + m^2}$ and $\mathbf{k}$ is the vector of magnitude k, rotated by the contour integration variable z: $\mathbf{k}(z) = (k\cos z, k\sin z)$. That (3.5) satisfies (3.3) is obvious; that also the boundary condition (3.4) is obeyed depends on the specific weight function in (3.5) and also on the contour, which is depicted in Fig. 2a.

The weight function has poles on the real axis at $z = z_n = 2\pi n\alpha$ and the contour C avoids them. However, the contour may be deformed. We can consider the equivalent, three segment contour $\tilde{C}$, depicted in Fig. 2b, where the poles are encircled and also there are integrals along the vertical lines. The contribution from the encircled poles is evaluated by Cauchy's theorem; it gives the incoming wave. The remaining integrals along the vertical lines give the scattered wave, but the integrations cannot be evaluated, so no closed form is available. Nevertheless, the large r asymptote is accessible, and the scattering amplitude is determined explicitly

$$\psi(r, \theta) = \psi^{\text{in}}(r, \theta) + \psi^{\text{sc}}(r, \theta)\,, \tag{3.6}$$

$$\psi^{\text{in}}(r, \theta) = \alpha \sum_n{}' e^{i\,\mathbf{k}(z_n)\cdot\mathbf{r}}\,, \tag{3.7a}$$

$$\psi^{sc}(r,\theta) = i \int_{-\infty}^{\infty} \frac{dy}{2\pi} e^{ikr\cosh y} \left[\frac{1}{1 - e^{i\frac{\pi}{\alpha}} e^{-\frac{1}{\alpha}(y+i\theta)}} - \frac{1}{1 - e^{-i\frac{\pi}{\alpha}} e^{-\frac{1}{\alpha}(y+i\theta)}} \right]$$

$$\xrightarrow[r\to\infty]{} \sqrt{\frac{i}{r}} f(\theta) e^{ikr}, \tag{3.7b}$$

$$f(\theta) = \frac{1}{2\sqrt{2\pi k}} \left[\left(\text{ctn}\frac{\theta - \pi}{2\alpha} - i \right) - \left(\text{ctn}\frac{\theta + \pi}{2\alpha} - i \right) \right]. \tag{3.8}$$

The prime on the sum in (3.7a) indicates that z_n must lie in the interval $[-\pi + \theta, \pi + \theta]$. Note that the incoming wave is not a plane wave, rather it is a superposition of variously rotated plane waves.

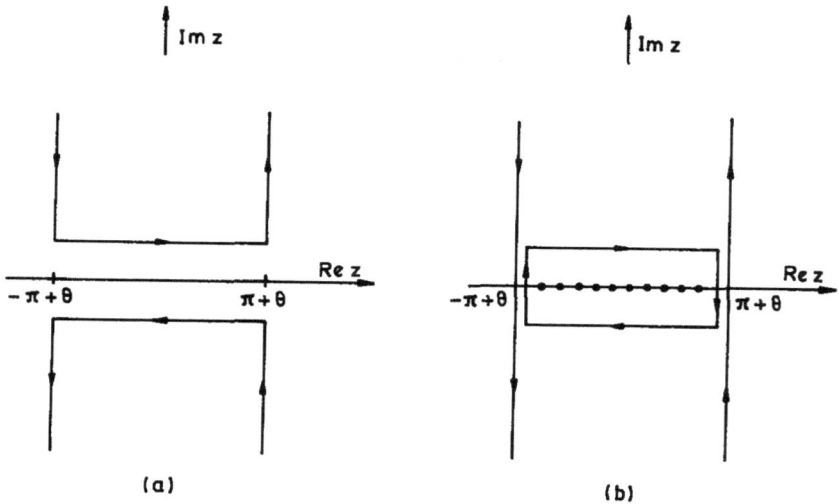

Fig. 2: (a) Integration contour C for the representation of $\psi(r,\theta)$ in (3.5). (b) Integration contour $\tilde{C}$ for the representation of $\psi(r,\theta)$ equivalent to that in (a) but giving rise to the decomposition $\psi = \psi^{\text{in}} + \psi^{\text{sc}}$. The incoming wave ψ^{in} is given by the [negative] Cauchy contour around the poles at $z = 2\pi n\alpha$, indicated by heavy dots. The integrals along the left and right vertical contours determine the scattered wave ψ^{sc}, whose large distance asymptote defines the scattering amplitude f.

We observe that the scattering amplitude $f(\theta)$ in (3.8) is real and vanishes when $1/\alpha$ is an integer. Also there are singularities at finite values of θ, where either of the two cotangents blows up. Finally, the optical theorem, which in two dimensions and with our normalization reads

$$\text{Im } f(0) = \sqrt{\frac{k}{4\pi}} \int d\theta |f(\theta)|^2 \tag{3.9}$$

fails because the left-hand side vanishes and the right-hand side diverges. Nevertheless, there is no loss of unitarity: one can verify from the exact solution (3.6)–(3.7) that the probability current is conserved. The peculiarities of the scattering amplitude are presumably related to the long-range nature of the "interaction": no matter how far the scattered particle is from the source, it remains on a cone. An interesting problem that here remains is the study of how a wave packet evolves in time.

Going beyond the simplest case, we consider the situation that arises when both the source and the test particle are spinning. The source spin S is arbitrary; for the test particle we consider spins 0 and 1/2, solving the Klein-Gordon and Dirac equations, respectively, but now with the more elaborate identification (2.13). One may again give a contour integral representation for the wave function, obtain the incoming wave by performing a Cauchy contour integral, and deduce an explicit formula for the scattering amplitude. The result is an elegant generalization of (3.8), which can be presented in universal form, provided the following definitions are made

S^s = spin of source [can be arbitrary, previously called S]

S^t = spin of test particle [actual calculations done only for $S^t = 0, 1/2$]

E^s = energy of source [taken to be M]

E^t = energy of test particle $\left(E^t = \sqrt{k^2 + m^2} \right)$. $\tag{3.10}$

The scattering amplitude is

$$f(\theta) = \frac{e^{-i[\omega]\theta/\alpha}}{2\sqrt{2\pi k}} \left[e^{-i\{\omega\}\pi/\alpha} \left(\text{ctn}\frac{\theta - \pi}{2\alpha} - i \right) - e^{i\{\omega\}\pi/\alpha} \left(\text{ctn}\frac{\theta + \pi}{2\alpha} - i \right) \right]. \tag{3.11}$$

Here ω is the symmetric cross product

$$\omega = E^s S^t + E^t S^s \tag{3.12}$$

while the square and curly brackets denote integer and fractional part, respectively

$$\omega = [\omega] + \{\omega\}. \tag{3.13}$$

For the spinless test particle, $S^t = 0$, one can determine from the phase shift $\delta(E)$ the time-delay by Wigner's formula. Agreement with the classical result (3.2) is found

$$\Delta T = 2\frac{\partial\delta(E)}{\partial E} . \tag{3.14}$$

We may understand the scattering amplitude as arising from diffraction effects [like in physical optics] which supplement the classical trajectories [whose analogy is geometrical optics]. These diffraction patterns are indicated by the dotted arcs in Fig. 1 and the two terms in (3.8) and (3.11) correspond to the two branches. We observe that scattering consists of a rotation through the angle $\pm\pi GM$, and we recall that in the presence of spin a rotation is accompanied by a phase change in the wave function. This explains the emergence of the additional phases in (3.11) as compared to (3.8).

The analysis of the Dirac equation is especially interesting owing to the fact that the Dirac Hamiltonian ceases to be self-adjoint on a conical, time-helical space time. [The same malady afflicts the Dirac equation in the presence of a vortex — the spinning Aharonov-Bohm effect.[5]] Of course, the derivatives are formally Hermitian, but consideration of the boundary conditions indicates that a self-adjoint extension, depending on parameters, must be made and different physical results emerge with different values for the parameters. [In deriving Eq. (3.11) a definite choice is made to insure universality — but other choices are possible.]

In physical terms, what is seen here is the failure of the point-particle description. Extended, smooth objects — described *eg.*, by fields — would lead to a self-adjoint Hamiltonian and in the point-particle limit various parameters, characterizing the extended object, survive as boundary terms on the particle surface and provide the missing information. The situation is similar to what is found for the Dirac equation with a [Dirac] point magnetic monopole. The Hamiltonian needs a one-parameter self-adjoint extension.[6] When a smooth 't Hooft-Polyakov monopole is considered, the parameter is identified as the QCD vacuum angle.[7] For the gravitational [and vortex] problems it remains an open question what model for the extended particle gives a physical origin to the mathematically necessary self-adjoint extension parameters.

The loss of self-adjointness appears to be related to the closed time-like curves that are present in a background metric arising from a spinning source.

We conclude this discussion of quantum motion by remarking that the true two-body problem — in contrast to its test particle source-particle equivalent description — is solved on a space with deficit angle given by the eigenvalues of the two-body Hamiltonian. This truly "Machian" behavior raises conceptual puzzles — for example it is impossible to superpose or compare energy eigensolutions. Moreover, the three- or more-body problem has thus far not been resolved [apart from a very easy special case] owing in part to difficulties in describing the multi-conical space on which the physical motion takes place.

4. TOPOLOGICAL ELABORATIONS

Up to now the discussion has been based on the three-dimensional version of the Einstein equation (2.2). However, in complete analogy to three-dimensional gauge theories, it is possible to modify (2.2) by an additional term, because in three dimensions there exists another second rank tensor that is symmetric and covariantly conserved. Sometimes called the *Cotton tensor*, its form is

$$C^{\mu\nu} = \frac{1}{2\sqrt{\det g_{\mu\nu}}} \varepsilon^{\mu\alpha\beta} D_\alpha R^\nu_\beta + \mu \leftrightarrow \nu. \tag{4.1}$$

Symmetry is manifest, covariant conservation follows from the Bianchi identities. $C^{\mu\nu}$ is traceless as also follows from (4.1) with the help of Bianchi identities

$$C^\mu_\mu = 0. \tag{4.2}$$

Moreover, $C^{\mu\nu}$ may be viewed as the three-dimensional conformal tensor — an odd-parity analog of the Weyl tensor, the latter vanishing identically in three dimensions. [That is why the Riemann tensor is determined by the Einstein tensor.] $C^{\mu\nu}$ is invariant against conformal redefinition of the metric tensor $g_{\mu\nu}(x) \to \lambda(x)g_{\mu\nu}(x)$ and vanishes if and only if space-time is conformally flat, $g_{\mu\nu}(x) = \lambda(x)\eta_{\mu\nu}$. We may supplement/replace the left-hand of (1.2) by the addition of a multiple of $C^{\mu\nu}$

$$G^{\mu\nu} + \frac{1}{\kappa}C^{\mu\nu} = 0, \tag{4.3a}$$

$$G^{\mu\nu} + \frac{1}{\kappa}C^{\mu\nu} = 2\pi G T^{\mu\nu}. \tag{4.3b}$$

[Also a cosmological constant can of course be added to the equation with or without sources, (4.3a) or (4.3b) respectively — we shall not do so.]

From its definition (4.1), we see that $C^{\mu\nu}$ is of one derivative order higher than $G^{\mu\nu}$, hence the dimension of κ is mass. Analysis of the linearized approximation yields dramatic results. While in the absence of the modification, there are no gravitational excitations, the addition of $C^{\mu\nu}$ "liberates" a previously "confined" graviton, which now becomes a single propagating mode; moreover, it is massive, while retaining covariance. The spin is ±2, the sign being correlated with the sign of κ. [The triple derivative nature of the differential equations (4.3) does not give rise to acausality; here, the conformal invariance comes into play, removing possibly dangerous terms from $C^{\mu\nu}$.] This theory is called *topologically massive gravity*.

$G^{\mu\nu}$ is obtained variationally from the Einstein-Hilbert action. Similarly, $C^{\mu\nu}$ may be obtained variationally from the Chern-Simons action, for the local Lorentz group in 2+1 dimensions — $SO(2,1)$. Constructing that quantity as in a gauge theory from the connection — either Christoffel or spin — but viewing the connection as a function of the fundamental dynamical variable — either the metric tensor or the *Dreibein*, respectively — and varying the dynamical variable gives $C^{\mu\nu}$.

No quantization condition need be imposed on κ. Non-trivial homotopies in a non-compact group like $SO(2,1)$ coincide with those of its maximal compact subgroup, here $SO(2)$; but $SO(2)$ is trivial in this respect, so the gravitational Chern-Simons action is invariant, just as the field equations are covariant, and κ is unrestricted.

It is not known whether topologically massive gravity is renormalizable.

Of course, a theory based solely on the Chern-Simons action/Cotton tensor field equation may also be considered. Here again, there are no propagating degrees of freedom, and due to the tracelessness of $C^{\mu\nu}$, only massless sources, with trace-free energy-momentum tensor can be coupled. However, owing to its triple derivative structure, the topological term is *not* natural for a low energy description, in contrast to the gauge theoretic Chern-Simons term. On the contrary, the Einstein/Hilbert theory is dominant at low energies, while the Chern-Simons/Cotton term dominates at high energy.

I conclude this discussion of topological elaborations on planar gravity by the following observations.

(a) Just like the gauge theoretic Chern-Simons term, the gravitational $SO(2,1)$ Chern-Simons term is induced by virtual fermions. This raises a puzzle about our treatment of quantum scattering, when the matter degrees of freedom are *second* quantized fermions and the "bare" gravitational action is just the conventional Einstein-Hilbert action. On the one hand, the bare gravitational action suggests that there are no propagating gravitational degrees of freedom. On the other hand, fermion loops induce a Chern-Simons action which when considered together with the bare action indicates the presence of massive, propagating gravitons. So which viewpoint is correct? Is the emergent "graviton" a fermion/anti-fermion bound state? How should perturbative calculations be organized?

(b) The Lagrangian for topologically massive gravity consists of $\mathcal{L}_{EH} + \frac{1}{\kappa}\mathcal{L}_{CS}$, the Einstein-Hilbert Lagrangian summed with κ^{-1} times the Chern-Simons term. Equivalently, we may write it as $\mathcal{L}_{CS} + \kappa\mathcal{L}_{EH}$, and view the higher derivative $\mathcal{L}_{CS}$ as the "kinetic" term and $\kappa\mathcal{L}_{EH}$ as the "mass" term. The former possesses more symmetry than the latter — it is conformally invariant. In some sense, that is "too much" symmetry and no propagation is possible with just the kinetic term. Inclusion of the less symmetric mass [Einstein-Hilbert] term lowers the symmetry and "liberates" the previously confined graviton. One may even promote κ to a scalar field ϕ with its own [unspecified] dynamics. The combination $\mathcal{L}_{CS} + \phi\mathcal{L}_{EH} + \mathcal{L}_\phi$ can be conformally invariant for suitably chosen $\mathcal{L}_\phi$. Then an expansion about $\langle\phi\rangle = 0$ contains no propagating gravitons, while the symmetry breaking starting point $\langle\phi\rangle = \kappa$ liberates the graviton.

(c) Some classical solutions to topologically massive gravity have been found. They are planar analogs of Gödel universes.

5. RELATION TO CHERN-SIMONS GAUGE THEORY

We have already stated that Einstein gravity may be modified by the addition of the $SO(2,1)$ Chern-Simons term, constructed as usual from connections, either Christoffel or spin, which are however viewed as functions of the metric or *Dreibein*. When this is varied with respect to the metric or *Dreibein* one obtains the Cotton tensor. The Einstein-Hilbert action, *viz.* the Riemann scalar, is also a function of the metric or the *Dreibein*, and when varied, gives the Einstein tensor.

Now we turn to a different role that Chern-Simons structures play in (2+1)-dimensional gravity. We shall here discuss solely Einstein gravity *without* the topological modification explained in the last Lecture, and alluded to again above.

As is well-known, the Einstein action [in any number of space-time dimensions greater than two] may be equivalently presented in first-order form, where the variables are spin connections and *Vielbein*, taken as independent quantities. When they are varied independently one obtains *two* sets of equations, which in combination reproduce the Einstein equations.

The spin connections and *Vielbein* may be viewed as vector potentials associated with the inhomogeneous Lorentz group [Poincaré group] with spin connections related to Lorentz rotations and the *Vielbein* to the translations.

The following remarkable fact has been established.[8] The Einstein-Hilbert action of (2+1)-dimensional gravity is precisely the Chern-Simons action for (2+1)-dimensional inhomogeneous Lorentz group [Poincaré group] $ISO(2,1)$ with gauge potentials as indicated above. To see this requires some explanation.

The [anti-Hermitian] generators of $ISO(2, 1)$ are J^a [Lorentz rotations] and P^a [translations]. They satisfy the algebra

$$[J^a, J^b] = \varepsilon^{abc} J_c, \qquad [J^a, P^b] = \varepsilon^{abc} P_c, \qquad [P^a, P^b] = 0. \qquad (5.1)$$

Latin indices are "internal." Various signs are hidden in the signature of their metric, which coincides with the signature of the metric for Greek indices labeling space-time.

The gauge potential for $ISO(2,1)$ takes its value in the Lie algebra

$$A_\mu = e_\mu{}^a P_a + \omega_\mu{}^a J_a. \qquad (5.2)$$

Here $e_\mu{}^a$ is the *Dreibein* and $\omega_\mu{}^a$ the spin connection. In the following, we shall often use the language of forms and thus omit the index μ.

In order to write down the action, we still have to define an invariant, non-degenerate, associative inner product, denoted by $\langle \, , \, \rangle$

$$\langle J_a, P_b \rangle = \eta_{ab}, \qquad \langle J_a, J_b \rangle = \langle P_a, P_b \rangle = 0. \qquad (5.3)$$

It replaces the trace operation in the more common gauge theories of simple Lie groups.

The action for pure Einstein gravity is then the Chern-Simons action for the Poincaré group

$$I_A = -\frac{1}{4\pi G} \int \left\langle A dA + \frac{2}{3} A^3 \right\rangle. \qquad (5.4)$$

The action in terms of the components reads

$$I_A = -\frac{1}{4\pi G} \int d^3 x \varepsilon^{\mu\nu\rho} e_\mu{}^a (\partial_\nu \omega_\rho{}^a - \partial_\rho \omega_\nu{}^a + \varepsilon_{abc} \omega_\nu{}^b \omega_\rho{}^c). \qquad (5.5)$$

It is well-known that invariance under homotopically non-trivial, "large", gauge transformations enforces a quantization condition on the Chern-Simons coupling. For a non-compact group, like $ISO(2,1)$, the non-trivial gauge transformations are those of the maximal compact subgroup, which for $SO(2,1)$ is

$SO(2)$, but this does not possess non-trivial gauge transformations in three-space, so the gravitational coupling constant G is not quantized.

For later reference, we also calculate the gauge field which is the variation of the Chern-Simons action with respect to the gauge potentials

$$F_{\mu\nu} = [D_\mu, \, D_\nu] = T^a_{\mu\nu} P_a + R^a_{\mu\nu} J_a \,. \tag{5.6}$$

Its components are the torsion

$$T^a \equiv \frac{1}{2} T^a_{\mu\nu} dx^\mu dx^\nu = de^a + \varepsilon^{abc} \omega_b e_c \tag{5.7}$$

and the Riemann curvature

$$R^a \equiv \frac{1}{2} F^a_{\mu\nu} dx^\mu dx^\nu = d\omega^a + \frac{1}{2} \varepsilon^{abc} \omega_b \omega_c \,. \tag{5.8}$$

That these equations coincide with familiar equations in differential geometry is recognized after defining $\omega^a = -\frac{1}{2}\varepsilon^{abc}\omega_{bc}$ and $R^a = -\frac{1}{2}\varepsilon^{abc} R_{bc}$

$$T^a \equiv de^a + \omega^a{}_b e^b \,, \tag{5.9a}$$

$$R_{ab} \equiv d\omega_{ab} + \omega_a{}^c \omega_{cb} \,. \tag{5.9b}$$

The variation of the action with respect to the gauge field A_μ or equivalently with respect to the [independent] components $e_\mu{}^a$ and $\omega_\mu{}^a$ leads to the field equation of vanishing curvature $F = 0$, or equivalently

$$T^a = 0 \,, \tag{5.10a}$$

$$R^a = 0 \,. \tag{5.10b}$$

Upon postulating invertibility of the *Dreibein*, the above equations are well-known to be an equivalent formulation of Einstein's equation in vacua: equation (5.10a) requiring vanishing torsion is equivalent to the symmetry of the Christoffel symbols, equation (5.10b) enforcing vanishing Riemann curvature implies the vanishing of the Einstein tensor and in (2+1) dimensions the reverse is also true as a consequence of (1.4).

When the gauge field vanishes the gauge potentials are pure gauges

$$A_\mu = U \partial_\mu U^{-1} \,, \tag{5.11}$$

where U is an element of the group, here $ISO(2,1)$. For point sources, the energy-momentum tensor $T_{\mu\nu}$ involves δ-functions, hence it vanishes almost everywhere and the vector potential is a pure gauge almost everywhere. In fact, the point particle vector potential can be presented as a pure gauge that is singular.

In this way, the situation is analogous to the electromagnetic vector potential arising from a point vortex [infinitely thin solenoid]: the magnetic field is a two-dimensional δ-function $B = \Phi \delta^2(\mathbf{r})$, where Φ is the flux through the plane; the vector potential that gives rise to this magnetic field $B = \nabla \times \mathbf{A}$, is $\mathbf{A} = \frac{\Phi}{2\pi} \nabla \theta$, where $\tan\theta = y/x$, $\mathbf{r} = (x,y)$. This requires $\nabla \cdot \nabla \times \theta = 2\pi \delta^2(\mathbf{r})$, or in terms of exterior derivatives $dd\theta = 2\pi \delta^2(\mathbf{r})dx\,dy$.

According to (5.2), specifying the vector potential is equivalent to specifying the spin connection and the *Dreibein*, from which the metric tensor is determined. It follows therefore that specifying the space-time created by point particles, which is what we described in Lecture II, is tantamount to specifying an element U of the Poincaré group — a fact that we already remarked upon.

It is interesting to find the group element corresponding to the space-time of a single spinning and massive particle at the origin. The line element was presented in (2.10) and equivalently in (2.12), where the angular range is reduced. Here we present it once again, after rescaling θ and r so that the range is conventional

$$(ds)^2 = (dt + GSd\phi)^2 - \frac{1}{1-GM}(d\rho)^2 - \rho^2(d\phi)^2$$

$$-\pi \le \phi = \frac{\theta}{1-GM} \le \pi \tag{5.12}$$

$$0 \le \rho = (1-GM)r \le \infty.$$

The *Dreibein* and spin connection for the above metric are

$$e^0 = dt + GSd\phi$$

$$e^1 = \frac{1}{1-GM}\cos\phi d\rho - \rho\sin\phi d\phi,$$
(5.15a)

$$e^2 = \frac{1}{1-GM}\sin\phi d\rho + \rho\cos\phi d\phi$$

$$\omega^0 = GMd\phi$$

$$\omega^1 = 0 = \omega^2. \tag{5.15b}$$

Outside the origin, these can easily be seen to satisfy the vacuum equation (5.10). By using the distributional identity mentioned earlier $dd\phi = 2\pi\delta^2(\rho)dx\,dy$ and recalling the definitions (5.9) and (5.10) for the torsion

and Riemann curvature one obtains the following non-vanishing components
for these two-forms

$$T^0 = 2\pi GS\delta^2(\rho)dx\,dy, \qquad R^0 = 2\pi GM\delta^2(\rho)dx\,dy. \qquad (5.16)$$

These equations provide the link between Einstein's theory and the gauge
theory of gravity. A very interesting fact is that angular momentum acts as a
source for torsion. [It has sometimes been suggested in four dimensions that the
interior solution for sources with angular momentum (*eg.*, for the Kerr metric)
might require torsion and we see that in three dimensions this is indeed the
case.]

We wish to present the group element U that corresponds to (5.15). We
use a 4×4 representation for $ISO(2,1)$ where an arbitrary group element V
has the form

$$V = \begin{pmatrix} \Lambda^a{}_b & q^a \\ 0 & 1 \end{pmatrix}, \qquad (5.17)$$

$\Lambda^a{}_b$ is a 3×3 Lorentz matrix and q^a is a three-vector. Denoting V by (Λ, q)
the matrix product $V_1 V_2 = (\Lambda_1 \Lambda_2, \Lambda_1 q_1 + q_2)$ reproduces the composition law
of the Poincaré group. The inverse of (Λ, q) is given by $(\Lambda^{-1}, -\Lambda^{-1}q)$. Also
we obtain

$$dV = d(\Lambda, q) = \begin{pmatrix} d\Lambda^a{}_b & dq^a \\ 0 & 0 \end{pmatrix} = \begin{pmatrix} d\Lambda^c(J_c)^a_b & dq^b(P_b)^a \\ 0 & 0 \end{pmatrix}. \qquad (5.18)$$

In the last expression, we separated the components of the generators

$$(J_a)^b{}_c = -\varepsilon_a{}^b{}_c, \qquad (P_a)^b = \delta_a{}^b. \qquad (5.19a)$$

These can be checked to satisfy the algebra (5.1). By a slight abuse of notation,
we use the same symbols for the quantities

$$J_a = \begin{pmatrix} (-\varepsilon_a)^b{}_c & 0 \\ 0 & 0 \end{pmatrix}, \qquad P_a = \begin{pmatrix} 0 & (\delta_a)^b \\ 0 & 0 \end{pmatrix}. \qquad (5.19b)$$

We remark that the derivative dV in (5.18) is not of the form (5.17), since
there is a zero in the lower right corner.

Equation (5.11) reads with $U^{-1} \equiv (\Lambda, q)$

$$A = U dU^{-1} \equiv \begin{pmatrix} \Lambda^{-1}d\Lambda & \Lambda^{-1}dq \\ 0 & 0 \end{pmatrix} = \begin{pmatrix} \omega^a J_a & e^b P_b \\ 0 & 0 \end{pmatrix}. \qquad (5.20)$$

With the *Dreibein* and the spin connection from Eq. (5.15) one can solve for Λ and q to obtain

$$
U^{-1} = \begin{pmatrix}
1 & 0 & 0 & t + GS\phi \\
0 & \cos GM\phi & \sin GM\phi & \dfrac{\rho}{1-GM}\cos(1-GM)\phi \\
0 & -\sin GM\phi & \cos GM\phi & \dfrac{\rho}{1-GM}\sin(1-GM)\phi \\
0 & 0 & 0 & 1
\end{pmatrix}. \tag{5.21}
$$

This matrix has a very simple interpretation. The vector q^a is simply the [local] coordinate transformation which transforms the metric $g_{\mu\nu}$ into η_{ab}. Such a coordinate transformation must exist, since the space is locally [Riemann] flat and is indeed given by $q^0 = t + GS\phi$, $q^1 = [\rho/(1-GM)]\cos(1-GM)\phi$ and $q^2 = [\rho/(1-GM)]\sin(1-GM)\phi$. The *Dreibein* in flat coordinates $e_a{}^b = \delta_a^b$ when transformed, becomes $e'_\mu{}^b = (dq^a/dx^\mu)\delta_a{}^b$. This however is not single valued, since the coordinate transformation is well-defined only locally. One thus has to make an additional Lorentz transformation Λ^{-1} to obtain $e_\mu{}^a = (\Lambda^{-1})^a{}_b e'_\mu{}^b$. In the above case Λ^{-1} is rotation by $GM\phi$.

This discussion also provides a prescription how to present U^{-1} in any other coordinate system.

An interesting quantity to compute in a gauge theory is the holonomy [Wilson line] of the field around these sources. We thus want to calculate

$$
W = P\exp\oint\{-A_\mu dx^\mu\} = P\exp\oint\{-(e_\mu{}^a P_a + \omega_\mu{}^a J_a)dx^\mu\}. \tag{5.22}
$$

Here the path encloses the world-line of the particle. The invariants of W with respect to gauge transformations at the end points should of course correspond to the mass and spin of the source. We shall give a direct description of these invariants below.

For a pure gauge vector potential (5.11) W is

$$
W = U(\phi = 2\pi)U^{-1}(\phi = 0). \tag{5.23}
$$

We may use U from (5.21); However for aesthetic reasons we prefer to define $U' = U^{-1}(\phi = 2\pi)U$, which amounts to a constant gauge transformation of A.

The transformed holonomy matrix is then given by

$$W = U'(\phi = 2\pi)U'^{-1}(\phi = 0) = U^{-1}(\phi = 0)U(\phi = 2\pi)$$

$$= \begin{pmatrix} 1 & 0 & 0 & -2\pi GS \\ 0 & \cos 2\pi GM & -\sin 2\pi GM & 0 \\ 0 & \sin 2\pi GM & \cos 2\pi GM & 0 \\ 0 & 0 & 0 & 1 \end{pmatrix}. \tag{5.24}$$

So W corresponds to a rotation by $-2\pi GM$ and time translation by $-2\pi GS$. The order is unimportant since the two operations commute.

We shall now show that M and S are the invariants of W under conjugation $W \to VWV^{-1}$. Every Lorentz matrix Λ can be characterized by a unit vector n^a [axis of rotation] and an angle α [amount of rotation]. For proper rotations n^a is time-like. It is well known from elementary geometry that conjugation "rotates the axis," but leaves the angle invariant. This angle corresponds to $-2\pi GM$ and is the first invariant of W. The second invariant of an element (Λ, q) is the component of q along the axis of rotation, *i.e.*, $q_a n^a$. This can be verified by an explicit calculation, using the composition law for the Poincaré group. In W from Eq. (5.24) the vector n^a points along the time-axis and the second invariant is indeed $-2\pi GS$.

This completes the group theoretic/Chern-Simons transcription of (2+1) gravity with a point-particle source. The utility of this framework for the quantum problem derives from the fact that it is known how to describe quantum mechanically point particles that interact through gauge fields that are singular pure gauges — a kind of Aharonov-Bohm / Ehrenberg-Siday problem, here in a non-Abelian realization. Taking over these techniques to the gravitational problem, reproduces precisely the point of view adopted in Lecture III, which had been arrived at heuristically.[9]

For details, the reader is referred to the literature. It is interesting to point out that the gauge theoretic framework allows performing gauge transformations which render the *Dreibein* in the gauge potential no longer invertible. This precludes an Einstein theory interpretation of the *Dreibein*. But it does not effect physical quantities that are gauge invariant, while by clever gauge choice, calculations may be enormously amplified.

Thus one may use an even simpler field configuration to study the physical effects of a point particle, to wit

$$e'^0 = GS d\phi, \qquad \omega'^0 = GM d\phi \tag{5.25}$$

all other components vanishing. The corresponding A' is a gauge transform of A from (5.15), $A' = VAV^{-1} + VdV^{-1}$ with V given by

$$V = \begin{pmatrix} 1 & 0 & 0 & t \\ 0 & 1 & 0 & \dfrac{\rho}{1 - GM} \cos\phi \\ 0 & 0 & 1 & \dfrac{\rho}{1 - GM} \sin\phi \\ 0 & 0 & 0 & 1 \end{pmatrix}. \qquad (5.26)$$

We emphasize that this is a globally well-defined, non-singular gauge transformation. Incidentally, V leaves Eq. (5.16) invariant $F' \equiv VFV^{-1} = F$. It is trivial to verify that A' of Eq. (5.25) indeed satisfies Eq. (5.16). Of course, the holonomy W is the same, characterized by the invariant M and S, but the line element corresponding to (5.25) is the ridiculous quantity

$$(ds)^2 = -G^2 S^2 (d\phi)^2$$

which not only is degenerate but has lost all information about the mass. However, within the gauge theoretic point of view, we have no reason to consider $(ds)^2$, which is not gauge invariant.

To conclude, let me remark that inclusion of a cosmological constant and of the topological mass term still permits a group theoretical, Chern-Simons description.[8]

REFERENCES

1. T. Kaluza, *Sitz. Ber. Preuss. Akad., Phys. Math. Kl.* 966 (1921); O. Klein, *Z. Physik* **37**, 895 (1929).

2. E. Abbott, *Flatland* (Dover, New York, NY, 1884).

3. Except where otherwise explicitly noted, these Lectures are based on the following papers by our group of collaborators: S. Deser, R. Jackiw and S. Templeton, *Phys. Rev. Lett.* **48**, 975 (1972), *Ann. Phys.* (NY) **140**, 372 (1982), (E) **185**, 406 (1988); S. Deser, R. Jackiw and G. 't Hooft, *Ann. Phys.* (NY) **152**, 220 (1984); S. Deser and R. Jackiw, *Ann. Phys.* (NY) **153**, 405 (1984); I. Vuorio, *Phys. Lett.* **B163**, 91 (1985), **B175**, 176 (1986); R. Percacci, P. Sodano and I. Vuorio, *Ann. Phys.* (NY) **176**, 344 (1987); R. Percacci, *Ann. Phys.* (NY) **177**, 27 (1987); G. 't Hooft, *Comm. Math. Phys.* **117**, 685 (1988); S. Deser and R. Jackiw, *Comm. Math. Phys.* **118**, 495 (1988); P. Gerbert and R. Jackiw, *Comm. Math. Phys.* **124**, 229 (1989); S. Deser and R. Jackiw, *Ann. Phys.* (NY) **192**, 352 (1989); G. Grignani and C. Lee, *Ann Phys.* **196**, 386 (1989). My earlier reviews of this and related subjects are in *Nucl. Phys.* **B252**, 343

(1985); *Group Theoretical Methods in Physics*, Y. St. Aubin and L. Vinet, eds. (World Scientific, Singapore, 1989); *Field Theory and Particle Physics*, O. Éboli, M. Gomes and A. Santoro, eds. (World Scientific, Singapore, 1990).

4. R. Jackiw, C. Teitelboim, in *Quantum Theory of Gravity*, S. Christensen, ed. (A. Hilger, Bristol, UK, 1984); C. Teitelboim, *Phys. Lett.* **126B**, 41 (1983); T. Fukuyama and K. Kamimura, *Phys. Lett.* **160B**, 259 (1985); A. Polyakov, *Mod. Phys. Lett.* **A2**, 893 (1987); K. Isler and C. Trugenberger, *Phys. Rev. Lett.* **63**, 834 (1989); A. Chamseddine and D. Wyler, *Phys. Lett.* **B228**, 75 (1989); see also "Gauge Theories for Gravity on a Line", reprinted in this volume on p. 197.

5. P. Gerbert, *Phys. Rev. D* **40**, 1346 (1989).

6. A. Goldhaber, *Phys. Rev. D* **16**, 1815 (1977); C. Callias, *Phys. Rev. D* **16**, 3068 (1977).

7. B. Grossman, *Phys. Rev. Lett.* **50**, 464 (1983); H. Yamagishi, *Phys. Rev. D* **27**, 2382 (1983).

8. A. Achúcarro and P. Townsend, *Phys. Lett.* **B180**, 89 (1986); E. Witten, *Nucl. Phys.* **B311**, 46 (1988/89).

9. The program of studying point-particle gravitational dynamics within the Chern-Simons formulation of (2+1)-dimensional gravity was begun by S. Carlip, *Nucl. Phys.* **B324**, 106 (1989) and completed by P. Gerbert, *Nucl. Phys.* **B346**, 440 (1990). My discussion is taken from Gerbert's paper.

Paper II.5

Planar gravity "in action" addressing a "physical" question: Is time travel possible in the presence of cosmic strings? The lecture was delivered at an anniversary celebration of MIT's Laboratory of Nuclear Science (May 1992) and memorializes G. Feinberg who showed that time travel can be analyzed scientifically. The text was prepared in collaboration with S. Deser.

TIME TRAVEL?

Comments Nucl. Part. Phys. **20**, *337 (1992)*

To travel into the past, to observe it, perhaps to influence it and correct mistakes of one's youth has been an abiding fantasy of mankind for as long as we have been aware of a past. Here are described some recent scientific investigations on this topic.

Before the twentieth century, time travel was discussed only in works of fiction; among innumerable instances, best known are surely the novels of Mark Twain and H. G. Wells. The latter marks a transition: Wells, a graduate of London's Imperial College of Science and Technology, couched his novel, *The Time Machine* in scientific language, giving us an early work of science fiction.

After 1900, special relativity made scientific discussion of time machines possible. The question may be posed in this way: it makes perfectly good sense to speak about travel that returns to the same point in *space*. But Einstein and Minkowski tell us that space and time are equivalent, so after a journey can we return to our starting position in *time*? Here the answer is well-known: the space-time of special relativity is flat and rigid; while time travel into the past is indeed possible, it requires faster-than-light velocities. This is seen as follows. The Lorentz transformation law of space (Δx) and time (Δt) intervals in some initial (inertial) reference frame into corresponding quantities (designated by an overbar) in another frame, moving relative to the original one with frame velocity v along say the x-axis, is given by

$$\overline{\Delta t} = \frac{1}{\sqrt{1 - v^2/c^2}} \left(\Delta t - \frac{v}{c^2} \Delta x \right), \tag{1a}$$

$$\overline{\Delta x} = \frac{1}{\sqrt{1 - v^2/c^2}} (\Delta x - v\Delta t). \tag{1b}$$

Since the velocity u of a moving object in the original frame is $\Delta x/\Delta t$, we also have

$$\overline{\Delta t} = \frac{1}{\sqrt{1 - v^2/c^2}} \Delta t \left(1 - \frac{uv}{c^2} \right). \tag{2}$$

For $\overline{\Delta t}$ to remain real, the frame velocity v must not exceed the velocity of light c, but $\overline{\Delta t}$ can have a sign opposite to Δt if

$$u/c > c/v > 1. \tag{3}$$

One can think of this result as a continuation of the familiar time-dilation story; the faster one travels starting from rest, the slower flows time; time stands still for travel at the velocity of light and runs backward once c is exceeded.

A time machine is now constructed in the following manner. When a faster-than-light object — called a *tachyon* — is emitted with $u > c$ by a source, moving backwards with speed $v < c$ and satisfying (3), the tachyon will arrive at its goal a time $|\overline{\Delta t}|$ before it was sent out; then if returned by a similarly backward-moving source, it will arrive at its point of origin at $2|\overline{\Delta t}|$ before emission. However, there are no known tachyons, and time machines cannot be constructed with the physics and engineering possibilities provided by special relativity. Nevertheless, it should be stressed that there is no *logical* prohibition against exceeding light velocity. Indeed, a hypothetical world in which tachyons exist is physically consistent. Although tachyons have been looked for experimentally none have ever been found; see Ref. 1 for a nice review of the entire subject.

But then we come to Einstein's general relativity, within which space-time geometry is no longer fixed; indeed it can take all kinds of unexpected configurations, depending on the matter content of the universe. In particular, there can be geometries containing paths along which one can travel into the past with velocity *less* than that of light — such paths are called *closed time-like curves*. The first solution to Einstein's theory with closed time-like curves was obtained in 1949 by Gödel and it permits construction of time machines.[2]

Gödel solved the Einstein gravity equations of general relativity

$$G_{\mu\nu} + \frac{\Lambda}{2} g_{\mu\nu} = \frac{8\pi G}{c^4} T_{\mu\nu}. \tag{4}$$

The left side is geometrical; it contains the Einstein tensor, $G_{\mu\nu}$, formed from the metric tensor $g_{\mu\nu}$ in which is encoded the geometry of space-time, together with a cosmological term. The right side describes matter, $T_{\mu\nu}$ being its local energy-momentum distribution; matter determines geometry through Eq. (4). (Newton's G is the appropriate dimensional proportionality constant.)

Gödel took $T_{\mu\nu}$ to be a space-time constant, not vanishing only in its time-time component (energy density)

$$T_{00} = \frac{c^4}{8\pi G} \Lambda > 0. \tag{5}$$

The metric tensor that solves Einstein's equations leads to the space-time interval

$$ds^2 = g_{\mu\nu}dx^\mu dx^\nu = \left(c\,dt - \sqrt{\frac{2}{\Lambda}}(\cosh\sqrt{\Lambda}\,r - 1)d\theta\right)^2$$

$$- dr^2 - \frac{1}{\Lambda}\sinh^2\sqrt{\Lambda}\,r d\theta^2 - dz^2\,,$$

(6)

where r, θ are planar circular coordinates, with $\theta = 0$ and 2π identified, and there is no interesting structure in the z-direction. A curve $x^\mu(\tau)$ is closed and time-like if both $x^\mu(0) = x^\mu(1)$ (closed) and $(ds/d\tau)^2 = g_{\mu\nu}(dx^\mu/d\tau)$ $(dx^\nu/d\tau) > 0$ (time-like). It is therefore clear that a circular path in the Gödel universe for which t, r and z remain constant, while θ varies from 0 to 2π, is closed and time-like provided $\cosh\sqrt{\Lambda}\,r > 3$, *i.e.*, $r > (2/\sqrt{\Lambda})\ln(1 + \sqrt{2})$.

This result caused great puzzlement, because first of all, there is no evidence for time travel — we know of no visitors from the future — and second, our classical notions of causality (which to be sure are already challenged by quantum mechanics) prejudice us against considering geometries with closed time-like curves, where effects precede causes. Here is Einstein's reaction[3]:

"Kurt Gödel's [time machine solution raises] the problem [that] disturbed me already at the time of the building up of the general theory of relativity, without my having succeeded in clarifying it. ... It will be interesting to weigh whether these [solutions] are not to be excluded on physical grounds."

A more recent comment is by Hawking,[4]

"Gödel presented a ... solution [which] was the first to be discovered that had the curious property that in it, it was possible to travel into the past. This leads to paradoxes such as 'What happens if you go back and kill your father when he was a baby?' It is generally agreed that this cannot happen in a solution that represents our universe, but Gödel was the first to show that it was not forbidden by the Einstein equations. His solution generated a lot of discussion of the relation between general relativity and the concept of causality."

Upon further reflection it comes as no surprise that Einstein's general relativity allows closed time-like curves: in Einstein's theory geometry is determined from matter by Eq. (4) whose schematic form is

$$\left.\begin{array}{r}\text{space-time}\\\text{geometry}\end{array}\right\} \Leftarrow \left\{\begin{array}{l}\text{distribution}\\\text{of matter}\,.\end{array}\right.$$

But Einstein's equations may be read in the opposite direction: pick an interesting geometry — no matter how strange — and in particular one containing closed time-like curves, then determine the (unphysical?) matter distribution that engenders it, thereby "finding" a time machine solution, *i.e.*, redirecting the arrow above:

$$\left.\begin{array}{c}\text{time machine}\\ \text{geometry}\end{array}\right\} \Rightarrow \left\{\begin{array}{l}\text{peculiar matter}\\ \text{"unphysical?"}\end{array}\right.$$

Our "defense" against these solutions and against the paradoxes they entail is to assert that the exotic matter distributions supporting time machines are unphysical. Thus Gödel's universe requires a constant and uniform energy density, which clearly is unphysical. The time machines studied by Thorne and his colleagues[5] at CalTech make use of wormholes, another exotic and presumably unphysical form of matter in which a narrow channel — the wormhole — connects distant regions of space-time and is threaded by a closed time-like curve.[6]

The reason for current interest in time travel ideas derives from the recent realization that infinitely long and arbitrarily thin cosmic strings can support closed time-like curves. Cosmic strings are hypothetical but entirely physical structures that may have survived from a cosmic phase transition, and that may even be responsible for the present-day large scale structure in the universe.[8] Only completely conventional physical ideas are relied upon in cosmic string speculations; cosmologists make use of cosmic strings in model building, and astrophysicists occasionally report sightings, although thus far evidence is inconclusive. Nevertheless, although infinite length and arbitrary thinness are idealizations, one would not view such cosmic strings as unphysical. Therefore, if they indeed give rise to closed time-like curves, there is something new that physicists must confront.

To recognize string-generated closed time-like curves, we need to record the space-time that is produced by a cosmic string, We use coordinates in which an infinitely long and thin cosmic string lies along the z-axis through the origin, Its mass per unit length is m and we also endow the string with intrinsic spin J per unit length. The cross-sectional area is assumed to be vanishingly small, so the mass density is proportional to a two-dimensional spatial δ-function, while the spin density is more singular, being produced by a momentum density proportion to derivatives of δ.

Solving Einstein's equation (4) (with vanishing cosmological constant Λ) for a stationary string, carrying the above mass and spin distributions, produces a metric described by the line element

$$ds^2 = g_{\mu\nu}dx^\mu dx^\nu = \left(c\,dt + \frac{4}{c^3}GJ\,d\theta\right)^2 - dr^2$$
$$- \left(1 - \frac{4}{c^2}Gm\right)^2 r^2 d\theta^2 - dz^2\,. \tag{7}$$

Indeed, this two-parameter (m, J) metric tensor is the general time-independent solution to (4) outside any matter distribution lying in a bounded region on the plane and having cylindrical symmetry. As in the Gödel line element (6), there is no structure in the z-direction, while in the perpendicular (r, θ) plane, with $\theta = 0$ and 2π identified, the non-trivial geometry supports closed time-like curves when J is non-vanishing: take constant t and r and describe a circle (θ ranging from 0 to 2π) with sufficiently small radius

$$r < \frac{4GJ}{c^3 - 4Gmc}\,. \tag{8}$$

By changing coordinates one can hide the presence of the string and make the line element appear locally Minkowskian: with the definitions

$$\tau = t + \frac{4}{c^4}GJ\theta,$$
$$\varphi = \left(1 - \frac{4}{c^2}Gm\right)\theta\,, \tag{9}$$

(7) becomes

$$ds^2 = (c\,d\tau)^2 - dr^2 - r^2 d\varphi^2 - dz^2\,. \tag{10}$$

Now, however, the ranges of these flat-looking coordinates are unconventional: that of φ is diminished to $(1 - 4Gm/c^2)2\pi$, while the new time variable τ, rather than flowing in a smooth and linear fashion jumps by $8\pi GJ/c^4$ whenever the string is circumnavigated, owing to the identification of $\theta = 0$ and 2π. The defect in the angular range turns the spatial plane into the surface of a cone with deficit angle determined by m, while the time-helical structure, where pitch is proportional to intrinsic spin J, is responsible for the closed time-like curves (8).

There is another entirely equivalent framework for understanding the geometry produced by our cosmic string: In the idealization of infinitely long and thin strings, there is no structure along the z-axis; hence we can suppress that direction altogether. Then the theory becomes (2+1)-dimensional "planar" gravity, governed by the same Einstein equations (4) but in this reduced space.

Cosmic string sources now become "point-particles" at the locations where the strings pierce the $z = 0$ plane.

Einstein's gravity theory in (2+1) dimensions is a much-studied model,[9] both for pedagogical reasons — one hopes that the dimensional reduction effects sufficient simplification to permit thorough analysis, while still retaining useful content to inform the physical (3+1)-dimensional problem — and also for practical calculation — as explained above, idealized cosmic strings are effectively described by (2+1)-dimensional gravity. (Analogously, motion of charged particles in magnetic fields that are constant along the z-axis is effectively governed by planar dynamics, as in Landau theory, quantum Hall effect and perhaps high-T_c superconductivity.) We shall use the name *cosmon* to refer uniformly to infinite cosmic strings in space and to particles on the plane.

A simplifying feature of the lower-dimensional interpretation (always without cosmological constant, although it can be included) is that space-time is locally flat whenever sources ($T_{\mu\nu}$) are absent. The reason for this is that the Riemann curvature tensor, which completely determines geometry, is linearly related to the Einstein tensor $G_{\mu\nu}$ in (and only in) three-dimensions (they are each other's double duals). Thus at any point where there are no sources, $G_{\mu\nu}$ vanishes by (4), and consequently, the three-dimensional Riemann tensor also vanishes: space-time is Minkowskian. Moreover, when sources are point-cosmons, the space-time is flat everywhere, except at these points, and this is why it is possible to transform the line element (7) to the flat one (10) almost everywhere; only the unconventional range and jump properties of the coordinates remind us that there are sources somewhere. Indeed, these coordinate defects contain all the physical information about sources that is accessible in regions outside them.[10]

Further examination of the geometry generated by cosmons shows that the mass of any individual one (= mass per unit length of a cosmic string) must not exceed $c^2/4G$, so that the deficit angle not exceed 2π ($m = c^2/4G$ corresponds to the cone closing into a cylinder). The *total* mass of an *assembly* of static cosmons, each with acceptable mass, may exceed $c^2/4G$, but then it must precisely equal $c^2/2G$ for space to be non-singular, and space is necessarily closed.[10] [This is a consequence of the fact that the total mass of static sources in (2+1)-dimensional gravity is proportional to the Euler number of the 2-space.] Note in particular: sources, which individually, *i.e.*, *locally*, are acceptable ($m < c^2/4G$) can give rise to configurations that are *globally* unacceptable ($c^2/4G < \Sigma_m m \neq c^2/2G$); this lesson will be essential in what follows.

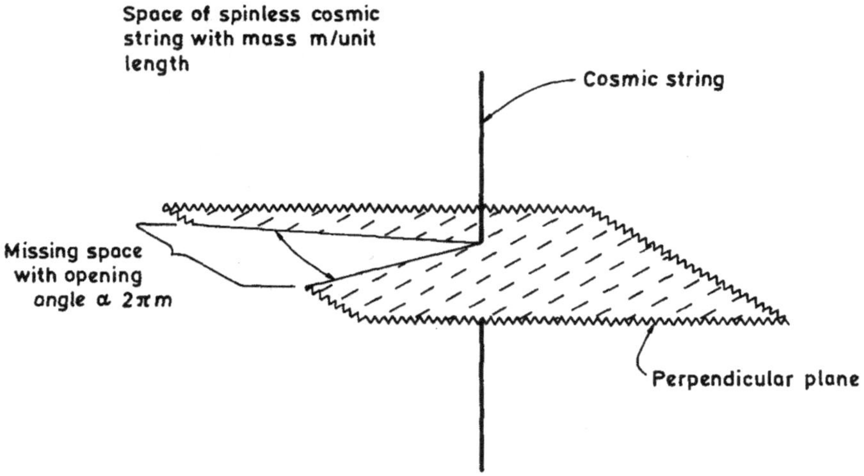

Space of spinless cosmic
string with mass m/unit
length

Cosmic string

Missing space
with opening
angle α 2π m

Perpendicular plane

Fig. 1. Space of a spinless cosmon.

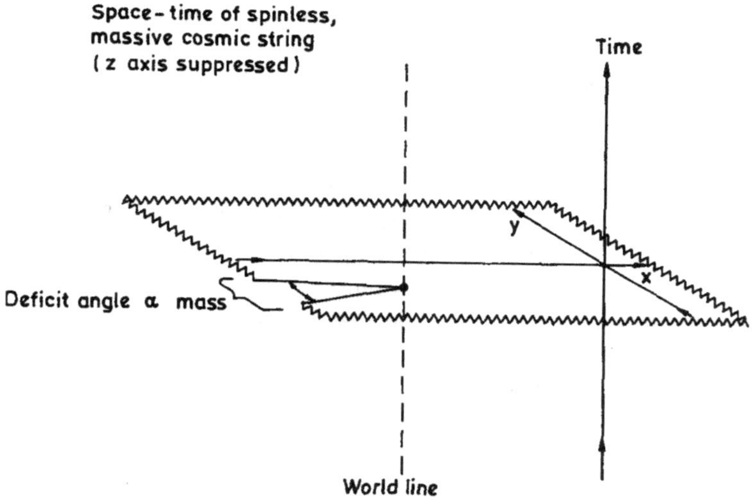

Space-time of spinless,
massive cosmic string
(z axis suppressed)

Time

y

x

Deficit angle α mass

World line

Fig. 2. Space-time of a spinless cosmon.

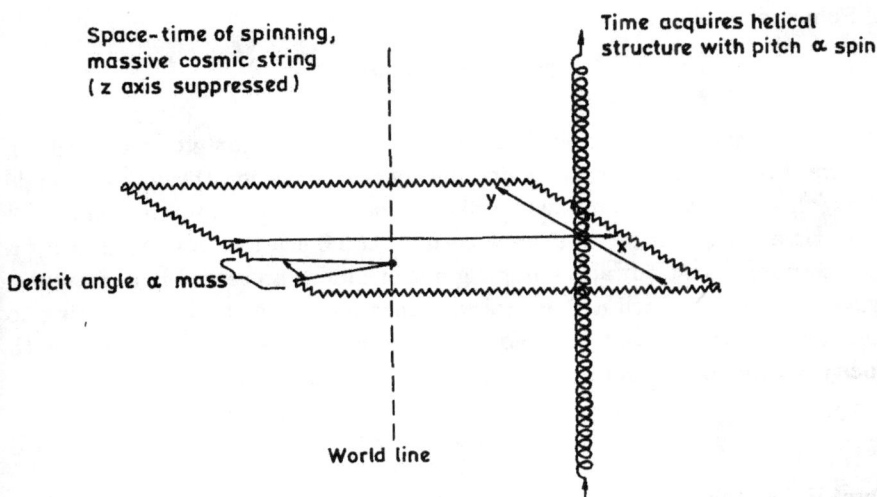

Fig. 3. Space-time of a spinless cosmon.

In Figs. 1–3 we depict the space and the space-time of a single cosmon using flat coordinates with unconventional ranges. In Fig. 1 the cosmon is spinless; the three-dimensional space is structureless along the z-direction (direction of the string); in the perpendicular plane an angular region, of magnitude proportional to $8\pi Gm/c^2$, is excised and the edges identified — the space is a cone. In Fig. 2, the z-axis is suppressed, the cosmon is a point-particle and the time axis as well as the world line of the stationary cosmon are indicated; since there is no spin, time flows smoothly from $-\infty$ to ∞. Figure 3 depicts the previous situation, but now the cosmon also carries spin, and time acquires a helical structure, as explained above. For more cosmons, the space is an assembly of cones, joined smoothly, as is possible provided each of their deficit angles and the sum are suitably bounded by the limits given above.

Another useful way of describing the space-time created by a cosmon is to notice that after the cosmon is circumnavigated, the flat space-time description requires an additional rotation in space and a jump in time, which effect the coordinate identifications that encode the presence of a massive, spinning cosmon. These identifications may be represented in terms of a Poincaré transformation (spatial rotation or Lorentz boost and space-time translation) on the three-dimensional space-time (since the invariance group of flat space-time is the Poincaré group),

$$\bar{x}^\mu = M^\mu_\nu x^\nu + a^\mu \,. \tag{11}$$

Here $\bar{x}$ and x are the identified three-vectors, M^μ_ν is a Lorentz transformation, specialized for the single static source to a spatial rotation through the angle $8\pi Gm/c^2$, and a^μ is the jump in coordinates, specialized to a jump of $8\pi GJ/c^4$ in the time component. A general cosmon configuration, not necessarily a single cosmon at rest but also a moving one or indeed an assembly of cosmons, is also described by such a Poincaré identification, with x and $\bar{x}$ referring to points exterior to the assembly. For example, a spinless cosmon moving with velocity $\mathbf{v}$ through the point $\mathbf{x}_0$ leads to the identification

$$\bar{x} = B_{\mathbf{v}}^{-1} M_m B_{\mathbf{v}} (x - x_0) + x_0 \,, \tag{12}$$

where $B_{\mathbf{v}}$ is the Lorentz boost that brings the cosmon to rest and M_m is the mass-determined rotation. Equation (12) is merely the statement that the coordinate $B_{\mathbf{v}}(x - x_0)$ is identified as in (11), with a vanishing when there is no spin.

As already remarked, owing to the time-helical structure that is required by the Einstein equation, a spinning $(J \neq 0)$ cosmon gives rise to closed time-like curves. Now *intrinsic* spin attached to a cosmic string may still be deemed unphysical (like any classical spin involving derivatives of δ-functions) so we need not worry about closed time-like curves supported by locally spinning cosmons. However, one may ask whether *two* or more spinless cosmons, moving relative to each other and thus also carrying *orbital* angular momentum, can still support closed time-like curves. This question was answered affirmatively by Gott at Princeton University.[11] He found that two spinless cosmons, each moving faster than a critical but subluminal $(v < c)$ velocity, do indeed support closed time-like curves.

In fact, the question of possible closed time-like curves in a many-spinless cosmon universe had been posed even earlier and answered negatively in the original investigation that first exhibited closed time-like curves in the presence of spinning cosmons.[10] Briefly, the reasoning there relied on the expectation that the perfectly good Cauchy development of two freely moving cosmons from $t = -\infty$ should not be spontaneously destroyed at some finite time, when they approach each other, and then be reinstated when they diverge as $t \to +\infty$. Resolution of the apparent contradiction between this expectation and Gott's result is the most recent development in the time machine story.

The disposition of cosmons that support closed time-like curves is portrayed in Fig. 4. Two spinless cosmons, each with equal (for simplicity) masses m are located at the center of the figure, and the excised wedges point up and down. Consider now the light path from A to B. It is credible that a path surrounding the cosmons and taking advantage of the "missing" space can be shorter than the direct, straight line AB. Then cosmons moving in opposite directions with speed v, in a manner already described above in connection with the tachyonic time machine, create closed time-like curves, provided v exceeds a critical value $v_{critical}$, which nevertheless is less than c, so that neither cosmon is tachyonic. Gott shows that

$$\frac{v}{c} > \frac{v_{critical}}{c} = \cos\left(\frac{4\pi Gm}{c^2}\right) < 1. \tag{13}$$

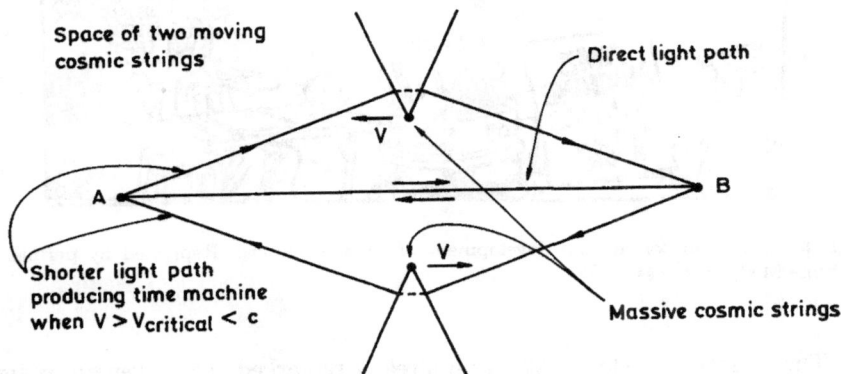

Fig. 4. Two-cosmon time machine.

This result, in what purports to be a physically acceptable situation, produced intense interest (not only in the physics community, as is seen from the list of semi-popular accounts[12]) since for the first time time-travel appeared to be consistent with unexceptional physical principles, awaiting only development of engineering possibilities. Fantastic applications obviously come to mind and were discussed, while at the same time, some pitfalls inherent in commercial development were noted in a nationally syndicated newspaper feature, published the same day as the analysis in Ref. 13 and reproduced in Fig. 5.

But there is a catch: since one needs a *pair* of moving strings, one may ask what is their combined energy and momentum. Now energy and momentum are conserved quantities that arise from invariance under time- and space-translation. However, when space-time is as globally complicated — "sewn-together" cones — as it becomes in the presence of strings, the addition rules for combining energy and momentum become non-trivial and non-linear. In particular, even though each of the two cosmons is moving sufficiently slowly to be non-tachyonic, as soon as the velocity of each exceeds the critical value (13) needed to support a time machine, their center-of-mass becomes tachyonic.

Fig. 5. Pitfall in commercial development of time machines. Reprinted by permission: Tribune Media Services.

This is seen as follows. We have already remarked that a general cosmon distribution is characterized by an element of the Poincaré group, as in (11). Concentrating now on the (2+1)-dimensional Lorentz transformation matrix M, we recall that it may be presented as the exponential of three generator matrices Σ_μ,

$$M = \exp\left\{ \frac{8\pi}{c^3} G P^\mu \Sigma_\mu \right\}. \tag{14}$$

(This is analogous to presenting a three-dimensional spatial rotation matrix as the exponential of angular momentum matrices.) The three-vector P^μ defines the energy-momentum of a cosmon assembly. For example, when a single cosmon is at rest, M is a pure rotation, which corresponds to

$$\mathbf{P} = 0, \qquad p^0 = mc, \qquad \Sigma_0 \equiv \begin{pmatrix} 0 & 0 & 0 \\ 0 & 0 & 1 \\ 0 & -1 & 0 \end{pmatrix}, \tag{15}$$

where the rows and columns are labelled by (t, x, y). For the cosmon assembly to be non-tachyonic, *i.e.*, physically constructible, we must require

$$P^\mu P_\mu \equiv m^2 c^4 > 0 \,. \tag{16}$$

Now Gott's time machine makes use of two equal-mass (m) cosmons, moving in opposite directions with speed v. Suitably extending (12), we see that the resulting Lorentz identification that describes it is

$$M_{\text{total}} = B_{\mathbf{v}}^{-1} M_m B_{\mathbf{v}} B_{-\mathbf{v}}^{-1} M_m B_{-\mathbf{v}} = B_{\mathbf{v}}^{-1} M_m B_{\mathbf{v}}^2 M_m B_{\mathbf{v}}^{-1} \tag{17a}$$

with total energy and momentum defined by

$$M_{\text{total}} = \exp\left\{ \frac{8\pi G}{c^3} P_{\text{total}}^\mu \Sigma_\mu \right\} \,. \tag{17b}$$

A straightforward calculation yields the formula for the total mass, m_{total},

$$\sin\left(\frac{2\pi G m_{\text{total}}}{c^2} \right) = \frac{1}{\sqrt{1 - v^2/c^2}} \sin\left(\frac{4\pi G m}{c^2} \right), \tag{18}$$

which for small Gm/c^2 recovers the special relativistic result

$$m_{\text{total}} = \frac{2m}{\sqrt{1 - v^2/c^2}} \,.$$

For real m_{total}, the left side of (18) is less than unity, so we must have

$$\frac{v}{c} < \cos\left(\frac{4\pi G m}{c^2} \right) \,. \tag{19}$$

This is precisely *opposite* to Gott's criterion (13) for the presence of closed time-like curves! Thus we learn that when (13) is satisfied, *each* cosmon travels with a velocity below that of light, but the *total* system possesses a tachyonic energy. It follows that only tachyons can give rise to Gott's time machine produced by sufficiently rapidly moving pair of cosmons.[13] In the absence of such tachyonic sources, energy is unavailable to produce the rapidly moving cosmons. Another perspective is given by formula (18). We see that when $v = 0$, $m_{\text{total}} = 2m$. As v increases from zero, $4G m_{\text{total}}/c^2$ increases and approaches unity when v reaches v_{critical}, corresponding to a deficit angle 2π in the surrounding space. As the critical velocity is exceeded, the deficit angle surpasses 2π, and the system becomes unphysical.

Thus the two opposite claims[10,11] about time machines are reconciled[13]: just like the special relativistic time machines, which could only be constructed if tachyons exist — but they do not — so also the cosmic string time machines require tachyonic center-of-mass velocities, and cannot be produced in the absence of tachyons, that is in our world.

Even without mentioning the tachyonic center-of-mass, we can give an equivalent geometrical description of the unphysical nature of Gott's system: Its exterior geometry necessarily contains closed time-like curves at spatial infinity. Existence of a time machine during that portion of the system's evolution when the cosmons are sufficiently close requires that closed time-like curves also be present at the boundary of space for *all* time. Space-times with this property have long been known in classical gravity, where they are produced by unusual identifications of space and time. An example is Misner space,[14] and one can show that the coordinate identifications by means of the Lorentz transformation (17a) when $v > v_{critical}$ give rise to such a space. Moreover, the *effective* energy-momentum tensor responsible for the exterior Gott metric is tachyonic, with energy density replaced by spatial stress.

The above analysis was confirmed in further investigations on the obstacles to constructing Gott's time machine. Carroll, Farhi and Guth[15] showed that it cannot be manufactured as a subsystem of a normal time-like matter assembly in an *open* universe. In a universe that is spatially *closed*, owing to a mass distribution with critical magnitude for closing space ($\Sigma_m m = c^2/2G$), a pair of cosmons traveling with velocity exceeding $v_{critical}$ can be produced[15]; nevertheless, a time machine cannot be built here either because the universe dramatically self-destructs just before the cosmons get close enough to engender a closed time-like curve — there is not enough time to travel into the past.[16]

While we are reassured, at least as concerns this most recent string-inspired attempt, that causality will not be violated by time travel, further questions remain and motivate further study. One wonders whether one can *a priori* and once-and-for-all characterize those mass distributions that *mathematically* support time travel and therefore should perhaps be deemed unphysical. Indeed, Hawking[7] has encapsulated these ideas in his "Chronology Protection Conjecture" that "the laws of physics do not allow the appearance of closed time-like curves." But establishing this conjecture requires knowing why the laws of physics prohibit certain forms of matter. That this question is difficult to answer is well-illustrated by the cosmon time machine: each individual cosmon is a perfectly acceptable subluminal and physical form of matter —

but fitting exactly two rapidly moving cosmons into a non-singular universe is impossible. Evidently global rather than local considerations come into the definition of "physical."

However, it should be remarked that as paradoxical as time-travel may appear ("killing one's infant father," *etc.*) we can be assured that mathematical analysis, correctly carried out, will provide a consistent physical picture — which of course need not correspond to reality if conditions for time-travel cannot be physically met. This had already been established for tachyonic effects within special relatively.[1] Recent investigations of physical processes in the presence of closed time-like curves, like those generated by spinning strings[17] or by other mechanisms,[18] confirm that unexpected but not self-contradictory results are found, *eg.*, perturbative violation of unitarity in interacting quantum field theories.

We study paradoxical aspects of physics — like time machines — to illuminate our understanding of fundamental issues. For example, the very concept of time remains mysterious in physical theory: What gives time its observed direction? Indeed what gives it a definition, in the first place, since general relativity requires space-time to be determined by matter? It should be apparent from the foregoing that here, there is still much to learn.

REFERENCES

1. G. Feinberg, *Scientific American* **222**(2), 68 (1970); see also O.-M. Bilaniuk and E. Sudarshan, *Phys. Today* **43** (May 1969).

2. K. Gödel, *Rev. Mod. Phys.* **21**, 447 (1949).

3. A. Einstein, in *Albert Einstein: Philosopher-Scientist*, P. Schilpp, ed. (Tudor, New York, NY, 1957).

4. S. Hawking, in *K. Gödel, Collected Works*, Vol. II, S. Feferman, ed. (Oxford University Press, New York, NY, 1990).

5. M. Morris and K. Thorne, *Am. J. Phys.* **56**, 395 (1988).

6. Specific, technical objections can also be raised to wormhole-supported time machines. The narrowness of the channel would prevent anything of significant size from squeezing through its opening. Moreover, Hawking[7] has argued that quantum fluctuations around a wormhole generate infinite energy and stress, thereby closing it, but the subject is still being investigated.

7. S. Hawking, in *The Sixth Marcel Grossmann Meeting*, H. Sato and T. Nakamura, eds. (World Scientific, Singapore, 1992).

8. For a review, see A. Vilenkin, *Phys. Rep.* **121**, 263 (1985).

9. For reviews, in *The Sixth Marcel Grossmann Meeting,* H. Sato and T. Nakamura, eds. (World Scientific, Singapore, 1992) and "Planar Gravity," reprinted in this volume on p. 155.

10. S. Deser, R. Jackiw and G. 't Hooft, *Ann. Phys.* (NY) **152**, 220 (1984).

11. J. Gott, *Phys. Rev. Lett.* **66**, 1126 (1991).

12. *Time*, 13 May 1991; *Science News*, 28 March 1992; *New Scientist*, 28 March 1992; *Discover*, April 1992; *Science*, 10 April 1992; *Nature*, 7 May 1992, *Scientific American*, February 1994. Only the *Nature*, *New Scientist* and *Science* articles convey accurate information.

13. S. Deser, R. Jackiw and G. 't Hooft, *Phys. Rev. Lett.* **68**, 267 (1992).

14. C. W. Misner, in *Relativity Theory and Astrophysics I: Relativity and Cosmology*, J. Ehlers, ed. (American Mathematical Society, Providence, RI, 1967).

15. S. Carroll, E. Farhi and A. Guth, *Phys. Rev. Lett.* **68**, 236, (E) 3368 (1992); see also D. Kabat, *Phys. Rev. D* **46**, 2720 (1992). The claim in the first of these papers that time machines can be constructed in closed universes was shown to be false in Ref. 16.

16. G.'t Hooft, *Class. Quant. Grav.* **9**, 1335 (1992).

17. P. Gerbert and R. Jackiw, *Comm. Math. Phys.* **124**, 495 (1988).

18. J. Friedman, M. Morris, I. Novikov, F. Echeverria, G. Klinkhammer, K. Thorne and U. Yurtsever, *Phys. Rev. D* **42**, 1915 (1990); D. Deutsch, *Phys. Rev. D* **44**, 3197 (1991); D. Boulware, *Phys. Rev. D* **46**, 4421 (1992); J. Friedman, N. Papastamatiou and J. Simon, *Phys. Rev. D* **46**, 4456 (1992); J. Hartle, *Phys. Rev. D* **49**, 6543 (1994).

Paper II.6

In Memoriam A. Polivanov, who was interested in lineal theories, I discuss how even gravity on a line can be interesting and realizes yet another example of a gauge theory that may provide research material for the nineteen-nineties.

GAUGE THEORIES FOR GRAVITY ON A LINE

Teor. Mat. Fiz. **92**, *404 (1992) [English Translation:*
Theor. Math. Phys. **92**, *979 (1992)]*

1. INTRODUCTION

Professor M. C. Polivanov and I met only a few times, during my infrequent visits to the-then Soviet Union in the 1970's and 1980's. His hospitality at the Moscow Steclov Institute made the trips a pleasure, while the scientific environment that he provided made them professionally valuable. But it is the human contact that I remember most vividly and shall now miss after his death. At a time when issues of conscience were both pressing for attention and difficult/dangerous to confront, Professor Polivanov made a deep impression with his quiet but adamant commitment to justice. I can only guess at the satisfaction he must have felt when his goal of gaining freedom for Yuri Orlov was attained, and even more so these days when human rights became defensible in his country; it is regrettable that he cannot now enjoy the future that he strived to attain.

One of our joint interests was the Liouville theory,[1,2] which in turn can be viewed as a model for gravity in two-dimensional space-time. Some recent developments in this field are here summarized and dedicated to Polivanov's memory, with the hope that he would have enjoyed knowing about them.

We study lower-dimensional gravity both for pedagogical reasons — one expects that the dimensional reduction effects sufficient simplification to permit thorough analysis, while still retaining useful content to inform the physical (3+1)-dimensional problem — and also, if one is lucky, there are practical applications — *eg.*, idealized cosmic strings are described by (2+1)-dimensional gravity, while the still lower-dimensional models are used in statistical mechanics.

The drastic dimensional reduction to (1+1) dimensions — gravity on a line, *i.e.*, *lineal* gravity — is not devoid of interest, provided dynamical equations are not based on the Einstein tensor $G_{\mu\nu} = R_{\mu\nu} - \frac{1}{2}g_{\mu\nu}R$, which vanishes identically in two dimensions.

In a proposal of several years ago,[3] it was suggested that gravity equations be based on the Riemann scalar R, the simplest entity that encodes in two dimensions all local geometric information about space-time. Moreover, in an action formulation it is necessary to introduce an additional scalar field, which acts as a Lagrange multiplier that enforces the equation of motion for R. Thus we are dealing with scalar-tensor theories, or — to use the contemporary string nomenclature — "dilaton" gravities.

Since the initial proposal, various models have been studied. Here I shall describe two that are selected by their group theoretical properties: they can be formulated as gauge theories based on groups relevant to space-time: de Sitter or anti-de Sitter [in (1+1)-dimensions both groups are $SO(2,1)$, although the geometries are different] and Poincaré. The first of these is the one proposed originally,[3] it is governed by the action

$$I_1 = \int d^2x \sqrt{-g}\, \eta (R - \Lambda).$$ (1)

The second is "string-inspired" and has been recently studied for purposes of modeling (on a line!) black hole physics[4]; its action is

$$\bar{I}_2 = \int d^2x \sqrt{-\bar{g}}\, e^{-2\varphi}(\bar{R} + 4\bar{g}^{\mu\nu}\partial_\mu\varphi\partial_\nu\varphi - \Lambda).$$ (2̄)

[Notation: time and space carry the metric tensor $\bar{g}_{\mu\nu}$ with signature $(1, -1)$. The two-vector $x^\mu = (t, x)$ will be frequently presented in light-cone components $x^\pm \equiv \frac{1}{\sqrt{2}}(t \pm x)$. Tangent space components are labeled by Latin letters a, b, ..., and the Minkowski metric tensor $h_{ab} = \mathrm{diag}\,(1, -1)$ raises/lowers these indices. Also we use the antisymmetric tensor $\varepsilon^{ab}, \varepsilon^{01} = 1$.]

In (1), R is the scalar curvature built from $g_{\mu\nu}$, η is a world scalar Lagrange multiplier related to the dilaton, while Λ is a cosmological constant. In (2̄) we temporarily use an overbar to denote a differently scaled metric tensor $\bar{g}_{\mu\nu}$ from which $\bar{R}$ is constructed, while φ is the dilaton. Formula (2̄) arises naturally from string theory, restricted to a two-dimensional target space, with the antisymmetric tensor field identically vanishing. In the string context, matter is taken to couple to $\bar{g}_{\mu\nu}$; for our purposes in the absence of matter it is convenient to redefine variables by $\bar{g}_{\mu\nu} = e^{2\varphi}g_{\mu\nu}$, $\eta = e^{-2\varphi}$. Then (2̄) becomes

$$I_2 = \int d^2x \sqrt{-g}\,(\eta R - \Lambda)$$ (2)

but it is to be remembered that because of the redefinition, the "physical" metric tensor is $g_{\mu\nu}/\eta$. Note that (2) is invariant against shifting η by a constant, because $\sqrt{-g}R$ is a total derivative.

It is seen that the two models (1) and (2) differ in the placement of the Lagrange multiplier with the cosmological term: in (1) η multiplies Λ, in (2) the η factor is absent from Λ. Of course in the limit $\Lambda = 0$, the difference disappears.

We now describe the interesting gauge group structure of (1) and (2) which we name *(anti) de Sitter gravity* and *extended Poincaré gravity*, respectively.

2. (ANTI) DE SITTER GRAVITY

The equations of motion that follow from varying η and $g_{\mu\nu}$ in (1) are

$$R = \Lambda, \tag{3}$$

$$(\mathcal{D}_\mu \mathcal{D}_\nu - g_{\mu\nu} \mathcal{D}^2)\eta - \frac{\Lambda}{2} g_{\mu\nu}\eta = 0. \tag{4a}$$

The second equation, with $\mathcal{D}_\mu$ the space-time covariant derivative, can be decomposed into traceless and trace parts

$$\left(\mathcal{D}_\mu \mathcal{D}_\nu - \frac{1}{2} g_{\mu\nu} \mathcal{D}^2\right)\eta = 0, \tag{4b}$$

$$(\mathcal{D}^2 + \Lambda)\eta = 0. \tag{4c}$$

The above geometric dynamics may be presented in a gauge theoretical fashion.[5] To this end, one uses the (anti) de Sitter group with Lorentz generator J and translation generators P_a satisfying the $SO(2,1)$ algebra (for $\Lambda \neq 0$)

$$[P_a, J] = \varepsilon_a{}^b P_b, \qquad [P_a, P_b] = \frac{\Lambda}{2} \varepsilon_{ab} J. \tag{5}$$

The gauge connection one-form is introduced $A = A_\mu dx^\mu$ and expanded in terms of the generators,

$$A = e^a P_a + \omega J, \tag{6}$$

where e^a_μ is the *Zweibein* and ω_μ is the spin-connection. The curvature two-form

$$F = dA + A^2 \tag{7}$$

becomes

$$F = f^a P_a + f J = (De)^a P_a + \left(d\omega + \frac{\Lambda}{4} e^a \varepsilon_{ab} e^b\right) J, \tag{8}$$

$$(De)^a \equiv de^a + \varepsilon^a{}_b \omega e^b. \tag{9}$$

It is seen that $d\omega$ is proportional to the scalar curvature density and $f^a = (De)^a$ is proportional to the torsion density, each expressed in terms of e^a and ω, which at this stage are independent variables.

The Lagrange density

$$\mathcal{L}'_1 = \sum_{A=0}^{2} \eta_A F^A = \eta_a (De)^a + \eta_2 \left(d\omega + \frac{\Lambda}{4} e^a \varepsilon_{ab} e^b\right)$$

$$F^A = (f^a, f), \qquad \eta_A = (\eta_a, \eta_2) \tag{10}$$

is gauge invariant: the three field strengths F^A transform covariantly according to the three-dimensional adjoint representation, while the Lagrangian multiplier triplet η_A transforms by the coadjoint representation.

The equation obtained from (10) by varying η_a gives the condition of vanishing torsion, and allows evaluating the spin connection in terms of the *Zweibein*

$$\omega = e^a(h_{ab}\varepsilon^{\mu\nu}\partial_\mu e^b_\nu)/\det e.\tag{11}$$

The equation which follows upon variation of η_2 regains (3) once (11) is used. Variation of e^a and ω produces equations for the Lagrange multipliers η_a and η_2, respectively, the latter coinciding with 2η in the geometric formulations (1), (3) and (4)

$$d\eta_a + \varepsilon_a{}^b\omega\eta_b + \frac{\Lambda}{2}\varepsilon_{ab}\eta_2 e^b = 0,\tag{12a}$$

$$d\eta_2 + \eta_a\varepsilon^a{}_b e^b = 0.\tag{12b}$$

Upon taking a space-time covariant derivative of (12b) and using (12a) to eliminate η_a, we recover (4). Finally, we see that when ω is eliminated from $\mathcal{L}'_1$ with the help of (11), so that the torsion (9) vanishes, what remains is the Lagrange density of (1), expressed in terms of *Zweibeine*.

Thus the geometric formulation of this gravity theory is contained within the (anti) de Sitter group theoretical framework for solutions with $\det e \neq 0$, but see below.

Explicit classical solutions to the equations are easy to find. Working within the geometric framework, we use coordinate invariance to choose a conformally flat metric tensor

$$g_{\mu\nu} = h_{\mu\nu}\exp 2\sigma.\tag{13}$$

Then (3) becomes the Liouville equation,

$$\Box\sigma = -\frac{\Lambda}{2}\exp 2\sigma\tag{14}$$

studied by Polivanov.[1] Its general solution depends on two arbitrary functions of the two light-cone variables, $F(x^+)$, $G(x^-)$,

$$\exp 2\sigma = \frac{F'(x^+)G'(x^-)}{\left(1 + \dfrac{\Lambda}{4}FG\right)^2}\tag{15}$$

whose derivatives fulfill the consistency condition $F'G' > 0$. But the residual coordinate invariance within the conformal gauge allows choosing $F(x^+) = x^+, G(x^-) = x^-$, hence

$$\exp 2\sigma = \frac{1}{\left(1 + \frac{\Lambda}{8}x^2\right)^2} \cdot \tag{16}$$

In conformal gauge, (4b) reduces to

$$\partial_\mu V_\nu + \partial_\nu V_\mu - h_{\mu\nu}h^{\alpha\beta}\partial_\alpha V_\beta = 0, \tag{17}$$

where V_μ is defined by

$$V_\mu \exp 2\sigma = \partial_\mu \eta. \tag{18}$$

Equation (17) is just the (flat-space) conformal Killing equation with solutions in terms of arbitrary functions of a single light-cone variable

$$V_- = V_-(x^+), \qquad V_+ = V_+(x^-). \tag{19}$$

Finally, the remaining equation (4c) together with (18) restricts these functions, so that the solution for η takes the form

$$\eta = \frac{\alpha_a x^a + \alpha_2\left(1 - \frac{\Lambda}{8}x^2\right)}{1 + \frac{\Lambda}{8}x^2}, \tag{20}$$

where α_a is a constant two-vector and α_2 is a constant scalar.

The *Zweibein* and spin connection of the gauge theoretical formulation are given by related formulas. The former, the "square root" of the metric tensor, becomes (apart from an arbitrary Lorentz transformation on the tangent-space indices),

$$e_\mu^a = \delta_\mu^a \exp \sigma = \frac{1}{1 + \frac{\Lambda}{8}x^2}\delta_\mu^a \tag{21}$$

while the latter is

$$\omega_\mu = -h_{\mu\alpha}\varepsilon^{\alpha\beta}\partial_\beta\sigma. \tag{22}$$

The Lagrange multiplier η_2 coincides with 2η, while Eq. (12) for η_a is solved by

$$\eta_a \exp \sigma = 2\varepsilon_a{}^\mu \partial_\mu \eta. \tag{23}$$

Of course, the general solution is an arbitrary coordinate transformation of the above.

Finally, we observe that the gauge theoretical formulation allows an alternative group theoretical presentation of solutions. The field equations following from (10), upon respective variation of η_A and A, are

$$F = 0\,,\tag{24}$$

$$dH + [A,\ H] = 0\,.\tag{25}$$

A, F and $H = \eta_a h^{ab} P_b - \frac{\Lambda}{2}\eta_2 J$ belong to the $SO(2,1)$ algebra (the factor $\Lambda/2$ is a consequence of the group metric). Equation (24) implies that A is a pure gauge given by an arbitrary element U of the $SO(2,1)$ group,

$$A = U^{-1}dU\tag{26}$$

while the Lagrange multiplier is then determined by (25) to be

$$H = U^{-1}\Phi U\,,\tag{27}$$

where Φ is a constant element in the algebra. The explicit group and algebra elements that correspond to the above solution, Eqs. (20)–(23), are

$$U = \exp\left(\frac{\tanh^{-1}\sqrt{-\dfrac{\Lambda}{8}x^2}}{\sqrt{-\dfrac{\Lambda}{8}x^2}}\right) x^a P_a\tag{28}$$

and

$$\Phi = -2\alpha_a \varepsilon^{ab} P_b - \Lambda\alpha_2 J\,.\tag{29}$$

U is unique up to a constant gauge transformation. Performing such a gauge transform allows setting two of the three constants (α_a, α_2) to zero, so that the invariant dependence is on a single quantity, which we can choose as $-4(\alpha^a\alpha_a + \frac{\Lambda}{2}\alpha_2\alpha_2) = \eta^a\eta_a - \frac{\Lambda}{2}\eta_2\eta_2$.

Within the gauge theoretical framework, an even simpler solution to (24) and (25) is available: $A = 0$, $H = \Phi$, which makes no sense geometrically: not only $\det e$, but both the connections e^a and ω vanish! But in fact, use can be made of such solutions: when presented with a geometrically singular configuration, perform any gauge transformation producing non-singular connections, for example with the group element U above. So we see that the group theoretical framework, even in its $\det e = 0$ sector, contains adequate information for encoding the gravity theory.

3. EXTENDED POINCARÉ GRAVITY

Equations of motion of the string-inspired gravitational theory (2) are, from varying η

$$R = 0 \tag{30}$$

and from varying $g_{\mu\nu}$

$$(\mathcal{D}_\mu \mathcal{D}_\nu - g_{\mu\nu}\mathcal{D}^2)\eta - \frac{\Lambda}{2}g_{\mu\nu} = 0 \tag{31a}$$

which is equivalent to

$$\mathcal{D}_\mu \mathcal{D}_\nu \eta = -\frac{\Lambda}{2}g_{\mu\nu} \, . \tag{31b}$$

Note that (31a) differs from (4a) by the absence of η in the last term.

To give a gauge theoretical formulation,[6] we make use of the *centrally extended* Poincaré group, whose algebra is

$$[P_a, J] = \varepsilon_a{}^b P_b, \qquad [P_a, P_b] = \varepsilon_{ab}I \, , \tag{32}$$

where the central element I commutes with P_a and J. Consequently, the connection A and curvature F now become

$$A = e^a P_a + \omega J + aI \, , \tag{33}$$

$$F = dA + A^2 = f^a P_a + fJ + gI$$

$$= (De)^a P_a + d\omega J + \left(da + \frac{1}{2}e^a \varepsilon_{ab}e^b\right)I \, . \tag{34}$$

Here a and g are the additional connection and curvature associated with the central element in the algebra.

This magnetic-like extension of the Poincaré group may be viewed as an unconventional contraction of the de Sitter group: The ordinary Poincaré algebra [Eq. (32) without the central element] is the $\Lambda \to 0$ contraction of the $SO(2,1)$ algebra (5). However, owing to the well-known ambiguity of two-dimensional angular momentum, in (5) one may replace J by $J - 2/\Lambda$ before taking the $\Lambda \to 0$ limit, which then leaves (32).

The extension reflects a 2-cocycle in the composition law for representatives of the Poincaré group. If the group acts on coordinates x^a by

$$x^a \longrightarrow \bar{x}^a = \mathcal{M}^a{}_b x^b + q^a, \tag{35a}$$

where $\mathcal{M}$ is a finite Lorentz transformation

$$\mathcal{M}^a{}_b = \delta^a{}_b \cosh\alpha + \varepsilon^a{}_b \sinh\alpha \tag{35b}$$

and q^a is a finite translation, the composition law for these is

$$\mathcal{M}_{(12)} = \mathcal{M}_1 \mathcal{M}_2, \tag{36a}$$

$$q_{(12)} = q_1 + \mathcal{M}_1 q_2. \tag{36b}$$

However, the composition law for a representation $G(\mathcal{M}, q)$ containing the extension (32) in its algebra acquires a 2-cocycle

$$G(\mathcal{M}_1, q_1) G(\mathcal{M}_2, q_2) = \exp\left\{\frac{i}{2} q_1^a \varepsilon_{ab} (\mathcal{M}_1 q_2)^b\right\} G(\mathcal{M}_1 \mathcal{M}_2, q_1 + \mathcal{M}_1 q_2). \tag{37}$$

(I is represented by $i = \sqrt{-1}$.)

A finite gauge transformation, generated by the gauge function Θ,

$$\Theta = \theta^a P_a + \alpha J + \beta I \tag{38}$$

produces the following transformations on the connections

$$e^a \to \bar{e}^a = (\mathcal{M}^{-1})^a{}_b (e^b + \varepsilon^b{}_c \theta^c \omega + d\theta^b)$$

$$\omega \to \bar{\omega} = \omega + d\alpha \tag{39}$$

$$a \to \bar{a} = a - \theta^a \varepsilon_{ab} e^b - \frac{1}{2}\theta^2 \omega + d\beta + \frac{1}{2} d\theta^a \varepsilon_{ab} \theta^b.$$

The multiplet of curvatures $F^A = (f^a, f, g)$ transforms by the adjoint 4×4 representation of the extended group,

$$f^a \to \bar{f}^a = (\mathcal{M}^{-1})^a{}_b (f^b + \varepsilon^b{}_c \theta^c f)$$

$$f \to \bar{f} = f \tag{40}$$

$$g \to \bar{g} = g - \theta^a \varepsilon_{ab} f^b - \frac{1}{2}\theta^2 f$$

or

$$F^A \to \bar{F}^A = \sum_{B=0}^{3} (U^{-1})^A{}_B F^B$$

$$U = \begin{pmatrix} \mathcal{M}^a{}_b & -\varepsilon^a{}_c \theta^c & 0 \\ 0 & 1 & 0 \\ \theta^c \varepsilon_{cd} \mathcal{M}^d{}_b & -\theta^2/2 & 1 \end{pmatrix}. \tag{41}$$

The upper left 3×3 block in U comprises the adjoint representation of the conventional Poincaré group with q^a of (35) identified with $-\varepsilon^a{}_c \theta^c$, while the fourth row and column arise from the extension. Note that in the above realization of the gauge action on F, the extension is not visible: I is represented by **O**. On the other hand, an additional connection and curvature by **O**. On the other hand, an additional connection and curvature (a, g) are present.

In this representation, the extended algebra possesses a non-singular metric, which is unavailable without the extension

$$h_{AB} = \begin{pmatrix} h_{ab} & 0 & 0 \\ 0 & 0 & -1 \\ 0 & -1 & 0 \end{pmatrix}. \tag{42}$$

It is true that $^T U h U = h$; this allows raising and lowering the indices (A, B). Additionally, there exists an invariant four-vector

$$i^A = \begin{pmatrix} 0 \\ 0 \\ -1 \end{pmatrix}, \qquad i_A = (0, 1, 0) \tag{43}$$

for which it is true that $(U^{-1})^A{}_B i^B = i^A$ and $i_B U^B{}_A = i_A$. [The occurrence of such invariant vectors is related to the fact that the algebra (32) is solvable.]

An invariant Lagrange density is now constructed with an extended multiplet of Lagrange multipliers η_A,

$$\mathcal{L}'_2 = \sum_{A=0}^{3} \eta_A F^A = \eta_a (De)^a + \eta_2 d\omega + \eta_3 \left(da + \frac{1}{2} e^a \varepsilon_{ab} e^b \right)$$

$$F^A = (f^a, f, g), \qquad \eta_A = (\eta_a, \eta_2, \eta_3) \tag{44}$$

which obey the coadjoint transformation law,

$$\eta_A \to \bar{\eta}_A = \sum_{B=0}^{3} \eta_B U^B{}_A \tag{45}$$

or in components

$$\eta_a \rightarrow \bar{\eta}_a = (\eta_b - \eta_3 \varepsilon_{bc} \theta^c) \mathcal{M}^b{}_a$$

$$\eta_2 \rightarrow \bar{\eta}_2 = \eta_2 - \eta_a \varepsilon^a{}_b \theta^b - \frac{1}{2} \eta_3 \theta^2 \qquad (46)$$

$$\eta_3 \rightarrow \bar{\eta}_3 = \eta_3 \,.$$

Using the invariant metric (42) and the invariant vector (43), other group invariants may be constructed

$$\mathcal{F}^2 = \sum_{A,B=0}^{3} {}^*F^A h_{AB} F^B \,, \qquad (47a)$$

$$M = -\frac{2}{\Lambda} \sum_{A,B=0}^{3} \eta_A h^{AB} \eta_B \,, \qquad (47b)$$

$$C = 2\eta_A i^A \,, \qquad (47c)$$

where ${}^*F^A$ is the 0-form $\frac{1}{2}\varepsilon^{\mu\nu} F^A_{\mu\nu}$, dual to the 2-form F^A.

We recognize in (43) the torsion $(De)^a$ and curvature $d\omega$ densities, which vanish as a consequence of varying η_a and η_2, respectively. Thus Eq. (30) is regained. The Lagrange multiplier η in (2) corresponds to $\frac{1}{2}\eta_2$ in the present formulas and the equation for it, obtained by varying ω, is as in the (anti) de Sitter model, (12b),

$$d\eta_2 + \eta_a \varepsilon^a{}_b e^b = 0 \qquad (48a)$$

while the equation for η_a, obtained by varying e^a, differs from (12a),

$$d\eta_a + \varepsilon_a{}^b \omega \eta_b + \eta_3 \varepsilon_{ab} e^b = 0 \,. \qquad (48b)$$

We need a value for η_3 to close the system (48). The equation for that multiplier is obtained by varying a,

$$d\eta_3 = 0 \qquad (48c)$$

and a constant, cosmological solution

$$\eta_3 = -\frac{\Lambda}{2} \qquad (48d)$$

renders (48b) similar to (12a),

$$d\eta_a + \varepsilon_a{}^b \omega \eta_b - \frac{\Lambda}{2} \varepsilon_{ab} e^b = 0 \qquad (48e)$$

except that there is no factor of η_2 in the last, cosmological term of (48e). This of course has the consequence that when (48a) and (48e) are combined as before, the second order equation that emerges for $\eta = \frac{1}{2}\eta_2$ reproduces (31).

The remaining equation of the gauge theoretical formulation, obtained by varying η_3

$$da = -\frac{1}{2}e^a \varepsilon_{ab} e^b \tag{49}$$

and allowing evaluation of a, has no counterpart in the geometric formulation. Equation (49) can always be locally integrated because the right side is a two-form, hence closed in two dimensions. However in general, there will be singularities in a, since upon integrating (49) over a two-space, the right side gives the total "volume," which could be a well-defined non-vanishing quantity, while the left side always integrates to zero if the manifold is closed and bounded, and a is non-singular.

Note that upon eliminating ω in $\mathcal{L}_2'$ with the zero-torsion equation $(De)^a = 0$ and evaluating η_3 at $-\Lambda/2$, $\mathcal{L}_2'$ coincides with the Lagrange density in (2), now expressed in terms of *Zweibeine*, apart from the total derivative $-\Lambda/2da$, which does not contribute to equations of motion.

Thus here again, the group theoretical formulation reproduces the geometric one, for solutions with det $e \neq 0$, but again see below. However, the former is more flexible: Eq. (48c) is satisfied with vanishing η_3; this corresponds to a vanishing cosmological constant. Thus the gauge theory built on the *extended* Poincaré group possesses as a solution a *non-extended* system. It is interesting therefore that here the cosmological term is an integration constant, and not inserted *a priori* into the theory.

Finding explicit solutions is straightforward. In the geometric formulation, (3) is solved by a flat metric tensor

$$g_{\mu\nu} = h_{\mu\nu}. \tag{50}$$

Then (31) immediately gives

$$2\eta = M - \frac{\Lambda}{2}(x - x_0)^2 \tag{51}$$

with M and x_0 being integration constants, the former reflecting the η-translation invariance mentioned earlier.

Interest in the model[4] derives precisely from the above "black-hole" solution with mass M [in terms of the "physical" metric $g_{\mu\nu}/\eta$], located at x_0. An arbitrary coordinate transformation of this configuration produces the general solution.

The gauge theoretical counterparts of the above are a flat *Zweibein* (apart from a constant tangent-space Lorentz transformation)

$$e_\mu^a = \delta_\mu^a \tag{52}$$

and a vanishing spin connection

$$\omega = 0. \tag{53}$$

Taking in (48c) the cosmological solution for η_3, allows solving (48e) for η_a

$$\eta_a = \frac{\Lambda}{2}\varepsilon_{a\mu}(x^\mu - x_0^\mu) \tag{54}$$

and from (48a) $\eta_2 = 2\eta$ is recovered to be as in (51). Finally (49) is solved for a

$$a_\mu = \frac{1}{2}\varepsilon_{\mu\nu}x^\nu \tag{55}$$

with a pure gauge contribution $\partial_\mu\chi$ left arbitrary. The potential in (55) corresponds to a constant "magnetic field," as is appropriate with our "magnetic-like" extension of translations.

Note the invariants defined in (47): $\mathcal{F}^2$ vanishes since F^A does, M is recognized as the "black hole" mass, while C is the cosmological constant.

The gauge theoretical solution may of course also be presented in a group theoretical fashion, since the equations are of the same form as in (24) and (25), with all quantities belonging to the *extended* algebra and group. The explicit formulas, corresponding to the "black hole" solution, Eqs. (50)–(55), are as follows. The group element U that leads to the pure gauge connection $A = U^{-1}dU$ is

$$U = \exp x^a P_a \tag{56}$$

up to a constant gauge transformation. The constant algebra element Φ that gives $H = \eta_a h^{ab} P_b - \eta_3 J - \eta_2 I = U^{-1}\Phi U$ is (placement of η_2 and η_3 dictated by the group metric (42), *viz.* $\eta^A = h^{AB}\eta_B$)

$$\Phi = \frac{\Lambda}{2}x_0^a \varepsilon_a{}^b P_b + \frac{\Lambda}{2}J + \left(\frac{M}{2} - \frac{\Lambda}{4}x_0^2\right)I. \tag{57}$$

The above-mentioned gauge transformation can be used to set two of the four constants (x_0^a, M, Λ) to zero, leaving an invariant dependence on the group scalars M and Λ, *i.e.*, on $\eta^a\eta_a - 2\eta_2\eta_3$ and $i^A\eta_A$.

As in the (anti) de Sitter model, we see that after a further gauge transformation we pass to the geometrically singular configuration $A = 0$, $H = \Phi$. This gives an especially compact account of the relevant geometric information: Φ encodes the integration constants, which characterize the intrinsic geometry: the cosmological constant Λ, the "black hole" mass M and location x_0. A geometry is built with these characteristics once a gauge transformation is performed, say with the above U, to obtain non-singular connections.

4. QUANTIZATION

The gauge theoretical formulation allows a succinct description of the quantum theory. Of course in the absence of matter, which is all that we here consider, there are no propogating degrees of freedom. Nevertheless, the quantal structure is interesting, albeit simple.

4.1 (Anti) de Sitter gravity

After a spatial integration by parts, the Lagrange density (10) is given by

$$\mathcal{L}_1'' = \eta_a \dot{e}_1^a + \eta_2 \dot{\omega}_1 + e_0^a \left(\eta_a' + \varepsilon_a{}^b \eta_b \omega_1 - \frac{\Lambda}{2} \eta_2 \varepsilon_{ab} e_1^b \right) + \omega_0(\eta_2' + \eta_a \varepsilon^a{}_b e_1^b). \quad (58)$$

Spatial end point contributions have been dropped, and the dot (dash) denotes differentiation with respect to time (space).

The form (58) exhibits the canonical, symplectic structure, where (η_a, η_2) are canonical "momenta", (e_a, ω_1) are conjugate "coordinates" and (e_0^a, ω_0) are Lagrange multipliers, enforcing the vanishing of the constraints,

$$G_a = \eta_a' + \varepsilon_a{}^b \eta_b \omega_1 - \frac{\Lambda}{2} \eta_2 \varepsilon_{ab} e_1^b, \quad (59a)$$

$$G_2 = \eta_2' + \eta_a \varepsilon^a{}_b e_1^b. \quad (59b)$$

One readily verifies with the help of canonical commutation relations between "momenta" and "coordinates" that the algebra of constraints follows the Lie algebra (5)

$$[G_a(x), G_2(y)] = i\varepsilon_a{}^b G_b(x)\,\delta(x-y)$$

$$[G_a(x), G_b(y)] = -i\frac{\Lambda}{2}\varepsilon_{ab}G_2(x)\,\delta(x-y)\,. \qquad (60)$$

The constraints annihilate physical states, which we take in a Schrödinger, momentum representation; *i.e.*, states are functional of (η_a, η_2) while (e_a, ω_1) act by functional differentiation. Thus physical states satisfy

$$\left(\eta_a' + i\varepsilon_a{}^b\eta_b\frac{\partial}{\partial\eta_2} - i\frac{\Lambda}{2}\eta_2\varepsilon_{ab}\frac{\partial}{\partial\eta_b}\right)\Psi(\eta_a, \eta_2) = 0\,, \qquad (61a)$$

$$\left(\eta_2' + i\eta_a\varepsilon_a{}^b\frac{\partial}{\partial\eta_b}\right)\Psi(\eta_0, \eta_2) = 0\,. \qquad (61b)$$

The solution of these equations is

$$\Psi(\eta_a, \eta_2) = \exp i\int dx\left(\eta_2\varepsilon^{ab}\eta_a\eta_b'/\eta_c\eta^c\right)\psi\,, \qquad (62)$$

where ψ is a function (not functional) of the position-independent part of the invariant $\eta^a\eta_a + (\Lambda/2)\eta_2\eta_2$.

4.2 Extended Poincaré gravity

The Lagrangian density (44) is equivalently given by

$$\mathcal{L}_2'' = \eta_a\dot{e}_1^a + \eta_2\dot{\omega}_1 + \eta_3\dot{a}_1 + e_0^a(\eta_a' + \varepsilon_a{}^b\eta_b\omega_1 + \eta_3\varepsilon_{ab}e_1^b)$$
$$+ \omega_0(\eta_2' + \eta_a\varepsilon^a{}_b e_1^b) + a_0\eta_3'\,. \qquad (63)$$

This formula identifies the canonical conjugates (η_a, η_2, η_3) and (e_1^a, ω_1, a_1) with (e_0^a, ω_0, a_0) enforcing vanishing of the constraints

$$G_a = \eta_a' + \varepsilon_a{}^b\eta_b\omega_1 + \eta_3\varepsilon_{ab}e_1^b\,, \qquad (64a)$$

$$G_2 = \eta_2' + \eta_a\varepsilon^a{}_b e_1^b\,, \qquad (64b)$$

$$G_3 = \eta_3'\,, \qquad (64c)$$

whose commutators follow the Lie algebra (32)

$$[G_a(x), G_2(y)] = i\varepsilon_a{}^b G_b(x)\,\delta(x-y)$$
$$[G_a(x), G_b(y)] = i\varepsilon_{ab}G_3(x)\,\delta(x-y)\,. \qquad (65)$$

Again, we use a functional momentum representation for the physical states which must satisfy

$$\left(\eta_a' + i\varepsilon_a{}^b\eta_b\frac{\partial}{\partial\eta_2} + i\eta_3\varepsilon_{ab}\frac{\partial}{\partial\eta_b}\right)\Psi(\eta_a,\eta_2,\eta_3) = 0\,, \qquad (65a)$$

$$\left(\eta_2' + i\eta_a\varepsilon^a{}_b\frac{\partial}{\partial\eta_b}\right)\Psi(\eta_a,\eta_2,\eta_3) = 0\,, \qquad (65b)$$

$$\eta_3'\Psi(\eta_a,\eta_2,\eta_3) = 0\,. \qquad (65c)$$

The general solution of these is

$$\Psi(\eta_a,\eta_2,\eta_3) = \exp i\int dx\left(\eta_2\varepsilon^{ab}\eta_a\eta_b'/\eta_c\eta^c\right)\psi\,, \qquad (66a)$$

where ψ depends on the constant part of the invariants $\eta^c\eta_c - 2\eta_2\eta_3$ and $\eta_3 = -\eta_A i^A = -\Lambda/2$. These are essentially the black hole mass M and the cosmological constant Λ. Observe further that the phase in the exponent of (66a) may be written as

$$\frac{1}{2}\int\frac{dx}{\eta_3}(2\eta_2\eta_3 - \eta^a\eta_a)\varepsilon^{ab}\eta_a\eta_b'/\eta_c\eta^c + \frac{1}{2}\int\frac{dx}{\eta_3}\varepsilon^{ab}\eta_a\eta_b'\,. \qquad (66b)$$

But both η_3 and $(2\eta_2\eta_3 - \eta^a\eta_a)$ contribute only their constant part and may be taken outside the integral. The remaining integrand in the first integral is a total derivative, hence that contribution to phase may be ignored. We are thus left with the wave functional

$$\Psi = e^{-i\Lambda\int dx\varepsilon^{ab}\eta_a\eta_b'}\psi(\Lambda,M)\,. \qquad (66c)$$

Thus the cosmological constant and the black hole mass characterize a physical state.

5. CONCLUSION

The two models here considered are special: their geometric dynamics possess a gauge theoretical formulation. The extended Poincaré model exhibits the intriguing possibility of a cosmological term that is an integration constant, as are the "black hole" mass M and location x_0; all there are encoded in the Lagrange multipliers of the theory.

Both models can also be obtained by dimensional reduction from (2+1) dimensions: To obtain (anti) de Sitter gravity in its geometric formulation one begins[3] with the Einstein theory/Hilbert action (with cosmological term), suppresses dependence on the third dimension, sets $g_{\mu 2}$ to zero for $\mu = 0, 1$ and g_{22} to η^2; for the gauge theoretical formulation one starts with the *Dreibein*-spin connection form of the theory, which also is equivalent to a Chern-Simons, $\mathcal{O}(2,2)$ or $\mathcal{O}(3,1)$ model.[7] Extended Poincaré gravity can be similarly constructed, but the higher-dimensional theory has to be suitably extended by an Abelian ideal. Indeed, it is found that *both* the (anti) de Sitter and extended Poincaré (1+1) dimensional theories arise as *different* dimensional reductions of a *single*, extended (2+1)-dimensional gravity.[8] This and another interesting topic — the coupling of matter consistently with the gauge principle[9] — are beyond the scope of our review. In yet a further investigation one could study non-topological theories in which invariants (46) and/or (47) are added to the Lagrange density (43).

In conclusion, we note that dynamics determined by a group has been familiar in physics since the invention of Yang-Mills theory. However, the examples described here offer a new possibility: in the Lie algebra that determines a gauge theory, one can allow an extension. This gives rise to richer dynamics within the same group theoretical structure, and in the gravity model studied above produces the cosmological constant.

ACKNOWLEDGEMENTS

The review was prepared with the assistance of D. Cangemi, particularly in finding the explicit solutions; this I gratefully acknowledge.

REFERENCES

1. G. Dzhordzhadze, A. Pogrebkov and M. Polivanov, *Dokl. Akad. Nauk. SSSR* **243**, 318 (1978) [English translation: *Sov. Phys. Dokl.* **23**, 828 (1978)]; *Teor. Mat. Fiz.* **40**, 221 (1979) [English translation: *Theor. Math. Phys.* **40**, 706 (1979)].

2. E. D'Hoker and R. Jackiw, *Phys. Rev. D* **26**, 3517 (1982); *Phys. Rev. Lett.* **50**, 1719 (1983); E. D'Hoker, D. Freedman and R. Jackiw, *Phys. Rev. D* **28**, 2583 (1983).

3. C. Teitelboim, *Phys. Lett.* **126B**, 41 (1983), and in *Quantum Theory of Gravity*, S. Christensen, ed. (A. Hilger, Bristol, UK, 1984); R. Jackiw, in *Quantum Theory of Gravity*, S. Christensen, ed. (A. Hilger, Bristol, UK, 1984), *Nucl. Phys.* **B252**, 343 (1985).

4. H. Verlinde, in *The Sixth Marcel Grossman Meeting*, H. Sato and T. Nakamura, eds. (World Scientific, Singapore, 1992); C. Callan, S. Giddings, A. Harvey and A. Strominger, *Phys. Rev. D* **45**, 1005 (1992). Quantization of the model is discussed in these papers.

5. T. Fukuyama and K. Kamimura, *Phys. Lett.* **160B**, 259 (1985); K. Isler and C. Trugenberger, *Phys. Rev. Lett.* **63**, 834 (1989); A. Chamseddine and D. Wyler, *Phys. Lett.* **B228**, 75 (1989). Quantization of the model is discussed in these papers.

6. D. Cangemi and R. Jackiw, *Phys. Rev. Lett.* **69**, 233 (1992).

7. A. Achúcarro and P. Townsend, *Phys. Lett.* **B180**, 89 (1986); E. Witten, *Nucl. Phys.* **B311**, 46 (1988/89).

8. D. Cangemi, *Phys. Lett.* **B297**, 261 (1992); A. Achúcarro, *Phys. Rev. Lett.* **70**, 1037 (1993).

9. G. Grignani and G. Nardelli, *Nucl. Phys.* **B412**, 320 (1994); D. Cangemi and R. Jackiw, *Phys. Lett.* **229**, 24 (1993), *Ann. Phys.* (NY) **225**, 229 (1994).

Section III

SYMMETRY BEHAVIOR

The conceptual revolutions in physics that accompanied establishment of relativity theory and quantum theory also engendered a return to the age-old fascination with symmetry behavior in Nature. Appreciating symmetries and conservation laws in natural processes satisfies an ancient aesthetic prejudice, also it aids unraveling complicated dynamics. Within quantum field theory, we recognized novel manifestations of symmetry: local gauge symmetries, infinite parameter symmetries like those arising from the two-dimensional conformal group. Also novel mechanisms of symmetry breaking were understood to occur in Nature: "spontaneous breaking" when energetically preferred configurations are asymmetric and "anomalous breaking" effected by the quantization procedure.

Paper III.1

Results of the MIT-SLAC experiments on high-energy electron-nucleon scattering, which established the physical reality of quarks, also aroused interest in the symmetries of theories for dynamical quantities that are independent of dimensional parameters.

INTRODUCING SCALE SYMMETRY

Phys. Today **25**, *No.1 (January) 23 (1972)*

When symmetries are present the solution of almost any physical problem is simplified, because we can get at the properties of a system without completely solving all the equations that describe the system. In high-energy physics, symmetry principles have been studied with an eye to circumventing two obstacles to the understanding of elementary particles: lack of precise knowledge about interparticle forces (with the exception of electromagnetism) and the mathematical intractability of any realistic models that propose to explain observed particle phenomena.

The useful symmetries, and their associated conservation laws, are of two kinds: "space-time" and "internal." In space-time symmetries, kinematic properties such as energy and momentum are conserved. For internal symmetries, properties such as charge or baryon number are conserved. High-energy theorists have found it profitable to consider *approximate* symmetries as well as *exact* symmetries. These approximate symmetries are not completely valid in Nature, but their physical predictions are moderately well satisfied. So far, only internal approximate symmetries have been successfully exploited. Examples are Werner Heisenberg's isotopic spin [SU(2)], which is conserved exactly only if we ignore electromagnetism, and chirality [SU(2) × SU(2)], a transformation, developed by Murray Gell-Mann, Feza Gürsey, Maurice Lévy, Yoichiro Nambu, Julian Schwinger and Steven Weinberg, that extends the concept of isospin by introducing both positive and negative parity isospin. Chirality becomes an exact symmetry only when the mass of the pion is zero. In that case, it corresponds to the possibility of changing the parity of a state without a change of mass, by emitting a massless pion.

Here I shall describe an approximate space-time symmetry, *scale* or *dilation* invariance that has recently become the focus of much attention because of results from inelastic electron-nucleon scattering experiments[1] done at the Stanford Linear Accelerator in collaboration with a group from the Massachusetts Institute of Technology. We shall see that the symmetry is exact only if there are massless particles present, so that it must be broken for strong interactions. If the symmetry were exact, it would be possible to rescale or "dilate" the space-time coordinates for a system, without changing the physical content. An understanding of the workings of this approximate invariance principle promises to help unravel the intricacies of physical systems at very high energies, such as those available at SLAC and at the National Accelerator Laboratory, where we can hope that the symmetry becomes exact, because in that kinematical domain it should be possible to ignore masses. This point of view has been advocated for a long time by H. A. Kastrup and Julius Wess.[2]

Nature of dilation symmetry

In its most intuitive form, dilation symmetry is related to dimensional analysis. Consider any dynamical quantity, for example a phase shift for particle scattering, computed in a quantum theory. It will depend on the appropriate kinematical variables, such as energy and momentum, as well as on the fundamental dimensional parameters present in the theory. Among these parameters are Planck's constant $\hbar$ (dimension of angular momentum) and the velocity of light c (dimension of velocity); for convenience, we can choose our units so that these quantities are set to one. Further dimensional parameters may, of course, also form a part of the theory; examples are masses of fundamental particles (if any), lengths describing various fundamental interactions (if any) and the like.

But suppose now that in fact no such additional parameters are present: This is the condition for dilation invariance. We may then find out how the quantity under consideration depends on the kinematical variables. Thus if we are discussing the phase shift, which in general depends on relative energy of the scattering particles, *exact* dilation invariance requires that there be no such dependence. Thus scale symmetry determines the phase shift completely except for an overall constant. The reason for this invariance is that the phase shift is a dimensionless quantity, and there is no way to build an energy dependence when there are no dimensional parameters available. Application of *approximate* dilation symmetry at high energies (where one assumes the dependence on dimensional parameters to be negligible) then implies that the phase shift approaches a constant in this energy region.

The elaboration of this simple feature, a characteristic of systems not depending on dimensional parameters, is the subject of scale symmetry. A class of dynamical models that possesses scale symmetry is easy to find. Consider the single-particle Hamiltonian $\mathcal{H}$ proportional to $p^n + ar^{-n}$, where a is a dimensionless constant, and $p(t)$, $r(t)$ are the time-dependent momentum and position of the particle. (An overall multiplicative constant has been ignored, as it is without consequence for our discussion.) Just as the invariance of this system under time translation

$$t \rightarrow t + \delta,\, r(t) \rightarrow r(t + \delta),\, p(t) \rightarrow p(t + \delta)$$

leads to the conservation of the total energy $E = \mathcal{H}$, so also a scale transformation

$$t \rightarrow \rho t,\, r(t) \rightarrow \rho^{-1/n} r(\rho t),\, p(t) \rightarrow \rho^{1/n} p(\rho t)$$

(this *defines* a dilation transformation for this model) leaves the system described by the above Hamiltonian unchanged. Because of this additional symmetry, a new conserved quantity D is present. The constant of motion is related to the virial pr, and its form in this model is

$$D = -\frac{1}{n}(p\,r) + Et .$$

Explicit evaluation of the scattering phase shift verifies the general consideration: It is energy independent, as can be seen even in the semi-classical WKB (Wentzel-Kramers-Brillouin) formula for the phase shift

$$\delta(E) = \int dr[(E - ar^{-n})^{1/n} - E^{1/n}] = \int dx[(1 - ax^{-n})^{1/n} - 1] .$$

In the second equality, the dummy integration variable r has been changed to $E^{-1/n}x$, and it is seen that all energy dependence has disappeared in this form of the equality.

A scale-invariant Hamiltonian (for a unit-mass particle) in any number of spatial dimensions is

$$\mathcal{H} = \frac{\mathbf{p}^2}{2} - \frac{1}{2\mathbf{r}^2} .$$

This corresponds to the case with $n = 2$ discussed in the text. The equations of motion of this particle are invariant under the transformation

$$t \rightarrow \rho t$$
$$\mathbf{r} \rightarrow \rho^{-1/2}\mathbf{r}$$
$$\mathbf{p} \rightarrow \rho^{1/2}\mathbf{p}$$

which leads to a new conservation law

$$\frac{dD}{dt} = 0$$

where

$$D = -\frac{1}{2}\mathbf{p} \cdot \mathbf{r} + \mathcal{H}t .$$

In nonrelativistic physics, where the kinetic part of the Hamiltonian is proportional to p^2 (that is, $n = 2$ in the above example), our discussion of scale-invariant systems indicates that only the r^{-2} potential possesses this symmetry. Because such a force law has no apparent physical significance, considerations of dilation symmetry in nonrelativistic particle physics are not especially fruitful.[3] On the other hand in relativistic physics, the kinetic part of the Hamiltonian is proportional to $(p^2 + m^2)^{1/2}$. When m vanishes exactly, or at high energies when m^2 may be neglected in comparison to p^2, the kinetic term goes as p, and the $1/r$ potential is scale invariant. Thus we conclude that for massless particles, or for massive particles at high energy, scale symmetry is present in relativistic systems if the forces are of the inverse-square form at short distances (that is, for large momentum). Electromagnetic and strong-interaction (Yukawa) force laws are indeed of this type, so that approximate scale symmetry may be a useful concept in relativistic particle physics if the discussion is confined to these interactions. (The short-distance form of the weak interaction is not yet known.)

More realistic models

The proper model for relativistic particle physics cannot of course be a single-particle Hamiltonian, as given above. However, the general features of scale invariance, which we have discussed, are not much different for a more appropriate framework, such as Lagrangian field theory. If the system is described by a field $\Phi(x)$, a scale transformation takes the position four-vector $\mathbf{x}^\mu$ into $\rho \mathbf{x}^\mu$, and the field $\Phi(x)$ into $\rho^d \Phi(\rho x)$. This scale transformation is a symmetry operation if the Lagrange density contains no dimensional parameters, and if we choose the scale dimension d of the field $\Phi(x)$ properly.

For the transformation to work in simple models, d should coincide with the natural dimension of the field, expressed in units of mass. The "natural dimension" is just the energy normalization of the field, as determined by ordinary dimensional analysis; for Fermi fields $d = 3/2$ and for Bose fields $d = 1$. We will look at possible modifications of this rule later.

To determine whether or not the transformation is a symmetry, we use a version of Noether's theorem and find[4] a dilation "current"

$$D^\mu(x) = (x)_\nu \theta^{\mu\nu}(x).$$

Here $\theta^{\mu\nu}$ is a symmetric energy-momentum tensor. When scale symmetry is present, $D^\mu(x)$ is a conserved "current," analogous to the conserved current $J^\mu(x)$ associated with charge conservation. $D^\mu(x)$ is also analogous to the conserved quantities $\theta^{\mu\nu}(x)$ and $x^\mu \theta^{\nu\alpha}(x) - x^\nu \theta^{\mu\alpha}(x)$ associated with the other space-time symmetries, the so-called "Poincaré symmetries":

translations, rotations and Lorentz transformations. In the usual fashion, the time-independent constant of motion D is constructed from the current

$$D = \int d^3x D^0(x).$$

Together with the (four) momentum

$$P^\mu = \int d^3\mathbf{x}\theta^{0\mu}(x)$$

and the (four) angular momentum

$$L^{\mu\nu} = \int d^3\mathbf{x}[x^\mu\theta^{\nu0}(x) - x^\nu\theta^{\mu0}(x)].$$

D satisfies the commutator algebra shown on the next page. The first three commutators are those of the usual Poincaré group, whereas the remaining three show how dilations combine with Poincaré transformations to form a larger group, if there is scale invariance, that is if D is a conserved quantity.

Dilations must be inexact

We now see that dilation symmetry, unlike Poincaré invariance, is not an exact symmetry of hadron (strong-interaction) physics. For if the commutator of D with P^μ is of the given form, then, from the usual rules of commutator algebra, it follows that

$$i[D, M^2] = 2M^2,$$

where

$$M^2 \equiv P^\mu P_\mu.$$

The eigenvalues of the mass operator M^2 (the masses) are either zero or continuous; neither case provides an acceptable mass spectrum for hadron physics (see box on the next page). Consequently, commutators with D must be modified, and the symmetry must be approximate, becoming exact only when the details of the mass spectrum are irrelevant, for example at high energies (if at all).

We expect therefore that in Nature the dilation current is not conserved; that is the divergence of $D^\mu(x)$ does not vanish:

$$\partial_\mu D^\mu(x) = \theta_\mu{}^\mu(x) \neq 0.$$

Poincaré and dilation commutators

$$i[P^\alpha, P^\beta] = 0$$
$$i[L^{\alpha\beta}, P^\mu] = g^{\alpha\mu}P^\beta - g^{\beta\mu}P^\alpha$$
$$i[L^{\alpha\beta}, L^{\mu\nu}] = g^{\alpha\mu}L^{\beta\nu} - g^{\beta\mu}L^{\alpha\nu} + g^{\alpha\nu}L^{\mu\beta} - g^{\beta\nu}L^{\mu\alpha}$$
$$i[D, D] = 0$$
$$i[D, L^{\mu\nu}] = 0$$
$$i[D, P^\mu] = P^\mu$$

Why must dilation symmetry be inexact?

We define the mass operator as

$$M^2 \equiv \sum P^\mu P_\mu \equiv (P^0)^2 - (\mathbf{P})^2$$

and look at the last commutator above

$$i[D, P^\mu] = P^\mu$$

That is

$$i[D, P^0] = P^0$$
$$i[D, P^1] = P^1$$

and so on, up to $\mu = 3$.

We see then that

$$i[D, M^2] = \sum(P_\mu i[D, P^\mu] + i[D, P^\mu]P_\mu) = \sum(P_\mu P^\mu + P^\mu P_\mu) = 2M^2$$

Given this result, $e^{i\alpha D}M^2 e^{-i\alpha D}$ must equal $e^{2\alpha}M^2$ where α is any arbitrary constant.

Then if $|m\rangle$ is an eigenstate of M^2 with eigenvalue m^2, the state $e^{-i\alpha D}|m\rangle$ is an eigenstate of M^2 with eigenvalue $e^{2\alpha}m^2$. Because α is arbitrary, we have either a continuous mass spectrum, or else m^2 is zero.

D then is not time independent, and one can show that the $[D, P^\mu]$ commutator is changed, so that the undesirable mass spectrum is avoided. Terms in the Lagrange density that are scale invariant and do not contribute to $\theta_\mu{}^\mu(x)$ are the boson or fermion kinetic energy $\partial_\mu\Phi(x)\partial^\mu\Phi(x)$ or $i\bar{\Psi}(x)\gamma^\mu\partial_\mu\Psi(x)$,

Yukawa-type fermion-boson couplings, [such as $\bar{\Psi}(x)\Psi(x)\Phi(x)$] and various self interactions [such as $\Phi^4(x)$, $(\bar{\Psi}(x)\Psi(x))^{4/3}$ and so on]. On the other hand, non-scale-invariant terms, which do contribute to $\theta_\mu{}^\mu(x)$, are boson or fermion mass terms, $m^2\Phi^2(x)$ or $m\bar{\Psi}(x)\Psi(x)$, and various interactions characterized by coupling constants with dimension of a power of length: $\lambda\Phi^3(x)$, $g\bar{\Psi}(x)\gamma^\mu \ \Psi(x)\bar{\Psi}(x)\gamma_\mu\Psi(x)$ and so on.

In applications, it is assumed that no dimensional coupling constants are present, and that $\theta_\mu{}^\mu(x)$ includes only mass terms. The hope that scale symmetry is a useful approximate symmetry is formalized in the statement that $\theta_\mu{}^\mu(x)$ (that is, the mass terms) is a "small" operator, whose matrix elements vanish at large energy. When the consequences of these ideas are developed in the abstract framework of quantum field theory, one obtains, without solving the theory, results like those discussed previously in connection with the phase shift. They may be summarized by the statement that, at high energy, dimensionless quantities become functions of dimensionless ratios of the available kinematic variables.

Application to MIT-SLAC work

The experimental verification of the above ideas has so far been confined largely to the MIT-SLAC results on inelastic electron scattering off nucleons.[1] This process is described in Fig. 1, where the lines labeled by k represent the incoming and outgoing electrons, the thick line is the target nucleon, with four-momentum p, and the spray n is the final hadronic state. Because the electrons have no direct strong interaction with the hadrons, the process of Fig. 1a is assumed to proceed through the exchange of a virtual photon, carrying four-momentum q as indicated by the wavy line in Fig. 1b. The interaction of the photon with the electrons is assumed to be known from lowest order quantum electrodynamics, and the interaction with the hadronic system is parametrized by the (unknown) matrix element $\langle p \mid J^\mu(0) \mid n \rangle$ where $J^\mu(x)$ is the electromagnetic current. Effectively, therefore, we may ignore the electrons entirely and view the process as the scattering of an unphysical photon with "mass" q^2 off a nucleon.

The total cross-section, that is, the sum over n, can be expressed by the forward, nucleon matrix element of the current commutator

$$C^{\mu\nu}(p,q) = \int d^4x e^{iqx} \langle p \mid [J^\mu(x), J^\nu(0)] \mid p \rangle .$$

This follows from the optical theorem, which relates the total cross-section to the imaginary part of the forward elastic-scattering amplitude; this amplitude involves a matrix element between target states.

Because the photon arises from an electron transition and is unphysical, it has longitudinal as well as transverse polarizations. Thus even though the target nucleon is not polarized, we may speak of two independent cross-sections: σ_L for longitudinally polarized photons and σ_T for transverse photons. The total cross-section σ equals $\sigma_L + \sigma_T$. The cross-sections are functions of the photon "mass" q^2, which is negative (spacelike), and of $\nu = p \cdot q$, the energy loss of the scattered electron in the target rest frame. James D. Bjorken[5] suggested that in the region where q^2 and ν became large at the same rate, so that their ratio is fixed at $\omega = -q^2/2\nu$, the quantities $\nu\sigma_L$ and $\nu\sigma$ should become functions *only of* ω. This is just the behavior subsequently observed by the MIT-SLAC experimentalists in the "deep inelastic" region, as the kinematical domain we have been discussing is called: $\nu\sigma_L$ and $\nu\sigma$ are described by functions of ω, conventionally called $F_L(\omega)$ and $F_2(\omega)$. They are the two "deep-inelastic structure functions" of inelastic electron-nucleon scattering.

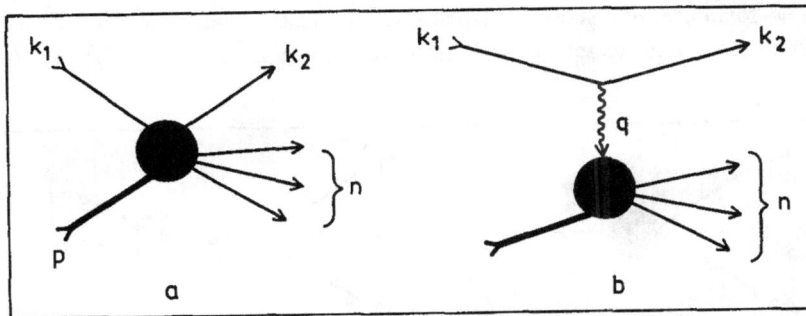

Fig. 1. Inelastic electron-nucleon scattering. Experimental verification of scale invariance has thus far been confined largely to the MIT-SLAC results on electron-nucleon scattering in the "deep-inelastic" region.

This unexpected high-energy behavior is easily explained in terms of approximate dilation invariance; indeed, most of the present interest in the theory derives from this experiment. Note that $\nu\sigma_L$ and $\nu\sigma$ are dimensionless quantities: The dimension of a cross-section is length squared, or in units that we are using, inverse mass squared, and the dimension of ν is mass squared. In a scale-invariant world, these dimensionless objects can depend on the appropriate kinematical variables q^2 and ν only through the dimensionless ratio $\omega = -q^2/2\nu$. Consequently, we can understand the regularities of the inelastic data if we assume that scale invariance becomes exact in this high-energy domain. An unanswered question remains: Why does scale invariance set in at as low an energy as is observed—a few GeV?

Other processes

Another process that lends itself to a similar analysis is the annihilation of electron pairs into hadrons. The reaction is described by Fig. 2, with the same notation as in figure 1. The total cross-section σ (which according to Fig. 2b depends on one kinematical variable, q^2, the "mass" of the final hadronic state) is determined by the matrix element of the current commutator between no-particle (that is, vacuum) states

$$C^{\mu\nu}(q) = \int d^4x e^{iqx} \langle 0 \, | [J^\mu(x), J^\nu(0)] | \, 0 \rangle \, .$$

This is analogous to the formula for the total electron-nucleon cross-section, except that the nucleon state $| \, p \rangle$ has been replaced by the vacuum state $| \, 0 \rangle$. Here we form $q^2\sigma(q^2)$ as our dimensionless quantity. Hence for large q^2, if scale invariance becomes exact, $q^2\sigma(q^2)$ approaches a constant and $\sigma(q^2) \propto 1/q^2$.

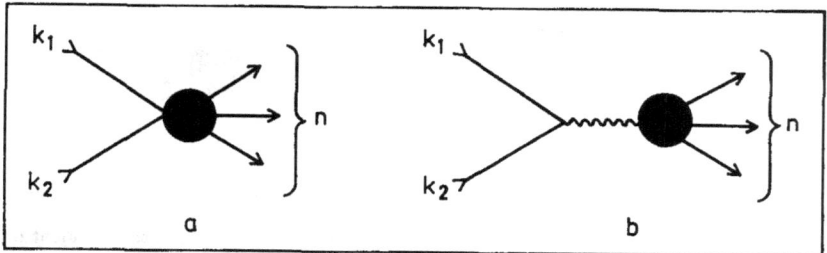

Fig. 2. Annihilation of electron pairs into hadrons is another process that could be analyzed in the same way as electron-nucleon scattering.

This behavior is not inconsistent with the data on pions from the storage rings at Frascati, although definite verification of the prediction must await further experimental study.[6]

In addition to the above *high-energy* applications of approximate dilation invariance, there are attempts to use these ideas at *low energies*. It was first noted by Gerhardt Mack[7] that there is a certain analogy between approximate scale invariance and approximate chiral invariance: Both symmetries are broken by nonvanishing masses. Approximate chiral symmetry led to an understanding of the low-energy dynamics of the *pseudoscalar* mesons, principally the pion. This happened because the axial vector current, which would be conserved if the symmetry were exact, has a nonvanishing divergence that can be used as the field describing a particle with the quantum numbers of the divergence: zero spin and odd parity, corresponding to the physical pion. Analogously, it is hoped that the nonvanishing divergence of the dilation current, a zero-spin even-parity operator, can similarly serve as the field for scalar mesons, and that the low-energy dynamics of such mesons can be understood.

So far, nothing particularly useful has emerged from the low-energy considerations about scale invariance because we need particles with mass sufficiently small to be approximated by zero to carry out these studies. This condition is satisfied for chirality, because the pion mass is negligible in the hadron mass spectrum. But for scale invariance at low energies, one would need a scalar meson with negligible mass and unfortunately, no such particle is seen experimentally.

Difficulties with models

The applications that I have described are rather abstract, and it would be gratifying to present a realistic relativistic field-theoretical model in which the general features of scale symmetry are realized *explicitly* in the solutions to that model. But there is apparently no nontrivial example of a theory whose solution exhibits scale symmetry, broken only by mass terms that are irrelevant at high energies. The reason is that all known relativistic quantum field theories, except free-field models, are beset by divergences and therefore undefined, at least in perturbation theory.

All we can do at the present time is introduce a cut-off and calculate, in perturbation theory, selected quantities that have a limit as the cut-off is removed. The cut-off, however, is always a dimensional quantity, typically a large mass, that breaks scale invariance in a large way. This breaking is sufficiently violent that the symmetry is not exact even at high energies. Furthermore, when we calculate the effect of a scale transformation on a field, we find, because of the divergences in the theory, that the scale dimension d, which characterizes this transformation, does not coincide with the natural dimension but becomes an arbitrary, dynamically determined quantity. This is called the "phenomenon of anomalous dimensions."

The consequence of these features of explicit models is that when one considers, for example, the deep-inelastic cross-sections, the computed formulas in perturbation theory do not possess the convergent high-energy behavior suggested by Bjorken and substantiated by experiment. Rather they diverge in the deep inelastic limit. Only theories without interactions are free of such difficulties. (Similar anomalies occur when one tries to verify some results of conventional current algebra.[8])

At the present time, we do not know how to assess our failure to find an explicit realization of approximate scale symmetry. It may be that the nonscaling behavior of the perturbative solutions is physically correct and that further experimental studies will modify the MIT-SLAC data. In this context, Kenneth Wilson has suggested that scale dimensions need not coincide with natural dimensions in Nature, but are in general, as in perturbation theory, anomalous. On the other hand, it may be that we have not as yet been sufficiently clever in choosing and solving the appropriate model, and that the present calculations are irrelevant to the real situation.[9]

Light-cone algebra

The free-field theory, as we have mentioned, is the only currently known model with asymptotic scale symmetry. This observation suggests that certain features of a noninteracting theory may be relevant in describing high-energy phenomena, specifically the deep-inelastic region. An example of this idea is Richard Feynman's parton model.[10] Here the target nucleon is considered to be a noninteracting gas of fundamental particles, and the observed deep-inelastic phenomena arise from an incoherent superposition of the scattering of photons from these partons.

Recently, another framework has been proposed for incorporating free-field features in a discussion of inelastic electron-nucleon scattering; this is the formalism of "light-cone commutators." With these methods, it has been possible to rederive some of the results of the parton model in a much more general context that does not rely on the specific dynamical assumptions of that model. The relation between the parton model and light-cone commutators is in this sense analogous to the relation between the quark model and current algebra; in both cases, the somewhat arbitrary results of the physical model are seen to come directly from the mathematical formalism.

To discuss light-cone algebra, we need some notations and definitions. For the position four vector $x^\mu = (x^0, x^1, x^2, x^3)$ as well as for any other Lorentz tensor, the $+$ and $-$ components are defined by $x^\pm = [(x^0 \pm x^3)]/\sqrt{2}$. The remaining components of x^μ are denoted by $\mathbf{x}_T$. We can see that the deep-inelastic region is attained by letting the minus component of the photon momentum, q^-, pass to infinity with q^+, $\mathbf{q}_T$ and p fixed: The reason is that q^2 and ν are linear in q^-:

$$q^2 = 2q^+q^- - \mathbf{q}_T^2;$$
$$\nu = p^+q^- + p^-q^+ - \mathbf{p}_T \cdot \mathbf{q}_T$$

so that when q^- gets large, q^2 and ν are also large, but their ratio remains finite. Consider next the expression that determines the cross-section

$$\int d^4x\, e^{iqx} \langle p \mid [J^\mu(x), J^\nu(0)] \mid p\rangle.$$

As $q^- \to \infty$, the exponential in general oscillates rapidly because

$$\exp(iqx) = \exp i(q^-x^+ + q^+x^- - \mathbf{q}_T \cdot \mathbf{x}_T).$$

The contributions to the x integral are therefore damped by this oscillation, unless $x^+ = 0$. Consequently, the limiting forms of the cross-sections, which are the structure functions are determined by the commutator on the light cone

$$[J^\mu(x), J^\nu(0)] \quad x^+ = 0.$$

The reason for the nomenclature is that when x^+ vanishes, the nontrivial contributions to the commutator come from $x^2 = 0$; that is, x^2 is on the light cone, so called because light propagates along it. (If $x^+ = 0$ and $x^2 \neq 0$, then $x^2 = -\mathbf{x}_T^2 < 0$, where causality forces the commutator to vanish.)

We have just seen that if we can provide a model for the light-cone commutator, we shall have a model for the deep-inelastic structure functions. For free fields, the commutator can be evaluated, and it is tempting to postulate that the same form holds even in an interacting field theory, just as the equal-time commutator

$$[J^\mu(x), J^\nu(0)] \quad x^0 = 0$$

is not very sensitive to the nature of the interaction. Several groups of theorists have successfully done a rather elegant investigation of the light-cone structure of current commutators, and it has been possible to determine completely the light-cone commutator in *an interacting theory*. The result indeed is largely interaction independent![11]

The light-cone commutator of currents involves operators that we have not previously met in physics. These are called "bilocal" currents. Unlike the usual current $J^\mu(x)$, the bilocal current depends on *two* different points, $J^\mu(x \mid y)$, where $(x - y)^2 = 0$; in the limit $x \to y$ it reduces to $J^\mu(x)$. For example, in a model where the current is constructed from fermion fields,

$$J^\mu(x) = \bar{\Psi}(x)\gamma^\mu\Psi(x)$$

the bilocal current is given by

$$J^\mu(x \mid y) = \bar{\Psi}(x)\gamma^\mu\Psi(y)\,.$$

The matrix elements of these bilocal currents determine the deep *inelastic* structure functions

$$p^\mu \int d\omega e^{i\omega x \cdot p} F(\omega) \propto \langle p \mid J(x \mid 0) \mid p \rangle$$

just as the matrix elements of the ordinary current determine the *elastic* form factors. This shows that the bilocal currents are operators with a physical significance quite analogous to that of the local currents.

At the present time, many theorists are studying the light-cone commutators, bilocal operators *etc.* very closely. An intriguing possibility is suggested by the fact that bilocal currents are conserved in free-field theory. Consequently, they correspond to a definite symmetry whose properties have not as yet been fully understood. We may speculate that in an interacting theory, where these bilocal currents are no longer conserved, there remain traces of the free-field symmetry, and that an *approximate,* as yet untried, symmetry is present. I cannot predict the direction that future research will take, but certainly the concepts of scale invariance and the light cone will dominate the discussion of high-energy processes, specifically the deep inelastic processes, for some time.

REFERENCES

1. H. W. Kendall, W. Panofsky, *Scientific American* **224**, June 1971, page 60.
2. H. A. Kastrup, *Ann. Physik* **7**, 388 (1962); J. Wess, *Nuovo Cim.* **18**, 1086 (1960).
3. Velocity-dependent forces arising from an electromagnetic vector potential that scales as r^{-1} are also dilation invariant. Thus dynamics of electromagnetic point-magnetic vortices and of (hypothetical) point-magnetic monopoles is scale invariant; see "Hidden Symmetry of Magnetic Point Monopole and Vortex," reprinted in this volume on p. 233.

4. C. Callan, S. Coleman, R. Jackiw, *Ann. Phys.* (NY) **59**, 42 (1970).

5. J. D. Bjorken, *Phys. Rev.* **179**, 1547 (1969).

6. *Phys. Today* **23**, December 1969, page 17.

7. G. Mack, *Nucl. Phys.* **B5**, 499, (1968).

8. S. Treiman, R. Jackiw, B. Zumino and E. Witten, *Current Algebra and Anomalies* (Princeton University Press, Princeton, NJ/World Scientific, Singapore, 1985).

9. K. Wilson, *Phys. Rev.* *D* **2**, 1473, 1478 (1970); C. Callan, *Phys. Rev.* *D* **2**, 1541 (1970); K. Symanzik, *Comm. Math. Phys.* **18**, 227 (1970); S. Coleman, R. Jackiw, *Ann. Phys.* (NY) **67**, 552 (1971).

10. *Phys. Today* **24**, February 1971, page 17.

11. J. M. Cornwall, R. Jackiw, *Phys. Rev.* *D* **4**, 367 (1971); H. Fritsch, M. Gell-Mann, in *Proceedings of the International Conference on Duality and Symmetry in Hadron Physics*, E. Gotsman, ed. (Weizmann Science Press, Jerusalem, 1971); D. J. Gross, S. Treiman, *Phys. Rev.* *D* **4**, 1059 (1971).

Paper III.2

The non-relativistic charged particle interaction, *i.e.*, the Kepler-Coulomb problem, possesses a "hidden" or "dynamical" symmetry that leads to the conservation of the Runge-Lenz vector. Interaction of a charged particle with a magnetic point-monopole or a magnetic point-vortex also is characterized by a hidden symmetry described by the SO(2,1) group — the *conformal symmetry group of non-relativistic physics*, which is also respected by the inverse square potential, *i.e.*, by forces proportional to r^{-3}. This result was first presented at the XX International Conference on High Energy Physics, Madison, WI, July 1980, and has engendered much subsequent research.

HIDDEN SYMMETRY OF MAGNETIC POINT MONOPOLE AND VORTEX

High Energy Physics – 1980, L. Durand and L. Pondrom, eds. (American Institute of Physics, New York, NY, 1981)

Please allow me to take a few minutes to acquaint you with an elementary property of the point magnetic monopole, and of the point magnetic vortex, which I recently found and which had not previously appeared in the literature that I thoroughly searched.

To begin with, I am concerned with the dynamics of a point particle with mass M and charge e, moving in the field of a strength-g point magnetic monopole, which is taken to be infinitely heavy and placed at the origin of the coordinate system. The equations for the dynamical variable $\mathbf{r}(t)$, the coordinate of our particle, are governed by the Lagrangian

$$L = L_{\text{kin}} + L_{\text{int}}, \quad L_{\text{kin}} = \frac{1}{2}Mv^2, \quad L_{\text{int}} = \frac{e}{c}\mathbf{v}\cdot\mathbf{A}, \tag{1}$$

$$\mathbf{v} = \frac{d}{dt}\mathbf{r}, \quad \mathbf{B} = \mathbf{\nabla}\times\mathbf{A} = g\frac{\mathbf{r}}{r^3} \tag{2}$$

which gives rise to the familiar Lorentz force law

$$M\frac{d^2}{dt^2}\mathbf{r} = \frac{e}{c}\mathbf{v}\times\mathbf{B} = -\frac{eg}{c}\frac{\mathbf{r}\times\mathbf{v}}{r^3}. \tag{3}$$

The symmetries that characterize this dynamics are the subject of my presentation.

Before describing the symmetries, let me remind you that there is a long story concerning the vector potential $\mathbf{A}$ which specifies the magnetic monopole. In order to accommodate (2), $\mathbf{A}$ must necessarily possess some singularities, and these give rise either to the "Dirac string,"[1] or to the "Wu-Yang sections."[2] Related to this is the fact that the angular momentum, which is a conserved quantity by virtue of the rotational invariance of the dynamics, has an unconventional form. In addition to the familiar kinematic contribution $\mathbf{r}\times M\mathbf{v}$, there is a further term, first given by Poincaré, which arises from the electromagnetic fields; $\mathbf{J} = \mathbf{r}\times M\mathbf{v} - (eg/c)(\mathbf{r}/r)$.

The rotational invariance of our problem, and the corresponding conserved quantity $\mathbf{J}$, are of course familiar. Also familiar is the symmetry of time-translation invariance. Under a time-translation, $t \to t + t_0$, the dynamical variable changes according to $\mathbf{r}(t) \to \mathbf{r}(t + t_0)$, which in infinitesimal form reads $\delta\mathbf{r} = \mathbf{v}$. The consequent conserved quantity is the Hamiltonian, which measures the energy; $H = \frac{1}{2}Mv^2$. Note that the Lagrangian is not invariant under a time-translation, but the action $I = \int dt\, L$, is.

The two further symmetries, to which I call your attention, are similar to the time-translation. They too describe invariance against further redefinitions of the time variable; they leave invariant not the Lagrangian but the action; and they lead to constants of motion.

Consider first time-dilation,

$$t \to \alpha t. \tag{4}$$

One verifies that transforming the dynamical variable as a density of weight $-\frac{1}{2}$

$$\mathbf{r}(t) \to \alpha^{-1/2} r(\alpha t), \quad \delta \mathbf{r} = t\mathbf{v} - \frac{1}{2}\mathbf{r} \tag{5}$$

leaves the action and the equations of motion invariant. The conserved quantity

$$D = tH - \frac{M}{4}(\mathbf{r} \cdot \mathbf{v} + \mathbf{v} \cdot \mathbf{r}) \tag{6}$$

involves the virial.

The second symmetry transforms the time by a special conformal transformation [translation of the reciprocal time]

$$\frac{1}{t} \to \frac{1}{t} - \frac{1}{t_0}. \tag{7}$$

The transformation law for $\mathbf{r}(t)$ which leaves the action and the equations of motion invariant, is

$$\mathbf{r}(t) \to \left(1 - \frac{t}{t_0}\right)\mathbf{r}\left(\frac{t}{1 - t/t_0}\right), \quad \delta \mathbf{r} = t^2 \mathbf{v} - t\mathbf{r} \tag{8}$$

and the corresponding constant of motion involves H and D

$$K = -t^2 H + 2tD + \frac{1}{2} M r^2. \tag{9}$$

The above invariances are realized both classically and quantum mechanically. In the latter context, $\mathbf{r}$ and $\mathbf{v}$ are operators that satisfy the equal time algebra

$$[r^i, r^j] = 0, \quad [r^i, v^j] = \frac{i\hbar}{M}\delta^{ij}, \quad [v^i, v^j] = \frac{i\hbar e}{Mc^2}\varepsilon^{ijk}B^k \tag{10}$$

and D is written in the symmetrized form already used in (6).

Even though D and K are constants of motion, their commutator [Poisson bracket] with H does not vanish; rather one finds, using (10) that

$$[H, D] = i\hbar H, \quad [H, K] = 2i\hbar D, \quad [D, K] = i\hbar K. \tag{11}$$

These allow verifying the time-independence of D and K by use of the formulas

$$\frac{dD}{dt} = \frac{i}{\hbar}[H, D] + \frac{\partial D}{\partial t} = 0, \quad \frac{dK}{dt} = \frac{i}{\hbar}[H, K] + \frac{\partial K}{\partial t} = 0. \tag{12}$$

The three-parameter set of transformations of the time: translations, dilations, and conformal transformations, form a three-parameter Lie group, with generator algebra (11). This group is recognized to be $O(2, 1)$. Hence when it is combined with the $O(3)$ group of rotations, we set that the problem admits an $O(3) \times O(2, 1)$ group of invariances.

The generators however are not all independent. The Casimir operator of the $O(2, 1)$ group is related to that of the rotation group

$$C = \frac{1}{2}[KH + HK] - D^2 = \frac{1}{4}\left[J^2 - \left(\frac{eg}{c}\right)^2 - \frac{3}{4}\hbar^2\right]. \tag{13}$$

[The last term is a quantal re-ordering contribution, absent in the classical mechanical context.] Equation (13) shows that at fixed angular momentum, the entire dynamics of the magnetic monopole-electric monopole interaction is characterized by a single, irreducible, unitary, hence infinite-dimensional, representation of $O(2, 1)$, labelled by the monopole strength. Within quantum mechanics, J^2 is quantized in units of $\hbar^2 j(j + 1)$, and so is eg/c in units of $\hbar\mu$ [$2j, 2\mu$ = integers].[1] Consequently C is also quantized.

We recognize that the situation here is quite analogous to that of the Coulomb/Kepler problem, which describes the dynamics of an electric monopole in the field of another infinitely heavy electric monopole. There too, the obvious rotational invariance, when combined with the hidden Runge-Lenz invariance, gives rise to an $O(3) \times O(2, 1)$ symmetry of the unbounded motion, and an $O(3) \times O(3) = O(4)$ symmetry of the bounded motion. [In the monopole problem there is no bounded motion.][3]

One may use the symmetry to give an algebraic treatment for monopole dynamics, either classically or quantum-mechanically. Since the answers have been available for many years, the algebraic approach does not produce any new results; rather it supplements the analytic treatment and, in a sense, explains why a simple, closed-form answer is attainable.[4]

One may also do something else: the $O(2, 1)$ invariance allows for quantization procedures, other than the conventional equal-time one. Rather one may quantize on other characteristic curves of the $O(2, 1)$ group. This is analogous to the situation in relativistic field theories where the $O(3, 1)$ Lorentz invariance [or in massless theories, the $O(4, 2)$ conformal invariance] allows for quantization on surfaces other than the fixed-time ones. That this should be an interesting alternative, which organizes the same physical content in novel ways, was first stressed by Dirac.[5] In our simple example, the idea can be worked out completely and explicitly.

Here there is no time to give details of the above calculations. You may find them in the published articles.[6,7] One obvious calculation does not however exist yet. In the Coulomb/Kepler problem, the higher symmetry allows for the introduction of a coordinate system which makes the hidden symmetry manifest and facilitates computation – this is the Fock sphere. An analogous formalism for the magnetic monopole, which would expose its hidden symmetry, has thus far not been indentified.

Let me remind that all the above symmetries are a property of the non-relativistic point monopole. How they may appear when the dynamics is relativistic, or when the monopole has additional structure is not known. However one may expect that some of our results may extend beyond the context here emphasized. The reason for this hope is that the *interaction* Lagrangian in (1) allows for an even larger set of symmetry transformations than the $O(2,1)$ group discussed here. One may verify that *any* reparametrization of the time, $t \to f(t)$, leaves $\int dt\, L_{\text{int}}$ unchanged. This general invariance is reduced to $O(2,1)$ by the fact that $\int dt\, L_{\text{kin}}$ is invariant only under the restricted transformations. Other kinetic terms may very well allow for other symmetries.

Finally, I record the fact that similar features characterize the point magnetic vortex problem.[7] A point vortex is most appropriately given a planar (two-dimensional) description, where the magnetic field is proportional to a two-dimensional δ-function and the vector potential is a singular pure gauge

$$\mathbf{A} = \frac{\Phi}{2\pi}\boldsymbol{\nabla}\Theta, \quad B = \boldsymbol{\nabla} \times \mathbf{A} = \Phi\delta^2(\mathbf{r}). \tag{14}$$

Here Φ is the flux through the plane, coordinized by (x,y) and $\Theta = \tan^{-1} y/x$. One verifies that (4), (5) as well as (7), (8) are still symmetries leading to the constants of motion (6) and (8) respectively. The commutators of $\mathbf{r}$ with $\mathbf{v}$ are as in (10) except that the indices range only over two values, and the last, velocity commutator equals $\frac{i\hbar e}{Mc^2}\epsilon^{ij}B$. Thus (11) and (12) still hold. The Casimir is

$$C = \frac{1}{2}[KH + HK] - D^2 = \frac{1}{4}\left[J - \frac{e}{2\pi c}\Phi\right]^2 - \frac{\hbar^2}{4}. \tag{15}$$

J is two-dimensional angular momentum quantized in integer multiples of $\hbar$, but the magnetic flux Φ of the vortex need not be quantized.

REFERENCES

1. P. A. M. Dirac, *Proc. Roy. Soc.* (London) **A133**, 60 (1931)
2. T. T. Wu and C. N. Yang, *Nucl. Phys.* **B107**, 365 (1976) and *Phys. Rev. D* **16**, 1018 (1977).
3. For a review of the hidden symmetries of the Kepler problem, see M. Bander and C. Itzykson, *Rev. Mod. Phys.* **38**, 330, 346 (1966).
4. The $O(2,1)$ group has been previously utilized to analyze various dynamical problems; for a review see B. Wybourne, *Classical Groups for Physicists* (Wiley, New York, NY, 1974). An application to the $1/r^2$ potential, very similar to the one I give here for the magnetic monopole, is by V. de Alfaro, S. Fubini, and G. Furlan, *Nuovo Cim.* **A34**, 569 (1976).
5. P. A. M. Dirac, *Rev. Mod. Phys.* **21**, 392 (1949).
6. R. Jackiw, *Ann. Phys.* (NY) **129**, 183 (1980).
7. R. Jackiw, *Ann. Phys.* (NY) **201**, 83 (1990).

Paper III.3

In a gauge theory, recognition of symmetry in physical situations is obscured by the fact that the formalism employs unphysical and redundant gauge variables. How to separate the invariant physics from the non-invariant mathematical description was described at the Schladming, Austria (February 1980) Ski School.

INVARIANCE, SYMMETRY AND PERIODICITY IN GAUGE THEORIES

Acta Phys. Austr. Suppl. XXII, 383 (1980)

I. INTRODUCTION

Studying symmetry properties of dynamical systems has always been a rewarding and useful activity. Not only is it aesthetically satisfying to identify simple, symmetric structures that lie within complicated formalisms, but also the appreciation of symmetry principles provides a framework for ordering and understanding an otherwise chaotic situation. Moreover, the universally shared belief that fundamental natural laws should possess a high degree of symmetry, which however may manifest itself in a subtle fashion, has stimulated the search for various mechanisms that may "hide" the action of a symmetry.

The reason that something new can still be said about this classical subject is the emergence of gauge theories as premier candidates for models of fundamental processes. Because gauge transformations can be performed at will, without affecting physical content, it is clear that they can obscure symmetry properties: a symmetric gauge-invariant configuration can be built from non-symmetric, gauge-variant objects. It is the interplay between gauge transformations and coordinate transformations that will concern us here and the nontrivial results that we shall present arise from the nontrivial nature of that relationship.

Our theory will aid in understanding the mixing of space-time and internal degrees of freedom, as observed in the well known "spin from isospin" phenomenon.[1] We shall systematize the frequently noticed circumstance that Yang-Mills fields that are invariant under some coordinate transformation are equivalent to a Yang-Mills-Higgs system.[2] It will be explained why some constants of motion, like the angular momentum of a charged point particle in the field of a monopole,[3] have unexpected non-kinematical additions. Finally, the concept of time-periodicity when combined with gauge variance will allow putting the Pontryagin index to new use in a description of magnetic monopole configurations.[4]

Appropriate references to relevant research papers will be given in the course of the lectures. Here let me recount the history of the subject. The problem of describing and constructing symmetric gauge fields, *i.e.*, gauge fields invariant under some specified transformation, was posed and solved completely in the context of the mathematical theory of fiber-bundles in the fifties. The principal work, due to Wang,[5] has found its way into the mathematical literature under the name of *Wang's Theorem*.[6] Needless to say, this had no impact on physics and physicists, who for a long time appreciated that symmetries operate in a subtle way in gauge theories. (It was known, for example, that a magnetic monopole gives rise to spherically symmetric dynamics, yet its vector potential is not rotationally invariant.) When non-Abelian gauge theories became the focus of attention, physicists quickly realized that interesting configurations like 't Hooft-Polyakov monopoles[7] and instantons[8] are invariant under coordinate transformations, yet the explicit vector potentials are not; a non-invariance is compensated by a gauge transformation. Systematic investigation began in ignorance of Wang's work,[9] with the most complete analysis being given by the physicists Forgàcs and Manton,[10] as well as by the mathematical physicists Harnad, Schnider and Vinet.[11] By now our knowledge has arrived to the level of the mathematicians but of course, our results are much more explicit and directly applicable to further physical speculation.

My own interest in the subject was aroused by a study of the instanton's symmetry properties. In establishing invariance under various conformal transformations, it was necessary to use gauge transformations, not only global ones — these had already been encountered in discussions of rotational invariance of the 't Hooft-Polyakov monopole[12] — but also local ones played a role.[13] Subsequent investigations[14] culminated in a recent paper with Manton[15] concerning the interplay of gauge transformations with conservation laws in physics. The work will be described below.

Let me conclude these introductory remarks by setting various conventional definitions. We shall be dealing with gauge potentials A_μ which are anti-Hermitian matrices in the space of the Lie algebra of the gauge group. The generators T^a satisfy

$$[T^a, T^b] = f^{abc}T^c \tag{1.1}$$

and the potentials may also be written in component notation

$$A_\mu = A_\mu^a T^a. \tag{1.2}$$

The field strength is related to the potential by

$$F_{\mu\nu} = \partial_\mu A_\nu - \partial_\nu A_\mu + [A_\mu, A_\nu], \tag{1.3}$$

where a possible coupling constant has been scaled to unity. The fields obey Bianchi identities

$$D_\mu {}^*F^{\mu\nu} = 0,$$
$$^*F^{\mu\nu} = \frac{1}{2}\varepsilon^{\mu\nu\alpha\beta}F_{\alpha\beta} \tag{1.4}$$
$$D_\mu = \partial_\mu + [A_\mu, \cdot$$

The field equation

$$D_\mu F^{\mu\nu} = 0 \tag{1.5}$$

is the Euler-Lagrange equation for stationarizing the action

$$S = \text{tr}\, \frac{1}{2} \int d^4x F^{\mu\nu} F_{\mu\nu}.$$ (1.6)

The theory is invariant under gauge transformations,

$$A_\mu \Rightarrow U^{-1} A_\mu U + U^{-1} \partial_\mu U,$$ (1.7a)

$$F_{\mu\nu} \Rightarrow U^{-1} F_{\mu\nu} U$$ (1.7b)

which in infinitesimal form

$$U = I + u + \dots$$ (1.8)

are given by the following formulas

$$A_\mu \Rightarrow A_\mu + D_\mu u,$$ (1.9a)

$$F_{\mu\nu} \Rightarrow F_{\mu\nu} + [F_{\mu\nu}, u].$$ (1.9b)

Note that equations (1.1) – (1.4) and (1.7) – (1.9) require no statement concerning the metric of the space; but in (1.5) and (1.6) a metric tensor must be used to raise indices. Mostly, we shall work in four-dimensional flat space, Euclidean or Minkowskian; for the latter

$$g_{\mu\nu} = \begin{pmatrix} 1 & 0 & 0 & 0 \\ 0 & -1 & 0 & 0 \\ 0 & 0 & -1 & 0 \\ 0 & 0 & 0 & -1 \end{pmatrix}.$$

Occasionally, I shall mention Yang-Mills fields in curved space; then (1.5) must be modified by the familiar methods of tensor calculus and the measure in (1.6) acquires the factor $\sqrt{g} = \sqrt{|\det g_{\mu\nu}|}$.

There are various questions that one may ask concerning the response of this theory to coordinate transformations. Firstly, one may wish to know what should be the rule for transforming A_μ and $F_{\mu\nu}$. Next, one may inquire how to characterize a symmetric gauge potential, *viz.* a gauge potential that is invariant under a coordinate transformation. A third question asks under which coordinate transformations are the dynamical equations of the theory (1.5) invariant; or equivalently which coordinate transformations leave the action (1.6) invariant. The answer to these and related questions comprises the content of my lectures. Because we are dealing with gauge fields, the material departs in a significant fashion from familiar results appropriate to ordinary fields.

II. COORDINATE TRANSFORMATION LAWS FOR ARBITRARY FIELDS

Consider a coordinate transformation which takes the four-vector x to another four-vector x', which is a specific function of x

$$x^\mu \Rightarrow x'^\mu(x). \tag{2.1}$$

We inquire how a field transforms under (2.1). The conventional answer is that a *scalar* field ϕ becomes a new scalar field ϕ', which is related to ϕ by

$$\phi'(x') = \phi(x). \tag{2.2a}$$

For a *vector* field v_μ, which goes to the new vector field v'_μ, the transformation rule is

$$v'_\mu(x') = v_\alpha(x)\frac{\partial x^\alpha}{\partial x'^\mu}. \tag{2.2b}$$

For a *2nd-rank tensor* field $t_{\mu\nu}$, we have

$$t'_{\mu\nu}(x') = t_{\alpha\beta}(x)\frac{\partial x^\alpha}{\partial x'^\mu}\frac{\partial x^\beta}{\partial x'^\nu} \tag{2.2c}$$

etc. It is useful to give an infinitesimal formulation of the above. Upon defining the first-order small quantities

$$\delta x^\mu = x'^\mu - x^\mu = -f^\mu(x) \tag{2.3}$$

for the coordinate transformation, and

$$\phi'(x) - \phi(x) = \delta\phi(x), \tag{2.4a}$$

$$v'_\mu(x) - v_\mu(x) = \delta v_\mu(x), \tag{2.4b}$$

$$t'_{\mu\nu}(x) - t_{\mu\nu}(x) = \delta t_{\mu\nu}(x) \tag{2.4c}$$

for the fields, we find from (2.2)

$$\delta_f\phi = f^\alpha\partial_\alpha\phi, \tag{2.5a}$$

$$\delta_f v_\mu = f^\alpha\partial_\alpha v_\mu + (\partial_\mu f^\alpha)v_\alpha, \tag{2.5b}$$

$$\delta_f t_{\mu\nu} = f^\alpha\partial_\alpha t_{\mu\nu} + (\partial_\mu f^\alpha)t_{\alpha\nu} + (\partial_\nu f^\alpha)t_{\mu\alpha}. \tag{2.5c}$$

The combination of derivatives appearing in the above occurs sufficiently frequently to merit a special name. The *Lie derivative* L_f of a general tensor field $t^{\alpha\beta\cdots}_{\mu\nu\cdots}$ is defined so that it produces a tensor of the same type

$$L_f t^{\alpha\beta\cdots}_{\mu\nu\cdots} = f^\lambda \partial_\lambda t^{\alpha\beta\cdots}_{\mu\nu\cdots}$$

$$+ (\partial_\mu f^\lambda) t^{\alpha\beta\cdots}_{\lambda\nu\cdots} + (\partial_\nu f^\lambda) t^{\alpha\beta\cdots}_{\mu\lambda\cdots} + \cdots \tag{2.6}$$

$$- (\partial_\lambda f^\alpha) t^{\lambda\beta\cdots}_{\mu\nu\cdots} - (\partial_\lambda f^\beta) t^{\alpha\lambda\cdots}_{\mu\nu\cdots} - \cdots .$$

The Lie derivative of products of tensors follows Leibniz rule and contractions are respected, so that the order of taking the Lie derivative and contracting indices is immaterial

$$L_f v^\mu = (L_f g^{\mu\nu}) v_\nu + g^{\mu\nu} L_f v_\nu. \tag{2.7}$$

With this definition, we see from (2.5) that the infinitesimal response of any field to an infinitesimal coordinate change f^μ is given by the Lie derivative with respect to f^μ.

$$\delta x^\mu = -f^\mu, \tag{2.8}$$

$$\delta_f \phi = L_f \phi, \tag{2.9a}$$

$$\delta_f v_\mu = L_f v_\mu, \tag{2.9b}$$

$$\delta_f t_{\mu\nu} = L_f t_{\mu\nu}. \tag{2.9c}$$

The above are the geometrical transformations laws. For translations $[f^\mu = a^\mu]$ and Lorentz transformations $[f^\mu = \omega^{\mu\nu} x, \ \omega^{\mu\nu} = -\omega^{\nu\mu}]$ they coincide with the familiar rules of field theory. For dilations $[f^\mu = a x^\mu]$ and special conformal transformations $[f^\mu = c^\mu x^2 - 2c \cdot x x^\mu]$ they coincide with the familiar rules only for those fields whose canonical scale dimension [in units of mass] is equal to the tensor rank [spin]. In particular for a 4-dimensional vector field, which has unit scale dimension, the geometrical transformation law is identical with the conventional one.[16]

A field is said to be *symmetric* or *form invariant* with respect to a definite coordinate transformation when it is invariant against that change of coordinates

$$\phi'(x) = \phi(x), \tag{2.10a}$$

$$v'_\mu(x) = v_\mu(x), \tag{2.10b}$$

$$t'_{\mu\nu}(x) = t_{\mu\nu}(x). \tag{2.10c}$$

In terms of the infinitesimal formulation, the invariance condition (2.10) is the statement that the Lie derivative vanishes

$$L_f \phi(x) = 0 \,, \tag{2.11a}$$

$$L_f v_\mu(x) = 0 \,, \tag{2.11b}$$

$$L_f t_{\mu\nu}(x) = 0 \,. \tag{2.11c}$$

Within these formulas we recognize familiar results: A translationally invariant field is one whose derivative vanishes: $a^\mu \partial_\mu \phi = 0$. A Lorentz invariant scalar field is one with vanishing antisymmetrized derivative: $\omega^{\mu\nu} x_\mu \partial_\nu \phi(x) = \frac{1}{2} \omega^{\mu\nu}(x_\mu \partial_\nu - x_\nu \partial_\mu)\phi(x) = 0$, etc. The condition for coordinate invariance of a 2nd-rank tensor is encountered in general relativity theory. If we take for $t_{\mu\nu}$ the metric tensor $g_{\mu\nu}$, then (2.11c) may also be written as

$$\nabla_\mu f_\nu + \nabla_\nu f_\mu = 0 \,, \tag{2.12}$$

where ∇_μ denotes coordinate covariant differentiation

$$\nabla_\mu f_\nu = \partial_\mu f_\nu - \Gamma^\lambda_{\mu\nu} f_\lambda \,, \tag{2.13a}$$

$$\Gamma^\lambda_{\mu\nu} = \frac{1}{2} g^{\lambda\alpha}[\partial_\mu g_{\alpha\nu} + \partial_\nu g_{\alpha\mu} - \partial_\alpha g_{\mu\nu}]. \tag{2.13b}$$

Equation (2.12) is the *Killing equation* which determines when a given metric is invariant against coordinate transformations. Infinitesimal coordinate transformations f^μ satisfying (2.12) are called *Killing vectors*. In flat space, the only solutions are the translations and Lorentz transformations, and the dynamical equations of field theories are invariant under these *Poincaré transformations*, provided the fields are transformed according to (2.9). This invariance persists for field theories formulated on a curved background geometry, when the infinitesimal coordinate transformations are the appropriate Killing vectors.

Finally, let us recall that for massless fields, dynamical field equations are frequently invariant under further coordinate transformations, the dilational and special conformal ones. In particular, Yang-Mills theory possesses this invariance, with the vector potentials transforming by (2.9b).[17] In curved space, one continues to find the larger invariance when the infinitesimal coordinate transformation satisfies the *conformal Killing equation*,

$$\nabla_\mu f_\nu + \nabla_\nu f_\mu = \frac{1}{2} g_{\mu\nu} \nabla_\alpha f^\alpha \tag{2.14}$$

whose only solutions in flat space are the already mentioned translations, Lorentz transformations, dilations and special conformal transformations.[16]

III. COORDINATE TRANSFORMATIONS FOR GAUGE FIELDS

The previous lecture recapitulated familiar ideas as applied to arbitrary fields. Let us now explore how these ideas should be modified and generalized when the discussion is restricted to gauge fields.

We begin by re-examining the condition for invariance of a field against coordinate transformations. In a gauge theory, it is correct and natural to speak of invariance, even when the vector potential is not manifestly invariant, but any non-invariance can be compensated by a gauge transformation. Thus for a gauge potential, the infinitesimal invariance condition (2.11b) is generalized to

$$L_f A_\mu = D_\mu W_f. \tag{3.1}$$

Indeed one may go further. Let us observe that the conventional transformation law

$$\delta_f A_\mu = L_f A_\mu \tag{3.2}$$

may be rewritten in the following way, by adding and subtracting certain terms

$$\delta_f A_\mu = f^\alpha (\partial_\alpha A_\mu - \partial_\mu A_\alpha + [A_\alpha, A_\mu])$$
$$+ f^\alpha \partial_\mu A_\alpha + [A_\mu, f^\alpha A_\alpha] + (\partial_\mu f^\alpha) A \tag{3.3}$$
$$= f^\alpha F_{\alpha\mu} + D_\mu (f^\alpha A_\alpha).$$

The infinitesimal coordinate transformation, when acting in its geometrical definition on the gauge potential, is given by one term involving the gauge-covariant field strength and a second term that is an infinitesimal gauge transformation. Since gauge transformations can be performed at will, it is suggestive that the action of an infinitesimal coordinate transformation on a gauge potential be defined as just the first term in (3.3). In this way, one is lead to a modified transformation law which I have called the *gauge-covariant coordinate transformation*[14]

$$\bar\delta_f A_\mu = f^\alpha F_{\alpha\mu}. \tag{3.4}$$

Clearly, nothing conceptually new is introduced by working with the gauge-covariant rule, rather than with the conventional, geometrical one (3.2). However, there are several advantages that (3.4) has over (3.2).

Firstly, for f^μ's that leave dynamical field equations invariant, [*i.e.*, conformal transformations or their curved-space generalizations] the action, which is gauge-invariant, remains invariant against (3.4), if it was invariant against (3.2).

Exercise: Verify directly that the curved space generalization for S in (1.6) is invariant against (3.4) when f^μ satisfies the conformal Killing equation.

By Noether's theorem a conserved current can be constructed, but the conventional formulation gives rise to a current involving the canonical energy-momentum tensor which is neither symmetric nor gauge-invariant. One has to embark upon the well-known Belinfante "improvement" of the stress-tensor by adding super-potentials — divergences of anti-symmetric tensors — which are by themselves conserved.[16] This awkward procedure is unnecessary when (3.4) is adopted — the conserved current which emerges from an application of Noether's theorem automatically involves the symmetric, gauge-invariant tensor

$$J_f^\mu = \theta^{\mu\nu} f_\nu, \tag{3.5a}$$

$$\theta^{\mu\nu} = \operatorname{tr} 2 \left(F^{\mu\alpha} F^\nu{}_\alpha - \frac{g^{\mu\nu}}{4} F^{\alpha\beta} F_{\alpha\beta} \right). \tag{3.5b}$$

Secondly, it is stated in discussions of super-symmetry that the composition law for two super-symmetry transformations is a coordinate transformation.[18] Nevertheless, when the formalism is examined in detail for gauge fields, one finds that the coordinate transformation in question is the gauge-covariant one (3.4) rather than the conventional one (3.2).

Thirdly, the gauge-covariant form has a very natural meaning in fiber-bundle language. It is the Lie derivative of the connection along the horizontal lift of f^μ, while the conventional transformation involves a Lie derivative along the fiber as well. The removal of the gauge transformation portion of $L_f A_\mu$ achieves this.[19]

Finally, it is worth mentioning, although I shall not elaborate on this here, that the finite version of (3.4) is also interesting. It leads in a natural way to the Wu-Yang nonintegrable phase factor[20] and to concepts related to the loop space of a gauge theory.[14]

Exercise: Compute the commutator of two gauge-covariant coordinate transformations and compare with that of two conventional transformations [see (4.8)]. Further discussion of the difference between the two is in Ref. 14.

An invariant gauge potential may also be characterized by reference to its behavior under a gauge-covariant coordinate transformation: it must be possible to compensate the coordinate change by a gauge transformation. In this way a criterion equivalent to (3.1) may be given

$$f^\alpha F_{\alpha\mu} = D_\mu \phi_f. \qquad (3.6)$$

For an invariant field, a projection of the field along the vector describing the coordinate transformation is a total gauge-covariant divergence of a scalar field ϕ_f. Comparison of (3.1) with (3.3) and (3.6) gives the relationship between the two generators of infinitesimal gauge transformations W_f and ϕ_f

$$W_f = \phi_f + f^\alpha A_\alpha. \qquad (3.7)$$

Formula (3.6) has an analogue in Riemannian geometry. Consider a metric tensor which admits Killing vectors; *i.e.*, it is invariant under coordinate transformations which satisfy (2.12). It therefore follows that

$$\nabla_\alpha(\nabla_\mu f_\nu) = \frac{1}{2}[\nabla_\nu, \nabla_\alpha]f_\mu + \frac{1}{2}[\nabla_\nu, \nabla_\mu]f_\alpha + \frac{1}{2}[\nabla_\alpha, \nabla_\mu]f_\nu. \qquad (3.8)$$

But the commutator of two coordinate covariant derivatives is given by the curvature tensor,

$$[\nabla_\mu \nabla_\nu]f_\alpha = R_{\nu\mu\alpha}{}^\lambda f_\lambda \qquad (3.9)$$

which possess a cyclic property under permutation of its first three indices. Consequently (3.8) becomes

$$R_{\omega\phi\alpha}{}^\lambda f_\lambda = \nabla_\alpha(\nabla_\omega f_\phi). \qquad (3.10a)$$

If we suppress the indices ω, ϕ by considering $R_{\omega\phi\alpha}{}^\lambda$ to be the 4×4 matrix $R_{..\alpha}^{\ \ \lambda}$, then (3.10a) reads

$$f^\lambda R_{..\alpha\lambda} = \nabla_\alpha(\nabla . f.). \qquad (3.10b)$$

This formula expresses the same fact as (3.6): in the invariant situation, certain projections of the curvature are given by a covariant gradient of some other quantity. In Yang-Mills theory that other quantity remains unspecified; in general relativity a formula for it is found: ∇f. This difference reflects the fact that general relativity, though analogous to a non-Abelian gauge theory, possess further structure that gives rise to crucial distinctions.[21]

IV. PROPERTIES OF SYMMETRIC GAUGE FIELDS

We have two equivalent characterizations of invariant gauge fields

$$L_f A_\mu = D_\mu W_f, \qquad (4.1)$$

$$f^\alpha F_{\alpha\mu} = D_\mu \phi_f. \qquad (4.2)$$

As will be seen, both formulations are useful and the connection between them is through

$$W_f = \phi_f + f^\alpha A_\alpha. \qquad (4.3)$$

Let us now explore the consequences. First we inquire how ϕ_f and W_f change under gauge transformations. Since (4.2) involves gauge covariant structures, we readily deduce that a gauge transformation on the potentials

$$A_\mu \Rightarrow U^{-1} A_\mu U + U^{-1} \partial_\mu U \qquad (4.4)$$

induces a covariant change on ϕ_f,

$$\phi_f \Rightarrow U^{-1} \phi_f U \qquad (4.5a)$$

and then (4.3) implies

$$W_f \Rightarrow U^{-1} W_f U + U^{-1} f^\mu \partial_\mu U = U^{-1} W_f U + U^{-1} L_f U. \qquad (4.5b)$$

Equation (4.5b) exhibits the important fact that when only *one* given transformation f^μ is under discussion, it is always possible to make a gauge transformation which transforms W_f to zero, and the transformed gauge potentials are explicitly invariant; they are annihilated by the appropriate Lie derivative. In order to achieve this, one needs to find the gauge function U_f, satisfying the equation

$$f^\mu \partial_\mu U_f + W_f U_f = 0 \qquad (4.6)$$

which always has solutions. Thus for one definite coordinate transformation, the generalization that we have here introduced is unnecessary since by a change of gauge, one may always achieve explicit invariance of the potentials.

The theory becomes non-trivial when more than one coordinate transformation is under discussion. Consider two such: f^α and g^α; one is naturally led also to a third h^α,

$$h^\alpha = f^\alpha \partial_\beta g^\alpha - g^\beta \partial_\beta f^\alpha = L_f g^\alpha = -L_g f^\alpha, \qquad (4.7)$$

which occurs in the commutator of two Lie derivatives

$$[L_f, L_g] = L_h. \qquad (4.8)$$

Invariance of the potentials under these three coordinate transformations gives rise to W_f, W_g, W_h, and all three cannot be made to vanish by one gauge transformation.

The W's satisfy an equation which follows from (4.1) and (4.8)

$$D_\mu W_h = D_\mu(L_f W_g - L_g W_f + [W_f, W_g]). \qquad (4.9a)$$

Apart from covariant constants that need be treated separately but will not be discussed here,[22] Eq. (4.9a) gives

$$L_f W_g - L_g W_f + [W_f, W_g] - W_h = 0. \qquad (4.9b)$$

This is a consistency requirement that must be satisfied by generators of gauge transformations W_f, W_g and W_h which are to compensate for the non-invariance of A_μ against coordinate transformations f^α, g^α and h^α.

Note that (4.9b) is a purely geometrical relation; no reference to the gauge potential occurs. Consequently, one may contemplate choosing a set of coordinate transformation functions, then solving (4.9b) for the most general gauge transformations, and finally determining from (4.1) the most general vector potentials invariant under the transformation. That this is always possible and the procedure for doing it is the content of Wang's theorem[5,6] which was recently rediscovered by physicists.[10,11]

I shall exemplify the analysis of (4.9b) in the next lecture, but first I wish to record the formula, analogous to (4.9b), which holds for the gauge-covariant functions ϕ. Upon substituting (4.3) into (4.9b), we get

$$f^\alpha D_\alpha \phi_g - g^\alpha D_\alpha \phi_f + [\phi_f, \phi_g] - \phi_h = f^\alpha g^\beta F_{\beta\alpha}. \qquad (4.10a)$$

However, the first two terms occurring on the left-hand side may be re-expressed, by virtue of (4.2), in terms of projections of the field strength. Hence, one further finds

$$f^\alpha g^\beta F_{\alpha\beta} = [\phi_f, \phi_g] - \phi_h. \qquad (4.10b)$$

Not only do certain projections of field strengths equal total covariant derivatives of scalars, as we saw before in (4.2), but also some further components are given by just the same scalars.

This fact allows for the following observation. Consider the Lagrangian of a pure Yang-Mills theory on some manifold M. Further make the *Ansatz* that the fields will be invariant under coordinate transformations on a sub-manifold M'. Then the Lagrangian, $L = \frac{1}{2}$ tr $F^{\mu\nu} F_{\mu\nu}$, which involves a summation over components μ, ν referring both to M' and its complement M'', $M = M' \cup M''$, may be split into terms in the direction parallel to the vectors specifying the symmetry transformations that leave the fields invariant, and in the directions perpendicular to these. The former lie in M', the latter in M''

$$L = \frac{1}{2} \operatorname{tr} F^{\parallel\,\parallel} F_{\parallel\,\parallel} + \operatorname{tr} F^{\parallel\,\perp} F_{\parallel\,\perp} + \frac{1}{2} \operatorname{tr} F^{\perp\,\perp} F_{\perp\,\perp}. \qquad (4.11a)$$

In the first term on the right-hand side, we may use (4.10b), in the second (4.2), to give symbolically

$$L = \frac{1}{2} \operatorname{tr} F^{\perp\,\perp} F_{\perp\,\perp} + (D_\perp \phi)(D^\perp \phi) - V(\phi)$$

$$V(\phi) = - \operatorname{tr} \frac{1}{2} \Big\{ [\phi, \phi] - \phi \Big\}^2. \qquad (4.11b)$$

The first term of L is recognized to be a Yang-Mills Lagrangian on M''; the second is a kinetic term for a Higgs field; the last is a Higgs potential. In this way, we see that a pure Yang-Mills theory on a manifold but with the *Ansatz* that all quantities be invariant against some coordinate transformation, is always equivalent to a Yang-Mills-Higgs system on a smaller manifold.

This result is familiar from individual instances. The 't Hooft-Polyakov monopole in the Prasad-Sommerfield limit is a solution to a Yang-Mills-Higgs system in Euclidean 3-space. But the equations are known to be formally equivalent to a 4-dimensional Yang-Mills theory in Euclidean 4-space with invariance against translations in the fourth direction. [The fourth component of the vector potential is identified with the Higgs field.][2] Another example is Euclidean 4-dimensional Yang-Mills theory with $O(3)$ symmetry, which is equivalent to a 2-dimensional Abelian gauge-Higgs theory.[2]

This remarkable and intriguing circumstance has prompted Manton and others[23] to speculate about the feasibility of obtaining the physically interesting Yang-Mills-Higgs structure of the Weinberg-Salam model[24] from a pure Yang-Mills theory in a space of dimensionality greater than four. Partial success has been achieved. One starts with a gauge theory based on a rank 2 Lie group, $SU(3)/Z^3$, $O(5)$ or G_2, in the 6-dimensional space $M^{(4)} \times S^{(2)}$, *i.e.*, Minkowski space enlarged with a 2-dimensional sphere, and demands invariance against rotations on the sphere. The resulting model is a 4-dimensional Yang-Mills-Higgs system based on $SU(2) \times U(1)$ with a complex doublet of Higgs fields. The Higgs kinetic term comes from the $F_{5\mu}^2 + F_{6\mu}^2$ contributions to the 6-dimensional action and the Higgs potential is the F_{56}^2 term. Remarkably the potential always leads to symmetry breaking. The number of parameters in the new approach is two [gauge coupling constant and radius of $S^{(2)}$] as compared to Weinberg's and Salam's four [two gauge coupling constants, coefficients of the quadratic and quartic terms in the Higgs potential]. Consequently, two predictions emerge: the mass of the Higgs particle is degenerate with that of the Z^0; θ_W is 60°, 45°, and 30°, respectively for each of the three groups. [These predictions are on the tree level, without quantum corrections.]

The appealing features of the idea, which is a modern reprise of Kaluza-Klein theory,[25] are obvious. Higgs fields, always an unattractive feature of models with spontaneous symmetry breaking, emerge naturally and are not introduced *ad-hoc*. Their couplings are not arbitrarily arranged to achieve desired results, but are completely determined; this leads to testable physical predictions. Unfortunately there are also difficulties. Fermions have not, as yet, been incorporated into the scheme. Very serious is the objection that in a quantum theory there is no place for a symmetry *Ansatz*. Quantum fluctuations cannot be arbitrarily limited to lie within any prescribed functional form. Moreover, if we really have to deal with a 6-dimensional space, it is hard to see how to renormalize. Also one wonders whether attempts to derive the Weinberg-Salam model are fruitless if that theory is merely provisional and should be supplanted by a grand unified one.

Thus it is likely that, just like its Kaluza-Klein predecessor, this attempt to go beyond 4-dimensional space-time will fail to find a useful place in physics, even though the idea is undenyably intriguing and elegant. I agree that Higgs fields should disappear from the theory, but I believe that they will be replaced by the bound-state mechanism,[26] and not by dimensionally reduced Yang-Mills fields.

V. CONSTRUCTION OF SYMMETRIC GAUGE FIELDS

In order to construct gauge potentials which are invariant, up to gauge transformations, against coordinate transformations, we proceed as follows. The coordinate transformations are specified by their infinitesimal form. One chooses f^α, g^α and $h^\alpha = L_f g^\alpha$, and this set is supplemented, if necessary, to form a *group* of transformations. Next one finds all W_f, W_g and W_h which satisfy

$$L_f W_g - L_g W_f + [W_f, W_g] - W_h = 0. \tag{5.1}$$

As we shall see, there are many distinct types of solutions to this equation. Of course, we are only interested in solutions modulo gauge transformations which leave (5.1) invariant

$$W_f \Rightarrow U^{-1} W_f U + U^{-1} L_f U. \tag{5.2}$$

The gauge potentials are determined from

$$L_f A_\mu = D_\mu W_f. \tag{5.3}$$

It turns out that (5.3) not only allows for the construction of general forms for A_μ, but further delimits W_f, beyond what is found from (5.1).

Before examplifying the procedure in a concrete application, let us note that a particularly simple result emerges if the W's are constant. In that special case (5.1) reduces to

$$W_h = [W_f, W_g] \tag{5.4}$$

which is the statement that the W's are those generators of the gauge group which follow the Lie algebra of the coordinate transformations. In other words, one solution to (5.1) is a set of constant W's which provide an embedding of the coordinate transformation group in the gauge group. In this case, the gauge transformation which compensates for the coordinate transformation is a global one, and this circumstance has been widely discussed.[12]

Less straightforward and, as we shall see, equally interesting is the general situation where the W's may be position dependent and the compensating gauge transformations become local. We illustrate this by examining in detail 3-dimensional rotations in a 4-dimensional Yang-Mills theory.[27]

For rotations, the infinitesimal coordinate transformations take the following forms

$$f^\mu = (0, \mathbf{n}_f \times \mathbf{r}), \quad g^\mu = (0, \mathbf{n}_g \times \mathbf{r}), \quad h^\mu = (0, \mathbf{n}_h \times \mathbf{r}). \tag{5.5a}$$

Here the $\mathbf{n}$'s are infinitesimal parameters specifying the rotation; they are the axes of rotation; the third $\mathbf{n}_h$ being determined by the other two

$$\mathbf{n}_h = \mathbf{n}_g \times \mathbf{n}_f. \tag{5.5b}$$

The W's, which generate infinitesimal gauge transformations that compensate for the infinitesimal rotations, must be linear in the parameters. Thus we set

$$W_f = \mathbf{W} \cdot \mathbf{n}_f, \quad W_g = \mathbf{W} \cdot \mathbf{n}_g, \quad W_h = \mathbf{W} \cdot \mathbf{n}_h. \tag{5.6}$$

Equation (5.1) now specializes to

$$[\mathbf{n}_f \times \mathbf{n}_g] \cdot [\nabla(\mathbf{r} \cdot \mathbf{W}) - \mathbf{r}(\nabla \cdot \mathbf{W}) + \mathbf{W} \times \mathbf{W}] = 0 \tag{5.7a}$$

which can be satisfied only by the vanishing of the vector quantity in the second bracket

$$\nabla(\mathbf{r} \cdot \mathbf{W}) - \mathbf{r}(\nabla \cdot \mathbf{W}) + \mathbf{W} \times \mathbf{W} = 0. \tag{5.7b}$$

The gauge transformation rule (5.2) becomes

$$\mathbf{W} \Rightarrow U^{-1}\mathbf{W}U + U^{-1}(\mathbf{r} \times \boldsymbol{\nabla})U. \tag{5.8}$$

The solution of (5.7b) is most easily presented in a radial coordinate system with the use of radial components

$$x = r\sin\theta\cos\phi, \quad y = r\sin\theta\sin\phi, \quad z = \cos\theta$$
$$\mathbf{W} = (W^r, W^\theta, W^\phi). \tag{5.9}$$

Note that W^r transforms homogeneously under gauge transformations, while W^θ and W^ϕ undergo inhomogeneous transformations

$$W^r \Rightarrow U^{-1}W^rU, \tag{5.10a}$$

$$W^\theta \Rightarrow U^{-1}W^\theta U - U^{-1}\frac{1}{\sin\theta}\frac{\partial}{\partial\phi}U, \tag{5.10b}$$

$$W^\phi \Rightarrow U^{-1}W^\phi U + U^{-1}\frac{\partial}{\partial\theta}U. \tag{5.10c}$$

This means that non-vanishing W^r is a gauge-invariant signal that $\mathbf{W}$ cannot be transformed to zero, and that the vector potentials cannot be made manifestly rotationally covariant. Correspondingly, a gauge may always be chosen such that W^θ or W^ϕ vanishes.

The solution is now straightforwardly obtained, by setting $W^\phi = 0$; a condition which may always be attained, according to (5.10c). One finds

$$W^r = \Omega, \quad W^\theta = \operatorname{ctn}\theta\,\Omega, \quad W^\phi = 0. \tag{5.11}$$

Here Ω is an arbitrary element of the Lie algebra, independent of θ and ϕ. The absence of any further ϕ dependence is also a consequence of the gauge freedom (5.10b). The above solution and gauge choices still allow for gauge transformations with gauge functions that do not depend on θ and ϕ, but vary with r and t.

To proceed, we need to implement Eq. (5.3), which here reduces to

$$(\mathbf{x} \times \boldsymbol{\nabla})A^0 = D_0\mathbf{W}, \tag{5.12a}$$

$$(\mathbf{r} \times \boldsymbol{\nabla})^i A^j = D_j W^i. \tag{5.12b}$$

These are most easily analyzed by utilizing the remaining gauge freedom to rotate Ω into the Cartan sub-algebra of the gauge group. One then finds that Ω must be totally constant, *i.e.*, independent of r and t. A^0 and A^r must be independent of θ and ϕ and must lie in the same Cartan sub-algebra as Ω; *viz.*

they are functions of r and t and commute with Ω. A^θ and A^ϕ are determined to be

$$A^\theta = \frac{1}{r} a^\theta(r,t), \tag{5.13a}$$

$$A^\phi = \frac{\operatorname{ctn}\theta}{r}\Omega + \frac{1}{r} a^\phi(r,t), \tag{5.13b}$$

where a^θ and a^ϕ are solutions to

$$[a^\theta, \Omega] = a^\phi, \tag{5.14a}$$

$$[a^\phi, \Omega] = -a^\theta. \tag{5.14b}$$

Exercise: By decomposing (5.7) and (5.12) into components, derive (5.11), (5.13) and (5.14).

Finally, let us observe that A^ϕ is singular at $\theta = 0$ and π. This singularity may be removed by the gauge transformation $U = e^{\phi\Omega}$ near $\theta = 0$ [north pole] and by $U = e^{-\phi\Omega}$ near $\theta = \pi$ [south pole]. At the equator, the two differ by $e^{2\phi\Omega}$, which must be single-valued. Hence, a global requirement on Ω emerges: in order that the singularity in the vector potential be a gauge artifact, we must have

$$e^{4\pi\Omega} = I. \tag{5.15}$$

The global requirement on Ω is relaxed when several quantities are identified with the identity. For example, for the group $SU(N)/Z^N$, the right-hand side of (5.15) is replaced by $e^{2\pi i (n/N)}I$, $n = 1,\dots,N$.]

The field strengths may now also be determined. Especially interesting is that component of $F_{\mu\nu}$ which is completely constrained by the symmetry, see (4.10b). We find

$$B^r = -\frac{1}{r^2}\{\Omega + [a^\theta, a^\phi]\}, \tag{5.16}$$

where the magnetic field is given by $B^i = -\frac{1}{2}\varepsilon^{ijk}F^{jk}$.

The procedure for constructing gauge potentials that are invariant under rotations is now clear. Most simply, Ω can be chosen to vanish. Then the only non-vanishing components are A^0 and A^r. They are arbitrary functions of r and t and rotational symmetry is realized in an explicit, trivial way.

Next, Ω is taken to be a constant element of the Cartan sub-algebra, consistent with (5.15) and A^0, A^r lie in the same Cartan sub-algebra. [In an explicit representation, Ω is a diagonal matrix.] If with the chosen Ω, the only solution to (5.14) is vanishing a^θ and a^ϕ, we are dealing with Abelian Dirac monopoles, as is seen from (5.16), with their strength quantized by virtue of (5.15). [In an Abelian theory a^θ and a^ϕ are of course zero.] It is clear that here, $\mathbf{W}$ is necessarily position dependent and cannot be made constant by a gauge transformation.

When the chosen Ω permits non-trivial forms for a^θ and a^ϕ, two cases can still be distinguished. Firstly, $-\Omega$ may have the proper normalization to be an element of the $SU(2)$ algebra. In that case a^θ and a^ϕ obviously are non-trivial; they are constructed from the remaining two $SU(2)$ generators. One can then perform a gauge rotation so that $\mathbf{W} = (W^r = \Omega,\ W^\theta = \operatorname{ctn}\theta\Omega,\ W^\phi = 0)$ becomes a constant matrix triplet $\mathbf{W} = (W^x, W^y, W^z)$. Indeed, the result is just the embedding of $SU(2)$ in the gauge group which was discussed in connection with (5.4). The relevant gauge transformation is

$$U = e^{-\theta\omega_y}\, e^{\theta\Omega}, \tag{5.17}$$

where ω_y forms an $SU(2)$ algebra with $-\Omega$

$$[\Omega, \omega_y] = +\omega_x, \tag{5.18a}$$

$$[\Omega, \omega_x] = -\omega_y, \tag{5.18b}$$

$$[\omega_y, \omega_x] = \Omega. \tag{5.18c}$$

Exercise: Prove that a necessary and sufficient condition for $\mathbf{W}$ to be gauge equivalent to a constant triplet is that $-\Omega$ be an element of the $SU(2)$ algebra.

From (5.14) and (5.18) it follows that B^r is entirely proportional to Ω, so that the singularity at $r = 0$ is absent when a^θ and a^ϕ satisfy appropriate boundary conditions at the origin. Therefore, within this case fall the non-singular, radially symmetric gauge field configurations. For $SU(2)$ this is the Witten *Ansatz*[28]; for $SU(3)$, two independent *Ansätze* are available corresponding to the two inequivalent ways of embedding $SU(2)$ in $SU(3)$: (T_1, T_2, T_3) and $(\sqrt{2}T_4 + \sqrt{2}T_7,\ \sqrt{2}T_5 + \sqrt{2}T_6,\ 2T_3)$, [with $T_a = \lambda^a/2i$].[29]

The second case obtains when $-\Omega$ is not a member of an $SU(2)$ algebra, but non-vanishing a^θ and a^ϕ do solve (5.14), with their commutator not closing on Ω. It is then impossible to find a gauge in which $\mathbf{W}$ is constant; the compensating gauge transformations are essentially local. Such configurations obviously do not exist for $SU(2)$, but in $SU(3)$ one can construct them. A choice for Ω, labelled by the integer n, and consistent with (5.15) is

$$-\Omega = T_3 + n\sqrt{3}T_8. \tag{5.19}$$

One verifies that non-vanishing a^θ and a^ϕ solve (5.14), but except in the cases $n = 0$ and $n = -1$, $-\Omega$ is not a member of an $SU(2)$ algebra. The magnetic field (5.16) necessarily possesses a singularity at the origin. It is precisely this type of symmetry *Ansatz*, but for $SU(3)/Z^3$ that Manton used in his derivation of the Weinberg-Salam-Higgs potential.[23]

In summary, radially symmetric gauge potentials come in various forms. There are the explicitly symmetric ones. There are the non-Abelian configurations which require a global gauge transformation to compensate the effect of a rotation. Next are the Abelian Dirac monopoles with a local gauge transformation compensating the rotational non-invariance of the vector potential. Lastly are non-Abelian configurations, whose occurrence has not been heretofore widely appreciated, which also use local gauge transformations. They lead to intriguing speculative applications.

Another place where essentially non-constant gauge transformations play a crucial role is in the analysis of the instanton's symmetries. That vector potential lives in Euclidean space; its form is

$$A_\mu = \frac{-2i\alpha_{\mu\nu}x^\nu}{x^2 + \lambda^2} \tag{5.19a}$$

with λ being a length scale, and $\alpha_{\mu\nu}$ is a matrix constructed from the Pauli matrices[8]

$$\alpha_{\mu\nu} = \frac{1}{4i}(\bar{\alpha}_\mu\alpha_\nu - \bar{\alpha}_\nu\alpha_\mu),$$
$$\alpha^\mu = (-i\boldsymbol{\sigma}, I), \quad \bar{\alpha}^\mu = (i\boldsymbol{\sigma}, I). \tag{5.19b}$$

The configuration is invariant against $O(4)$ rotations generated by

$$f^\mu = x_\alpha\omega^{\alpha\mu}, \quad \omega^{\alpha\mu} = -\omega^{\mu\alpha} \tag{5.20a}$$

and gauge compensated by

$$W_f = \frac{i}{2}\alpha_{\mu\nu}\omega^{\mu\nu}. \tag{5.20b}$$

The instanton is also invariant against $O(5)$ transformations, which are generated by the sum of special conformal transformations and translations

$$f^\mu = c^\mu(x^2 + \lambda^2) - 2c \cdot xx^\mu. \tag{5.21a}$$

However, the gauge compensation involves a local gauge transformation[13]

$$W_f = 2i\alpha_{\mu\nu}x^\mu c^\nu. \tag{5.21b}$$

Exercise: Verify that the W's of (5.20) and (5.21) satisfy the consistency condition (5.1).

VI. PHYSICAL SIGNIFICANCE OF GAUGE TRANSFORMATIONS

The generators of infinitesimal gauge transformations, defined by

$$L_f A_\mu = D_\mu W_f, \tag{6.1a}$$

$$f^\alpha F_{\alpha\mu} = D_\mu \phi_f \tag{6.1b}$$

entered the theory in a formal, mathematical way through the requirement of symmetry. However, they also have a very direct, physical meaning which I shall now explain.

Consider first a "matter" system of fields and/or particles propagating through space-time either freely or with self-interactions, but with no gauge fields. When the dynamical equations for the matter objects are invariant against various space-time transformations, there will be various constants of motion. Examples are the energy as a consequence of time-translational invariance; the angular momentum arising from rotational invariance; *etc.* We call these constants C^f, where f labels the infinitesimal coordinate transformation which leaves the dynamics unchanged, and we write the subscript "matter" to remind that the constant consists of contributions from matter fields and particles

$$C^f = C^f_{\text{matter}}. \tag{6.2a}$$

Next let the matter system interact with an external, prescribed gauge field A_μ. The interaction Lagrangian is

$$L_{int} = 2\,\mathrm{tr}\,j^\mu A_\mu \tag{6.2b}$$

where j^μ is the matter charge current which mediates the interaction.

Since a generic external field removes all symmetries, the invariance will be absent in the general case, and there will no longer be any constants of motion. However, in the special case when the external gauge fields themselves are invariant under the infinitesimal coordinate transformation f^μ, in the sense of (6.1), then the matter dynamics continues to be characterized by constants, whose form will now be

$$C^f = C^f_{\text{matter}} + C^f_{\substack{\text{external} \\ \text{gauge field}}} . \qquad (6.3)$$

The second term on the right-hand side allows for a contribution to the total conserved quantity from the external gauge field. As I shall show, such contributions are indeed present; they are essentially determined by ϕ_f, or equivalently by W_f. In other words, for invariant fields, the generators of the gauge transformations have the physical significance of carrying the external gauge field contribution to the total constant of motion. This provides another example of the familiar identity between constants of motion and generators of symmetry transformations.

The proof of the assertion proceeds through an application of Noether's theorem; the canonical method for determining conservation laws in dynamical systems. This study has been carried through by Manton and me,[15] but I shall not present it here, owing to the tedium that details of the argument entail. Rather, I shall give a simple, heuristic discussion, which gets to the result quickly, even though it leaves some things muddled. Those interested can consult the cited literature for a complete derivation.

Let us consider first the matter quantities in interaction with dynamical gauge fields [as opposed to external gauge fields], where the entire dynamical system is invariant under the infinitesimal coordinate transformation f^μ. The argument is simplest if f^μ is taken without a time component; I consider here changes of coordinates that do not modify the time, the general case is treated in Ref. 15. A constant of motion will be present; its form is

$$C^f = \tilde{C}^f_{\text{matter}} + C^f_{\substack{\text{dynamical} \\ \text{gauge field}}} . \qquad (6.4)$$

The first term is the contribution from the matter variables and from the matter-gauge field interaction terms. The latter enter in order to make the matter contribution gauge-invariant, for example by changing coordinate derivatives of charged fields into gauge-covariant derivatives. The second term arises from the dynamical gauge fields. It is gauge-invariant and its form is

$$C^f_{\substack{\text{dynamical} \\ \text{gauge fields}}} = \int d\mathbf{r}\, \theta^{0\alpha}_{YM} f_\alpha, \qquad (6.5)$$

where $\theta^{\mu\nu}_{YM}$ is the Yang-Mills portion of the total energy-momentum tensor; see (3.5)

$$\theta^{\mu\nu}_{YM} = \text{tr}\, 2\left(F^{\mu\alpha} F^\nu_\alpha - \frac{g^{\mu\nu}}{4} F^{\alpha\beta} F_{\alpha\beta} \right). \qquad (6.6)$$

Equation (6.5) is now rewritten in the following equivalent ways

$$\int d\mathbf{r}\, \theta^{0\alpha}_{YM} f_\alpha = \int d\mathbf{r}\, \text{tr}\, 2F^{0\alpha} F_{\beta\alpha} f^\beta$$

$$= \int d\mathbf{r}\, \text{tr}\, 2F^{0\alpha} (F_{\beta\alpha} f^\beta - D_\alpha \phi)$$

$$+ \int d\mathbf{r}\, \text{tr}\, 2F^{0\alpha} D_\alpha \phi$$

$$= \int d\mathbf{r}\, \text{tr}\, 2F^{0\alpha}(f^\beta F_{\beta\alpha} - D_\alpha \phi) + \int d\mathbf{r}\, \text{tr}\, 2j^0\phi. \qquad (6.7a)$$

In the last step, we have integrated by parts and used the Yang-Mills field equation in the presence of charged matter. Thus far (6.7a) is true for an arbitrary ϕ; let us suppose now that the dynamical gauge field "freezes" into a prescribed external and symmetric configuration characterized by (6.1b). Then the next-to-last term in (6.7a) vanishes, and the gauge field contribution to the total constant of motion is determined by ϕ_f.

$$C^f = \tilde{C}^f_{\text{matter}} - \int d\mathbf{r}\, \rho_a \phi^a_f. \qquad (6.7b)$$

Here ρ_a is the charge density, the time component of j^μ_a, exhibited in component notation.

The decomposition in (6.7b) is gauge-invariant, since the gauge-covariant generator ϕ_f enters. Alternatively, one may rewrite this in terms of W_f

$$C^f = \tilde{C}^f_{\text{matter}} + \int d\mathbf{r}\, \rho_a A^\mu_a f_\mu - \int d\mathbf{r}\, \rho_a W^a_f. \qquad (6.8a)$$

It can be shown that the second term combines with the first and cancels away all gauge field-matter interaction terms, leaving the gauge-non-invariant, pure matter contribution

$$C^f = C^f_{\text{matter}} - \int d\mathbf{r}\, \rho_a W^a_f. \tag{6.8b}$$

This as well as other missing parts of the proof are supplied in Ref. 15.

Equations (6.7b) and (6.8b) are the desired results. They take on especially simple forms in various interesting special cases. When W^a_f is constant, the integral may be evaluated to give T^a, the total matter isospin [or its generalizations for non-$SU(2)$ theories]. Then (6.8b) becomes

$$C^f = C^f_{\text{matter}} - T^a W^{\,a}_f \tag{6.9}$$

and the constants W^a_f describe the embedding of the coordinate transformation group into the gauge group. Equation (6.9) is a generalization of the "spin-from-isospin" phenomenon that was described a few years ago.[1] That effect is recognized in (6.9) when we specialize to rotations. Then $-W^a_f = n^a_f$, the axis of rotation, and (6.9) becomes

$$\mathbf{J} = \mathbf{L} + \mathbf{T}. \tag{6.10}$$

The total angular momentum consists of a kinematical matter part $\mathbf{L}$ and a matter isospin part $\mathbf{T}$. When the matter carries half-integral isospin, the total angular momentum has half-integer eigenvalues even if the matter is purely bosonic.

For another interesting and familiar application, consider the matter to be a classical point particle

$$\rho_a(\mathbf{r}) = T^a \delta(\mathbf{r} - \mathbf{r}_0). \tag{6.11}$$

Then (6.7b) becomes

$$C^f = \tilde{C}^f_{\text{matter}} - T^a \phi^a_f(\mathbf{r}_0). \tag{6.12}$$

An example is the motion of a charged point particle in the field of an Abelian monopole

$$\mathbf{B} = g\, \frac{\hat{r}}{r^2}. \tag{6.13}$$

The criterion of rotational invariance, (6.1b), becomes in this simple case

$$(\mathbf{n} \times \mathbf{r}) \times B = \boldsymbol{\nabla}\phi \tag{6.14a}$$

and is solved by

$$\phi = -g\mathbf{n} \cdot \hat{r}. \tag{6.14b}$$

Removing the axis of rotation and deleting the isospin in this Abelian application, we regain the famous result of Poincaré concerning the non-kinematical contribution to the angular momentum of this dynamical system[3]

$$\mathbf{J} = \mathbf{r} \times m\dot{\mathbf{r}} - g\hat{\mathbf{r}}. \tag{6.15}$$

When the compensating gauge transformation is local, and the matter charge density distributed in space, one must remain with the formulas (6.7b) and (6.8b) which provide the proper extension to complicated general cases of the above familiar situations.

VII. MAGNETIC MONOPOLE TOPOLOGY WITHOUT HIGGS FIELDS

In this last lecture, I depart somewhat from our previous subject, and shall describe work done in collaboration with N. Christ concerning the Pontryagin index of magnetic monopole configurations.[4] Monopoles are among the first topological objects that were studied in the contemporary semi-classical explorations of quantum field theory. After their initial discovery as interesting classical solutions,[7] and quantization in the context of soliton physics,[30] their role was eclipsed somewhat by instantons and by the apparent circumstance that monopoles arise in unphysical models, for example, in the $SU(2)$ Georgi-Glashow[31] theory and not in the physically relevant Weinberg-Salam model.

In the present time, we are witnessing a resurgence of magnetic monopole investigations. Not only have the instanton calculations almost reached their conclusion, but also it has been realized that monopoles do occur in physically interesting theories. The various grand unification proposals are typically based on a simple gauge group, yet in order to describe Nature the symmetry must break down spontaneously to $SU(3) \times U(1)$ [color$\times$ electromagnetism]. Such a pattern is a rather reliable clue to the existence of monopoles, and those who are committed to the physical picture given by these theories must also accept the presence of magnetic monopoles in Nature.[32]

Monopole solutions require Higgs fields. Moreover, the entire discussion of topological behavior in gauge theories, that may give rise to the monopoles, is tied to Higgs fields at spatial infinity, where they become gauge-covariant constants.[33] This is unfortunate, since many of us believe that the scalar fields, which initiate symmetry breakdown, are a provisional part of theory, eventually to be replaced by a dynamical picture based on bound states,[26] or some other mechanism.[23] It behooves us therefore to see how much of the monopole story can be told without referring to scalar Higgs fields.

Of course in the absence of scalars, there are no classical solutions, and no semi-classical description of quantal monopoles can be given. However, if we believe that symmetry breaking should occur through a deep quantum mechanical mechanism, we should also acknowledge that monopole formation requires the full quantum theory.

But one still faces the question of how to describe the monopole sector and its topological properties in the absence of Higgs', which are crucial to the conventional analysis.[33] The purpose of our investigation, which I am here describing, is to expose the topological information that is encoded in the Yang-Mills fields themselves and to use properties of the gauge fields as a monopole signal. The discussion will be limited to an $SU(2)$ gauge theory. A generalization to other groups is given in Ref. 4; but this generalization does not provide as detailed a classification of monopoles in larger groups as does the usual theory based on Higgs fields. Clearly the subject requires further research, and it remains an open question how to achieve a detailed description of magnetic monopoles for groups larger than $SU(2)$, solely in terms of Yang-Mills fields.

Another motivation for our investigation is to provide a setting for the Pontryagin index in Minkowski space. Its role in Euclidean space in connection with the instanton phenomena is well-known.[34] While it certainly has been possible in Minkowski space to exhibit gauge fields with non-trivial index, these configurations do not possess any distinguishing characteristics [other than the non-zero index] which might make them interesting objects of study. It is therefore useful to establish a link between the Pontryagin index on the one hand, and monopole strength on the other.

What I shall show is that configurations of gauge fields that are periodic in time, and that satisfy some further conditions detailed below, possess a Pontryagin index that coincides with the monopole strength. The index, for Minkowski space fields with temporal period T, is defined by

$$q = -\frac{1}{16\pi^2} \int_0^T dt \int d\mathbf{r} \, \mathrm{tr} \, {}^*F^{\mu\nu} F_{\mu\nu}. \tag{7.1}$$

The time integration does not range to infinity, but is limited to one period, as it must be in order to avoid obvious divergence for periodic fields. Effectively, our fields live on $S^{(1)} \times R^{(3)}$ rather than on Minkowski space.

The proper mathematical setting for the Pontryagin index is on a compact space; thus our use of $R^{(3)}$ is somewhat unconventional. The standard move is to compactify, a procedure familiar from the instanton story where $R^{(4)}$ is replaced by $S^{(4)}$.[34] Alternatively, and more relevant physically, one may posit large-distance fall-off conditions on the fields. This we shall do by demanding that potentials drop off as r^{-1} and fields as r^{-2} at spatial infinity

$$A^\mu \xrightarrow[r\to\infty]{} O(r^{-1}), \tag{7.2a}$$

$$F^{\mu\nu} \xrightarrow[r\to\infty]{} O(r^{-2}). \tag{7.2b}$$

The asymptotes (7.2) are also physically motivated. That the scalar potential A^0 should vary as $1/r$ at large distances is merely the reflection that electric charges may be present; a vector potential $\mathbf{A}$ that persists as $1/r$ for large r is appropriate when monopoles are present. Note that this behavior is more singular than had been allowed in the discussion of vacuum periodicity and vacuum angle in Yang-Mills theory. In that analysis, we insisted that the vector potentials $\mathbf{A}$ disappear faster than $1/r$ at large r.[35] The difference is that in the earlier discussion, we were concerned with the vacuum sector of the theory; here we are dealing with the soliton-monopole sector. Some further aspects of the vacuum angle in the monopole sector will be touched on later.

A second condition imposed on the fields under discussion is that their entire time dependence, and the periodicity in time, reside in a gauge transformation. In other words, the fields are physically static, and the potentials with which we are dealing are gauge transformations of static potentials

$$A^\mu = U^{-1}A^\mu_S U + U^{-1}\partial^\mu U. \tag{7.3a}$$

Here A^μ_S is static, and the periodic time-dependence of A^μ is in U

$$U = e^u, \qquad u(\hat{r}, t) = -\omega(\hat{r})\frac{2\pi}{T}t$$

$$\omega = \frac{\omega_a \sigma^a}{2i}, \qquad \omega_a \omega_a = 1. \tag{7.3b}$$

Before proceeding, we need some discussion of the further conditions (7.2) and (7.3). Note that for the magnetic field $\mathbf{B}$, (7.2a) implies (7.2b)

$$B^i = -\frac{1}{2}\varepsilon^{ijk}F^{jk} = \frac{1}{2}\varepsilon^{ijk}(\partial_j A^k - \partial_k A^j - [A^j, A^k]). \tag{7.4a}$$

If A^i is $O(r^{-1})$, clearly F^{ij} is $O(r^{-2})$. But for the electric field E^i, the two are independent requirements

$$E^i = F^{i0} = D^i A^0 - \frac{d}{dt}A^i. \tag{7.4b}$$

While the first term on the right-hand side is $O(r^{-2})$ as consequence of (7.2a), the second term, which is also given by

$$-\frac{d}{dt}A^i = D_i\dot{u} = -D_i\omega\frac{2\pi}{T} \tag{7.5}$$

will be $O(r^{-2})$ at larger r if the covariant derivative of ω satisfies

$$\lim_{r\to\infty} rD_i\omega = 0. \tag{7.6}$$

In other words, the fall-off conditions (7.2) require that the infinitesimal gauge function ω be a covariant constant at spatial infinity.

[Here we are encountering for the first time some of the elements of the conventional analysis: the Higgs field is a covariant constant at spatial infinity.[33] Moreover, the present circumstances are reminiscent of the ideas in the earlier lectures where also we found Higgs-like structures in gauge fields.]

Why is it that I am working in a gauge where A^μ is time-dependent, rather than using A_S^μ which is static? The reason is that the static potentials do not satisfy the fall-off requirements (7.2). For when A^μ vanishes at large r, it is clear that A_S^0 tends to $\omega(2\pi/T)$, hence in general is non-vanishing. The gauge transformation, which introduces the time periodicity, is needed to remove a "singularity" at spatial infinity. Note however that the gauge function U introduces a singularity in A^μ at the origin, owing to the presence of $\omega(\hat{r})$ which is ill-defined there.

It is possible to give a description which is everywhere non-singular. If we begin with the static potentials A_S^μ, we can gauge transform them to the everywhere-regular potentials $\tilde{A}^\mu$ by a gauge transformation U

$$\tilde{A}^\mu = \tilde{U}^{-1}A_S^\mu U + \tilde{U}^{-1}\partial^\mu\tilde{U} \tag{7.7}$$

which has the following properties. In order that the singularity at infinity of A_S^μ be removed

$$\tilde{U} \xrightarrow[r\to\infty]{} U. \tag{7.8a}$$

In order that no new singularity be introduced at the origin

$$\tilde{U} \xrightarrow[r\to 0]{} I. \tag{7.8b}$$

[We are of course assuming that A_S^μ is non-singular in the finite plane.] An explicit form for $\tilde{U}$ might be

$$\tilde{U} = \exp\{uf(r)\} = \exp\left\{-\omega(\hat{r})\frac{2\pi}{T}tf(r)\right\}, \tag{7.9a}$$

$$f(0) = 0, \quad f(\infty) = 1. \tag{7.9b}$$

The price one pays for a formulation in terms of everywhere-regular potentials is that they cease to be periodic, but are periodic only up to a gauge transformation[36]

$$\tilde{A}(t+T) = \text{gauge transform } \tilde{A}(t). \tag{7.10}$$

Both the singular, periodic description as well as the non-singular gauge-periodic formulation are useful and will be employed in the subsequent.

We are now in a position to prove the main assertion. The Pontryagin index (7.1) is rewritten in terms of the magnetic and electric fields where the latter is derived from the periodic potentials

$$
\begin{aligned}
q &= \frac{1}{4\pi^2} \int_0^T dt \int d\mathbf{r}\, \text{tr}\, \mathbf{B} \cdot \mathbf{E} \\
&= -\frac{1}{4\pi^2} \int_0^T dt \int d\mathbf{r}\, \text{tr}\, \mathbf{B} \cdot \frac{d\mathbf{A}}{dt} \\
&\quad - \frac{1}{4\pi^2} \int_0^T dt \int d\mathbf{r}\, \text{tr}\, \mathbf{B} \cdot \mathbf{D}A^0 .
\end{aligned}
\tag{7.11a}
$$

In the last formula, an integration by parts shows that term to vanish: the time-component of the Bianchi identity gives $\mathbf{D} \cdot \mathbf{B} = 0$; no surface contributions survive since A^0 is well behaved at infinity according to (7.2a) and the singularity at the origin is too weak to interfere with the result. In the next-to-last term, we use (7.5)

$$q = -\frac{1}{2\pi} \int_0^T \frac{dt}{T} \int d\mathbf{r}\, \text{tr}\, \mathbf{B} \cdot \mathbf{D}\omega. \tag{7.11b}$$

Another integration by parts and use of the Bianchi identity allows only a possible surface term to survive

$$
\begin{aligned}
q &= \int_0^T \frac{dt}{T} \int \frac{d\mathbf{S}}{4\pi} \cdot (-2\,\text{tr}\,\mathbf{B}\omega) \\
&= \int_0^T \frac{dt}{T} \int \frac{d\mathbf{S}}{4\pi} \mathbf{B}^a \omega_a .
\end{aligned}
\tag{7.11c}
$$

The surface integral is recognized as the magnetic monopole strength m: the average over the sphere at infinity of the magnetic field contracted with a covariant constant. Thus we find that Pontryagin index coincides with the monopole strength

$$q = m . \tag{7.12}$$

An alternate derivation will show that (7.12) is in fact an integer; also additional insight into the problem will be provided. For the second derivation, we evaluate q, a gauge-invariant object, by writing $\mathrm{tr}\ ^*F^{\mu\nu}F_{\mu\nu}$ as total 4-divergence of a gauge-variant quantity, for which we use the everywhere-regular potentials $\tilde{A}$

$$q = -\frac{1}{4\pi^2} \int_0^T dt \int d\mathbf{r}\, \partial_\mu \varepsilon^{\mu\alpha\beta\gamma}\, \mathrm{tr}\left(\frac{1}{2}\tilde{A}_\alpha \partial_\beta \tilde{A}_\gamma + \frac{1}{3}\tilde{A}_\alpha \tilde{A}_\beta \tilde{A}_\gamma\right). \qquad (7.13a)$$

The spatial part of the divergence can be integrated to zero, since the potentials, being everywhere regular, do not leave a surface integral at the origin or at infinity. Thus we are left with

$$q = \int_0^T dt \frac{d}{dt} W(\tilde{A}) = W(\tilde{A}(T)) - W(\tilde{A}(0)) \qquad (7.13b)$$

$$= W(\text{gauge transform } \tilde{A}(0)) - W(\tilde{A}(0)),$$

where we have used (7.10) and $W(A)$, is defined by

$$W(A) = -\frac{1}{4\pi^2} \int d\mathbf{r}\, \varepsilon^{ijk}\, \mathrm{tr}\left(\frac{1}{2}A_i\, \partial_j A_k + \frac{1}{3}A_i A_j A_k\right). \qquad (7.14)$$

To proceed, we must determine how $W(A)$ behaves under gauge transformations. The answer is well known from the investigation which was first performed when the vacuum structure was analyzed.[35] Let us recall those results.

When only those gauge transformations which tend to a unique global limit at spatial infinity are considered,

$$U \xrightarrow[r\to\infty]{} \quad \text{global limit independent of angles} \qquad (7.15)$$

they may be categorized into "small" ones which are continuously deformable to the identity, and "large" ones that are not so deformable [with (7.15) maintained during the deformation]. The latter are classified into homotopy classes labelled by the integers, with a representative of the first class being

$$U_1 = e^{i\sigma\cdot\hat{r}\pi f(r)}$$

$$f(0) = 1, \quad f(\infty) = 1 \qquad (7.16a)$$

and of the nth class

$$U_n = (U_1)^n. \qquad (7.16b)$$

The integers are called winding numbers of the gauge transformation.

With this classification of gauge transformations in mind, we can determine how $W(A)$ changes under gauge transformations. The answer is that $W(A)$ is invariant against small gauge transformations, while the large ones change it by the winding number of the gauge transformation

$$W(\text{gauge transform } A) = W(A) + n. \tag{7.17}$$

The restriction to those gauge transformations which tend to a global limit at infinity is quite proper in the vacuum sector. We recall that there the potentials, by hypothesis, fall off faster than $1/r$. To allow gauge transformations that have a directionally dependent limit would interfere with this asymptotic requirement; hence they are excluded. But for the monopole sector of the theory, where we are now working, physical effects give rise to $O(r^{-1})$ potentials, and there is no basis for excluding gauge transformations which do not tend to a global limit. Indeed, we see that the gauge transform $\tilde{U}(t)$ of (7.9) does not tend to a unique global limit for arbitrary times.

However, observe that when $t = T$, $\tilde{U}(T)$ ceases to be afflicted with a directionally dependent limit; it attains a unique value at large distance, hence is classifiable according to the scheme outlined above. It follows from (7.13) and (7.17) that q is an integer; it is equal to the winding number of the gauge transformation.

For times different from integral multiples of the period, $\tilde{U}(t)$ is not classifiable; indeed it provides a homotopy between the distinct equivalence classes. In the vacuum sector, this homotopy is forbidden; in the monopole sector evidently it is allowed. As a consequence, effects associated with the vacuum angle can only arise through quantum tunnelling in the vacuum sector, but are more accessible in the monopole sector. This feature is exemplified by Witten's recent observation that quantum dyons, the charged excitations of the monopoles, carry a charge which differs from an integer by an amount proportional to the vacuum angle.[37]

Let us illustrate the calculations in this section by reference to the Julia-Zee dyon, which is described by the static potentials

$$(A_a^i)_S = \varepsilon^{aij} \frac{\hat{r}^j}{r} [K(r) - 1], \tag{7.18a}$$

$$(A_a^0)_S = \hat{r}^a \frac{J(r)}{r}. \tag{7.18b}$$

While $\mathbf{A}_S$ satisfies (7.2), because $K(r)$ tends to unity at large distances, A_S^0 does not since $J(r)/r \to M \neq 0$.[38] Consequently, its long-range tail must be removed by a gauge transformation. We may pass to the periodic description with

$$U = \exp i\boldsymbol{\sigma} \cdot \hat{r} \frac{M}{2} t \qquad (7.19)$$

which leads to the period

$$T = \frac{2\pi}{M}. \qquad (7.20)$$

The calculation of the spatial integral in the Pontryagin index makes use of a published result[39]

$$q = \int_0^T dt \left(-\frac{1}{16\pi^2} \int d^3\mathbf{r} \, \mathrm{tr} \, {}^*F^{\mu\nu} F_{\mu\nu} \right) = \int_0^T dt \frac{M}{2\pi}. \qquad (7.21)$$

That the above is indeed unity follows from (7.20), thus confirming that the Julia-Zee dyon, which carries unit monopole strength, has the proper Pontryagin index.

Alternatively, we may use the nowhere singular gauge with the gauge transformation

$$\tilde{U} = \exp i\boldsymbol{\sigma} \cdot \hat{r} \frac{M}{2} t f(r), \quad f(0) = 0, \quad f(\infty) = 1. \qquad (7.21')$$

At $t = T$, $\tilde{U}$ coincides with (7.16a); thus it has unit winding number.

The original 't Hooft-Polyakov monopole fits into this discussion in a limiting way. The electric field is zero, as is $\mathbf{E} \cdot \mathbf{B}$. But the period is infinite, since A^0 vanishes. In order to regulate the $(0) \times (\infty)$ ambiguity of (7.21), we may view this monopole as the $M \to 0$ limit of the dyon. Thus the previous results hold as well.

VIII. CONCLUSION AND SUGGESTIONS FOR FURTHER RESEARCH

Understanding the action of symmetries in gauge theories allows us to unify the various theoretical topics mentioned in the Introduction: mixing of space-time and internal degrees of freedom, emergence of Higgs fields from Yang-Mills fields, modification of constants of motion. Also a practical advantage is gained. In solving field equations, one frequently wants to make a symmetry *Ansatz* to simplify the algebra. To do this, one obviously needs to know how to construct symmetric fields.

There are various further investigations that suggest themselves. Several "study" problems were indicated in the course of the lectures, additional ones are the following: The conformal invariance of Yang-Mills theory gives rise to an $O(4,2)$ group of symmetry transformations in Minkowski space, and $O(5,1)$ in Euclidean space. Configurations that are invariant under subgroups of these have been studied exhaustively only for the $O(3)$ case, in connection with monopoles. But other subgroups also give rise to interesting solutions. [$O(5)$ invariance in Euclidean space gives instantons[13]; $O(4) \times O(2)$ invariance in Minkowski space gives merons[40]; $O(4)$ invariance generalizes the merons.[41]] However, symmetry *Ansätze*, based on invariances other than 3-dimensional rotations, have not been thoroughly examined. To do this would be a useful application of the theory.

Another simple exercise would be to classify the known $SU(5)$ dyons[32] in terms of the Pontryagin index, as has already been done for the $SU(2)$ case.[4]

More ambitious, research problems would address the questions raised by the intriguing attempts to get Higgs fields by dimensional reduction.[23] How should fermions be incorporated? Are the models renormalizable, even though they are formulated on 6-dimensional space-time? What is the meaning of a symmetry *Ansatz* for quantum fields?

It would be especially useful to find an answer to the last question, since it is known that various dimensional reductions of conventional 4-dimensional Yang-Mills theory yield models that are sufficiently simple to be completely analyzable and even solvable.[34] While the resulting quantal systems clearly share features with the full theory [multiple vacua, for example], it has never been clear what information, if any, about the complete Yang-Mills model survives in these simplified descendants.

If the above questions can be adequately answered, one may try to derive the grand unified Higgs potentials of $SU(5)$ or $O(10)$.

I emphasized the distinction between a symmetric potential that requires a global gauge transformation to compensate for non-invariance under a coordinate transformation, and one where the compensating gauge transformation is local. This difference was exemplified by the $SU(2)$ instanton whose $O(4)$ invariance is of the first kind, and $O(5)$ invariance of the second. Yet when we established the $O(5)$ symmetry, we extended the gauge group to $SU(2) \times SU(2) = O(4)$ and then to $O(5)$; also we compactified R^4 to S^4.[13] On this manifold the gauge transformation is global; it becomes position dependent only when the problem is projected back to the original group and manifold. The question therefore presents itself whether or not a symmetry *Ansatz* always involves global gauge transformations, not necessarily on the original space, but on some "higher" one.

A last problem, which I already mentioned, is to refine the categorization of magnetic monopoles in groups higher than $SU(2)$, without using Higgs fields.

I trust that you will find these investigations interesting and challenging, and that carrying them out to a successful conclusion will contribute to our understanding of gauge theories.

REFERENCES

1. R. Jackiw and C. Rebbi, *Phys. Rev. Lett.* **36**, 1116 (1976); P. Hasenfratz and G. 't Hooft, *Phys. Rev. Lett.* **36**, 1119 (1976); A. Goldhaber, *Phys. Rev. Lett.* **36**, 1122 (1976).

2. The static equations of a Yang-Mills-Higgs system, solved by M. Prasad and C. Sommerfield, *Phys. Rev. Lett.* **35**, 760 (1975), are formally equivalent to pure Yang-Mills equations; see W. Marciano and H. Pagels, *Phys. Rev. D* **14**, 531 (1976) and M. Lohe, *Phys. Lett.* **70B**, 325 (1975). That $SU(2)$ Yang-Mills fields in 4-dimensional Euclidean space, but with an $O(3)$ invariance, are equivalent to an Abelian gauge-Higgs system in two dimensions was shown by E. Witten, *Phys. Rev. Lett.* **38**, 121 (1977).

3. This classical result of H. Poincaré, *Compt. Rend.* **123**, 530 (1896), is frequently discussed in the contemporary literature, for example by J. Schwinger, *Particles, Sources and Fields* (Addison-Wesley, Reading, MA, 1970).

4. N. Christ and R. Jackiw, *Phys. Lett.* **91B**, 228 (1980).

5. H. Wang, *Nagoya Math. J.* **13**, 1 (1958).

6. S. Kobayashi and K. Nomizu, *Foundations of Differential Geometry* (Interscience, New York, NY, 1963).

7. G. 't Hooft, *Nucl. Phys.* **B79**, 276 (1974); A. Polyakov, *Zh. Eksp. Teor. Fiz., Pis'ma Redakt.* **20**, 430 (1974) [English translation: *JETP Lett.* **20**, 194 (1974)].

8. A. Belavin, A. Polyakov, A. Schwartz and Y. Tyupkin, *Phys. Lett.* **59B**, 85 (1975).

9. P. Bergmann and E. Flaherty, *J. Math. Phys.* **19**, 212 (1978); A. Trautman, *Bull. Acad. Polon., Ser. Sci. Phys. et Astron.* **27**, 7 (1979).

10. P. Forgàcs and N. Manton, *Comm. Math. Phys.* **72**, 15 (1980).

11. J. Harnad, S. Shnider and L. Vinet, *J. Math. Phys.* **21**, 2719 (1980).

12. For a review see P. Goddard and D. Olive, *Rep. Prog. Phys.* **41**, 1357 (1978).

13. R. Jackiw and C. Rebbi, *Phys. Rev. D* **14**, 517 (1976).

14. R. Jackiw, *Phys. Rev. Lett.* **41**, 1635 (1979).

15. R. Jackiw and N. Manton, *Ann. Phys.* (NY) **127**, 257 (1980). The sign conventions in this paper differ from those used here.

16. For a summary of conformal transformations in field theory, see S. Treiman, R. Jackiw, B. Zumino and E. Witten, *Current Algebra and Anomalies* (Princeton University/World Scientific, Princeton NJ/Singapore, 1985).

17. G. Mack and A. Salam, *Ann. Phys.* (NY) **53**, 174 (1969).
18. J. Wess and B. Zumino, *Nucl. Phys.* **78B**, 1 (1974); B. de Wit and D. Freedman, *Phys. Rev. D* **12**, 2286 (1975).
19. G. Horowitz (private communication).
20. T. T. Wu and C. N. Yang, *Phys. Rev. D* **12**, 3845 (1975).
21. I thank D. Eardley for helping me develop the analogy.
22. An investigation which makes use of the difference between (4.9a) and (4.9b) is by L. Brown and W. Weisberger, *Nucl. Phys.* **B157**, 285 (1979).
23. N. Manton, *Nucl. Phys.* **B158**, 141 (1979).
24. S. Weinberg, *Phys. Rev. Lett.* **19**, 1264 (1967); A. Salam in *Relativistic Groups and Analyticity*, N. Svartholm, ed. (Interscience, New York, NY, 1969).
25. T. Kaluza, *Sitz. Preuss. Akad. Wiss. Phys. Math. Klasse* (1921) 966; O. Klein, *Z. Phys.* **37**, 895 (1926). For a description, see P. Bergman, *An Introduction to the Theory of Relativity* (Prentice-Hall, New York, NY, 1942).
26. R. Jackiw and K. Johnson, *Phys. Rev. D* **8**, 2386 (1973); J. Cornwall and R. Norton, *Phys. Rev. D* **8**, 3338 (1973).
27. My approach here follows the infinitesimal formulation and solution of the general problem by Forgàcs and Manton, Ref. 10; and some of the results given below are stated by Manton in Ref. 23. An alternative global analysis has also been given; see Ref. 11.
28. Witten, Ref. 2.
29. A. C. T. Wu and T. T. Wu, *J. Math. Phys.* **15**, 53 (1974); A. Chakrabarti, *Ann. Inst. Henri Poincaré* **23**, 235 (1975). These authors use for the second triplet $(2T_7, -2T_5, 2T_2)$ which is related to our formula by an $SU(3)$ rotation.
30. R. Jackiw in *Gauge Theories and Modern Field Theory*, R. Arnowitt and P. Nath, eds. (MIT Press, Cambridge, MA, 1976); E. Tomboulis and G. Woo, *Nucl. Phys.* **B107**, 221 (1976); N. Christ, A. Guth and E. Weinberg, *Nucl. Phys.* **B114**, 61 (1976).
31. H. Georgi and S. Glashow, *Phys. Rev. D* **6**, 2977 (1972).
32. The following papers address the question of physical abundance of monopoles: Y. Zeldovich and M. Khlopov, *Phys. Lett.* **79B**, 239 (1978); J. Preskill, *Phys. Rev. Lett.* **43**, 1365 (1979); A. Guth and S.-H. Tye, *Phys. Rev. Lett.* **44**, 631 (1980).
33. For a review, see S. Coleman in *New Phenomena on Sub-Nuclear Physics*, A. Zichichi, ed. (Plenum, New York, NY, 1977).
34. For a review see R. Jackiw, C. Nohl and C. Rebbi in *Particles and Fields*, D. Boal and A. Kamal, eds. (Plenum, New York, NY, 1979).
35. R. Jackiw and C. Rebbi, *Phys. Rev. Lett.* **37**, 172 (1976).
36. Euclidean-space potentials that are periodic up to a gauge transformation have also been used by G. 't Hooft, *Nucl. Phys.* **B153**, 141 (1979).
37. E. Witten, *Phys. Lett.* **86B**, 283 (1979).

38. B. Julia and A. Zee, *Phys. Rev.* D **11**, 2227 (1975).
39. Marciano and Pagels, Ref. 2.
40. V. de Alfaro, S. Fubini and G. Furlan, *Phys. Lett.* **65B**, 163 (1976).
41. M. Lüscher, *Phys. Lett.* **B70**, 321 (1977); B. Schechter, *Phys. Rev.* D **16**, 3015 (1977).

Paper III.4

While the mechanisms of spontaneous and/or anomalous symmetry breaking allow describing non-symmetric reality by symmetric theories, one wonders whether symmetric solutions to the dynamical equations possess any physical relevance. For the spontaneously broken symmetries, the phenomenon of phase transitions at finite temperature suggests that Nature may find itself in a symmetric phase at sufficiently high temperature. Calculational methods for establishing this within the standard model for elementary particles were presented at the Kyoto meeting (January 1975) devoted to quantum field theory and statistical mechanics. This opened a new and still on-going field of research.

SYMMETRY RESTORATION AT FINITE TEMPERATURE

Springer Lecture Notes in Physics **39**, *319 (1975)*

Some field theories, possessing a symmetry, admit only solutions which do not respect the symmetry. This is the familiar Goldstone-Nambu phenomenon. However, when the theory is examined in an external environment, the hidden symmetry may be restored. We show how spontaneously broken global and local symmetries are restored in a temperature environment. The methods for studying this question are explained, and the critical temperature for symmetry restoration is computed in several models.

The formalism, which we shall use, is that of the generalized effective potential $V(\phi, G)$. That object, an extension of the familiar effective potential $V(\phi)$, is defined as follows, for a theory described by a Lagrangian $\mathcal{L}(\Phi)$[1]. Consider the vacuum persistence amplitude $Z(J, K)$ in the presence of space-time varying sources $J(x)$ and $K(x, y)$; $J(x)$ couples to $\Phi(x)$ and $K(x, y)$ to $\frac{1}{2}\Phi(x)\Phi(y)$. The generalized effective action $\Gamma(\phi, G)$ is defined by a double Legendre transform of $-i \ln Z(J, K) = W(J, K)$

$$\frac{\delta W(J, K)}{\delta J(x)} = \phi(x), \quad \frac{\delta W(J, K)}{\delta K(x, y)} = \frac{1}{2}[\phi(x)\phi(y) + G(x, y)], \tag{1}$$

$$\Gamma(\phi, G) = W(J, K) - \int dx \, \phi(x) J(x),$$

$$- \frac{1}{2} \int dx dy \, [\phi(x)\phi(y) + G(x, y)] K(x, y). \tag{2}$$

In the physical theory the sources are absent, hence $\Gamma(\phi, G)$ satisfies

$$\frac{\delta \Gamma(\phi, G)}{\delta \phi(x)} = -J(x) - \int dy K(x, y)\phi(y) = 0$$

$$\frac{\delta \Gamma(\phi, G)}{\delta G(x, y)} = -\frac{1}{2}K(x, y) = 0. \tag{3}$$

It is clear from (1) that $\phi(x)$ is the expectation of the quantum field and $G(x, y)$ is the propagator of the theory. However, translation invariance which we do not expect to be spontaneously broken, implies that the field expectation is constant: $\phi(x) = \phi$, and that the propagator depends only on the coordinate difference: $G(x, y) = G(x - y)$. The generalized effective potential $V(\phi, G)$ is defined from $\Gamma(\phi, G)$ by using these translation invariant form

$$\Gamma(\phi, G) \Big|_{\substack{\text{translation} \\ \text{invariant}}} = -V(\phi, G) \int d^4x. \tag{4}$$

It is a function of ϕ and a functional of $G(x - y)$; for later convenience G is expressed in the momentum representation:

$$G(x - y) = \int \frac{d^4k}{(2\pi)^4} e^{-ik(x-y)} G(k).$$

The stability equations (3) now become

$$\frac{\partial V(\phi, G)}{\partial \phi} = 0$$

$$\frac{\delta V(\phi, G)}{\delta G(k)} = 0.$$

(5)

For a definite example, we chose a theory described by

$$\mathcal{L}(\Phi) = \frac{1}{2} \partial_\mu \Phi \partial^\mu \Phi - U(\Phi)$$

$$U(\Phi) = \frac{\lambda_0}{4!} \left(\frac{6\mu_0^2}{\lambda_0} - \Phi^2 \right)^2$$

(6)

and compute $V(\phi, G)$ in the Hartree-Fock approximation[1]

$$V(\phi, G) = U(\phi) + \frac{1}{2} \int \frac{d^4k}{(2\pi)^4} \ln G(k)$$

$$- \frac{i}{2} \int \frac{d^4k}{(2\pi)^4} G(k) \left[k^2 + \mu_0^2 - \frac{\lambda_0 \phi^2}{2} \right] + \frac{\lambda_0}{8} \left[\int \frac{d^4k}{(2\pi)^4} G(k) \right]^2$$

(7)

The first term on the right-hand side of (7) is the no-loop tree approximation, the next two terms represent the one-loop quantum correction, the last term arises from a two-loop graph. From (5), we find

$$\left(-\mu_0^2 + \frac{\lambda_0}{2} \int \frac{d^4k}{(2\pi)^4} G(k) + \frac{\lambda_0 \phi^2}{6} \right) \phi = 0,$$

(8a)

$$iG^{-1}(k) = k^2 + \mu_0^2 - \frac{\lambda_0}{2} \int \frac{d^4k}{(2\pi)^4} G(k) - \frac{\lambda_0 \phi^2}{2}.$$

(8b)

Upon defining the renormalized mass parameter

$$-m^2 = -\mu_0^2 + \frac{\lambda_0}{2} \int \frac{d^4k}{(2\pi)^4} G(k)$$

the above equation becomes

$$\left(-m^2 + \frac{\lambda_0 \phi^2}{6}\right)\phi = 0$$

$$iG^{-1}(k) = k^2 + m^2 - \frac{\lambda_0 \phi^2}{2}$$

(9)

and the only consistent solution is the one which breaks spontaneously the $\Phi \leftrightarrow -\Phi$ symmetry of (6)

$$\phi^2 = \frac{6m^2}{\lambda_0}$$

$$iG^{-1}(k) = k^2 - 2m^2 \,.$$

(10)

[The solution $\phi = 0$ is unacceptable, since it leads to a propagator with imaginary mass.] Similar calculations can be performed in an $O(N)$ invariant Φ^4 theory, [for which the Hartree-Fock approximation dominates when N is large] in different dimensions.[2] The behavior of spontaneous symmetry breaking with varying dimensionality is thereby exposed: a continuous symmetry can be broken only when the dimension of spacetime is greater than 2.

Another interesting question is whether a spontaneously broken symmetry can be restored at finite temperature. Qualitative arguments indicating that this should happen were given by Kirzhnits and Linde,[3] and subsequent detailed computations have established the phenomenon.[4] This topic can be readily analyzed by a straightforward extension of the formalism.

A field theory at a finite temperature, proportional to $1/\beta$, is conveniently described by its finite-temperature Green's functions, which are defined by a statistical average: $\operatorname{tr} e^{-\beta H} T\Phi(x_1) \cdots \Phi(x_n)/\operatorname{tr} e^{-\beta H}$. These Green's functions satisfy the same differential equations as the corresponding ones at zero temperature. However, the boundary conditions are different: in the complex time interval $[0, -i\beta]$ there are periodicity requirements. It follows that in a theory with interactions, the formulas for temperature Green's functions in terms of the elementary free-field propagator and vertices are the same as at zero temperature, except that the free-field propagators have a "momentum" representation consistent with the periodicity conditions

$$D_\beta(x) = \int_k e^{-ikx} \frac{i}{k^2 - m^2} \,,$$

(11)

where $\int_k$ stands for $\frac{1}{-i\beta} \sum_n \int \frac{d^3k}{(2\pi)^3}$, $n = 0, \pm 1 \cdots$; $k^2 = k_0^2 - \mathbf{k}^2$; and k_0 equals

$\frac{2\pi n}{-i\beta}$. Consequently, symmetry behavior at finite temperature can be studied in terms of solution to Eqs. (8) which are also valid at finite temperature.[5]

We are thus led to the system of equations

$$\left[-\mu_0^2 + \frac{\lambda_0}{2}\int_k G_\beta(k) + \frac{\lambda_0\phi_\beta^2}{6}\right]\phi_\beta = 0, \tag{12a}$$

$$iG_\beta^{-1}(k) = k^2 + \mu_0^2 - \frac{\lambda_0}{2}\int_k G_\beta(k) - \frac{\lambda_0\phi_\beta^2}{2} \tag{12b}$$

$$\phi_\beta = \operatorname{tr} e^{-\beta H}\Phi(x)/\operatorname{tr} e^{-\beta H}$$

$$\phi_\beta^2 + \int_k e^{-ik(x-y)}G_\beta(k) = \operatorname{tr} e^{-\beta H}T\Phi(x)\Phi(y)/\operatorname{tr} e^{-\beta H}.$$

The symmetric solution $\phi_\beta = 0$ will be acceptable if the mass parameter in the propagator is positive.

Equation (12b) is solved by $G_\beta^{-1}(k) = i[-k^2 + m_\beta^2]$ where the temperature dependent mass m_β satisfies [at $\phi_\beta = 0$] the gap equation

$$m_\beta^2 = -\mu_0^2 + \frac{\lambda_0}{2}\int_k \frac{i}{k^2 - m_\beta^2}$$

$$= -\mu_0^2 + \frac{\lambda_0}{2\beta}\sum_n \int \frac{d^3k}{(2\pi)^3}\frac{1}{\frac{(2\pi n)^2}{\beta^2} + \mathbf{k}^2 + m_\beta^2}, \tag{13a}$$

$$= -\mu_0^2 + \frac{\lambda_0}{2}\int \frac{d^3k}{(2\pi)^3}\frac{1}{2\sqrt{\mathbf{k}^2 + m_\beta^2}} + \frac{\lambda_0}{\beta^2}f(m_\beta^2\beta^2)$$

$$f(a^2) = \frac{1}{4\pi^2}\int_0^\infty \frac{x^2 dx}{\sqrt{x^2 + a^2}}\left[e^{\sqrt{x^2 + a^2}} - 1\right]^{-1}. \tag{13b}$$

The integral occurring in (13a) is divergent; the gap equation must be renormalized. It is important that renormalization does not involve any counterterms beyond those of the zero temperature theory. To carry out the renormalization we evaluate the integral in (13a) with a cutoff, and rewrite that expression

$$m_\beta^2 = -m^2 + \frac{\lambda}{\beta^2}f(m_\beta^2\beta^2) + \frac{\lambda}{32\pi^2}m_\beta^2\ln\frac{m_\beta^2}{m^2}. \tag{13c}$$

Here m^2 is the renormalized mass parameter, and λ is the renormalized coupling constant; they are given in terms of the bare parameters and cutoff Λ

$$-m^2 = -\mu_0^2 + \frac{\lambda_0}{32\pi^2}\left[\frac{\Lambda^2}{2} + m^2 \ln\frac{\Lambda^2}{m^2} - m^2\right]$$

$$\frac{1}{\lambda} = \frac{1}{\lambda_0} + \frac{1}{32\pi^2}\left[\ln\frac{\Lambda^2}{m^2} - 1\right].$$

Equation (13c) is the renormalized gap equation for m_β^2 and we seek solutions for positive m_β^2. At low temperature $[\beta \to \infty]$ there is symmetry breaking and no positive solution exists. At high temperature $[\beta \to 0]$ one can satisfy (13c) with $m_\beta^2 > 0$. The critical temperature β_c^{-1} is that value of the temperature for which m_β^2 vanishes. Therefore from (13b) and (13c) we have

$$0 = -m^2 + \frac{\lambda}{\beta_c^2}f(0)$$

$$\frac{1}{\beta_c^2} = \frac{24m^2}{\lambda}. \tag{14}$$

For $\beta^{-1} > \beta_c^{-1}$ symmetry is manifest, for $\beta^{-1} < \beta_c^{-1}$ the symmetry is broken. In this calculation we have ignored higher loop corrections, which are proportional to higher powers of the coupling constant. Hence the entire approach is correct only if the coupling is small, in which case β_c^{-1}, given by (14), is indeed large. One can estimate that for small λ, the corrections are $O(\lambda^{1/2})$ i.e., $\beta_c^{-2} = \frac{24m^2}{\lambda}[1 + O(\lambda^{1/2})]$. Equation (13c) can be solved for m_β when $\beta \approx \beta_c$

$$m_\beta = \frac{2\pi}{3}(\beta^{-1} - \beta_c^{-1}). \tag{15}$$

The computation may be extended to an $O(N)$ invariant Φ^4. [The Hartree-Fock approximation then dominates for large N.] Once again (14) and (15) are found, except that λ is replaced by $\lambda\frac{(N+2)}{3}$. Also if gauge fields are included, so that the theory is locally $O(N)$ invariant, one gets

$$\frac{1}{\beta_c^2} = \frac{m^2}{\lambda\dfrac{(N+2)}{72} + e^2\dfrac{(N-1)}{4}}, \tag{16}$$

where e is the gauge coupling strength.[6]

What is the significance of phase transitions in field theory which restore a spontaneously broken symmetry? One answer addresses itself to questions of principle. It may be thought that a theory with a hidden, spontaneously broken symmetry is equivalent to a theory without any symmetry at all, and that the various relationships that exist between masses, coupling constants *etc.* are merely consequences of perturbative unitary or renormalizability. However, if the physical environment can be arranged so that the symmetry becomes manifest, one cannot doubt the existence of the symmetry. Practical application of this phenomenon must be confined to speculations about the early universe, since only in that environment were there temperatures sufficiently high to effect a phase transition. [The order of magnitude of β_c^{-1} may be estimated as follows. For global symmetries, we identify the vacuum expectation value of the field with f_π, the pion decay constant. Hence $f_\pi^2 \approx \frac{6m^2}{\lambda}$ and, $\beta_c^{-1} = O(f_\pi) \approx 100$ MeV. For local symmetries, recall that the spontaneously generated vector meson mass is $\sqrt{2}ef_\pi$ and the mass of the Higgs particle is $\sqrt{2}m^2$. Hence (16) is also given by

$$\beta_c^{-2} = \left[\frac{e^2}{m_W^2} \frac{N+2}{6} + \frac{e^2}{m_H^2} \frac{N-1}{2} \right]^{-1}.$$

With $e^2 \approx 10^{-2}$ and $m_H < m_W$, we get $\beta_c^{-1} \approx O(10m_H)$.] Thus a temperature environment for field theoretic phase transitions is not readily available. However, one may study other environments which effect phase transitions described by critical parameters that have more immediate experimental consequences.[7]

REFERENCES

1. J. Cornwall, R. Jackiw and E. Tomboulis, *Phys. Rev.* D **10**, 2428 (1974).
2. L. Dolan and R. Jackiw, *Phys. Rev.* D **9**, 3320 (1974); H. Schnitzer, *Phys. Rev.* D **10**, 1800 (1974); S. Coleman, R. Jackiw and H. Politzer, *Phys. Rev.* D **10**, 2491 (1974).
3. D. Kirzhnits and A. Linde, *Phys. Lett.* **42B**, 471 (1972).
4. L. Dolan and R. Jackiw, *Phys. Rev.* D **9**, 3320 (1974); S. Weinberg, *Phys. Rev.* D **9**, 3357 (1974).
5. This brief synopsis of field theory at finite temperature is of course an inadequate summary of the beautiful work of Martin, Schwinger and others. For a fuller account, see for example, L. P. Kadanoff and G. Baym, *Quantum Statistical Mechanics* (W. A. Benjamin, Menlo Park, CA, 1962). Discussions of this topic which focus on the present application are also given in Refs. 4.
6. The approximation and results of this section are familiar in statistical mechanics, where they are known as "spherical model", "mean field theory", *etc.* For a review, see E. Stanley, *Introduction to Phase Transitions and Critical Phenomena* (Oxford University Press, New York, NY, 1971).

7. That phase transitions can be induced by high matter density has been shown by T. D. Lee and G. C. Wick, *Phys. Rev.* D **9**, 2291 (1974); A. Salam and J. Strathdee, *Nature* **252**, 569 (1974) have studied field theoretic phase transitions in electromagnetic environments.

DISCUSSION

K. Symanzik: Could you comment on the gauge dependence or independence of your results for a gauge theory?
The critical temperature was calculated in a variety of renormalizable gauges parametrized by gauge constants (covariant gauges, R_ξ gauges, *etc.*) and the answer comes out independent of these constants, hence gauge invariant within that class. For non-renormalizable gauges (the unitary gauge and others) a well defined answer does not emerge. We attribute this to a failure of perturbation theory.

J. Zittartz: Could you not get your answers just by dimensional arguments?
We give a numerical value for the critical temperature. Dimensional arguments, which were used initially by Kirzhnits and Linde, will yield the dependence of β_c^2 on m^2 and λ, but will not provide the numerical coefficient.

K. R. Ito: You consider an equation for the temperature dependent propagator. On the other hand, we can define phase transitions by the order parameter. Are these equivalent?
Yes; indeed Dolan has rederived our results by calculating the order parameter.

T. Kugo: What happens in massless scalar electrodynamics, where Coleman and Weinberg find symmetry breaking at zero temperature?
We have not done calculations for this massless theory, since our perturbation theory requires a mass parameter. However, I would speculate that finite temperature again restores the symmetry, and that a mass is generated for the charged particle, but not for the photon.

Paper III.5

While equilibrium statistical mechanics was easily transcribed into the context of relativistic quantum field theory, the non-equilibrium situation is much more challenging to deal with. A simple proposal was made at First Thermal Field Theories Workshop in Cleveland, OH (October 1988).

MEAN FIELD THEORY FOR NON-EQUILIBRIUM QUANTUM FIELDS

Physica **A158**, *269 (1989)*

I. INTRODUCTION

When a quantum system is in a pure state, its time evolution is governed by the time-dependent Schrödinger equation. The initial data needed for a definite solution is the wave function at initial time, and then the first-order [in time] differential equation allows following the time development. But for systems that are in a mixed state, *i.e.*, those described by a density matrix, the description of time evolution is more involved. A mixed state typically describes a system in an external environment. While time evolution is certainly governed by some microscopic Hamiltonian for *all* the dynamical degrees of freedom [system *and* environment], the separation of systemic degrees of freedom and the description of time evolution solely in terms of them is a non-trivial matter.

The density matrix is given by

$$\rho(x_1, x_2) = \sum_n p_n \psi_n(x_1) \psi_n^*(x_2)$$

$$\operatorname{tr} \rho = \sum_n p_n \int dx \psi_n^*(x) \psi_n(x) = \sum_n p_n = 1 \tag{1.1}$$

where $\{\psi_n\}$ is a complete set of normalized wave functions, and p_n is the probability that the system is in the state n. The entropy is determined by these probabilities,

$$S = -k \sum_n p_n \ln p_n \tag{1.2}$$

where k is Boltzmann's constant. [We are actually interested in quantum field theoretic, not quantum mechanical, applications. But, as is explained below, in the Schrödinger representation for quantum field theory,[1] one can take over the quantum mechanical formalism. So for simplicity, we shall first use the notation of the latter, and only when needed for specific calculations will the full quantum field theoretic apparatus come into play.]

Average values of physical quantities, described by operators $\mathcal{O}$ in abstract Hilbert space, and realized by operation with kernels $\mathcal{O}(x_1, x_2)$ on Schrödinger wave functions, are given by

$$\langle \mathcal{O} \rangle = \operatorname{tr} \rho \, \mathcal{O} = \int dx_1 dx_2 \, \rho(x_1, x_2) \mathcal{O}(x_2, x_1) \,. \tag{1.3}$$

Hence, knowledge of the time-dependence of ρ determines the time-dependence of these averages.

In the simplest situation, the density matrix is time-independent and we say that the system is in *equilibrium* with the environment. This happens, for example, when the wave functions ψ_n are energy eigenfunctions whose pure-phase time-dependence, $e^{-iE_n t}$, disappears from $\psi_n \psi_n^*$, and also when the occupation probabilities are constant in time, as is the entropy. Moreover, when the constant occupation probabilities are Boltzmann distributed,

$$p_n = e^{-\beta E_n} \Big/ \sum_{n'} e^{-\beta E_{n'}} \tag{1.4}$$

we say that the system is in *thermal equilibrium*, at temperature $T = 1/k\beta$, but more general, *non-thermal equilibria* are also possible with constant occupation probabilities different from (1.4).

For over ten years now, particle physicists have been familiar with the idea of a quantum field in thermal equilibrium with a heat bath. More than a decade ago, techniques for studying such problems were taken over from non-relativistic many-body theory and developed for relativistic quantum field theory.[2] In equilibrium, there is no significant change in time and the time variable, which is not needed to describe dynamical evolution can be profitably used for other purposes; for example, for continuation to imaginary time, where calculations at equilibrium and finite temperature are similar to Feynman-Dyson calculations at zero temperature.

However, these old techniques and results have been frequently applied to situations for which there *is* rapid change in time, like in the inflationary universe where equilibrium should not be assumed. There is a need to do better!

This then is the problem that Oscar Èboli, So-Young Pi and I[3] decided to research: How to study the time evolution of a field theory in an environment that is itself changing. We call this subject "Quantum Fields out of Thermal Equilibrium," which is also the title of our recent paper.[3]

The topic of non-equilibrium behavior is vast, touching not merely all branches of physics, but actually all physical sciences. Even our paper, which deals with only a small corner of this wide field, is lengthy. But this lecture's duration is limited, so here I can only outline what we did, and those of you who are interested can find more details in the paper.[3]

II. EVOLUTION EQUATION FOR THE DENSITY MATRIX

The first question we need to answer is which differential equation in time does the density matrix satisfy — ideally it should be a first-order equation so that an initial density matrix determines the subsequent evolution.

Simply differentiating with respect to time the defining equation (1.1) for ρ gives

$$\frac{d}{dt}\rho(x_1, x_2) = \sum_n p_n \left(\dot{\psi}_n(x_1)\psi_n^*(x_2) + \psi_n(x_1)\dot{\psi}_n^*(x_2) \right)$$
$$+ \sum_n \dot{p}_n \psi_n(x_1)\psi_n^*(x_2) . \tag{2.1}$$

[The over-dot signifies differentiation in time.] The first assumption that we shall make is that the time evolution of ψ_n is governed by a Hamiltonian H, which enters in a time-dependent Schrödinger equation for ψ_n

$$\begin{aligned} \dot{\psi}_n &= -iH\psi_n \qquad \text{(acting to the right)} \\ \dot{\psi}_n^* &= i\psi_n^*H \qquad \text{(acting to the left)} . \end{aligned} \tag{2.2}$$

It must be stressed that this is indeed an assumption and involves approximation. Observe that $\psi_n(x)$ refers only to systemic variables x, while a microscopic Hamiltonian for *all* the variables contains both systemic and environmental variables; moreover, the environment is changing and influencing our system. Nevertheless, we suppose that there exists an effective H, depending only on systemic dynamical variables, and that *the effect of the changing environment can be coded in H by prescribing a time variation for the parameters in H*. Examples of such parameters, which we assume change in time owing to the changing environment, are masses, coupling strengths and [in cosmological applications] the background geometry of space-time. In other words, the effective Hamiltonian describes an isolated system in the sense that reference is made only to systemic variables, but the influence of the environment is still felt through the explicit time-dependence. As a consequence of the time-dependence, energy is not conserved; there is energy exchange between the system and the environment.

The next question concerns $\dot{p}_n$: some model must be adopted for the way that the occupation probabilities vary with time, as a consequence of the changing environment. Presumably for a specific many-body system with known microscopic interactions, the physical situation may be carefully analyzed, at least in principle, and something sensible can be said about $\dot{p}_n$. According to (1.2), this is equivalent to adopting a model for the rate of change of entropy. In the cosmological application that we have in mind, such analysis does not appear feasible since we do not have an adequate description of the physical "environment" in which our cosmos evolved. Therefore, we shall assume that p_n *does not vary in time* and correspondingly, that the entropy is constant.

With these two assumptions, the equation satisfied by the density matrix is the Liouville [von Neumann] equation

$$\frac{d\rho}{dt} = i\,[\rho, H\,].$$
(2.3)

[Note that the sign is opposite from the Heisenberg equation of motion.]

Our two assumptions restrict our considerations to *isoentropic* but *energy non-conserving* evolution. While we are led to these assumptions by necessity — we do not know how the probabilities vary in time — the end result describes physically plausible processes in which energy is not conserved, but entropy is. Precisely such processes are thought to take place in the early universe. In the language of statistical mechanics, we are speaking of *adiabatic* evolution — closely related to but not identical with the *quantum adiabatic theorem* discussed later.

In order to obtain further insight into the approximations that we are using, let us relate ρ and its exact time derivative to a microscopic description in terms of a pure state for a larger, closed system. The density matrix $\rho(x_1, x_2)$ arises when we consider a system in a pure state described by two sets of variables $\{P, X\}$ and $\{p, x\}$ and ignore the former

$$\rho(x_1, x_2) = \int dX\,\Psi(X, x_1)\Psi^*(X, x_2)\,,$$
(2.4a)

$$\operatorname{tr}\rho = \int dx\,\rho(x, x) = \int dX\,dx\,|\Psi(X, x)|^2 = 1\,.$$
(2.4b)

The Hamiltonian for the entire system is taken as

$$\mathbb{H}\,(P, X; p, x) = \frac{P^2}{2m} + \frac{p^2}{2m} + V(X, x)\,.$$
(2.5)

The time-dependent Schrödinger equation gives a differential equation satisfied by ρ

$$
\begin{aligned}
i\frac{d}{dt}\rho = &\int dX\Big\{\Big(-\frac{1}{2M}\frac{d^2}{dX^2}\Psi(X, x_1)\Big)\Psi^*(X, x_2)\\
&+ \Psi(X, x_1)\Big(\frac{1}{2M}\frac{d^2}{dX^2}\Psi^*(X, x_2)\Big)\Big\}\\
&+ \int dX\Big\{\Big(-\frac{1}{2M}\frac{d^2}{dx_1^2}\Psi(X, x_1)\Big)\Psi^*(X, x_2)\\
&+ \Psi(X, x_1)\Big(\frac{1}{2M}\frac{d^2}{dx_2^2}\Psi^*(X, x_2)\Big)\Big\}\\
&+ \int dX\{V(X, x_1) - V(X, x_2)\}\Psi(X, x_1)\Psi^*(X, x_2)\,.
\end{aligned}
$$
(2.6a)

The first term on the right-hand side of (2.6a) vanishes, because it receives contributions only from the surface boundary of X space

$$\int dX \left\{ \left(-\frac{1}{2M} \frac{d^2}{dX^2} \Psi(X, x_1) \right) \Psi^*(X, x_2) + \Psi(X, x_1) \left(\frac{1}{2M} \frac{d^2}{dX^2} \Psi^*(X, x_2) \right) \right\}$$

$$= \frac{1}{2M} \int dX \frac{d}{dX} \left\{ \Psi(X, x_1) \frac{d}{dX} \Psi^*(X, x_2) - \frac{d}{dX} \Psi(X, x_1) \Psi^*(X, x_2) \right\}.$$

The second and third terms in (2.6a) may be rewritten as

$$i \frac{d}{dt} \rho = [H, \rho]$$

$$+ \int dX \left\{ \left(V(X, x_1) - V(x_1) \right) - \left(V(X, x_2) - V(x_2) \right) \right\} \Psi(X, x_1) \Psi^*(X, x_2).$$
$$(2.6b)$$

Here

$$H = \frac{p^2}{2m} + V(x) \tag{2.7}$$

and $V(x_i)$ is some suitable average of $V(X, x_i)$ over the complete wave function $\Psi(X, x_1) \Psi^*(X, x_2)$.

Comparison of (2.3) and (2.6) shows that the former follows from the microscopic theory of the latter, provided there exists an average potential $V(x)$ which permits dropping the last term in (2.6b)

$$V(x_1) - V(x_2) \approx \frac{1}{\rho(x_1, x_2)} \int dX \left(V(X, x_1) - V(X, x_2) \right) \Psi(X, x_1) \Psi^*(X, x_2).$$
$$(2.8)$$

Note that such a $V(x)$ will in general be time-dependent, since $\Psi(X, x)$ is.

Another perspective on the circumstances in which the approximations that we employ are realized is gotten by assuming that the interactions between the $\{P, X\}$ and $\{p, x\}$ systems are sufficiently weak so that the complete Hamiltonian (2.5) may be approximated

$$\mathbb{H} \approx \frac{P^2}{2M} + U(X) + \frac{p^2}{2m} + V(x). \tag{2.9}$$

$U(X) + V(x)$ in (2.9) should be viewed as an approximate and average replacement for the exact $V(X, x)$ of (2.5) as in (2.8); hence $U(X) + V(x)$ may carry a time-dependence that represents the residual influence of one system on the other. With (2.9) the last term in (2.6b) vanishes. When the Hamiltonian is a sum as in (2.9), the complete wave function $\Psi(X, x)$ is a superposition of factorized orthonormalized wave functions for each subsystem

$$\Psi(X, x) = \sum_n c_n \Theta_n(X) \psi_n(x).$$ (2.10)

Here c_n are constant and $\frac{P^2}{2M} + U(X)$ governs the time evolution of $\Theta_n(X)$, while $\frac{p^2}{2m} + V(x)$ governs $\psi_n(x)$. Thus the density matrix (2.4) becomes

$$\rho(x_1, x_2) = \sum_n p_n \psi_n(x_1) \psi_n^*(x_2)$$ (2.11a)

with constant probabilities

$$p_n = |c_n|^2.$$ (2.11b)

To summarize, for isoentropic, energy non-conserving processes, we need to solve the Liouville equation (2.3), with a time-dependent Hamiltonian.

The time dependence that we consider can take various forms. When interest focuses on time evolution through a phase transition, we shall take the time-dependence in H to reside in the quadratic term: in the mass squared for a field theory, in the square of the harmonic frequency for quantum mechanical examples; we shall allow this quantity to vary in time in a prescribed fashion, eg., passing from one positive value to another, or from positive [stable] to negative [unstable] values. When interest is in quantum field theory of the early universe, the time-dependence resides in the background geometry, eg., in the metric describing de Sitter space.

We take the time-dependence to occur in an interval $t_i < t < t_f$. For times earlier than t_i, the Hamiltonian is assumed constant, and the initial data for the Liouville equation will always be specified in this static regime, where we shall take ρ to be given by $\rho_i(\beta_i)$ — the initial Hamiltonian's Boltzmann distributed density matrix at some initial β_i, or an approximation thereto. The solution to (2.3) is then examined in the late period, $t > t_f$, where the Hamiltonian is again static but perhaps with different parameters. We wish to determine whether at late times ρ is static or not, and if static whether it is given by a Boltzmann distribution but perhaps at some other temperature. [In some examples $t_{i,f}$ may be $\mp\infty$.]

Therefore we are considering the problem of a system in thermal equilibrium, which becomes disturbed by the environment so that Hamiltonian parameters change. We wish to know whether there is a return to equilibrium, in particular to thermal equilibrium, after the disturbance ceases, and also we wish to follow the behavior of various interesting quantities through the disturbance. Our methods allow considering an arbitrary initial distribution, not necessarily in thermal equilibrium, and we could calculate the time evolution in this more general situation. However, we do not examine such problems here.

III. VARIATIONAL PRINCIPLES

Our goal is now settled: obtain solutions to the Liouville equation (2.3) with an "interesting" time-dependent Hamiltonian and a canonical distribution density matrix as initial condition. But the method for achieving this goal must still be developed because (2.3) cannot be integrated directly, except for linear problems, without self-interactions, described by a quadratic Hamiltonian. To this end, we use a variational principle first stated in the many-body context by Balian and Vénéroni,[4] which yields, under arbitrary variation, the Liouville equation. An approximate application of this principle with a restricted variational *Ansatz*, in the Rayleigh-Ritz manner, leads to approximate but tractable equations for the density matrix.

Let me begin with an overview of variational principles in physics. In classical mechanics, one encounters time-independent and time-dependent variational principles for the relevant dynamical equations of mechanics. Static solutions stationarize the Hamiltonian [energy] as a function of p and q; this is seen from the Hamiltonian equations

$$\dot{q} = 0 = \frac{\partial H(p,q)}{\partial p}, \quad -\dot{p} = 0 = \frac{\partial H(p,q)}{\partial q}. \tag{3.1}$$

The full time-dependent Newtonian equations are derived by Hamilton's variational principle, which requires stationarizing the classical action I_{cl} — the time integral of the Lagrangian L, and a functional of $q(t)$

$$I_{\mathrm{cl}} = \int_{t_i}^{t_f} dt\, L$$
$$\delta I_{\mathrm{cl}}(q) = 0. \tag{3.2}$$

Notice that static variations require no boundary conditions, while the time-dependent ones must vanish at the endpoints of the time interval that defines the action.

Both the static and the time-dependent variation principles of classical physics have their analogs in quantum physics. The former, (3.1), translates into the requirement that normalized expectation values of the Hamiltonian be stationary

$$\delta\frac{\langle\psi|H|\psi\rangle}{\langle\psi|\psi\rangle} = 0. \tag{3.3}$$

This yields the time-independent Schrödinger equation. The quantum analog of the time-dependent Hamilton's variational principle of classical physics, (3.2), is Dirac's little known time-dependent variational principle,[5] which results in the time-dependent Schrödinger equation. One is asked to stationarize the *quantum action*,

$$I_q = \int_{t_i}^{t_f} dt \left\langle \psi; t \left| i\frac{d}{dt} - H \right| \psi; t \right\rangle \tag{3.4}$$

with suitable initial (final) conditions on the time-dependent states $|\psi; t\rangle$ and $\langle\psi; t|$, so that $\delta|\psi; t\rangle$ and $\delta\langle\psi; t|$ vanish there.

In statistical physics, static problems are summarized by the variational principle that the Helmholtz free energy $U - TS$ be stationary. U is the average of H and T is a Lagrange multiplier insuring constant entropy when U is varied [equivalently $\frac{1}{T}$ is the Lagrange multiplier insuring constant U when S is varied]. In the quantum theory, we form

$$\beta\,\mathrm{tr}\rho H + \mathrm{tr}\,\rho\ln\rho - \lambda\,\mathrm{tr}\,\rho$$

where λ enforces $\mathrm{tr}\,\rho = 1$. Varying ρ yields the Boltzmann distribution (1.3) for ρ, which is seen to correspond to maximum entropy at fixed energy. For time-dependent statistical problems, we seek a variational principle giving rise to the Liouville equation for ρ. Such a principle has been given by Balian and Vénéroni[4] who define the action-like quantity

$$I = -\int_{t_i}^{t_f} dt\,\mathrm{tr}\,\rho\Big(\frac{d}{dt}\Lambda + i[H,\Lambda]\Big) - \mathrm{tr}\,\rho\Lambda\,\Big|_{t=t_i}. \tag{3.5}$$

Λ is a Lagrange multiplier that will disappear from the discussion. Varying Λ and ρ leaves

$$\delta_\Lambda I = \int_{t_i}^{t_f} \mathrm{tr}\,\Big(\frac{d}{dt}\rho + i[H,\rho]\Big)\delta\Lambda - \mathrm{tr}\,(\rho\,\delta\Lambda)\,\Big|_{t=t_f}, \tag{3.6a}$$

$$\delta_\rho I = -\int_{t_i}^{t_f} dt\,\mathrm{tr}\,\Big(\frac{d}{dt}\Lambda + i[H,\Lambda]\Big)\delta\rho - \mathrm{tr}\,(\Lambda\delta\rho)\,\Big|_{t=t_i}. \tag{3.6b}$$

We impose the initial condition that $\rho|_{t=t_i}$ is a Boltzmann distribution with $H|_{t=t_i}$ and that $\Lambda|_{t=t_f}$ is equal to the identity $\mathbf{1}$; this means that the corresponding variations vanish at $t = t_{i,f}$, and the last terms in (3.6a,b) may be dropped. Consequently, in order that I be stationary, the Liouville equation must be satisfied by both ρ and Λ. However, since $\Lambda|_{t=t_f} = 1$, it follows that $\Lambda = 1$ at all times.

IV. GAUSSIAN DENSITY MATRIX

A. Variational *Ansatz* for ρ

A variational formulation for the Liouville equation provides a useful starting point for approximating it; this is necessary because the equation cannot be solved, except for linear problems governed by quadratic Hamiltonians. The approximation consists of making a [variational] *Ansatz* for ρ and Λ, with specified x dependence governed by time-dependent parameters. The action (3.5) is then evaluated with a specific Hamiltonian and the posited expressions for ρ and Λ. The arbitrary variations (3.6) are replaced by variations of the parameters, and equations are obtained that determine them. In other words, a Rayleigh-Ritz type variational approximation replaces the exact variational principle, and captures some of the actual non-linearities that would not be seen in a straightforward perturbation expansion.

For the density matrix we choose a Gaussian *Ansatz*

$$
\rho(x_1, x_2) = e^{-\gamma} \exp -\frac{1}{2} \left\{ x_1 \left(\frac{G^{-1}}{2} - 2i\Pi \right) x_1 + x_2 \left(\frac{G^{-1}}{2} + 2i\Pi \right) x_2 \right.
$$
$$
\left. - x_1 G^{1/2} \xi G^{-1/2} x_2 \right\}. \tag{4.1}
$$

Here the parameters are γ, G, Π and ξ; they all depend on time; the constraints of hermiticity $\rho(x_1, x_2) = \rho^*(x_2, x_1)$ are satisfied when they all are real. An operator notation for (4.1) is

$$
\rho(x_1, x_2) = \langle x_1 | \hat{\rho} | x_2 \rangle, \tag{4.2a}
$$

$$
\hat{\rho} = N \exp -\frac{1}{2} (Ap^2 + Bx^2 + C(px + xp)) \tag{4.2b}
$$

for suitably chosen real functions of time A, B, C and N which are related to the previous γ, G, Π and ξ.

The reasons for choosing a Gaussian for ρ are (a) one can explicitly evaluate all integrals that arise [this is especially important in the field theory application, where the integrals are functional] and (b) the Gaussian form for ρ is exact for quadratic Hamiltonians, *i.e.*, linear dynamics [see below].

We make the following observations about the parameters. The prefactor involving γ assures proper normalization: $\mathrm{tr}\,\rho = 1$. Hence, we expect that the variational equation for γ is solved by

$$\gamma = \frac{1}{2}\ln 2\pi G^{1/2}(1-\xi)^{-1}G^{1/2}\,. \tag{4.3}$$

The parameter ξ that governs the last, x_1/x_2 cross term in the exponent, (4.1) is a measure of the degree of mixing. For $\xi = 0$, the density matrix of (4.1) reduces to that of a pure state, described by a Gaussian wave function

$$\rho(x_1, x_2)|_{\xi=0} = \psi(x_1)\psi^*(x_2)\,, \tag{4.4a}$$

$$\psi(x) = e^{-\gamma/2}\exp -\frac{1}{2}\left\{x\left(\frac{G^{-1}}{2} - 2i\Pi\right)x\right\}\,. \tag{4.4b}$$

Thus ξ is related to the occupation probabilities and to the entropy, and we expected that the variational equation for it implies that ξ is constant. The role of the remaining parameters is brought out by computing averages of bilinears [Averages of x and p vanish.]

$$\langle x^2\rangle = G^{1/2}\frac{1}{1-\xi}G^{1/2}\,, \tag{4.5a}$$

$$\langle p^2\rangle = \frac{1}{4}G^{-1/2}(1+\xi)G^{-1/2} + 4\Pi G^{1/2}\frac{1}{1-\xi}G^{1/2}\Pi\,, \tag{4.5b}$$

$$\langle xp\rangle = \frac{i}{2} + 2G^{1/2}\frac{1}{1-\xi}G^{1/2}\Pi\,. \tag{4.5c}$$

Within the parametrization, (4.1), we can describe the classical limit as the situation when the next-to-last term in (4.5b) is dominated by the last term, and when the next-to-last imaginary term in (4.5c) is dominated by the real last term. The reason is that the next-to-last terms are $\mathcal{O}(\hbar)$, while the last terms are $\mathcal{O}(1)$. [Remember that in our units $\hbar = 1$, hence invisible in the above formulas.] Indeed the next-to-last imaginary term in (4.5c) gives rise to the Heisenberg commutator. When the classical limit holds, then

$$\langle x^2\rangle = (\Delta x)^2 = \frac{G}{1-\xi} \equiv \langle x^2\rangle_c\,, \tag{4.6a}$$

$$\langle p^2\rangle = (\Delta p)^2 \approx \frac{4\Pi^2 G}{1-\xi} \equiv \langle p^2\rangle_c = 4\Pi^2\langle x^2\rangle_c\,, \tag{4.6b}$$

$$\langle xp\rangle \approx \frac{2\Pi G}{1-\xi} \equiv \langle xp\rangle_c = 2\Pi\langle x^2\rangle_c\,, \tag{4.6c}$$

and $\triangle p \triangle x = \langle xp \rangle$. The imaginary contribution to $\langle xp \rangle$ that is ignored in the classical limit has magnitude $1/2$. Hence, the classical limit is alternatively characterized by $\triangle p \triangle x \gg 1/2$, which expresses the fact that one is far from the minimal quantum uncertainty. Evidently, Π must not vanish in the classical limit.

A classical distribution function $f(p, x)$ can be constructed to reproduce the classical expectations (4.6)

$$f(p, x) = \delta(p - 2\Pi x)\, \rho(x, x) . \tag{4.7}$$

One easily verifies that $\langle x^2 \rangle_c = \int dp\, dx\, x^2 f(p, x)$, $\langle p^2 \rangle_c = \int dp\, dx\, p^2 f(p, x)$, $\langle xp \rangle_c = \int dp\, dx\, xp f(p, x)$. For another perspective on this matter, we calculate the Wigner distribution function f_W

$$f_W(p, x) = \int \frac{dy}{2\pi} e^{ipy} \rho\left(x - \frac{1}{2}y, x + \frac{1}{2}y\right) . \tag{4.8a}$$

With the Gaussian ρ (4.1), this becomes

$$f_W(p, x) = \sqrt{\frac{2G}{\pi(1 + \xi)}}\, e^{-\frac{2G}{1+\xi}(p - 2\Pi x)^2}\, \rho(x, x) \tag{4.8b}$$

and for large $2G(1 + \xi)^{-1}$ coincides with (4.7).

It still remains to choose an *Ansatz* for Λ, select a Hamiltonian, evaluate I in (3.5), and vary all the parameters to obtain equations for them. Before doing this, we observe that the Gaussian density matrix exactly solves the Liouville equation for quadratic systems, hence we discuss these first.

B. Quadratic Hamiltonians and Linear Dynamics

The Gaussian density matrix is an exact solution to Liouville's equation when the Hamiltonian is quadratic and dynamics is linear, without self-interaction. It is instructive to examine these solvable cases, but first we must assess their relevance to problems of equilibrium/non-equilibrium behavior.

When an external agent acts on a system and destroys its equilibrium, it is natural to believe that after the disturbance ceases, equilibrium will be re-established owing to self-interactions. Conversely, in the absence of self-interactions, one would not expect a return to equilibrium. It would therefore appear that linear systems, without self-interaction, though solvable, never exhibit initial equilibrium, out-of-equilibrium evolution in the presence of a disturbance, return to equilibrium after the disturbance. Our explicit calculations support the above observations in most, *but not all*, cases.

For a quantum mechanical example, we chose a harmonic oscillator with time-dependent frequency

$$H = \frac{1}{2}p^2 + \frac{1}{2}\omega^2(t)x^2 \, . \tag{4.9}$$

The frequency squared is a constant equal to ω_i^2 in the past, and again constant, ω_f^2, in the future. Substituting the Gaussian *Ansatz* (4.1) for ρ into (2.3) shows that the Liouville equation is satisfied — the exact solution is Gaussian provided, as anticipated above, γ is given by (4.3), ξ is constant and G and Π obey

$$\dot{G} = 4\Pi G \, , \tag{4.10a}$$

$$\dot{\Pi} = \frac{1 - \xi^2}{8G^2} - 2\Pi^2 - \frac{\omega^2}{2} \, . \tag{4.10b}$$

[As a check, we consider the static case when ω^2 is time-independent. Then (4.10) are solved by setting $\dot{G}$, $\dot{\Pi}$ and Π to zero,

$$G = \frac{\sqrt{1 - \xi^2}}{2\omega} = \frac{1}{2\omega}\tanh\beta\omega \, . \tag{4.11}$$

The definition $\xi^{-1} = \cosh\beta\omega$ relates the mixing parameter ξ to the temperature $T = 1/k\beta$. The density matrix (4.1), with parameters taken as above, reproduces the Boltzmann distributed harmonic oscillator.]

Here is not the place, nor do we have the time, to discuss the [numerical] solutions to (4.10) with various profiles in time for $\omega^2(t)$. We summarize: generically as expected equilibrium is not restored when $\omega^2(t)$ settles to the constant ω_f^2 for $t > t_f$. However, the slower the variation in $\omega^2(t)$, as it passes from ω_i^2 to ω_f^2, the smaller the departure from equilibrium at $t > t_f$. Moreover, for a class of very special functional forms for the time-dependence of $\omega^2(t)$, which bear a curious mathematical relation to reflectionless potentials, the system *does* settle into thermal equilibrium. For details, see Ref. 3.

C. Non-Linear Dynamics

For non-quadratic Hamiltonians, we cannot solve the Liouville equation exactly, and our previously described variational approximation comes into play. In order to implement it, we must make a variational *Ansatz* for Λ, to supplement the one for ρ.

Choice of a variational expression for Λ is dictated by several considerations. Because the action to be varied is linear in Λ and ρ, variation of the former produces equations determining the latter. Consequently, there should be as many parameters in the trial expression for Λ as in ρ, so that a sufficient number of equations determining the parameters of ρ is obtained. From the boundary condition $\Lambda|_{t=t_f} = 1$, we know that the exact solution for Λ is $\Lambda = 1$. Therefore, the trial form for Λ must accommodate this possibility. Λ should be parametrized in such a way that $\dot{\xi} = 0$ emerges as a variational equation; this expression of constant entropy should characterize the approximation. Finally, note that while the exact variational equations for Λ and ρ decouple, this need not be the case for the approximate equations. But it would be most convenient to preserve decoupling, and this also influences the selection of Λ.

To motivate the *Ansatz* that we make for Λ, let us first recall the static variational principle of statistical mechanics, which was described earlier: Maximizing entropy at fixed average H, produces the Boltzmann distribution formula for ρ. Clearly, a density matrix proportional to $e^{-\beta\Theta}$, where Θ is an arbitrary operator, maximizes the entropy with fixed average Θ.

We now recognize from (4.2) that the Gaussian density matrix (4.1) is of the maximum entropy form, where the constrained operator Θ is a linear superposition of x^2, p^2, $px+xp$, and the identity operator 1, with time-dependent coefficients related to γ, G, Π and ξ. Therefore, it is natural to take Λ in the same space of operators

$$\hat{\Lambda} = \Lambda^{(1)}1 + \Lambda^{(p^2)}p^2 - \Lambda^{(xp)}(xp + px) + \Lambda^{(x^2)}x^2 \tag{4.12a}$$

$$\Lambda(x_1, x_2) = \langle x_1|\hat{\Lambda}|x_2\rangle,$$

$$\begin{aligned} &= \Lambda^{(1)}\delta(x_1 - x_2) - \Lambda^{(p^2)}\delta''(x_1 - x_2) \\ &\quad + i\Lambda^{(xp)}(x_1 + x_2)\delta'(x_1 - x_2) + \Lambda^{(x^2)}x_1^2\delta(x_1 - x_2). \end{aligned} \tag{4.12b}$$

Here $\Lambda^{(a)}$ are variational parameters. They all depend on time, and satisfy the condition at final time $t = t_f$ that $\Lambda^{(1)}$ becomes 1, and the others vanish. The advantage of (4.12) is that the action (3.5) to be varied is linear in the $\Lambda^{(a)}$, so the equations for the parameters in ρ do not involve $\Lambda^{(a)}$. The equations for $\Lambda^{(a)}$ remain coupled to the parameters of ρ, but we need not solve them.

For an example with non-linear dynamics, we consider the anharmonic oscillator with time-dependent frequency

$$H = \frac{1}{2}p^2 + \frac{1}{2}\omega^2(t)x^2 + \frac{\lambda x^4}{6}. \tag{4.13}$$

We evaluate the action with ρ as in (4.1) and Λ as in (4.12). Upon setting the $\Lambda^{(a)}$ variations to zero, the resulting equations that determine ρ are as before, except that (4.10) are replaced by

$$\dot{G} = 4\Pi G, \tag{4.14a}$$

$$\dot{\Pi} = \frac{1 - \xi^2}{8G^2} - 2\Pi^2 - \frac{\omega^2}{2} - \lambda \frac{G}{1 - \xi}. \tag{4.14b}$$

[In the static problem, ω^2 is time-independent — this is not our primary concern, but presents an interesting exercise, since a variational approximation to the density matrix for the anharmonic oscillator does not seem to appear in the literature. To solve (4.14) in that case, we set to zero $\dot{G}$, $\dot{\Pi}$ and Π. With G replaced by $\frac{\sqrt{1-\xi^2}}{2\omega} g \equiv \frac{g}{2\omega} \tanh \beta\omega$, ξ related to the temperature by $\xi^{-1} = \cosh \beta\omega$ as in (4.11), g satisfies

$$\lambda \left(\frac{g}{\omega} \right)^3 \mathrm{ctnh} \frac{\beta\omega}{2} = 1 - g^2. \tag{4.15}$$

This cubic equation may be solved: for weak coupling it gives

$$g \approx 1 - \frac{\lambda}{2\omega^3} \mathrm{ctnh} \frac{\beta\omega}{2} \tag{4.16a}$$

for strong coupling

$$g \approx \frac{\omega}{\left(\lambda \mathrm{ctnh} \frac{\beta\omega}{2} \right)^{1/3}} - \frac{\omega^3}{3\lambda \mathrm{ctnh} \frac{\beta\omega}{2}} \tag{4.16b}$$

while in the absence of the harmonic term, $\omega = 0$, we find

$$G = \frac{(1 - \xi)^{2/3}(1 + \xi)^{1/3}}{2\lambda^{1/3}} \tag{4.16c}$$

but ξ may no longer be related to a temperature, and ρ — though time-independent — is not Boltzmann distributed.]

Given a profile for $\omega^2(t)$ and values for λ, (4.14) may be solved numerically. For a detailed discussion, see Ref. 3. Here we merely record that, consistent with physical intuition, departures from equilibrium at late times decrease with increasing λ [strengthening interaction] and also decrease as the rate of change of $\omega^2(t)$ between ω_i^2 and ω_f^2 decreases. The latter behavior, already seen in linear systems, can also be established on general principles, with the help of the quantum adiabatic theorem. Consider the defining expression for the

Boltzmann distributed density matrix at early times

$$\rho_i(x_1, x_2) = \sum_n p_n \psi_n(x_1; \omega_i) \psi_n^*(x_2; \omega_i), \qquad (4.17a)$$

$$p_n = e^{-\beta_i E_n(\omega_i)} \Big/ \sum_{n'} e^{-\beta_i E_{n'}(\omega_i)}. \qquad (4.17b)$$

Dependence on the parameter ω that will vary with time for $t_i < t < t_f$ is explicitly indicated. If the variation is sufficiently slow, the quantum adiabatic theorem states that apart from position-dependent phases, the wave functions evolve into $\psi_n(x; \omega(t))$; hence at intermediate times, the density matrix is

$$\rho(x_1, x_2; t) = \sum_n p_n \psi_n(x_1; \omega(t)) \psi_n^*(x_2; \omega(t)) \qquad (4.17c)$$

and in late times attains equilibrium

$$\rho_f(x_1, x_2) = \sum_n p_n \psi_n(x_1; \omega_f) \psi_n^*(x_2; \omega_f). \qquad (4.17d)$$

But this need not be thermal equilibrium. The occupation probabilities remain given by (4.17b) and involve $\beta_i E_n(\omega_i)$ which in general cannot be written as $\beta_f E_n(\omega_f)$. Only for energy eigenvalues of the form $E_n(\omega) = f(\omega)e_n$, as in the harmonic oscillator, can a final temperature be defined by $\beta_f = \frac{f(\omega_i)}{f(\omega_f)}\beta_i$. [Of course the argument holds for arbitrary slowly varying parameters in the Hamiltonian, not just frequency.]

V. FIELD THEORETIC APPLICATION

Finally, we come to the topic of our principal interest: the application of the above ideas to a quantum field theory. In order to extend the formalism, which has been presented in the context of a single quantum mechanical variable, to the infinite number of degrees of freedom that comprise a quantized field, we make use of the [functional] field theoretic Schrödinger representation,[1] which we now review briefly.

A. Schrödinger Representation

The field theoretic Schrödinger representation is well-known,[1] at least for bosonic fields, but it is not as widely used as the Green's function method. The reason is that perturbative divergences can be isolated and renormalized more conveniently within the latter framework. However, by now we have become experienced in dealing with field theoretical infinities, so their occurrence is not an obstacle. Indeed, recent results establish renormalizability of the Schrödinger picture, both for static[6] and time-dependent problems.[7] Moreover, since the formalism is a generalization from ordinary quantum mechanics to the infinite number of degrees of freedom that comprise a field, it offers the possibility of using physical/mathematical intuition gained in quantum mechanics to analyze approximately field theoretic problems.

In the field theoretic Schrödinger representation and picture, states are described by wave functionals $\Psi(\varphi)$ of a c-number field $\varphi(\mathbf{r})$ at fixed time. The inner product is defined by functional integration,

$$\langle \Psi_1 | \Psi_2 \rangle = \int \mathcal{D}\varphi \, \Psi_1^*(\varphi) \Psi_2(\varphi) \tag{5.1}$$

while operators are represented by functional kernels

$$\mathcal{O}|\Psi\rangle \longleftrightarrow \int \mathcal{D}\bar{\varphi} \, \mathcal{O}(\varphi, \bar{\varphi}) \Psi(\bar{\varphi}) \,. \tag{5.2}$$

For the canonical field operator at fixed time, $\Phi(\mathbf{r})$, [the time argument is common to all operators in the Schrödinger picture, so it is suppressed] we use a diagonal kernel $\Phi(\mathbf{r}) \longleftrightarrow \varphi(\mathbf{r})\delta(\varphi - \bar{\varphi})$: the canonical commutation relations determine the canonical momentum kernel as $\Pi(\mathbf{r}) \longleftrightarrow \frac{1}{i}\frac{\delta}{\delta\varphi(\mathbf{r})}\delta(\varphi - \bar{\varphi})$: both kernels involve a functional δ-function. Evidently, Φ acts by multiplication on functionals of φ, while Π acts by functional differentiation. In this way, the action of any operator constructed from Π and Φ is

$$\mathcal{O}(\Pi, \Phi)|\Psi\rangle \longleftrightarrow \mathcal{O}\left(\frac{1}{i}\frac{\delta}{\delta\varphi}, \varphi\right)\Psi(\varphi) \,. \tag{5.3}$$

The fundamental dynamical equation is the time-dependent Schrödinger equation for a time-dependent functional $|\Psi; t\rangle \longleftrightarrow \Psi(\varphi; t)$. The equation takes definite form, once a Hamiltonian operator $H(\Pi, \Phi)$ is specified

$$i\frac{\partial}{\partial t}\Psi(\varphi; t) = H\left(\frac{1}{i}\frac{\delta}{\delta\varphi}, \varphi\right)\Psi(\varphi; t) \,. \tag{5.4}$$

For time-independent Hamiltonians, the usual separation of variables leads to a functional eigenvalue problem

$$\Psi(\varphi; t) = e^{-iEt}\Psi_E(\varphi) \,, \tag{5.5}$$

$$H\left(\frac{1}{i}\frac{\delta}{\delta\varphi}, \varphi\right)\Psi_E(\varphi) = e\Psi_E(\varphi) \,. \tag{5.6}$$

In particular, for quadratic Hamiltonians

$$H = \frac{1}{2} \int (\Pi^2 + \Phi h \Phi)$$

(5.7)

the ground state [vacuum] is a Gaussian functional,

$$\Psi_0(\varphi) = \det^{1/4} \frac{\omega}{\pi} \exp -\frac{1}{2} \int \varphi \omega \varphi$$

(5.8a)

with covariance ω determined by the "first quantized" Hamiltonian h,

$$\omega^2 = h$$

(5.8b)

and vacuum energy

$$E_0 = \frac{1}{2} \operatorname{tr} \omega .$$

(5.9)

[Throughout, a self-evident functional/matrix notation is used: ω and h are kernels, $\int \varphi \omega \varphi \equiv \int dr\, dr'\, \varphi(\mathbf{r}) \omega(\mathbf{r}, \mathbf{r}') \varphi(\mathbf{r}')$, $\operatorname{tr} \omega \equiv \int dr\, \omega(\mathbf{r}, \mathbf{r})$, and the determinant in (5.8a) is functional.]

Equation (5.8) represents the conventional Fock vacuum in the Schrödinger picture. Higher excited [multi-particle] Fock states are represented by polynomials of φ multiplying the vacuum Gaussian (5.8a); they are orthonormalized if taken in linear combinations corresponding to functional Hermite polynomials.

For translationally invariant theories, kernels depend only on differences of coordinates and diagonalization may be achieved by Fourier transformation: $h(\mathbf{r}, \mathbf{r}') = \int_{\mathbf{p}} e^{-i\mathbf{p} \cdot (\mathbf{r}-\mathbf{r}')} h(\mathbf{p})$, $\omega(\mathbf{r}, \mathbf{r}') = \int_{\mathbf{p}} e^{-i\mathbf{p} \cdot (\mathbf{r}-\mathbf{r}')} \omega(\mathbf{p})$, and (5.8b) enforces $\omega^2(\mathbf{p}) = h(\mathbf{p})$. [$\int_{\mathbf{p}}$ denotes $\int \frac{d^3\mathbf{p}}{(2\pi)^3}$.] As a consequence our free-field formulas are superpositions over all $\mathbf{p}$ of corresponding quantum mechanical formulas for harmonic oscillators with frequencies $\omega(\mathbf{p})$, which for a conventional bosonic free field theory, describing particles with mass m, is given by $\omega(\mathbf{p}) = \sqrt{\mathbf{p}^2 + m^2}$.

In the translationally invariant case, there are infrared divergences owing to the infinite volume V of space, *eg.*, $\frac{1}{2} \operatorname{tr} \omega = \frac{1}{2} V \int_{\mathbf{p}} \omega(\mathbf{p})$. These we shall ignore. Also there are ultraviolet divergences when the integral over momenta diverges. The fact that the normalization factor of (5.8a) may also diverge $\left[\det^{1/4} \frac{\omega}{\pi} = \exp \frac{1}{4} \operatorname{tr} \ln \frac{\omega}{\pi} = \exp \frac{V}{4} \int_{\mathbf{p}} \ln \frac{\omega(\mathbf{p})}{\pi} \right]$ is ignored, because the norm disappears from all matrix elements — it is chosen precisely so this will be the case. More intricate questions of ultraviolet finiteness will be addressed as they arise in our discussion of interacting field theories.

The analogous Schrödinger picture for fermion theories has also been recently developed,[8] but we shall not make use of it here.

The [functional] density matrix is constructed as

$$\rho(\varphi_1, \varphi_2) = \sum_n p_n \Psi_n(\varphi_1) \Psi_n^*(\varphi_2), \tag{5.10a}$$

$$\operatorname{tr} \rho = \int \mathcal{D}\varphi \rho(\varphi, \varphi) = 1. \tag{5.10b}$$

Here $\{\Psi_n\}$ is a complete set of wave functionals, and p_n is the probability that the system is in the state n. Average values of physical quantities described by operators $\mathcal{O}$, which in turn are represented by kernels $\mathcal{O}(\varphi, \tilde{\varphi})$, are determined by the density matrix

$$\langle \mathcal{O} \rangle = \operatorname{tr} \rho \mathcal{O} = \int \mathcal{D}\varphi \mathcal{D}\tilde{\varphi} \rho(\varphi, \tilde{\varphi}) \mathcal{O}(\tilde{\varphi}, \varphi). \tag{5.11}$$

With the same approximations as in the quantum mechanical case, ρ satisfies the Liouville equation (2.3) with a time-dependent Hamiltonian and describes isoentropic, energy non-conserving processes.

B. Gaussian Density Matrix

It is now evident that we can take over all the quantum mechanical formulas into the quantum field theory, merely by replacing the independent variable x by a field variable $\varphi(\mathbf{r})$; functions of x, $\psi(x)$, by functionals of $\varphi(\mathbf{r})$, $\Psi(\varphi)$; derivatives with respect to x by functional derivatives with respect to $\varphi(\mathbf{r})$; and integrals over x by functional integrals over $\varphi(\mathbf{r})$.

In particular, a Gaussian *Ansatz* for the functional density matrix is

$$\rho(\varphi_1, \varphi_2) =$$
$$e^{-\gamma} \exp -\frac{1}{2} \int \left\{ \varphi_1 \left(\frac{G^{-1}}{2} - 2i\Pi \right) \varphi_1 + \varphi_2 \left(\frac{G^{-1}}{2} + 2i\Pi \right) \varphi_2 - \varphi_1 G^{-1/2} \xi G^{-1/2} \varphi_2 \right\}. \tag{5.12}$$

G, Π and ξ are real, time-dependent kernels in $\mathbf{r}$ space. Translation invariance allows diagonalizing the kernels

$$G(\mathbf{r}_1, \mathbf{r}_2) = \int_{\mathbf{p}} e^{-i\mathbf{p} \cdot (\mathbf{r}_1 - \mathbf{r}_2)} G(\mathbf{p}), \tag{5.13a}$$

$$\Pi(\mathbf{r}_1, \mathbf{r}_2) = \int_{\mathbf{p}} e^{-i\mathbf{p}\cdot(\mathbf{r}_1-\mathbf{r}_2)}\Pi(\mathbf{p}), \qquad (5.13b)$$

$$\xi(\mathbf{r}_1, \mathbf{r}_2) = \int_{\mathbf{p}} e^{-i\mathbf{p}\cdot(\mathbf{r}_1-\mathbf{r}_2)}\xi(\mathbf{p}). \qquad (5.13c)$$

Averages of field operators follow (4.5). Linear averages vanish, bilinear averages are

$$\langle \Phi(\mathbf{r})\Phi(\mathbf{r}')\rangle = (G^{1/2}(1-\xi)^{-1}G^{1/2})(\mathbf{r},\mathbf{r}'), \qquad (5.14a)$$

$$\langle \Pi(\mathbf{r})\Pi(\mathbf{r}')\rangle = \frac{1}{4}(G^{-1/2}(1+\xi)G^{-1/2})(\mathbf{r},\mathbf{r}')$$
$$+ 4(\Pi G^{1/2}(1-\xi)^{-1}G^{1/2}\Pi)(\mathbf{r},\mathbf{r}'), \qquad (5.14b)$$

$$\langle \Phi(\mathbf{r})\Pi(\mathbf{r}')\rangle = \frac{i}{2}\delta(\mathbf{r}-\mathbf{r}') + 2(G^{1/2}(1-\xi)^{-1}G^{1/2}\Pi)(\mathbf{r},\mathbf{r}'). \qquad (5.14c)$$

As in quantum mechanics, γ is a normalization factor and ξ is the mixing kernel. When ξ vanishes, the density matrix describes a pure state

$$\rho(\varphi_1, \varphi_2)|_{\xi=0} = \Psi(\varphi_1)\Psi^*(\varphi_2), \qquad (5.15a)$$

$$\Psi(\varphi) = e^{-\gamma/2}\exp{-\frac{1}{2}\int\left\{\varphi\left(\frac{G^{-1}}{2} - 2i\Pi\right)\varphi\right\}}. \qquad (5.15b)$$

The choice for the Lagrange multiplier kernel Λ is a functional generalization of (4.12)

$$\hat{\Lambda} = \Lambda^{(1)}\mathbf{1} + \Pi\Lambda^{(\Pi^2)}\Pi - \Phi\Lambda^{(\Phi\Pi)}\Pi - \Pi\Lambda^{(\Phi\Pi)}\Phi + \Phi\Lambda^{(\Phi^2)}\Phi, \qquad (5.16a)$$

$$\Lambda(\varphi_1, \varphi_2) = \langle \varphi_1|\hat{\Lambda}|\varphi_2\rangle$$

$$= \Lambda^{(1)}\delta(\varphi_1 - \varphi_2) - \int_{\mathbf{r}_1,\mathbf{r}_2} \Lambda^{(\Pi^2)}(\mathbf{r}_1,\mathbf{r}_2)\frac{\delta^2}{\delta\varphi_1(\mathbf{r}_1)\delta\varphi_1(\mathbf{r}_2)}\delta(\varphi_1 - \varphi_2)$$

$$+ i\int_{\mathbf{r}_1,\mathbf{r}_2} \Lambda^{(\Phi\Pi)}(\mathbf{r}_1,\mathbf{r}_2)(\varphi_1(\mathbf{r}_1) + \varphi_2(\mathbf{r}_1))\frac{\delta}{\delta\varphi_1(\mathbf{r}_2)}\delta(\varphi_1 - \varphi_2)$$
$$\qquad (5.16b)$$

$$+ \int_{\mathbf{r}_1,\mathbf{r}_2} \Lambda^{(\Phi^2)}(\mathbf{r}_1,\mathbf{r}_2)\varphi_1(\mathbf{r}_1)\varphi_1(\mathbf{r}_2)\delta(\varphi_1 - \varphi_2).$$

Differential equations [in time] for the kernels determining ρ are obtained by evaluating the field theoretic generalization of the action (3.5) with ρ and Λ as in (5.12) and (5.16), and a definite Hamiltonian H governing the dynamics of interest.

We consider a scalar field theory in de Sitter space-time, described by the interval

$$(ds)^2 = (dt)^2 - e^{2\chi t}(dr)^2 \qquad (5.17)$$

wherein the universe expands at constant rate χ. The dynamics of the scalar field is summarized by the Lagrange density

$$\mathcal{L} = e^{3\chi t}\left[\frac{1}{2}\dot{\Phi}^2 - \frac{1}{2}e^{-2\chi t}(\nabla\Phi)^2 - \frac{1}{2}(m^2 + gR)\Phi^2 - \frac{\lambda}{6}\Phi^4\right], \qquad (5.18a)$$

where g controls a non-minimal coupling to the Ricci scalar R, given by $R = 12\chi^2$ in de Sitter space. [The scalar field is conformally coupled when $g = 1/6$.] It is convenient to introduce a new field by $\tilde{\Phi} = e^{\frac{3}{2}\chi t}\Phi$. Apart from a total time derivative, which we drop, the Lagrange density (5.18a), expressed in terms of $\tilde{\Phi}$ [which we rename Φ] reads

$$\mathcal{L} = \frac{1}{2}\dot{\Phi}^2 - \frac{1}{2}e^{-2\chi t}(\nabla\Phi)^2 + \frac{1}{2}\left(\frac{9}{4}\chi^2 - m^2 - gR\right)\Phi^2 - \frac{\lambda}{6}e^{-3\chi t}\Phi^4 \qquad (5.18b)$$

and leads to the Hamiltonian

$$H = \int_r \mathcal{H}$$

$$\mathcal{H} = \frac{1}{2}\Pi^2 + \frac{1}{2}e^{-2\chi t}(\nabla\Phi)^2 + \frac{1}{2}\left(-\frac{9}{4}\chi^2 + m^2 + gR\right)\Phi^2 + \frac{\lambda}{6}e^{-3\chi t}\Phi^4 . \qquad (5.18c)$$

The functional integrals arising in the definition of the action are all Gaussian and can be evaluated. The variational equations can thus be derived; they involve spatial integrals over the kernels $G(\mathbf{r}, \mathbf{r}')$, $\Pi(\mathbf{r}, \mathbf{r}')$ and $\xi(\mathbf{r}, \mathbf{r}')$. But by passing to momentum space and diagonalizing the kernels, the equations decouple and one finds that

$$\dot{G}(\mathbf{p}) = 4\Pi(\mathbf{p})G(\mathbf{p}), \qquad (5.19a)$$

$$\dot{\Pi}(\mathbf{p}) = \frac{1 - \xi^2(\mathbf{p})}{8G^2(\mathbf{p})} - 2\Pi^2(\mathbf{p})$$

$$- \frac{1}{2}\left(e^{-2\chi t}\mathbf{p}^2 - \frac{9}{4}\chi^2 + m^2 + gR\right) - \lambda e^{-3\chi t}\int_{\mathbf{k}} \frac{G(\mathbf{k})}{1 - \xi(\mathbf{k})} . \qquad (5.19b)$$

Also $\xi(\mathbf{k})$ must be time-independent and the normalization factor γ is given by

$$\gamma = \frac{1}{2}\ln \det 2\pi G^{1/2}(1 - \xi)^{-1}G^{1/2} . \qquad (5.20)$$

In the presence of the interaction $[\lambda \neq 0]$ the equations have to be renormalized, owing to the divergence of $\int_k \frac{G(\mathbf{k})}{1-\xi(\mathbf{k})}$. This has been done.[3]

The equations (5.19) have been solved numerically for the following specializations.[3] In flat space $(\chi = 0 = R)$, m^2 is allowed to vary time, with and/or without interaction. In the former case, the renormalized equations are solved. In de Sitter space $(\chi \neq 0 \neq R)$ the time-dependence of the Hamiltonian is contained in the geometry $(e^{\chi t})$. While the renormalized equations in the presence of the interaction are well-defined and coincide with those previously obtained by Mazenko,[9] the exponential growth of the geometry overwhelmed our computational resources and no definite results are available. In the absence of interaction, analytic solution is possible, which validates and generalizes earlier analysis by Guth and Pi[10] of the same problem. The flat space model, with varying mass, follows qualitatively behavior seen in quantum mechanics. For details, see Ref. 3.

VI. CONCLUSION

Non-equilibrium phenomena are drawing the attention of an ever widening circle of particle physicists. Analysis of early universe processes, like inflation, must confront the fact of a rapidly changing environment. Quark plasma physics is another subject for which non-equilibrium phenomena are important. We expect that our theory provides the initial steps in the development of a formalism within which these issues can be analyzed.

Of course, much more must be done. It is recognized that ours is a mean-field theory approach to the problem — similar to the Hartree-Fock, large N and Gaussian approximations in equilibrium problems. As such it suffers from well-known shortcomings. At the very least, one should find a systematic procedure for going beyond the initial Gaussian [mean-field theory] approximation. Moreover, considerations based on the Liouville equation necessarily are restricted to describing isoentropic phenomena. It is important to allow entropy change but this requires a more detailed model for the environment-system interaction.

REFERENCES

1. For a review see "Analysis on Infinite-Dimensional Manifolds — Schrödinger Representation for Quantized Fields", reprinted in this volume on p. 383.
2. "Symmetry Restoration at Finite Temperature", reprinted in this volume on p. 275.

3. O. Èboli, S.-Y. Pi and R. Jackiw, *Phys. Rev. D* **37**, 3557 (1988).

4. R. Balian and R. Vénéroni, *Phys. Rev. Lett.* **47**, 1353, (E) 1765 (1981); *Ann. Phys.* (NY) **164**, 334 (1985).

5. See *eg.*, R. Jackiw and A. Kerman, *Phys. Lett.* **A71**, 158 (1979); R. Jackiw, *Intl. Jnl. Quantum Chem.* **17**, 41 (1980).

6. K. Symanzik, *Nucl. Phys.* **B190** [FS3], 1 (1983); M. Lüscher, *Nucl. Phys.* **B254**, 52 (1985).

7. F. Cooper and E. Mottola, *Phys. Rev. D* **36**, 1114 (1987); S.-Y. Pi and M. Samiullah, *Phys. Rev. D* **36**, 3128 (1987); O. Èboli, S.-Y. Pi and M. Samiullah, *Ann. Phys.* (NY) **193**, 102 (1989).

8. R. Floreanini and R. Jackiw, *Phys. Rev. D* **37**, 2206 (1988), for a review see Ref. 1.

9. G. Mazenko, *Phys. Rev. Lett.* **54**, 2163 (1985). Mazenko presents three equations, while we have only two non-trivial ones, (5.19a) and (5.19b). Apparently, one linear combination of his three merely expresses the constancy of our ξ.

10. A. Guth and S.-Y. Pi, *Phys. Rev. D* **32**, 1899 (1985).

Section IV

APPROACHES TO QUANTUM THEORIES FOLLOWING DIRAC

While unsolvable quantum mechanical problems can be analyzed by a variety of approximations, for a long time quantum field theory was studied only by its own analog of the quantum mechanical Born series: the Dyson-Feynman-Schwinger perturbative expansion. Eventually, other quantum mechanical techniques were brought to quantum field theory, and a wealth of new physical phenomena was uncovered. Flexibility of formalism when confronting a complex quantum mechanical system vividly characterized Dirac's work; adopting and developing his techniques for modern quantum field theory provided new insights.

Paper IV.1

Dirac called attention to the possibility of quantizing in ways different from the usual fixed-time method. One of these, "light-cone" quantization, arose in the infinite momentum frame of current algebra, and came into its own in the analysis of high-energy ("deep inelastic") electron-nucleon scattering. Recently, it has been revived in attempts to analyze low energy dynamics of the standard model. My lectures on this subject were given at the DESY summer school in Hamburg (July 1971).

CANONICAL LIGHT-CONE COMMUTATORS AND THEIR APPLICATIONS

Springer Tracts in Modern Physics **62**, 1 *(1972)*

I. INTRODUCTION

I shall report here on recent work done by several colleagues* and me where we succeeded in exhibiting the canonical structure of current commutators on the light cone, and applied it to several interesting physical problems. Two reasons may be advanced at the present time for studying this topic.

(1) It has been appreciated already in the early days of current algebra, that the fixed mass sum rules, the most famous of which is due to Dashen, Fubini and Gell-Mann,[1,2] which are usually derived with the help of the unreliable $p \to \infty$ technique,[3] are in fact equivalent to appropriate light-cone commutators.[4]

(2) Recently, it has been shown that the dramatic MIT-SLAC electroproduction experiments[5] measure directly the commutator of electromagnetic currents on the light cone.[6]

The following notation will be used. For the coordinate x^μ and for all other vector or tensor quantities, we define the $\pm$ components by

$$x^\pm = 2^{-1/2}(x^0 \pm x^3). \qquad (\text{I-1})$$

The remaining components will be denoted by $i, i = 1, 2$, or by the subscript $\perp$. The metric tensor now is $g^{++} = g^{--} = g^{12} = g^{21} = 0$; $g^{+-} = g^{-+} = -g^{11} = -g^{22} = 1$. Consequently $x^2 = 2x^+x^- - x_\perp^2$.

Consider the commutator of two currents

$$C_{ab}^{\mu\nu}(x) = [J_a^\mu(x), J_b^\nu(0)]. \qquad (\text{I-2})$$

In a given field theory, the usual canonical formalism permits one to evaluate the *equal time commutator*, $C_{ab}^{\mu\nu}(x)|_{x^0=0}$, without solving the theory. I shall show how similarly one can compute the *light cone commutator*, $C_{ab}^{\mu\nu}(x)|_{x^+=0}$, without unraveling the dynamics. The reason for the nomenclature can be given. Since $C_{ab}^{\mu\nu}(x)$ vanishes for negative x^2, and since $x^2 = 2x^+x^- - x_\perp^2 = -x_\perp^2$ when $x^+ = 0$, knowledge of $C_{ab}^{\mu\nu}(x)$ at the point $x^+ = 0$ is equivalent to the knowledge of $C_{ab}^{\mu\nu}(x)$ on the light cone, $x^2 = 0$. The relation of the present results to other investigations is as follows. Fritsch and Gell-Mann[7] independently and simultaneously suggested that the *most singular* contribution to $C_{ab}^{\mu\nu}(x)$ near $x^2 = 0$ might be as in the free-field theory, where of course, it is completely calculable. Subsequently, Gross and Treiman[8] verified the Fritsch-Gell-Mann[7] conjecture in various interacting theories. For purposes of studying the deep-inelastic, "scaling" region,[9] both our approach and

---

*This work was done in collaboration with J. M. Cornwall, D. Discus and V. Teplitz.

the Fritsch-Gell-Mann[7] technique give identical results. However, as will be seen below, the determination of the fixed mass sum rules requires the *full* commutator at $x^+ = 0$, not just the most singular part at $x^2 = 0$.

II. LIGHT CONE CURRENT COMMUTATORS

Since, as we have argued, light cone current commutators are of interest both for experimental and theoretical physics, we now show how one might calculate them.[10] Consider first a conserved, internal symmetry current J_a^μ. The time independent charge is given by $Q_a = \int d^3x \, J_a^0(x)$, and the assumption that the current transforms in a known fashion in group space implies

$$[Q_a, J_a^\mu(0)] = i f_{abc} J_c^\mu(0). \tag{II-1}$$

It is not hard to show that for *conserved* currents an alternate formula for Q_a may be given

$$Q_a = \int d^2x_\perp dx^- J_a^+(x). \tag{II-2}$$

Of course Q_a is independent of the unintegrated variable x^+. From (II-1) and (II-2) it now follows that

$$\begin{aligned}[J_a^+(x), J_b^\mu(0)]_{x^+=0} &= i f_{abc} J_c^\mu(0) \delta^2(x_\perp)\delta(x^-) \\ &\quad + \partial_- S_{ab}^\mu(x) + \partial_i S_{ab}^{i\mu}(x).\end{aligned} \tag{II-3}$$

The terms in (II-3) which disappear upon integration over x^- and $x_\perp$ are not unlike the Schwinger terms that are present in selected equal time commutators. Indeed by taking vacuum expectation values, one can show that they *must* be present, though of course the vacuum matrix element is not sensitive to their c or q number character. However, in one important respect these structures are different from what has been encountered previously. $S_{ab}^\mu(x)$ and $S_{ab}^{i\mu}(x)$ are not, in general, local in x^-, though they do possess this properly in $x_\perp$. The reason for this is the following. The commutator must vanish for $x^2 = 2x^+x^- - x_\perp^2 < 0$. When $x^+ = 0$, as it is in (II-3), $x^2 = -x_\perp^2$, and by causality, the commutator must be local on $x_\perp$ (*i.e.*, it has support only at $x_\perp = 0$). However, no constraint on x^- is imposed. Therefore, we expect the objects on the right-hand side of (II-3) to have an arbitrary dependence on x^-. Such quantities are called *bilocal* operators.

The bilocal operators are model dependent, and at the present time there is no *a priori* way of calculating them. However, as we shall demonstrate repeatedly, they are measurable, physical quantities, and their form must be specified, if we wish to understand various and diverse physical phenomena. As will be demonstrated below, they are known to have non-zero connected matrix elements; hence they are q numbers. It will be shown presently that the $p \to \infty$ technique of current algebra would predict their vanishing; hence we have the possibility of rectifying that unreliable method, if we can specify the bilocal operators. It is this task which we now embark upon.

A. Light-Cone Quantization

The reason it was possible to give a plausible model for *equal time* commutators without solving any of the dynamical equations of quantum field theory, was of course due to the fact that equal time commutators for fields form a boundary condition on the theory. They are assumed to be of a definite canonical form which can be given without solving the theory. Consequently, for any function of the fields, like a current, the appropriate commutators can be computed. It turns out that something similar may be done for light cone commutators. Rather than quantizing the theory on a space like surface — which corresponds to the usual equal time quantization — one may equivalently quantize on a light-like surface. With the latter technique, light cone commutators emerge canonically.[11]

In order to obtain familiarity with these ideas, let us begin by examining a simple scalar theory given by the Lagrange density

$$\mathcal{L} = \frac{1}{2}\partial_\mu \varphi \partial^\mu \varphi - \frac{1}{2}\mu^2 \varphi^2 + \frac{1}{4}\lambda \varphi^4. \tag{II-4}$$

In the usual way, this leads to the equation of motion

$$\Box \varphi + \mu^2 \varphi = \lambda \varphi^3. \tag{II-5}$$

However, the quantum theory is not specified only by (II-5). One also postulates the following equal time commutation relations

$$i[\varphi(x), \varphi(0)]_{x^0=0} = 0, \tag{II-6a}$$

$$i[\partial_0 \varphi(x), \varphi(0)]_{x^0=0} = \delta^3(x), \tag{II-6b}$$

$$i[\partial_0 \varphi(x), \partial_0 \varphi(0)]_{x^0=0} = 0, \tag{II-6c}$$

where the special role of $\partial_0 \varphi$ follows from the fact that it is the canonical momentum $\partial_0 \varphi = \partial \mathcal{L}/\delta \partial_0 \varphi$. This scheme is the quantum analogue of the classical (non-quantized) procedure of solving the partial differential (II-5) with the specification of the Cauchy initial value data: $\varphi(x)$ and $\partial_0\varphi(x)$ on the space-like surface $x^0 = 0$.

Let us now rewrite things in terms of the light cone variables. Of course, the Lagrange function and the equation of motion are equivalent

$$\mathcal{L} = \partial_-\varphi\partial_+\varphi + \frac{1}{2}\partial_i\varphi\partial^i\varphi - \frac{1}{2}\mu^2\varphi^2 + \frac{1}{4}\lambda\varphi^4, \qquad \text{(II-7)}$$

$$2\partial_+\partial_-\varphi + \partial_i\partial^i\varphi + \mu^2\varphi = \lambda\varphi^3. \qquad \text{(II-8)}$$

The new element comes in if we define the canonical momentum to be $\delta\mathcal{L}/\delta\,\partial_+\varphi = \partial_-\varphi$. Consequently, we postulate that

$$i[\partial_-\varphi(x),\varphi(0)]_{x^+=0} = \frac{1}{2}\delta(x^-)\delta^2(x_\perp). \qquad \text{(II-9a)}$$

This may be integrated and gives the basic light cone commutator

$$i[\varphi(x),\varphi(0)]_{x^+=0} = \frac{1}{4}\varepsilon(x^-)\delta^2(x_\perp), \qquad \text{(II-9b)}$$

where $\varepsilon(x) = -\varepsilon(-x) = 1$ for $x > 0$. This procedure is the quantum analog of solving the classical theory by specifying the initial data on the light like surface $x^+ = 0$. It can be shown that classically a unique solution is obtained by specifying only φ. The knowledge of derivatives of φ is unnecessary.[12]

Therefore (II-9b) is the fundamental light cone commutator in this theory, and light cone commutators of all other functions of $\varphi(x)$ may be now computed without solving the theory. There are peculiar features of this quantization scheme that should be commented upon. Note that the commutator between field and momentum (II-9a) has an unexpected factor of $\frac{1}{2}$. Furthermore, according to (II-9b) fields do not commute with themselves, and it is also clear that canonical momenta are noncommuting. This difference from the state of affairs in equal time quantization arises from the fact that the canonical momentum $\pi(x)$ is related by an equation of *constraint* to the canonical field: $\pi(x) = \partial_-\varphi(x)$. In the present scheme, equations of motion involve $+$ derivatives; the $+$ coordinate plays the role of time. Consequently, the canonical formalism is more complicated than at equal times, and careful investigation shows that the *Ansatz* (II-9) is correct. A related phenomenon is that the light cone method has changed a second-order differential equation in time, to a first-order equation in $+$. Therefore, the present scheme is somewhat analogous to the conventional quantization of fermion theories, which are also first-order.

The question now arises whether or not the light cone theory is the same as the equal time theory, or whether it is different; *i.e.* is the S matrix the same for both schemes. This is an important question, for if the two theories are different, then we cannot assume the *simultaneous* validity of the conventional equal time commutators and the new light cone commutators. To answer this question with complete certainty would require solving the two theories, and comparing the results — a task impossible to carry out at present. Hence, we must content ourselves with a partial answer which relies on various formal properties of the two theories. The following four facts are offered as evidence that the two methods of quantization result in the same physical theory.

1. It can be shown that the free theories are the same.[11]

2. If the light cone theory is developed to the point of computing S matrix elements, then one encounters instead of the usual Feynman-Dyson rules, the Weinberg "$p \to \infty$"[13] rules, which are supposed to be equivalent.[11]

3. In the light cone theory, Green's functions are ordered along the $+$ direction, rather than along the time direction. However, causality requires the two to be the same, up to seagulls. To see this, consider

$$T_0(x) = [A(x), B(0)]\theta(x^0),\tag{II-10a}$$

$$T_+(x) = [A(x), B(0)]\theta(x^+).\tag{II-10b}$$

The only place that (II-10a) and (II-10b) appear to be unequal is when $x^0 > 0$ and $x^+ < 0$, or $x^0 < 0$ and $x^+ > 0$. However in this region, x^2 is spacelike and the commutator function vanishes by causality.[14]

4. The *vacuum expectation value* of light cone commutators can be calculated in the usual framework. The results are consistent with the canonically postulated forms. For example, consider

$$i\langle 0|[\varphi(x), \varphi(0)]|0\rangle = i \int_0^\infty d\,a^2 \varrho(a^2)\triangle(x|a^2)\,,\tag{II-11}$$

where $\varrho(a^2)$ is some spectral function, and $\triangle(x|a^2)$ is the free-field commutator function.

$$\triangle(x|a^2) = (2\pi)^{-3} \int d^4k\varepsilon(k^0)\delta(k^2 - a^2)e^{-ikx}\,.\tag{II-12}$$

In the usual way, we have from the equal time commutators the result that

$$\int_0^\infty d\,a^2 \varrho(a^2) = 1.\tag{II-13}$$

To calculate the light cone commutator, we observe that from (II-12) it follows that

$$\Delta(x|a^2)|_{x^+=0} = (-i/4)\varepsilon(x^-)\delta^2(x_\perp). \qquad \text{(II-14)}$$

Consequently, we have with the help of (II-13)

$$i\langle 0|[\varphi(x), \varphi(0)]|0\rangle|_{x^+=0} = \frac{1}{4}\varepsilon(x^-)\delta^2(x_\perp) \qquad \text{(II-15)}$$

which reproduces the canonically postulated (II-9b), as far as vacuum expectation values are concerned. Other evaluations of vacuum expectation values always lead to the same result — the light cone method is consistent with and equivalent to the equal time method.

Supported by the above four partial arguments, as well as by the fact that no conflict has been found between the two methods, we shall assume that the form of light cone commutators evaluated in the light cone quantization scheme is valid simultaneously with the equal time commutators.

To give a form for the $SU(3) \times SU(3)$ current commutators, we turn to a model which has previously served as an inspiration for equal time commutators: the quark model. Since we wish to present results which are possibly true even in an interacting field theory, we shall assume that the quarks interact with a vector gluon — the end result does not depend on the nature of the gluon-quark interaction, as long as it is a nonderivative one. Thus we are led to the Lagrange density

$$\mathcal{L} = \bar{\psi}[((i/2)\overset{\leftrightarrow}{\partial}_\mu - g\,B_\mu)\gamma^\mu - M]\psi - \frac{1}{4}F^{\mu\nu}F_{\mu\nu} \qquad \text{(II-16)}$$

$$F^{\mu\nu} = \partial^\mu B^\nu - \partial^\nu B^\mu.$$

We have set the gluon (B^μ) mass to zero in order that the computations be simple — the end results with which we deal are insensitive to this simplification. The quark field ψ has mass M. If we wish to introduce $SU(3)$ symmetry breaking we would allow M to be a mass matrix, though for the present, we do not do this. The light cone quantization of this model may now be taken over from the published literature.[11] Various peculiar features of this scheme are the following. The 4 component fermion field ψ cannot be viewed as a canonically independent field. Rather a projection of it $\psi_+ \equiv P_+\psi$, $P_+ = \frac{1}{2}\gamma^-\gamma^+$ is independent; while the other projection $\psi_- \equiv P_-\psi$, $P_- = \frac{1}{2}\gamma^+\gamma^-$ is a dependent field. The canonical dependence is expressed by the following equation of constraint, which may be shown to hold in this theory

$$(i\partial_- - g\,B^+)\psi_- = \frac{1}{2}[(i\partial_j - g\,B_j)\gamma^j + M]\gamma^+\psi_+. \qquad \text{(II-17a)}$$

Since B^μ is a *massless* vector meson field, there exists a gauge freedom, which may be exploited to set B^+ to zero.[15] Then (II-17a) is easily integrated, to give a formula, at fixed x^+, for ψ_- in terms of the canonically independent fields B_j and ψ_+

$$\psi_-(x) = (-i/4) \int d\xi \varepsilon(x^- - \xi)$$
$$\cdot \left\{ \left[i\partial_j - g\, B_j(x^+, \xi, \boldsymbol{x}_\perp) \right] \gamma^j + M \right\} \gamma^+ \psi_+(x^+, \xi, \boldsymbol{x}_\perp). \qquad \text{(II-17b)}$$

The canonical commutators are

$$[B^i(x), B^j(0)]_{x^+=0} = (i/4)g^{ij}\varepsilon(x^-)\delta^2(\boldsymbol{x}_\perp), \qquad \text{(II-18)}$$

$$\{\psi_+(x), \psi_+^*(0)\}_{x^+=0} = 2^{-1/2}P_+\delta(x^-)\delta^2(\boldsymbol{x}_\perp). \qquad \text{(II-19)}$$

The operators ψ_+ and ψ_+^* anticommute with themselves, and commute with the B^i. Another commutator which we shall need is the one between ψ_+^* and ψ_-. We have from (II-17) and (II-19)

$$\{\psi_-(x), \psi_+^*(x')\}_{x^+=x'^+}$$
$$= -\left(i/4\sqrt{2} \right) \varepsilon(x^- - x'^-) \left\{ [i\partial_j - g\, B_j(x')] \gamma^j + M \right\} \gamma^+ \delta^2(\boldsymbol{x}_\perp - \boldsymbol{x}'_\perp). \qquad \text{(II-20)}$$

It is also possible to give other commutators, *eg.*, between ψ_- and ψ_-^*. These are extremely complicated, and fortunately we have no need of them. Note that the commutator (II-20) makes reference to the interaction.

B. Derivation of Current Commutators

The current commutators may now be calculated. For the present, considerations of internal symmetry are ignored, and the current is defined by

$$J^\mu = \bar{\psi}\gamma^\mu\psi. \qquad \text{(II-21a)}$$

In terms of the fields ψ_+ and ψ_- the formula for the current is as follows

$$J^+ = 2^{1/2}\psi_+^*\psi_+, \qquad \text{(II-21b)}$$

$$J^- = 2^{1/2}\psi_-^*\psi_-, \qquad \text{(II-21c)}$$

$$J^i = 2^{-1/2}\psi_-^*\gamma^+\gamma^i\psi_+ + 2^{-1/2}\psi_+^*\gamma^-\gamma^i\psi_-. \qquad \text{(II-21d)}$$

The commutator of J^+ with itself is quite simple. Only the ψ_+ field is involved whose commutator, (II-19), contains no interaction terms. We find

$$[J^+(x), J^+(y)]_{x^+=y^+} = 0. \tag{II-22}$$

For the $[J^+, J^-]$ commutator, we need the more complicated formula (II-20). A rather lengthy computation yields

$$[J^+(x), J^-(y)]_{x^+=y^+}$$

$$= \left(-\frac{i}{2}\right)\varepsilon(x^- - y^-)\bar{\psi}(x)\{\gamma^j\left[i\overleftrightarrow{\partial}_j + g\,B_j\right] + M\}P_-\psi(y)\delta^2(\boldsymbol{x}_\perp - \boldsymbol{y}_\perp) \tag{II-23a}$$

$$- \frac{1}{2}\varepsilon(x^- - y^-)\partial_i^x\left\{\bar{\psi}(x)\gamma^i P_-\psi(y)\delta^2(\boldsymbol{x}_\perp - \boldsymbol{y}_\perp)\right\} - \text{h.c.}$$

[The abbreviation h.c. stands for Hermitian conjugate which must be subtracted from the right-hand side of (II-23a).] The commutator is not expressible in terms of the current itself. Like the commutator of the fields it appears to depend explicitly on the coupling to $g\,B^\mu$. However this dependence may be eliminated. By use of the equation (II-17a), (II-23a) may be re-expressed by

$$[J^+(x), J^-(y)]_{x^+=y^+}$$

$$= \partial_-^x\left\{-\frac{1}{2}\varepsilon(x^- - y^-)\delta^2(\boldsymbol{x}_\perp - \boldsymbol{y}_\perp)\bar{\psi}(x)\gamma^-\psi(y)\right\}$$

$$+ \partial_i^x\left\{-\frac{1}{4}\varepsilon(x^- - y^-)\delta^2(\boldsymbol{x}_\perp - \boldsymbol{y}_\perp)[\bar{\psi}(x)\gamma^i\psi(y) - \varepsilon^{ij}\bar{\psi}(x)\gamma_j\gamma_5\psi(y)]\right\} - \text{h.c.} \tag{II-23b}$$

We have used the relations $\gamma^i P_- = \frac{1}{2}\gamma^i - \frac{1}{2}\varepsilon^{ij}\gamma_j\gamma_5$,

$$\gamma_5 \equiv \gamma^0\gamma^1\gamma^2\gamma^3, \quad \varepsilon^{ij} = -\varepsilon^{ji} = 1 \quad \text{for} \quad i=1,\, j=2.$$

The remarkable feature of this final formula is its elegant simplicity. The light cone commutator of these two currents is expressible in terms of bilocal operators, which are straight-forward generalizations of the local currents. These bilocal operators are $\bar{\psi}(x)\gamma^\mu\psi(y)$ and $\bar{\psi}(x)\gamma^\mu\gamma_5\psi(y)$.

It is clear* that the operators enter with $(x-y)^2 = 0$, since $x^+ = y^+$ and $\boldsymbol{x}_\perp = \boldsymbol{y}_\perp$. Consequently, the non-locality is confined to the $-$ direction. The bilocal terms do not contribute when an integration over x^- and $\boldsymbol{x}_\perp$ is performed. Thus the structure (II-23b) is of the type (II-3), which we arrived at by general considerations.

Finally, a completely similar argument gives

$$\left[J^+(x), J^i(y)\right]_{x^+=y^+}$$

$$= \partial_-^x \left\{ -\frac{1}{4}\varepsilon(x^- - y^-)\delta^2(\boldsymbol{x}_\perp - \boldsymbol{y}_\perp)\left[\bar\psi(x)\gamma^i\psi(y) + \varepsilon^{ij}\bar\psi(x)\gamma_j\gamma_5\psi(y)\right]\right\}$$

$$+ \partial_j^x \left\{ \frac{1}{4}\varepsilon(x^- - y^-)\delta^2(\boldsymbol{x}_\perp - \boldsymbol{y}_\perp)\left[g^{ij}\bar\psi(x)\gamma^+\psi(y) + \varepsilon^{ij}\bar\psi(x)\gamma^+\gamma_5\psi(y)\right]\right\} - \text{h.c.}$$

$$\text{(II-24)}$$

Again, the light cone commutator is expressible in terms of bilocal generalizations of the vector and axial vector currents.

Commutators of current components, not involving at least one + component, are considerably more complicated as they require the commutators between the ψ_- and ψ_-^* fields. We shall not concern ourselves with these here.

There exists an alternate method for computing commutators, due to Schwinger.[16] Schwinger's original arguments apply to ordinary quantization, but it is not hard to generalize them to light cone quantization. This "action principle" procedure gives the following formula for the *equal time* commutator of the *time* component of a current, with the current itself

$$\left[J^0(x), J^\mu(y)\right]_{x^0=y^0} = -i\partial_\alpha\delta J^\alpha(x)/\delta A_\mu(y). \qquad \text{(II-25)}$$

Here it is assumed that the current is conserved, and that it couples to an external field A^μ. At the end of the computation A^μ is set to zero. The variation in (II-25) is performed with fixed canonical variables, *i.e.*, J^α is first expressed in terms of canonical coordinates and momenta and the variation is

---

*Our formulas for the bilocal currents are gauge dependent, since they were calculated in the special gauge $B^+ = 0$. An explicitly gauge independent form is

$$\bar\psi(0, x^-, 0)\gamma^\mu\psi(0) \exp ig \int_0^{x^-} dy^- B^+(0, y^-, 0);$$

see Gross and Treiman.[8]

performed only with respect to the residual dependence on A^μ. The analogous formula arising from light cone quantization is

$$[J^+(x), J^\mu(y)]_{x^+=y^+} = -i\partial_\alpha \delta J^\alpha(x)/\delta A_\mu(y). \qquad \text{(II-26)}$$

In many models, J^+ has no dependence on external fields. (In our theory for example, $J^+ = \sqrt{2}\,\psi_+^*\psi_+$ and ψ_+ is a canonical variable which is held fixed in the explicit variation.) In that case (II-26) becomes

$$[J^+(x), J^\mu(y)]_{x^+=y^+} = -i\partial_- \delta J^-(x)/\delta A_\mu(y) - i\partial_i \delta J^i(x)/\delta A_\mu(y). \qquad \text{(II-27)}$$

This formula is exactly of the general form (II-3). An abstract representation for the bilocal operators is given by (II-27): $-i\delta J^-(x)/\delta A_\mu(y)$ and $-i\delta J^i(x)/\delta A_\mu(x)$, and it is seen that when an integration over x^- and $x_\perp$ is performed, the bilocal terms cease to contribute. Moreover, we again recognize that these structures are light-cone generalizations of the ordinary Schwinger term. That object, it will be recalled, is given by $\delta J^\nu(x)/\delta A_\mu(x)$, where the variation is performed at a common value of time.[16]

A specific form for the bilocal operators is obtainable only by specifying a model, *i.e.*, by specifying the dependence on A^μ. It is not hard to show that in the quark-gluon model this dependence is such that (II-22), (II-23) and (II-24) are true, regardless of the nature of the gluoan, provided it is not coupled derivatively.

C. Non-Canonical Contributions

Let us take the vacuum expectation value of the current commutator function

$$\langle 0|[J^\mu(x), J^\nu(0)]|0\rangle = (g^{\mu\nu}\Box - \partial^\mu\partial^\nu)\int_0^\infty d\lambda^2 \sigma(\lambda^2)\Delta(x|\lambda^2). \qquad \text{(II-28)}$$

The representation is conserved, as a consequence of current conservation, and a non-negative spectral function $\sigma(\lambda^2)$ has been introduced. Consider now the $++$ components of (II-28) when $x^+ = 0$. We have from (II-14), a nonvanishing result

$$\langle 0|[J^+(x), J^+(0)]|0\rangle_{x^+=0}$$

$$= -\partial_-\partial_-\int_0^\infty d\lambda^z \sigma(\lambda^2)\Delta(x|\lambda^2)|_{x^+=0} \qquad \text{(II-29)}$$

$$= \partial_-\partial_-[(i/4)\varepsilon(x^-)\delta^2(x_\perp)]\int_0^\infty d\lambda^2 \sigma(\lambda^2) \neq 0.$$

On the other hand, the canonical light cone commutator (II-22) would lead us to expect zero for the right-hand side of (II-29). Of course, this is just the ancient problem that naive canonical commutators do not yield the conventional Schwinger term in fermion theories. Recall that the Schwinger term is given by

$$\langle 0|[J^0(x), J^i(0)]|0\rangle_{x^0=0} = i\partial^i \delta^3(\boldsymbol{x}) \int_0^\infty d\lambda^2 \sigma(\lambda^2) \qquad \text{(II-30)}$$

which involves the same spectral integral as (II-29). Hence, we learn that the light cone commutator (II-22) must be modified by a non-canonical contribution, just as the equal time commutator. We shall make the very important assumption that all these non-canonical additions are c numbers, and therefore given by their vacuum expectation value. This assumption is *not* true in perturbation theory, as will be seen in Section III below. Thus we set

$$\left[J^+(x), J^+(y)\right]_{x^+=y^+} = (-i/4)\partial^x_- \partial^y_- \left[\varepsilon(x^- - y^-)\delta^2(\boldsymbol{x}_\perp - \boldsymbol{y}_\perp)S\right]$$

$$S = \int_0^\infty d\lambda^2 \sigma(\lambda^2). \qquad \text{(II-31)}$$

It is instructive to examine this commutator in theories where the usual Schwinger term emerges canonically *eg.*, in scalar electro-dynamics. A straight forward calculation yields (II-31), except S becomes a bilocal operator $S(x|y)$, with $S(x|x)$ coinciding with the usual Schwinger term.[17]

Similar modifications are present in other commutators. For the $+-$ components (II-28) gives

$$\langle 0|[J^+(x), J^-(0)]|0\rangle_{x^+=0}$$

$$= (i/8) \int_0^\infty d\lambda^2 (\lambda^2 - \partial_i \partial^i)\sigma(\lambda^2)\varepsilon(x^-)\delta(\boldsymbol{x}_\perp). \qquad \text{(II-32)}$$

Apart from the term not involving derivatives, which can be reproduced by the canonical commutators, we find in (II-32) a contribution proportional to $S\partial_i\partial^i \delta^2(\boldsymbol{x}_\perp)$. This second derivative structure has no analogue in the canonical result, hence the latter must be modified

$$[J^+(x), J^-(0)]|_{x^+=y^+}$$

$$= \partial^x_- \left\{-\frac{1}{2}\varepsilon(x^- - y^-)\delta^2(\boldsymbol{x}_\perp - \boldsymbol{y}_\perp)\bar{\psi}(x)\gamma^-\psi(y)\right\}$$

$$+ \partial^x_i \left\{-\frac{1}{4}\varepsilon(x^- - y^-)\delta^2(\boldsymbol{x}_\perp - \boldsymbol{y}_\perp)[\bar{\psi}(x)\gamma^i\psi(y) - \varepsilon^{ij}\bar{\psi}(x)\gamma_j\gamma_5\psi(y)]\right\} - \text{h.c.}$$

$$- (i/8)\partial^x_i \partial^i_x \{\varepsilon(x^- - y^-)\delta^2(\boldsymbol{x}_\perp - \boldsymbol{y}_\perp)S\}. \qquad \text{(II-33)}$$

For the $+i$ component (II-28) gives

$$\langle 0|[J^+(x), J^i(0)]|0\rangle_{x^+=0} = (i/4)\partial_-\partial^i\varepsilon(x^-)\delta^2(\boldsymbol{x}_\perp)S \neq 0 \qquad \text{(II-34)}$$

while the canonical form gives zero. Hence this commutator is also modified

$$[J^+(x), J^i(y)]_{x^+=y^+}$$

$$= \partial_-^x\left\{-\frac{1}{4}\varepsilon(x^- - y^-)\delta^2(\boldsymbol{x}_\perp - \boldsymbol{y}_\perp)[\bar{\psi}(x)\gamma^i\psi(y) + \varepsilon^{ij}\bar{\psi}(x)\gamma_j\gamma_5\psi(y)]\right\}$$

$$+ \partial_j^x\left\{\frac{1}{4}\varepsilon(x^- - y^-)\delta^2(\boldsymbol{x}_\perp - \boldsymbol{y}_\perp)[g^{ij}\bar{\psi}(x)\gamma^+\psi(y) + \varepsilon^{ij}\bar{\psi}(x)\gamma^+\gamma_5\psi(y)]\right\} - \text{h.c.}$$

$$+ (i/4)\partial_-^x\partial_x^i\{\varepsilon(x^- - y^-)\delta^2(\boldsymbol{x}_\perp - \boldsymbol{y}_\perp)S\}.$$

$$\text{(II-35)}$$

In this discussion, we have pretended that S is finite. If it is quadratically divergent, which is the scale invariant result, then in addition to the exhibited, quadratically divergent vacuum singularities, there are also well defined, non-infinite terms involving higher derivatives of $\delta^2(\boldsymbol{x}_\perp)$. We shall not concern ourselves with these objects.

D. $SU(3)$ Generalization

The $SU(3)$ generalization of the previous results is obtained easily. The current is given by

$$V_a^\mu(x) = \bar{\psi}(x)\gamma^\mu\frac{1}{2}\lambda_a\psi(x). \qquad \text{(II-36)}$$

We shall also need the axial current

$$A_a^\mu(x) = i\bar{\psi}(x)\gamma^\mu\gamma_5\frac{1}{2}\lambda_a\psi(x). \qquad \text{(II-37)}$$

The internal symmetry matrices satisfy

$$\lambda_a\lambda_b = (if_{abc} + d_{abc})\lambda_c. \qquad \text{(II-38)}$$

In deriving the current commutators, we allow for non-conservation of the currents by introducing the mass matrix M in the Lagrangian (II-16). The commutators are the following

$$[V_a^+(x), V_b^+(y)]_{x^+=y^+} = if_{abc}V_c^+(x)\delta(x^- - y^-)\delta^2(\boldsymbol{x}_\perp - \boldsymbol{y}_\perp)$$

$$- (i/4)\delta_{ab}\partial_-^x\partial_-^y\left[\varepsilon(x^- - y^-)\delta^2(\boldsymbol{x}_\perp - \boldsymbol{y}_\perp)S\right], \qquad \text{(II-39)}$$

$$[V_a^+(x), V_b^-(y)]_{x^+=y^+} - i f_{abc} V_c^-(x) \delta(x^- - y^-) \delta^2(\boldsymbol{x}_\perp - \boldsymbol{y}_\perp)$$

$$= -\frac{1}{4} \{i f_{abc} + d_{abc}\} \{\partial_-^x [\varepsilon(x^- - y^-) \delta^2(\boldsymbol{x}_\perp - \boldsymbol{y}_\perp) V_c^-(x|y)]$$

$$+ \frac{1}{2} \partial_i^x [\varepsilon(x^- - y^-) \delta^2(\boldsymbol{x}_\perp - \boldsymbol{y}_\perp)(V_c^i(x|y) + i\varepsilon^{ij} A_{jc}(x|y))]\}$$

$$+ \frac{1}{16} i\varepsilon(x^- - y^-) \delta^2(\boldsymbol{x}_\perp - \boldsymbol{y}_\perp) \bar{\psi}(x) \gamma^+ \gamma^- \Lambda_{ab} \psi(y) - \text{h.c.}$$

$$- \frac{1}{8} i\delta_{ab} \partial_i^x \partial_x^i \{\varepsilon(x^- - y^-) \delta^2(\boldsymbol{x}_\perp - \boldsymbol{y}_\perp) S\}, \tag{II-40}$$

$$[V_a^+(x), V_b^i(y)]_{x^+=y^+} - i f_{abc} V_c^i(x) \delta(x^- - y^-) \delta^2(\boldsymbol{x}_\perp - \boldsymbol{y}_\perp)$$

$$= -\frac{1}{8} \{i f_{abc} + d_{abc}\} \{\partial_-^x [\varepsilon(x^- - y^-) \delta^2(\boldsymbol{x}_\perp - \boldsymbol{y}_\perp)(V_c^i(x|y) - i\varepsilon^{ij} A_{jc}(x|y))]$$

$$+ \partial_j^x [\varepsilon(x^- - y^-) \delta^2(\boldsymbol{x}_\perp - \boldsymbol{y}_\perp)(-g^{ij} V_c^+(x|y) + i\varepsilon^{ij} A_c^+(x|y))]\}$$

$$+ \frac{1}{16} i\varepsilon(x^- - y^-) \delta^2(\boldsymbol{x}_\perp - \boldsymbol{y}_\perp) \bar{\psi}(x) \gamma^+ \gamma^i \Lambda_{ab} \psi(y) - \text{h.c.}$$

$$+ \frac{1}{4} i\delta_{ab} \partial_-^x \partial_x^i \{\varepsilon(x^- - y^-) \delta^2(\boldsymbol{x}_\perp - \boldsymbol{y}_\perp) S\}. \tag{II-41}$$

We have introduced the bilocal generalizations of the vector and axial vector current

$$V_a^\mu(x|y) = \bar{\psi}(x) \gamma^\mu \frac{1}{2} \lambda_a \psi(y) \tag{II-42}$$

$$A_a^\mu(x|y) = i\bar{\psi}(x) \gamma^\mu \gamma_5 \psi(y). \tag{II-43}$$

These enter the commutation relations with $(x - y)^2 = 0$, $x^- - y^- \neq 0$. The term involving Λ_{ab} is an internal symmetry operator, given by $\Lambda_{ab} = [M, \lambda_a]\lambda_b$, *i.e.*, it probes the nonconservation of the current V_a^μ. It is not hard to see how the commutators with the axial current work out, but we do not present them here. In Section IV the model dependence of these formulas will be discussed.

E. BJL Theorem on the Light Cone

We conclude this Section by presenting a method for computing light cone commutators directly from Green's functions. The method is analogous to the usual BJL[18,19] technique which allows one to determine an equal time commutator. Consider the x^+ ordered product of two operators

$$T_+(q) = \int d^4x e^{iqx} \langle \alpha | T_+ O_1(x) O_2(0) | \beta \rangle. \tag{II-44}$$

An integration by parts gives

$$T_+(q) = (i/q^-) \int d^4x e^{iqx} \partial_+ \langle \alpha | T_+ O_1(x) O_2(0) | \beta \rangle$$

$$= (i/q^-) \int d^2x_\perp dx^- e^{iq^+ x^-} e^{-iq_\perp \cdot x_\perp} \langle \alpha | [O_1(x), O_2(0)] | \beta \rangle_{x^+=0} + O[(1/q^-)^2]. \tag{II-45}$$

Not unexpectedly, the $1/q^-$ term in a x^+ ordered product is the light-cone commutator, just as the $1/q^0$ term in a time ordered product is the equal time commutator. The non-trivial aspect of the theorem emerges when we recall that a covariant Green's function (T^* product) is equal to the T_+ product, apart from seagulls. Consequently, we have

$$T^*(q) \equiv \int d^4x e^{iqx} \langle \alpha | T^* O_1(x) O_2(0) | \beta \rangle \xrightarrow[q^- \to \infty]{} \text{polynomials}$$

$$+ (i/q^-) \int d^2x_\perp dx^- e^{iq^+ x^-} e^{-iq_\perp \cdot x_\perp} \langle \alpha | [O_1(x), O_2(0)] | \beta \rangle_{x^+=0} + \cdots. \tag{II-46}$$

A more careful analysis indicates that q^- should become large away from the real axis. In the usual way, if the limit diverges, we interpret this as the statement that the commutator has divergent matrix elements.

As an example, consider the free propagator of Boson fields. For large q^- it goes as $i/2q^+ q^-$ and $1/2q^+$ is indeed the Fourier transform of the canonical (free field) light cone commutator:

$$-\frac{1}{4} i\varepsilon(x^-) \delta^2(x_\perp).$$

III. APPLICATIONS TO DEEP INELASTIC PROCESSES

The theory that we have developed will now be applied to the electroproduction processes studied by the MIT-SLAC experiment. As is well known, that investigation gives an experimental determination of the quantity

$\int d^4 x e^{iqx} \langle p|[J^\mu(x), J^\nu(0)]|p\rangle$, where $|p\rangle$ is the spin-averaged nucleon target state, and J^μ is the electro-magnetic current. (In the next section, we generalize the present results to include spin and internal symmetry.)

A. Scaling Representation

One of the remarkable experimental observations is that the invariant functions which determine the commutator defined above have a convergent limit in the "deep inelastic region", q^2 and $\nu = p \cdot q$ both large, and $\omega = -q^2/2\nu$ fixed. We may inquire what must be the form of $\langle p|[J^\mu(x), J^\nu(0)]|p\rangle$ in *position* space, such that its Fourier transform exhibit the properly convergent form in the deep inelastic region. The desired form, which incorporates this regularity, as well as causality and current conservation is the following[20]:

$$i\langle p|[J^\mu(x), J^\nu(0)]|p\rangle = [g^{\mu\nu}\Box - \partial^\mu\partial^\nu]\varepsilon(x^0)$$

$$\times [\delta(x^2)(1/8\pi^2)\int_{-1}^{+1} d\omega \omega^{-2}\cos\omega xp F_L(\omega) + \theta(x^2)f_1(x^2, x\cdot p)]$$

$$+ [p^\mu p^\nu \Box - p\cdot\partial(\partial^\mu p^\nu + \partial^\nu p^\mu) + g^{\mu\nu}(p\cdot\partial)^2]\varepsilon(x^0)\theta(x^2) \qquad \text{(III-1)}$$

$$\times [(1/8\pi^2)\int_{-1}^{+1} d\omega(\omega xp)^{-1}\sin\omega xp F_2(\omega) + f_2(x^2, x\cdot p)],$$

$$x^2 f_1(x^2, x\cdot p) \xrightarrow[x^2\to 0]{} 0, \qquad \text{(III-2a)}$$

$$f_2(x^2, x\cdot p) \xrightarrow[x^2\to 0]{} 0. \qquad \text{(III-2b)}$$

The non-trivial aspect of this representation is that the degree of singularity at $x^2 = 0$ is limited, and the most singular terms at that point are parametrized by the deep inelastic structure functions of Bjorken,[9] $F_L(\omega)$ and $F_2(\omega)$. The $f_i(x^2, x\cdot p)$ determine the subdominant light cone singularity, as is seen from (III-2). The degree of singularity at $x^2 = 0$ exhibited by (III-1), is no worse than in free-field theory. Also the singularity is such that the light cone commutator, *i.e.*, the restriction to $x^+ = 0$, exists.

B. Equal Time Sum Rules

The representation (III-1) is especially convenient for deriving the usual equal-time sum rules appropriate to deep-inelastic processes. Thus by going to $x^0 = 0$ in the $0i$ components of (III-1) we obtain the Schwinger-term sum rule.[20]

$$\langle p|[J^0(x), J^i(0)]|p\rangle|_{x^0=0} = (i/4\pi)\partial^i \delta^3(x) \int_{-1}^{+1} d\omega \omega^{-2} F_L(\omega). \tag{III-3}$$

Similarly, by differentiating the ij components with respect to x^0, the Callan-Gross[21] sum rule emerges, at $x^0 = 0$. We have after some manipulation

$$\lim_{p_0 \to \infty} (p_0^2)^{-1} \int d^3x \langle p|[\partial_0 J^i(x), J^k(0)]|p\rangle|_{x^0=0}$$

$$= \int_{-1}^{+1} d\omega [F_L(\omega)\delta^{jk} - F_2(\omega)(\delta^{jk} - \hat{p}^j\hat{p}^k)]. \tag{III-4}$$

Clearly, this procedure may be continued to obtain relations between

$$\int_{-1}^{+1} d\omega \cdot \omega^{2n} F_i(\omega)(i = L, 2) \text{ and}$$

$$\lim_{p_0 \to \infty} (p_0^2)^{-n-1} \int d^3x \langle p|[\partial_0^{2n+1} J^j(x), J^k(0)]|p\rangle|_{x^0=0}.$$

One disadvantage of these relations is that one needs to compute more and more obscure commutators, which depend on dynamics in a complicated way. Furthermore, it would be preferable to have *one* relation determining $F_i(\omega)$, rather than the *infinite* number of moment relations for $\int d\omega \omega^{2n} F_i(\omega)$. The light-cone commutators provide these desirable results.

C. Light Cone Sum-Rules

Rather than restricting the representation (III-1) to equal times, $x^0 = 0$, we may just as well restrict it to the light cone, $x^+ = 0$. It is easy to show that

$$\langle p|[J^+(x), J^+(0)]|p\rangle|_{x^+=0}$$

$$= \partial_-\partial_-\left\{(i/16\pi)\varepsilon(x^-)\delta^2(x_\perp) \int_{-1}^{+1} d\omega \omega^{-2} \cos \omega x^- p^+ F_L(\omega)\right\}. \tag{III-5}$$

Comparing this with the model for the light-cone commutator, given in (II-31) or (II-39), we have[22]

$$(4\pi)^{-1} \int_{-1}^{+1} d\omega \omega^{-2} \cos \omega x^- p^+ F_L(\omega) = \langle p|S|p\rangle = 0. \tag{III-6}$$

The last equality follows from the fact that the Schwinger term is taken to be a c number; hence its connected matrix element vanishes.

Eq. (III-6) contains the Schwinger term sum rule, (III-3); the Callan-Gross sum rule for F_L (III-4); as well as *all* the further moment sum rules for $\int_{-1}^{1} d\omega$ $\omega^{2n} F_L(\omega)$. It is very important to note that (III-6) may be inverted to obtain the *unintegrated* result

$$F_L(\omega) = 0 . \tag{III-7}$$

Thus Eq. (III-7) may be arrived at *without* use of positivity for $F_L(\omega)$. This happens for all the deep inelastic current algebra sum rules: relations which in that context involved an integration over ω, are now rederived in an unintegrated form.

Further results are obtained from the $+-$ components. For simplicity, we set F_L to zero as indicated by (III-7). It then follows from (III-1) that

$$\langle p|[J^+(x), J^-(0)]|p\rangle|_{x^+=0}$$

$$= -i\varepsilon(x^-)\delta^2(\boldsymbol{x}_\perp)\left[f(x^-p^+) + (\boldsymbol{p}_\perp^2/8\pi)\int_{-1}^{1} d\omega \, \cos \omega x^- p^+ F_2(\omega)\right]$$

$$-i\varepsilon(x^-)p^i\partial_i\delta^2(\boldsymbol{x}_\perp)(1/8\pi)\int_{-1}^{1} d\omega \, \sin \omega x^- p^+ F_2(\omega)/\omega. \tag{III-8}$$

Here $f(\alpha)$ parametrizes corrections to scaling, and will not concern us here; but see section IV.

$$f(\alpha) = (M^2/8\pi)\int_{-1}^{1} d\omega \, \cos \omega\alpha F_2(\omega) + \partial/\partial\alpha[\alpha f_1(0,\alpha)]. \tag{III-9}$$

The expression (III-8) may be compared with the operator formula (we ignore c number Schwinger terms)

$$[J^+(x), J^-(y)]_{x^+=y^+}$$

$$= -\frac{1}{2}\partial_-^x\{\varepsilon(x^- - y^-)\delta^2(\boldsymbol{x}_\perp - \boldsymbol{y}_\perp)V^-(x|y)\}$$

$$-\frac{1}{4}\partial_i^x\{\varepsilon(x^- - y^-)\delta^2(\boldsymbol{x}_\perp - \boldsymbol{y}_\perp)[V^i(x|y) + i\varepsilon^{ij}A_j(x|y)]\} - \text{h.c.} \tag{III-10}$$

In a fermion model, from which (III-10) is abstracted, we have

$$J^\mu(x) = \bar\psi(x)\gamma^\mu Q\psi(x); \qquad V^\mu(x|y) = \bar\psi(x)\gamma^\mu Q^2\psi(y),$$
$$A^\mu(x|y) = i\bar\psi(x)\gamma^\mu\gamma_5 Q^2\psi(y),$$

where Q is the charge matrix. (In the *quark* model, a more specific formula may be obtained since Q is given by the $SU(3)$ λ matrices.) Consequently, (III-6) and (III-10) imply[22]

$$-(p^i/2\pi)\int_{-1}^{1} d\omega \, \sin \omega x^- p^+ F_2(\omega)/\omega = \langle p|iV^i(x|0) + \text{h.c.}|p\rangle_{x^2=0} \tag{III-11}$$
$$= \langle p|i\bar\psi(x)\gamma^i Q^2\psi(0) + \text{h.c.}|p\rangle_{x^2=0}.$$

As always, the bilocal operator is evaluated on the light cone, $x^2 = 0$ with $x^+ = 0$, $x_\perp = 0$.

This then is the final, remarkably simple result, which summarizes *all* the moment sum rules. It is very gratifying that the bilocal operator, whose matrix elements determine $F_2(\omega)$ has no dependence on the gluon field (in the gauge $B^+ = 0$). As was noted before, the bilocal current is a simple generalization of the usual local current. Now we see that just as the matrix elements of the *local* current (the form factors) are relevant to *elastic* scattering, so analogously the matrix elements of the *bilocal* current determine the form factors appropriate to *deep inelastic* scattering. Of course, we do not have at the present time an evaluation of

$$\langle p|i\bar\psi(x)Q^2\gamma^\mu\psi(0) + \text{h.c.}|p\rangle,$$

just as at the present time, one cannot evaluate the elastic form factors $\langle p'|\bar\psi(0)\,Q\gamma^\mu\psi(0)|p\rangle$. The importance of the result (III-11) lies in the fact that it demonstrates the measurability of the matrix elements of bilocal operators. Thus these operators should be considered on the same footing as local currents in their relevance for practical physics.

Eq. (III-11) may be considered as a starting point for the calculation of the deep inelastic structure functions. Rather than computing the cross-sections for all q^2 and ν, and then passing to the deep inelastic limit, one needs to compute only the matrix elements in (III-11).

It is clear that similar investigations can be carried out for the deep inelastic neutrino processes. All previous current algebraic sum rules which provided relations between *moments* of the appropriate structure functions are now replaced by relations between *Fourier* transforms of these functions; or equivalently between the functions themselves.

We conclude this section with several further (unrelated) observations about the bilocal operators. If $i\bar\psi(x)\gamma^\mu\psi(0)|_{x^2=0}$ is expanded at small x^- in powers of x^-, one encounters local structures of the form

$$\bar\psi\gamma^\mu \overleftrightarrow{\partial}^{\nu_1}\ldots\overleftrightarrow{\partial}^{\nu_n}\psi.$$

In particular, the *fermion* part of the energy-momentum tensor occurs in this expansion — though the *boson* part is missing.[22] For large x^-, only small ω, *i.e.*, the Regge region, contributes. Thus the bilocal operator at large x^- parametrizes Regge poles. Finally, we repeat once more the *caveat* that in perturbation theory, the commutators possess anomalies and are not of the canonical form given here. One consequence of this is that $F_L \neq 0$, in perturbation theory, is contrary to (III-7).[23] Evidently, we are ignoring here, these perturbative q number anomalies.

IV. FIXED-MASS SUM RULES

It is known that ordinary current algebra is insufficient to derive fixed mass sum rules, like the one of Dashen, Fubini and Gell-Mann.[1,2] One possible technique, which *does* yield fixed mass sum rules, is the $p \to \infty$ procedure. Unfortunately, that method frequently fails because of illegitimate interchange of limit and integral and many of the results are invalid even in free-field models. On the other hand, it has been known for some time that the fixed mass sum rules are equivalent to appropriate light cone commutators and that the $p \to \infty$ technique is an attempt (not always succeessful) to determine the (then unknown) light cone commutator from the (then known) equal time commutator.[4] However, since we now have a model for the light cone commutator, we can dispense with the unreliable $p \to \infty$ method. We derive in this section corrected versions of various fixed mass sum rules.[17]

A. Kinematic Preliminaries

We consider the diagonal matrix element of the commutator of two vector currents V_a^μ, assumed to be conserved, between fermion states, which are *not* spin averaged

$$C_{ab}^{\mu\nu}(p,q) = \int d^4x\, e^{iqx} \langle p|[V_a^\mu(x), V_b^\nu(0)]|p\rangle. \tag{IV-1}$$

The fermion state $|p\rangle$ is characterized by a spin (pseudo) vector $s^\alpha \equiv \bar{u}(p)\, i\gamma^\alpha\gamma_5 u(p)$. This vector is orthogonal to p and has the form $s^\mu = (s^0, \mathbf{s})$, $s^0 = \mathbf{p} \cdot \hat{n}$, $\mathbf{s} = m\hat{n} + \mathbf{p}\mathbf{p} \cdot \hat{n}/[E + m]$. Here, $\hat{n}$ is an arbitrary unit vector specifying the rest frame spin direction, $\hat{n} = \langle\sigma\rangle$. Note that $p^2 = -s^2 = m^2$.

Considerations of parity, time inversion invariance, Lorentz invariance and current conservation give the following expression for $C^{\mu\nu}_{ab}(p,q)$

$$C^{\mu\nu}_{ab}(p,q) = [-g^{\mu\nu} + q^\mu q^\nu/q^2]W^{ab}_L$$
$$+ [p^\mu p^\nu - (\nu/q^2)(p^\mu q^\nu + p^\nu q^\mu) + g^{\mu\nu}\nu^2/q^2]W^{ab}_2 \qquad \text{(IV-2)}$$
$$+ i\varepsilon^{\mu\nu\alpha\beta}s_\alpha q_\beta W^{ab}_3 + i\varepsilon^{\mu\nu\alpha\beta}p_\alpha q_\beta q\cdot sW^{ab}_4.$$

The invariants are functions of q^2, ν. We shall decompose them into parts symmetric in ab, denoted by (ab); and antisymmetric in ab, denoted by $[ab]$:

$$W^{ab}_i = W^{(ab)}_i + iW^{[ab]}_i, \quad i = L, 2, 3, 4. \qquad \text{(IV-3)}$$

Crossing now implies that

$$W^{(ab)}_i(q^2, \nu) = -W^{(ab)}_i(q^2, -\nu). \quad i = L, 2, 3, \qquad \text{(IV-4a)}$$
$$W^{(ab)}_4(q^2, \nu) = W^{(ab)}_4(q^2, -\nu) \qquad \text{(IV-4b)}$$

and the opposite symmetry obtains for the invariants which are antisymmetric in a and b. Hermiticity insures that the invariants occurring in the right hand side of (IV-3) are real.

Since the current has dimension 3, and since our states are covariantly normalized, the following functions are dimensionless: W^{ab}_L, νW^{ab}_2, νW^{ab}_3 and $\nu^2 W^{ab}_4$. Hence these objects approach a limit as $\nu \to \infty$ with fixed $-q^2/2\nu \equiv \omega^*$. In the present context, this is *not* a hypothesis, but a consequence of our use of the light cone commutators. In the preasymptotic region, we shall frequently consider the above functions to depend on ω and q^2. In that case, W^{ab}_L will be denoted by $(-1/2\omega)F^{ab}_L(\omega, q^2)$, and the others by $F^{ab}_i(\omega, q^2)$. In the scaling limit, the $F^{ab}_i(\omega, q^2)$ become $F^{ab}_i(\omega)$, $i = L, 2, 3, 4$. The quark model commutator structure which we employ implies that $F^{ab}_L(\omega)$ vanishes, as will be seen below. We shall need the correction to scaling for this function. Hence we define $\lim\limits_{-q^2 \to \infty} q^2 F^{ab}_L(\omega, q^2) \equiv G^{ab}_L(\omega)$. One can show that $F^{ab}_i(\omega)$ and $G^{ab}_L(\omega)$ vanish for $|\omega| > 1$.

B. Infinite Momentum Derivation of Sum Rules

We summarize the $p \to \infty$ sum rules which emerge from the 0ν components of (IV-1). Some of these results are well known, others presumably have been

---

*This is in complete analogy with the electroproduction results discussed above.

recorded in the literature. Eq. (IV-1) and the usual current algebra imply

$$(2\pi)^{-1} \int_{-\infty}^{\infty} dq^0 C_{ab}^{0\alpha}(p,q) = i f_{abc} p^\alpha \Gamma_c$$

$$\langle p|V_a^\alpha(0)|p\rangle \equiv p^\alpha \Gamma_a.$$

(IV-5)

We have taken the Schwinger term to be a c number. Throughout subsection B we set $\mathbf{p}\cdot\mathbf{q}=0$. A change of variable is now performed $q^0 = \nu/p^0$

$$(2\pi)^{-1} \int_{-\infty}^{\infty} (d\nu/p^0) C_{ab}^{0\alpha}(p,q)\Big|_{\substack{q^0=\nu/p^0, \\ \mathbf{p}\cdot\mathbf{q}=0}} = i f_{abc} p^\alpha \Gamma_c.$$

(IV-6)

The above is exact, however it is not a *fixed mass* sum rule since $q^2 = \nu^2/p_0^2 - q^2$ varies with ν. To obtain fixed mass sum rules, p_0 is set to ∞ in the usual fashion. Then by equating independent tensors in (IV-6b), the following nontrivial relations are obtained.

From $\alpha = 0$ in (IV-6), the Dashen, Fubini, Gell-Mann[1,2] result emerges

$$\int_0^\infty d\nu W_2^{[ab]}(q^2,\nu) = \pi f_{abc}\Gamma_c \quad q^2 \leq 0.$$

(IV-7a)

From $\alpha = i$ in (IV-6) the "good-bad" sum rules are a consequence

$$\int_0^\infty d\nu W_3^{[ab]}(q^2,\nu) = 0,$$ (IV-7b)

$$\int_0^\infty d\nu\nu W_4^{[ab]}(q^2,\nu) = 0,$$ (IV-7c)

$$\int_0^\infty d\nu W_L^{[ab]}(q^2,\nu) = 0,$$ (IV-7d)

$$\int_0^\infty d\nu W_4^{(ab)}(q^2,\nu) = 0,$$ (IV-7e)

$$\int_0^\infty d\nu\nu W_2^{(ab)}(q^2,\nu) = 0, \quad q^2 \leq 0.$$ (IV-7f)

The sum of (IV-7b) and (IV-7c) is the Bég sum rule,[24] which when evaluated at $q^2 = 0$ gives a relation for anomalous magnetic moments.

If the results (IV-7) are compared with the explicit calculations in the *free* quark model, it is then found that (IV-7b) and (IV-7f) are not valid, while (IV-7c), (IV-7d) and (IV-7e) are trivially satisfied, in the sense that both sides of the equality vanish. Numerical evaluation,[25] which checks the validity of these in Nature is consistent with (IV-7a), in that the Cabbibo-Radicatti relation[26] which is a consequence of (IV-7a), appears to be satisfied. However, the Bég sum rule[24] is not verified experimentally. Finally, (IV-7f) cannot be true since it can be rewritten in terms of the scaling variables as $\int_{-1}^{1} d\omega \omega^{-2} F_2^{(ab)}(\omega, q^2) = 0$. When q^2 is then set to ∞, one gets $\int_{-1}^{1} d\omega \omega^{-2} F_2^{(ab)}(\omega) = 0$ which is inconsistent with the MIT-SLAC experiments, which show that $F_2^{(ab)}(\omega) \neq 0$.[5]

In the free-field model all integrals converge, since the invariants are δ functions. Hence, the failures in that context cannot be attributed to divergences. In a Regge pole model, the relation which diverge are (IV-7d) and (IV-7f). Note that the Bég sum rule converges in a Regge pole model.[27]

C. Light Cone Derivation of Sum Rules

To derive the fixed mass sum rules with light cone techniques, we set q^+ to zero in the $+\nu$ components of (IV-1), and integrate over q^-

$$(2\pi)^{-1} \int_{-\infty}^{\infty} dq^- C_{ab}^{+\alpha}(p, q)|_{q^+=0}$$

$$= \int d^2 x_\perp e^{-iq_\perp \cdot x_\perp} \langle p | [\int dx^- V_a^+(x), V_b^\alpha(0)] | p \rangle |_{x^+=0}.$$

(IV-8)

With this procedure, we obtain directly a fixed mass sum rule, since $q^2 = -q_\perp^2$ when $q^+ = 0$; *i.e.*, q^2 does not vary in the course of the integration. However, the right-hand side of (IV-8) involves a light cone commutator, rather than an equal time commutator. We assume that this object can be computed from our *Ansatz*, (II-39), (II-40), and (II-41)

$$\langle p | [\int dx^- V_a^+(x), V_b^\alpha(0)] | p \rangle |_{x^+=0}$$
$$= \int dx^- \langle p | [V_a^+(x), V_b^\alpha(0)] | p \rangle |_{x^+=0}.$$

(IV-9)

The nature of this assumption shall be discussed further in the Conclusion. Finally, we change variables in (IV-8) by setting $\nu = p^+ q^- - \boldsymbol{p}_\perp \cdot \boldsymbol{q}_\perp$. This yields

$$(2\pi)^{-1} \int_{-\infty}^{\infty} (d\nu/p^+) C_{ab}^{+\alpha}(p,q) \Bigg|_{\substack{q^+ = 0 \\ q^- = (\nu + \boldsymbol{p}_\perp \cdot \boldsymbol{q}_\perp)/p^+}}$$

(IV-10)

$$= \int d^2\boldsymbol{x}_\perp \, dx^- e^{-i\boldsymbol{q}_\perp \cdot \boldsymbol{x}_\perp} \langle p|[V_a^+(x), V_b^\alpha(0)]|p\rangle|_{x^+=0}.$$

We now use the model given previously to evaluate the commutator in (IV-10) for $\alpha = +, -$. (No new information emerges from $\alpha = i$). The Schwinger terms are ignored as they are taken to be c numbers. Also since the current is conserved, the commutators do not have symmetry breaking terms in them. Clearly, the right-hand side of (IV-10) will involve matrix elements of the bilocal operators. These can be given by the following form factor decomposition. Define

$$\mathcal{V}_a^\mu(x|y) \equiv \frac{1}{2} V_a^\mu(x|y) + \frac{1}{2} V_a^\mu(y|x),$$

(IV-11a)

$$\bar{\mathcal{V}}_a^\mu(x|y) \equiv \frac{1}{2i} V_a^\mu(x|y) - \frac{1}{2i} V_a^\mu(y|x),$$

(IV-11b)

$$\mathcal{A}_a^\mu(x|y) \equiv \frac{1}{2} A_a^\mu(x|y) + \frac{1}{2} A_a^\mu(y|x),$$

(IV-11c)

$$\bar{\mathcal{A}}_a^\mu(x|y) \equiv \frac{1}{2i} A_a^\mu(x|y) - \frac{1}{2i} A_a^\mu(y|x).$$

(IV-11d)

The form factors are by definition

$$\langle p|\mathcal{V}_a^\mu(x|0)|p\rangle = p^\mu V_a^1(x^2, x \cdot p) + x^\mu V_a^2(x^2, x \cdot p),$$

(IV-12a)

$$\langle p|\bar{\mathcal{V}}_a^\mu(x|0)|p\rangle = p^\mu \bar{V}_a^1(x^2, x \cdot p) + x^\mu \bar{V}_a^2(x^2, x \cdot p),$$

(IV-12b)

$$\langle p|\mathcal{A}_a^\mu(x|0)|p\rangle = s^\mu A_a^1(x^2, x \cdot p) + p^\mu x \cdot s A_a^2(x^2, x \cdot p)$$
$$+ x^\mu x \cdot s A_a^3(x^2, x \cdot p),$$

(IV-12c)

$$\langle p|\bar{\mathcal{A}}_a^\mu(x|0)|p\rangle = s^\mu \bar{A}_a^1(x^2, x \cdot p) + p^\mu x \cdot s \bar{A}_a^2(x^2, x \cdot p)$$
$$+ x^\mu x \cdot s \bar{A}_a^3(x^2, x \cdot p).$$

(IV-12d)

T inversion invariance eliminates a possible structure of the form $\varepsilon^{\mu\alpha\beta\gamma} x_\alpha p_\beta s_\gamma$ in (IV-12a) and (IV-12b). It is clear that (IV-12a) and (IV-12c) are even in x, while (IV-12b) and (IV-12d) are odd in x.

It is now straightforward to extract the sum rules by equating independent tensors in (IV-10). The results are[17]

$$\int_0^\infty d\nu W_2^{[a,b]}(q^2,\nu) = \pi f_{abc}\Gamma_c, \qquad \text{(IV-13a)}$$

$$\int_0^\infty d\nu W_3^{[ab]}(q^2,\nu) = \frac{1}{2}\pi f_{abc}\int_0^\infty d\alpha \bar{A}_c^1(0,\alpha), \qquad \text{(IV-13b)}$$

$$\int_0^\infty d\nu\nu W_4^{[ab]}(q^2,\nu) = \frac{1}{2}\pi f_{abc}\int_0^\infty d\alpha \bar{A}_c^2(0,\alpha), \qquad \text{(IV-13c)}$$

$$\int_0^\infty d\nu W_L^{[ab]}(q^2,\nu) = 0, \qquad \text{(IV-13d)}$$

$$\int_0^\infty d\nu W_4^{(ab)}(q^2,\nu) = 0, \qquad \text{(IV-13e)}$$

$$\int_0^\infty d\nu(\nu/-q^2)W_2^{(ab)}(q^2,\nu) = \frac{1}{2}\pi d_{abc}\int_0^\infty d\alpha \bar{V}_c^1(0,\alpha), \quad q^2 \leqq 0. \quad \text{(IV-13f)}$$

The Dashen, Fubini, Gell-Mann sum rule[1,2] (IV-13a), as well as (IV-13d), and (IV-13e) are rederived, but the Bég sum rules[24] (IV-13b) and (IV-13c) as well as (IV-13f) are found to have corrections. This is extremely gratifying, since it is precisely these which fail both in free-field theory and in Nature.[25] The corrections are expressed in terms of integrals over matrix elements of the bilocal operators. In the next sub-Section, we shall show that these matrix elements are measurable, even when an integration is not performed. Here, we merely demonstrate how (IV-13b), ((IV-13c) and (IV-13f) can be exploited.

Observe that the right-hand sides of (IV-13b), (IV-13c) and (IV-13f) are independent of q^2. Let us rewrite the left-hand sides in terms of the scaling functions

$$\int_0^\infty d\nu W_3^{[ab]}(q^2,\nu) = \int_0^1 d\omega\omega^{-1}F_3^{[ab]}(\omega,q^2), \qquad \text{(IV-14a)}$$

$$\int_0^\infty d\nu\nu W_4^{[ab]}(q^2,\nu) = \int_0^1 d\omega\omega^{-1}F_4^{[ab]}(\omega,q^2), \qquad \text{(IV-14b)}$$

$$\int_0^\infty d\nu(\nu/-q^2)W_2^{(ab)}(q^2,\nu) = \int_0^1 d\omega(2\omega^2)^{-1}F_2^{(ab)}(\omega,q^2), \quad q^2 \leqq 0 \quad \text{(IV-14c)}$$

According to the sum rules, the integrals are q^2-independent, and may be evaluated by letting $-q^2 \to \infty$. Thus the right-hand sides of (IV-14) are expressible in terms of the deep inelastic structure functions $F_i^{ab}(\omega)$. Performing a similar analysis of the remaining sum rules in (IV-13) gives finally the following results[17]

$$\int_0^\infty d\nu\, W_2^{[ab]}(q^2,\nu) = \int_0^1 d\omega\, \omega^{-1} F_2^{[ab]}(\omega) = \pi f_{abc} \Gamma_c, \tag{IV-15a}$$

$$\int_0^\infty d\nu\, W_3^{[ab]}(q^2,\nu) = \int_0^1 d\omega\, \omega^{-1} F_3^{[ab]}(\omega), \tag{IV-15b}$$

$$\int_0^\infty d\nu\, \nu W_4^{[ab]}(q^2,\nu) = \int_0^1 d\omega\, \omega^{-1} F_4^{[ab]}(\omega), \tag{IV-15c}$$

$$\int_0^\infty d\nu\, W_L^{[ab]}(q^2,\nu) = \int_0^1 d\omega\, \omega^{-3} F_L^{[ab]}(\omega), \tag{IV-15d}$$

$$\int_0^\infty d\nu\, W_4^{(ab)}(q^2,\nu) = \int_0^1 d\omega\, F_4^{(ab)}(\omega), \tag{IV-15e}$$

$$\int_0^\infty d\nu\, \nu(-q^2)^{-1} W_2^{(ab)}(q^2,\nu) = \int_0^1 d\omega\, (2\omega^2)^{-1} F_2^{(ab)}(\omega). \tag{IV-15f}$$

The sum rule (IV-15f) has already been derived by an entirely different method, by Cornwall, Corrigan and Norton.[28] Note that this relation diverges in a Regge pole model. Cornwall, Corrigan and Norton discuss a truncation technique, which possibly might give meaning to this divergent sum rule.

D. Measuring Bilocal Operators

It was seen in our discussion of deep inelastic scattering that matrix elements of the bilocal operators, with light-like separated arguments are completely measurable in terms of deep inelastic structure functions. Clearly, the same is true here. This can be established in one of two ways. In complete analogy to the discussion of Section IV, one may exhibit a position space representation for $(2\pi)^{-1} \int d^4q\, e^{-iqx} C_{ab}^{\mu\nu}(p,q)$ which incorporates scaling as in (III-1). Alternatively, one may work directly in momentum space and use the light-cone version of the BJL limit discussed in Section II. The results are as follows[17]

$$d_{abc}\bar{V}_c^1(0,\alpha) + f_{abc} V_c^1(0,\alpha) = (2\pi i)^{-1} \int_{-1}^1 d\omega\, \omega^{-1} e^{i\omega\alpha} F_2^{ab}(\omega), \tag{IV-16a}$$

$$d_{abc} A_c^1(0,\alpha) - f_{abc} \bar{A}_c^1(0,\alpha) = \pi^{-1} \int_{-1}^1 d\omega\, e^{i\omega\alpha} F_3^{ab}(\omega), \tag{IV-16b}$$

$$d_{abc}\alpha A_c^2(0,\alpha) - f_{abc}\alpha \bar{A}_c^2(0,\alpha) = \pi^{-1}\int_{-1}^{1} d\omega e^{i\omega\alpha} F_4^{ab}(\omega), \qquad \text{(IV-16c)}$$

$$d_{abc}\alpha \bar{V}_c^2(0,\alpha) + f_{abc}\alpha V_c^2(0,\alpha) = \frac{1}{16}i\pi^{-1}\int_{-1}^{1} d\omega \omega^{-3} e^{i\omega\alpha} G_L^{ab}(\omega). \qquad \text{(IV-16d)}$$

This is the generalization to spin and internal symmetry of (III-11). Also one finds[17]

$$F_L^{ab}(\omega) = 0 \qquad \text{(IV-17)}$$

which is the generalization of (III-7).

Additional results follow from (IV-16). Setting α to zero in (IV-16a) we have, since $\bar{V}_c^1(0,\alpha)$ is odd in α,

$$f_{abc}V_c^1(0,0) = \pi^{-1}\int_0^1 d\omega \omega^{-1} F_2^{[ab]}(\omega) \qquad \text{(IV-18a)}$$

with $F_i^{ab}(\omega) = F_i^{(ab)}(\omega) + iF_i^{[ab]}(\omega)$. The right-hand side can be evaluated from the Dashen, Fubini, Gell-Mann[1,2] sum rule, evaluated in the scaling region, (IV-15a). Thus (IV-18a) becomes

$$f_{abc}V_c^1(0,0) = f_{abc}\Gamma_c. \qquad \text{(IV-18b)}$$

Let us recall now the definitions of $V_c^1(0,0)$ and Γ_c

$$p^\mu f_{abc}V_c^1(0,0) = f_{abc}\langle p|\mathcal{V}_c^\mu(0|0)|p\rangle = f_{abc}\langle p|V_c^\mu(0|0)|p\rangle, \qquad \text{(IV-19)}$$

$$p^\mu \Gamma_c = \langle p|V_c^\mu(0)|p\rangle. \qquad \text{(IV-20)}$$

Thus (IV-18) shows that the proton matrix element of the bilocal operator $f_{abc}V_c^\mu(x|y)$ reduces to that of the vector current as $x \to y$. This of course is obvious in the field theoretic model considered where $V_c^\mu(x|y) = \bar{\psi}(x)\frac{1}{2}\lambda_c\gamma^\mu\,\psi(y)$. However, if we generalize and abstract from the model, we see that (IV-18) places an important model independent constraint on the bilocal operator $V_c^\mu(x|y)$.

We now assume that the same is true for the axial vector bilocal operator, as is of course the case in our model. We have from[17] (IV-16b)

$$\begin{aligned} d_{abc}A_c^1(0,0)s^\mu &= d_{abc}\langle p|A_c^\mu(0|0)|p\rangle \\ &= d_{abc}\langle p|A_c^\mu(0)|p\rangle = d_{abc}\Gamma_c^A s^\mu \\ &= 2s^\mu\pi^{-1}\int_0^1 d\omega F_3^{(ab)}(\omega) \end{aligned} \qquad \text{(IV-21)}$$

$[\bar{A}_c^1(0,0)$ vanishes by symmetry.] The sum rule, which has emerged, relating Γ_c^A to the spin odd, isospin even deep inelastic cross section is similar to a relation first obtained in the quark model by Bjorken.[29]

V. CONCLUSION

It is clear that the approach, which we have here developed, yields very informative and elegant results. A question which naturally arises is how generally valid are the conclusions of our investigation. It is this topic which we shall discuss in this Section. We shall not elaborate again on the canonical, formal nature of the present results, which invalidates them in perturbation theory.

The bilocal operators which occur in the light cone commutators are certainly model dependent, even when they are interaction independent as in (II-39), (II-40) and (II-41). As we have already mentioned, scalar electrodynamics gives entirely different expressions, and the same is true of a Yang-Mills theory. These various commutator models may be said to differ among themselves in the space-time tensor structure. In addition to this, there is the internal symmetry model dependence, which we have taken to be the $SU(3)$ triplets of the quark model.

It is plausible to suppose that nature favors the space-time tensor structure of the fermion model, since the other models have nonvanishing $F_L(\omega)$ or $F_2(\omega) = F_L.(\omega)$[17,20,21] In this connection, it would be most interesting to explore what *model independent* information can be obtained about the $[V_a^+, V_b^\alpha]$ light cone commutator, if it is assumed that for $\alpha = +$ the form is as in (II-39)* The internal symmetry structure that we have used does not, at the present time, possess any convincing experimental verification. Our results are true only in the triplet realization of $SU(3)$ where $\lambda_a\lambda_b = (if_{abc} + d_{abc})\lambda_c$. To generalize beyond this, one could for example, set $\lambda_a\lambda_b = if_{abc}\lambda_c + d_{(ab)}$ where $d_{(ab)}$ is no longer $d_{abc}\lambda_c$, but contains all the symmetric parts of the $8 \otimes 8$ representation.

With these *internal symmetry* generalization, the commutators would retain the same form as before, except that the $SU(3)$ content of the bilocal operators would be more complicated. One would need to replace $d_{abc}V_c^\mu(x|y)$ and $d_{abc}A_c^\mu(x|y)$ by $V_{(ab)}^\mu(x|y)$ and $A_{(ab)}^\mu(x|y)$. Consequently, (IV-16) would

---

*This problem is analogous to the model-independent determination of the $[V_a^0, V_b^i]$ equal-time commutator from the $[V_q^0, V_b^0]$ equal-time commutator.[30]

possess left-hand sides with more complicated $SU(3)$ structures. Thus the measurement of the internal symmetry of the deep inelastic structure functions will provide important information concerning the validity of specific models. For example, the inequality of the proton and neutron data[5] shows that the bilocal operators are not $SU(2)$ singlets. In this context, we see that (IV-18) is an important, model independent constraint.

The corrected deep inelastic sum rules (IV-15b), (IV-15c) and (IV-15f), as well as the first equality in (IV-15a), (IV-15d) and (IV-15e) are quite model independent, since they are consequences merely of (IV-9), causality and scaling. The point is that according to (IV-8) and (IV-9) a fixed mass sum rule is given by an integral over x^- and a Fourier transform with respect to $q_\perp$ of a light-cone commutator. This commutator is local in $x_\perp$ due to causality, *i.e.*, it is composed of δ functions of $x_\perp$ and derivatives of δ functions. Consequently, in Fourier space the integral over ν of an invariant function $W(-q_\perp^2, \nu)$ must be a polynomial in $q_\perp^2$. But the degree of the polynomial is fixed by scaling. Specifically, for example for (IV-15b) we can conclude quite generally from (IV-9) and locality

$$\int_0^\infty d\nu\, W_3^{[ab]}(-q_\perp^2, \nu) = \int_0^1 d\omega\, \omega^{-1} F_3^{[ab]}(\omega, -q_\perp^2)$$

$$= \sum_{n=0}^M (q_\perp^2)^n C_n^{[ab]}. \tag{V-1a}$$

However, scaling requires that the limit $q_\perp^2 \to \infty$ of (V-1a) exists; hence we learn that

$$\int_0^\infty d\nu\, W_3^{[ab]}(-q_\perp^2, \nu) = C_0^{[ab]} = \int_0^1 d\omega\, \omega^{-1} F_3^{[ab]}(\omega) \tag{V-1b}$$

which is (IV-15b). The second equality in (IV-15a), (IV-15d) and (IV-15e) is however a consequence of the specific model considered.

Finally, we come to the question of the validity of (IV-9), which is seen to involve an interchange of integration over x^- with the limit $x^+ \to 0$. A careful investigation of this has been given elsewhere[17] and the result is the following. In the terminology of Adler and Dashen[31] the class 2 graphs, *i.e.*, the graphs which have fixed mass singularities in the current lines, must satisfy the same superconvergence relation which is required for the validity of the $p \to \infty$ technique. Thus as far as the class 2 graphs are concerned, the present considerations have nothing new to offer. On the other hand, the Z

graphs, which were improperly treated by the $p \to \infty$ method, seem to be now accurately included, as is seen from the free-field example.

Put in another way, the $p \to \infty$ technique for (IV-7b) to (IV-7f) requires the absence of all fixed poles; while the present method can account for fixed poles with residues which are polynomials in q^2; see (IV-15b) to (IV-15f). If there are non-polynomial residues, then the light-cone method also fails due to the invalidity of (IV-9)*

REFERENCES

1. R. Dashen and M. Gell-Mann, in *Symmetry Principles at High Energy*, A. Perlmutter, J. Wojtaszek, G. Sudarshan and B. Kursunoğlu, eds. (W. Freeman, San Francisco, CA, 1966).

2. S. Fubini, *Nuovo Cim.* **34A**, 475 (1966).

3. G. Furlan, N. Paver and C. Verzegnassi, *Springer Tracts Mod. Phys.* **62**, 118 (1972).

4. K. Bardakci and G. Segre, *Phys. Rev.* **153**, 1263 (1967); J. Jersak and J. Stern, *Nuovo Cim.* **59**, 315 (1969); H. Leutwyler, *Springer Tracts Mod. Phys.* **50**, 29 (1969).

5. For a summary, see R. E. Taylor, *Rev. Mod. Phys.* **63**, 573 (1991); H. Kendall, *Rev. Mod. Phys.* **63**, 597 (1991); J. Friedman, *Rev. Mod. Phys.* **63**, 615 (1991).

6. R. Jackiw, R. Van Royen and G. B. West, *Phys. Rev. D* **2**, 2473 (1970); Other investigations include B. L. Ioffe, *Zh. Eksp. Teor. Fiz., Pis'ma Redakt.* **9**, 163 (1969) [English translation: *JETP Lett.* **9**, 97 (1969)]; L. Brown, in *Lectures in Theoretical Physics*, K. Mahanthappa and W. E. Brittin eds. (Gordon Breach, New York, NY, 1971); R. Brandt, *Phys. Rev. Lett.* **23**, 1260 (1969); H. Leutwyler and J. Stern, *Nucl. Phys.* **B20**, 77 (1970).

7. H. Fritsch and M. Gell-Mann, *Duality and Symmetry in Hadron Physics*, E. Gotsman, ed. (Weizmann Science Press, Jerusalem 1971).

8. D. J. Gross and S. B. Treiman, *Phys. Rev. D* **4**, 1059 (1971).

9. That scaling should occur in the electroproduction experiments was first suggested by J. D. Bjorken, *Phys. Rev.* **179**, 1547 (1969).

10. The derivation is due to J. M. Cornwall and R. Jackiw, *Phys. Rev. D* **4**, 367 (1971).

11. Quantization of a quantum theory on the light cone was carried out by J. B. Kogut and D. E. Soper, *Phys. Rev. D* **1**, 2901 (1970).

---

*The considerations in this paragraph were developed with D. J. Gross, whose comments are much appreciated.

12. F. Rohrlich, *Acta Phys. Aust.* **32**, 87 (1970), examines the quantization of theories on the light cone, with particular emphasis on the associated boundary value problem.

13. S. Weinberg, *Phys. Rev.* **150**, 1313 (1966).

14. This argument was shown to us by Prof. S. Coleman, whom we thank.

15. This freedom also exists for massive Abelian vector meson theories, since the propagator of such particles may be chosen to be without a + component as a consequence of the conserved current interaction.

16. J. Schwinger, *Phys. Rev.* **130**, 406 (1963).

17. D. Dicus, R. Jackiw and V. Teplitz, *Phys. Rev. D* **4**, 1733 (1971).

18. J. D. Bjorken, *Phys. Rev.* **148**, 1467 (1966).

19. K. Johnson and F. E. Low, *Prog. Theor. Phys.* (Kyoto) Suppl. **37–38**, 74 (1966).

20. R. Jackiw, R. Van Royen and G. B. West, *Phys. Rev. D* **2**, 2473 (1970).

21. C. Callan and D. J. Gross, *Phys. Rev. Lett.* **22**, 156 (1969).

22. This was derived by Cornwall and Jackiw, Ref. 10.

23. R. Jackiw and G. Preparata, *Phys. Rev. Lett.* **22**, 975 (1969); S. L. Adler and W.-K. Tung, *Phys. Rev. Lett.* **22**, 978 (1969).

24. M. A. B. Bég, *Phys. Rev. Lett.* **17**, 333 (1966).

25. G. C. Fox and D. Z. Freedman, *Phys. Rev.* **182**, 1628 (1969).

26. N. Cabibbo and L. A. Radicatti, *Phys. Lett.* **19**, 697 (1966).

27. A detailed Regge analysis is given in Ref. 17.

28. J. M. Cornwall, J. D. Corrigan and R. E. Norton, *Phys. Rev. Lett.* **24**, 1141 (1970) and *Phys. Rev. D* **3**, 536 (1971).

29. J. D. Bjorken, *Phys. Rev.* **148**, 1467 (1966).

30. D. J. Gross and R. Jackiw, *Phys. Rev.* **163**, 1688 (1967); R. Jackiw, *Phys. Rev.* **175**, 2015 (1968).

31. S. L. Adler and R. Dashen, *Current Algebra* (W. A. Benjamin, New York, NY, 1968).

Paper IV.2

In addition to "light-cone" quantization, Dirac suggested quantizing at fixed x^2. This idea was used in the early days of string theory to derive the Virasoro algebra in field theory, and was described at the Erice summer school (July 1973). The rebirth of string theory, reawakened interest in this "radial" quantization method.

INVARIANT QUANTIZATION, SCALE SYMMETRY AND EUCLIDEAN FIELD THEORY

Laws of Hadronic Matter, A. Zichichi ed.
(Academic Press, New York, NY, 1975)

I. INTRODUCTION

A method for extracting interesting results from field theoretic models, without explicitly solving the equations of motion which govern these models — an impossible task at the present time — is to study the algebraic relations which follow from canonical commutators. In the past, equal-time canonical commutators were used to derive the very important and successful *algebra of currents at equal times.*[1] More recently, canonical light-cone commutators led to *current algebra on the light cone*, which supplements the old equal time algebra and also gives a succinct description of Bjorken scaling in deep inelastic processes.[2]

The fortunate circumstance that canonical commutators can be determined *a priori* without solving the equations of motion, is a consequence of the quantization procedure. In the conventional approach, time is selected as the direction of evolution of the system, while the quantization conditions — the canonical commutators — are specified *a priori* on the space-like surface $t = $ constant. In light-cone quantization, the system is allowed to develop along the x^+ direction,[3] and commutators are given their *a priori* values on the surface $x^+ = $ constant. It is natural to inquire whether yet other quantization procedures exist which provide useful information.

In these lectures, I shall report on a new method of quantization which was developed by S. Fubini, A. Hanson and me.[4] The surface of quantization is chosen to be the space-like hyperboloid $x^2 = $ positive constant, and propagation of the system proceeds along the normal direction, from one hyperboloid to the next. Of course, the physical content of the theory is not changed by the choice of quantization surface, and we shall show that the formula for the S matrix in our approach, coincides with the conventional Feynman-Dyson expression. Nevertheless, as with light-cone quantization, the present technique organizes the theory in a novel fashion and offers new insights into the structure of field theory.

The results that have been obtained up to now are the following. In addition to exhibiting new canonical commutators, we shall give an unexpected interpretation for anomalous dimensions. An operator basis for Euclidean field theory has been discovered which differs from the one used in the past.[5] Especially interesting is the application of our methods to 2-dimensional field theories where we find that the present ideas provide a bridge between the formalism of conventional field theory and that of the dual resonance model. Explicitly, we can derive the Virasoro algebra for a large class of 2-dimensional field theories, and with further hypotheses, the Hagedorn spectrum is observed.

II. REVIEW OF QUANTIZATION METHODS

Let me begin by giving a general discussion of our quantization procedure, comparing and contrasting it with other approaches. In the conventional method, one considers the planes $x^0 = $ constant. The *evolution operator*, that is the operator which takes the physical system from one surface to the next, is the time translation generator P^0 (the Hamiltonian). Also there exist six *kinematical operators*, that is operators generating motions which leave the surface invariant. These are space translations generated by P^i, and space rotations generated by $M^{ij} = -M^{ji}$ ($i, j = 1, 2, 3$). The quantum conditions, that is the commutators, are specified on the selected surface; in the conventional case this is at $x^0 = $ constant, *i.e.*, at equal times.

In light-cone quantization, the selected surfaces are $x^+ = $ constant. The evolution operator is P^-, while the six kinematical operators are P^+, P^i, M^{ij}, M^{i+} ($i, j = 1, 2$). Canonical commutators are specified at equal x^+.

A third possibility, which is the basis of our theory, is a family of hyperboloids $x^2 = $ constant > 0. Clearly the six kinematical operators are the six Lorentz generators $M^{\mu\nu}$ ($\mu, \nu = 0, 1, 2, 3$), while the evolution operator is the dilation generator D. In other words if we wish to proceed from one hyperboloid $x^2 = \tau_1^2$ to another $x^2 = \tau_2^2$, this can be achieved by a dilation $x^\mu \to \rho x^\mu$. Canonical commutators are given their *a priori* values on surfaces of equal x^2.

Some attractive and interesting aspects of our quantization are already apparent. First, note that the method is explicitly Lorentz invariant at every stage. Unlike in the previous cases, the quantization surface does not select a preferred direction in space-time; clearly the surface $x^2 = $ constant transforms into itself under a Lorentz transformation. Second, and more striking, is the crucial role that dilations play in the present discussion: the evolution of the system is governed by the dilation generator. Currently, dilation transformations are the focus of much theoretical activity, as a consequence of the famous MIT-SLAC deep inelastic experiments. In the conventional approach to quantum field theory, scale transformations are somewhat peripheral to the main lines of development; thus we are pleased that these interesting transformations are central to the present theory.

An object of great importance in the discussion of the evolution of the system is the propagator function (Green's function). This is given by the vacuum expectation value of the ordered product of two fields, where the ordering is along the direction of propagation: time ordering or x^+ ordering in the familiar examples. For us, the ordering will be along the surfaces $x^2 = \tau^2$.

When quantizing a field, it is convenient to expand it on the quantization surface in a complete set of functions associated with *three* of the *six* kinematical operators. In the usual equal time case, the three kinematical operators are chosen to be the translation generators P^i ($i = 1, 2, 3$), and the expansion functions are eigenfunctions of $\frac{\partial}{\partial x^i}$, i.e., they are $e^{-i\mathbf{k}\cdot\mathbf{x}}$. For a free theory, one further expands in eigenfunctions associated with the evolution operator. This is P^0 in the conventional approach and one expands in eigenfunctions of $\frac{\partial}{\partial t} : e^{ik^0 t}$. Thus a free field is expanded in e^{ikx}.

In the light-cone quantization, the three kinematical operators are P^+, P^i ($i = 1, 2$) and the evolution operator is P^-. Hence, a free field is again expanded in terms of e^{ikx}.

The present theory is characterised by the kinematical operators $M^{\mu\nu}$; hence we shall expand in eigenfunctions associated with three objects constructed from $M^{\mu\nu}$. The evolution operator is D; thus the free field will further be expanded in eigenfunctions of $x^\mu \frac{\partial}{\partial x^\mu}$. Consequently, our expansion functions, which replace the conventional exponentials are powers and 4-dimensional harmonics.

All this may appear bizzare; yet in fact our approach is quite common in Schrödinger theory. The Schrödinger wave function, which satisfies

$$-\frac{1}{2}\boldsymbol{\nabla}^2\psi = (E - V)\psi,$$

can be expanded in momentum eigenfunctions

$$\psi(\mathbf{r}) = \int \frac{d^3k}{(2\pi)^3} e^{i\mathbf{k}\cdot\mathbf{x}} \phi(\mathbf{k})$$

$$\left(\frac{k^2}{2} - E\right)\phi(\mathbf{k}) = -\int \frac{d^3q}{(2\pi)^3} V(\mathbf{k} - \mathbf{q})\phi(\mathbf{q}).$$

This would be the analogy to the conventional approach in field theory. Yet practical calculations are not performed in momentum space. Rather one expands the wave function in terms of harmonics

$$\psi(\mathbf{r}) = \sum_{\ell m} \frac{R_{\ell m}(r)}{r} Y_{\ell m}(\Omega)$$

and furthermore, the radial wave function is frequently expanded in powers of r

$$R_{\ell m}(r) = \sum_n C_{n\ell m} r^n.$$

This is quite similar to the expansion procedure in our approach to field theory. Moreover, the Green's function for the Laplacian is given by

$$\frac{1}{|\mathbf{r} - \mathbf{r}'|} = \theta(r - r') \sum_{\ell m} \frac{1}{r'} \left(\frac{r}{r'}\right)^\ell \frac{4\pi}{2\ell + 1} Y_{\ell m}(\Omega) Y_{\ell m}^*(\Omega')$$

$$+ \theta(r' - r) \sum_{\ell m} \frac{1}{r'} \left(\frac{r}{r'}\right)^\ell \frac{4\pi}{2\ell + 1} Y_{\ell m}(\Omega) Y_{\ell m}^*(\Omega') .$$

It is seen that this has a representation as an ordered product, where the ordering is along the radial direction, in complete analogy to the ordering which we shall employ. Thus it is recognized that our method of quantization has already been widely used in Schrödinger theory, and one may view our work as an attempt to use in field theory techniques which proved themselves so successful in Schrödinger theory.

III. EUCLIDEAN FIELD THEORY

A technical difficulty presents itself in carrying out quantization on hyperboloids. The problem is that the space-like surfaces $x^2 = \tau^2 > 0$ do not span all of space-time; as τ^2 varies from 0 to ∞; the region outside the light-cone $x^2 < 0$ is not reached. Thus the evolution of the system in that region must be discussed separately, a task made difficult by the fact that the surface $x^2 = -\tau^2$ are time-like.[6] In order to overcome this obstacle, we simply continue our field theory to Euclidean space, where of course the surfaces $x^2 = \tau^2$ span all of space.

The continuation in no way excludes application of the general method to physical Minkowski-space theories. Indeed, even in the conventional approach, all practical computations are always performed in Euclidean space. Furthermore, I shall show explicitly that our formula for the S matrix coincides with the conventional Feynman-Dyson expression, continued to Euclidean space.

Our definition of Euclidean field theory will now be presented since it differs in all respects from other approaches.[5] We replace x^0 by ix^4, and for every field in Minkowski space $\phi(x_0, \mathbf{x})$, we define a corresponding field in Euclidean space $\phi_E(x_4, \mathbf{x})$

$$\phi_E(x_4, \mathbf{x}) = \phi(ix_4, \mathbf{x}) . \tag{3.1}$$

Thus the free, massless equation in Minkowski space

$$\Box \phi(x) = (\partial_0^2 - \boldsymbol{\nabla}^2)\phi(x) = 0 \tag{3.2}$$

goes over in Euclidean space to[7]

$$\Box_E\phi_E(x) = (\partial_4^2 + \mathbf{\nabla}^2)\phi_E(x) = 0. \tag{3.3}$$

The Green's function for a free massless and spinless field in Minkowski space is

$$G(x,y) = \frac{1}{4\pi^2}\frac{1}{(x-y)^2 - i\varepsilon}. \tag{3.4}$$

In Euclidean space this becomes

$$G_E(x,y) = \frac{1}{4\pi^2}\frac{1}{(x-y)^2}. \tag{3.5}$$

It is required that this Green's function be given by the ordered vacuum-expectation value of two Euclidean fields

$$\begin{aligned}G_E(x,y) &= \langle 0|R\phi_E(x)\phi_E(y)|0\rangle\\ &\equiv \theta(x^2 - y^2)\langle 0|\phi_E(x)\phi_E(y)|0\rangle\\ &+ \theta(y^2 - x^2)\langle 0|\phi_E(y)\phi_E(x)|0\rangle.\end{aligned} \tag{3.6}$$

Clearly, the insistence on ordering makes sense only if the Euclidean fields are non-commuting operators.[7] In the subsequent, I shall suppress the subscript E, since the discussion will be confined to Euclidean fields exclusively. Also vectors will be frequently written as

$$x^\mu = (r, \alpha^\mu)$$
$$x^2 = r^2, \quad \alpha^\mu = x^\mu/r.$$

IV. QUANTIZATION OF FREE MASSLESS FIELD

The form of canonical field commutators can be deduced from the Green's function, (3.5) and (3.6), which satisfies

$$\Box_x\langle 0|R\phi(x)\phi(y)|0\rangle = -\delta^4(x-y). \tag{4.1}$$

However, since the field itself satisfies $\Box_x\phi(x) = 0$, the nonvanishing result (4.1) can arise only from the effect of $\Box_x$ on the step functions which define the R ordering. The differentiation can be carried out with the help of the identity

$$\Box = \frac{1}{r^3}\left[\left(\frac{d}{d\log r}\right)^2 - (L^2 + 1)\right]r, \tag{4.2}$$

where L^2 is a purely angular derivative operator defined by

$$L^{\mu\nu} = i(x^\mu \partial^\nu - x^\nu \partial^\mu)$$

$$L^2 = \frac{1}{2} \ell^{\mu\nu} \ell_{\mu\nu} \,. \tag{4.3}$$

Eq. (4.1) may be regained, provided we set

$$[\chi(r, \alpha_1), \chi(r, \alpha_2)] = 0 \,, \tag{4.4a}$$

$$[\chi(r, \alpha_1), \chi(r, \alpha_2)] = -\delta^3(\alpha_1 - \alpha_2) \,, \tag{4.4b}$$

$$[\chi(r, \alpha_1), \chi(r, \alpha_2)] = 0 \,. \tag{4.4c}$$

Here the notation is

$$\chi(r, \alpha) = r\phi(x)$$

$$\dot{\chi}(r, \alpha) = \frac{d}{d \log r} \chi(r, \alpha) \,.$$

The δ function is defined on the sphere

$$\int d\alpha_1 f(\alpha_1)\delta^3(\alpha_1 - \alpha_2) = f(\alpha_2) \,.$$

The commutators (4.4) can also be obtained from the action principle. The action is

$$I = \frac{1}{2} \int d^4x \, \partial_\mu \phi \partial^\mu \phi = -\frac{1}{2} \int d^4x \, \phi \Box \phi \,. \tag{4.5a}$$

With the help of (4.2), this may be rewritten

$$I = -\frac{1}{2} \int_0^\infty \frac{dr}{r} \int d\alpha \chi(r, \alpha) \left[\left(\frac{d}{d \log r} \right)^2 - (L^2 + 1) \right] \chi(r, \alpha)$$

$$= \frac{1}{2} \int_{-\infty}^\infty d \log r \int d\alpha \left[\left(\frac{d}{d \log r} \chi \right)^2 + \chi(L^2 + 1)\chi \right] \,. \tag{4.5b}$$

Clearly, the canonical coordinate should be identified with χ, while the canonical momentum is $\dot{\chi}$. The results (4.4) now follow.

Next, I wish to discuss the properties of the field under Hermitian conjugation. We notice that the right-hand side of Eq. (4.4b) is real. This means that $\chi(r, \alpha)$ can not be an Hermitian operator, which is not surprising when it is recalled that the passage from Minkowski space to Euclidean space introduces factors of i. It is postulated that the field $\chi(r, \alpha)$ satisfies the relation

$$\chi^\dagger(r, \alpha) = \chi\left(\frac{1}{r}, \alpha\right) \tag{4.6}$$

which is consistent with (4.4b). Thus another transformation, which is currently much studied — inversion and its relative, the conformal transformation — plays a central role in our theory. The *Ansatz* (4.6) will have far reaching consequences. As will be seen later, the momentum operator P^μ will be the negative of the Hermitian conjugate of the conformal operator K^μ.

Finally, an expansion for the fields is postulated in terms of 4-dimensional harmonics, which are complete and orthonormal[8]

$$\chi(r, \alpha) = \sum_{\ell nm} \left\{ g_{\ell nm}^{(+)}(r) Y_{\ell nm}(\alpha) + g_{\ell nm}^{(-)}(r) Y_{\ell nm}^*(\alpha) \right\}. \tag{4.7}$$

The dependence of the operators $g_{\ell nm}^{(\pm)}(r)$ on r is deduced from $\Box\phi = 0$. This gives an expansion for $g_{\ell nm}^{(\pm)}(r)$ in terms of powers

$$\chi(r, \alpha) = \sum_{\ell nm} \left\{ \frac{a_{\ell nm}^{(+)} r^{\ell+1}}{\sqrt{2\ell + 2}} Y_{\ell nm}(\alpha) + \frac{a_{\ell nm}^{(-)} r^{-(\ell+1)}}{\sqrt{2\ell + 2}} Y_{\ell nm}^*(\alpha) \right\}. \tag{4.8}$$

The Hermiticity *Ansatz* (4.6) implies that

$$a_{\ell nm}^{(+)} = [a_{\ell nm}^{(-)}]^\dagger. \tag{4.9}$$

Let me now determine the algebraic properties of our operators. We shall postulate the fundamental commutators

$$[a_{\ell nm}^{(\pm)}, a_{\ell' n' m'}^{(\pm)}] = 0$$
$$[a_{\ell nm}^{(-)}, a_{\ell' n' m'}^{(+)}] = \delta_{\ell\ell'} \delta_{nn'} \delta_{mm'}. \tag{4.10}$$

It is now easy to verify that the field χ satisfies the commutation relations (4.4). We introduce a vacuum state $|0\rangle$, annihilated by $a_{\ell nm}^{(-)}$

$$a_{\ell nm}^{(-)} |0\rangle = 0. \tag{4.11a}$$

The Hermiticity conditions (4.6) and (4.9) insure the existence of a dual vacuum state $\langle 0|$ annihilated by $a_{\ell nm}^{(+)}$

$$\langle 0| a_{\ell nm}^{(+)} = 0. \tag{4.11b}$$

We are now in a position to check whether the above formalism correctly reproduces the Green's function according to (3.5) and (3.6). From (4.8), (4.10) and (4.11) it follows that

$$\langle 0|R\phi(x_1)\phi(x_2)|0\rangle = \sum_{\ell nm} \frac{1}{2\ell+2} Y^*_{\ell nm}(\alpha_1) Y_{\ell nm}(\alpha_2) x$$

$$\cdot \left\{ \frac{\theta(r_1 - r_2)}{r_1^2}\left(\frac{r_2}{r_1}\right)^\ell + \frac{\theta(r_2 - r_1)}{r_2^2}\left(\frac{r_1}{r_2}\right)^\ell \right\}. \tag{4.12}$$

With the help of the addition formula for 4-dimensional spherical harmonics,[8] (4.12) is recognised to be $\frac{1}{4\pi^2}\frac{1}{(x_1-x_2)^2}$. This shows that we have succeeded in giving a consistent quantization scheme, and have provided a new and intuitively attractive operator basis for Euclidean field theory.

The expansion (4.12) can be used to give a spectral representation for the propagator. By use of the identity

$$\frac{1}{\pi}\int_{-\infty}^{\infty} d\gamma \frac{z^{i\gamma}}{\gamma^2 + (\ell+1)^2} = \frac{z^{\ell+1}\theta(1-z) + z^{-(\ell+1)}\theta(z-1)}{\ell+1} \tag{4.13}$$

$$z > 0, \quad \ell > -1.$$

Eq. (4.12) can be cast in the form

$$G(x_1, x_2) = \frac{1}{4\pi^2}\frac{1}{(x_1-x_2)^2} = \frac{1}{2\pi}\int_{-\infty}^{\infty} d\gamma \sum_{\ell nm} \frac{F^*_{\gamma\ell nm}(x_1)F_{\gamma\ell nm}(x_2)}{\gamma^2 + (\ell+1)^2},$$

where the functions

$$F_{\gamma\ell nm}(x) = r^{i\gamma-1}Y_{\ell nm}(\alpha) \tag{4.14}$$

define the transformation between the coordinate space and the conjugate $(\gamma\ell nm)$ space in which the *dimensionality* γ and the angular momentum variables (ℓnm) are diagonalized. In this new space, particularly convenient for dilation invariant problems, the propagator is diagonal. It is

$$D(\gamma, \ell, n, m) = \frac{1/2\pi}{\gamma^2 + (\ell+1)^2}. \tag{4.15}$$

This 4-dimensional $(\gamma\ell nm)$ space plays the same role in the present context as 4-dimensional momentum space in the usual quantization schemes.

Although the above has been derived for the free massless case, the introduction of masses poses no special difficulty. The main difference is that the expansion functions are no longer simple powers; rather they are Bessel functions.[9] How to deal with interactions will be explained below.

V. CONSERVED QUANTITIES AND GENERATORS OF MOTION

As always, symmetries of the action lead to currents J^μ, which are conserved according to Noether's theorem

$$\partial_\mu J^\mu = \frac{\partial}{\partial t} J^0 + \nabla \cdot \mathbf{J} = 0 \,. \tag{5.1}$$

In order to obtain the "charge" associated with this symmetry, the current is integrated over the quantization surface

$$Q = \int d\tau^\mu J_\mu(x) \,, \tag{5.2}$$

where $d\tau^\mu$ is the infinitesimal surface element. For us, the surface is given by $x^2 = \tau^2$, and

$$d\tau^\mu = d^4 x \delta(x^2 - \tau^2) 2x^\mu \tag{5.3}$$

(The normal to the surface is $2x^\mu$). Hence the charge is

$$Q = \int d^4 x \delta(x^2 - \tau^2) 2x_\mu J^\mu(x) \,. \tag{5.4a}$$

Of course, as a consequence of current conservation, Q does not depend on τ, which may be set equal to 1. Thus (5.4a) may be cast into a purely angular integral

$$Q = \int d\alpha \, \alpha_\mu J^\mu(1, \alpha) \,. \tag{5.4b}$$

An especially important object in our theory is the dilation generator

$$D = i \int d\tau^\mu x^\nu \theta_{\mu\nu} \,, \tag{5.5a}$$

where $x_\nu \theta^{\mu\nu}(x)$ is the dilation current, expressed in terms of the new improved energy momentum tensor.[10] For the free massless theory governed by the action (4.5) this current is conserved. Upon expressing $\theta^{\mu\nu}$ in terms of the field ϕ, expanding ϕ in creation and annihilation operators as in (4.8), and performing the integral indicated in (5.5a), we get

$$D = -i \sum_{\ell n m} (\ell + 1) a_{\ell n m}^{(+)} a_{\ell n m}^{(-)} \,. \tag{5.5b}$$

Thus the dilation generator is diagonal and anti-Hermitian. Frequently, it will be convenient to define a Hermitian generator

$$\Delta = iD \,. \tag{5.6}$$

(If the theory is not scale invariant, D will no longer be independent of the surface over which $x_\nu \theta^{\mu\nu}$ is integrated; *i.e.*, it will depend on τ. This in no way changes the fact that $D(\tau)$ describes the evolution of the system. The analogous situation in the ordinary theory occurs when time-dependent interactions are present. The Hamiltonian is then time-dependent, yet it still governs evolution in time.)

The remaining generators of the conformal group can also be constructed

$$M^{\mu\nu} = i \int d\tau_\alpha (x^\mu \theta^{\nu\alpha} - x^\nu \theta^{\mu\alpha}), \tag{5.7a}$$

$$P^\mu = i \int d\tau_\nu \theta^{\mu\nu}, \tag{5.7b}$$

$$K^\mu = i \int d\tau_\nu (2x^\mu x_\alpha - g^\mu_\alpha x^2) \theta^{\nu\alpha}. \tag{5.7c}$$

Their expression in terms of creation and annihilation operators is quite complicated; I shall not exhibit it here. However, the Hermiticity properties can be easily deduced. We find

$$(M^{\mu\nu})^\dagger = M^{\mu\nu}, \tag{5.8a}$$

$$(P^\mu)^\dagger = -K^\mu, \tag{5.8b}$$

$$(K^\mu)^\dagger = -P^\mu. \tag{5.8c}$$

As promised, conformal transformations are as important in our theory as translations, since the two are related by Hermitian conjugation. The geometrical meaning of (5.8b) and (5.8c) is clear, once it is remembered that Hermitian conjugation is just coordinate inversion, see (4.6): a translation in the inverted coordinate system is equivalent to a conformal transformation.

VI. STATES OF THE SYSTEM

A "single particle" state is defined by

$$|\ell n m\rangle = a^{(+)}_{\ell n m} |0\rangle. \tag{6.1}$$

Naturally, this state does not coincide with the physical one particle state defined by Minkowski space quantization. Observe that $|\ell n m\rangle$ is an eigenstate of D with eigenvalue $-i(\ell+1)$. Furthermore, it is $(\ell+1)^2$-fold degenerate: for fixed value of ℓ, there are $(\ell+1)^2$ different values of n and m.[8]

The singularity structure of the dilation propagator, (4.15), may now be understood. Note that the propagator has poles at $\gamma = \pm i(\ell + 1)$. This is just the dilation eigenvalue of the single particle state. Thus the propagator in our conjugate $(\gamma \ell n m)$ space has poles at the eigenvalues of the evolution operator. This is in complete analogy with the conventional situation in momentum space, where the propagator has poles at $p^0 = \pm\sqrt{\mathbf{p}^2 + m^2}$, *i.e.*, at the eigenvalues of the Hamiltonian.

In the usual theory, interactions modify the position of the pole in the propagator — a phenomenon known as mass renormalization. We expect that in the present theory a similar shift should occur, once interactions are taken into account. Clearly, this is the origin of anomalous dimensions in our formalism. Thus the intriguing possibility presents itself of computing anomalous dimensions as the poles of the complete dilation propagator.

VII. PERTURBATION THEORY AND THE S MATRIX

The above considerations give a complete solution to the problem of quantizing a *free* theory. Now we must describe how interactions should be included. Suppose the equation of motion for the interacting field ϕ is

$$\Box \phi = \eta\,, \tag{7.1}$$

where η is the source. Free field canonical commutators (4.4) may still be imposed on $\phi(x)$ and $\frac{d}{d \log r}[r\phi(x)]$, at equal r.

To solve the equations of motion (7.1) we define the operator $U(r, r_0)$ by the equations

$$\frac{d}{d \log r} U(r, r_0) = -iD^{\text{int}}(r)U(r, r_0)$$

$$U(r_0, r_0) = 1\,, \tag{7.2}$$

where $D^{\text{int}}(r)$ is the interaction part of the dilation generator, constructed from the *free* fields ϕ. In a familiar fashion, one proves that $U(r, r_0)\phi(x)\,U^{-1}(r, r_0)$ is a free field. Since we have solved the free-field problem, all that remains is to integrate (7.2).

Integration of (7.2) gives

$$U(r, r_0) = R \exp -i \int_{r_0}^{r} \frac{dr'}{r'} D^{\text{int}}(r')\,, \tag{7.3}$$

where the R ordering is along the spheres. Recalling the definition of D^{int}, we see that

$$-i \int_{r_0}^{r} \frac{dr'}{r'} D^{\text{int}}(r') = -i \int_{r_0}^{r} \frac{dr'}{r'} \int d\tau^\mu i x^\nu \theta^{\text{int}}_{\mu\nu}$$

$$= -i \int_{r_0}^{r} \frac{dr'}{r'} \int d^4 x \delta(x^2 - r'^2) 2 x^\mu (i x^\nu \theta^{\text{int}}_{\mu\nu}(x))$$

$$= - \int_{r_0}^{r} \frac{dr'}{r'} \int d^4 x \delta(x^2 - r'^2) 2 r'^2 L^{\text{int}}(x)$$

$$= - \int_{r_0}^{r} d^4 x L^{\text{int}}(x)$$

$$U(r, r_0) = R \exp - \int_{r_0}^{r} d^4 x L^{\text{int}}(x). \tag{7.4}$$

The integration is over all space bounded by the spheres $r^2 > x^2 > r_0^2$.

We have regained the Schwinger-Tomonaga result that the evolution of the system from $x^2 = r_0^2$ to $x^2 = r^2$ is governed by the ordered exponential of the integral of the interaction Lagrangian. Clearly, the S matrix is $U(\infty, 0)$

$$S = R \exp - \int d^4 x L^{\text{int}}(x)$$

$$= \sum_{n=0}^{\infty} \frac{(-1)^n}{n!} \int d^4 x_1 \dots d^4 x_n R L^{\text{int}}(x_1) \dots L^{\text{int}}(x_n). \tag{7.5}$$

Applying Wick's theorem to this, we regain the familiar Feynman-Dyson perturbation theory. Thus we see that our quantization procedure is entirely successful and consistent with the conventional approach.

The evaluation of the perturbation series may be carried out in position space, or in our conjugate $(\gamma \ell n m)$ space. Calculations in the $(\gamma \ell n m)$ space are especially convenient for conformally invariant interactions. We have not yet researched questions of divergences and renormalization in this new conjugate space.

VIII. COUNTING STATES, ASYMPTOTIC DEGENERACY

It has already been observed that the single particle state $|\ell n m\rangle$ is an eigenstate of $\Delta = iD$, with eigenvalue $\ell + 1$ and degeneracy $(\ell + 1)^2$. Consider now the N particle state

$$|N\rangle = a^{(+)}_{\ell_1 n_1 m_1} \dots a^{(+)}_{\ell_N n_N m_N} |0\rangle. \tag{8.1}$$

This is an eigenstate of $\triangle$ with eigenvalue d_N, and degeneracy $g(N)$ $(d_1 = \ell+1$, $g(1) = (\ell+1)^2)$. We wish to compute $g(N)$ for large N in terms of d_N.

The computation is facilitated by introducing a generating function

$$F(\beta) = \text{Tr} \ \exp \ -\beta\triangle . \tag{8.2}$$

From its definition, two expressions for $F(\beta)$ can be given. Clearly, from (8.2) it follows that

$$F(\beta) = \sum_N g(N) \ \exp \ -\beta d_N . \tag{8.3}$$

An alternate expression is got for $F(\beta)$ by using the formulas (5.5) and (5.6) for $\triangle$

$$\begin{aligned}
F(\beta) &= \text{Tr} \ \exp \ -\beta \sum_{\ell nm}(\ell+1)a^{(+)}_{\ell nm}a^{(-)}_{\ell nm} \\
&= \text{Tr} \ \prod_{\ell nm} \exp \ -\beta(\ell+1)a^{(+)}_{\ell nm}a^{(-)}_{\ell nm} \\
&= \prod_{\ell nm} \text{Tr} \ \exp \ -\beta(\ell+1)a^{(+)}_{\ell nm}a^{(-)}_{\ell nm} .
\end{aligned} \tag{8.4a}$$

Recalling that for each mode $\text{Tr} \ x^{a^+a} = (1-x)^{-1}$, and that each mode is $(\ell+1)^2$-fold degenerate, it follows from (8.4a) that

$$F(\beta) = \prod_\ell \left\{1 - e^{-\beta(\ell+1)}\right\}^{-(\ell+1)^2}$$

$$\log F(\beta) = -\sum_{\ell=0}^\infty (\ell+1)^2 \log\left[1 - e^{-\beta(\ell+1)}\right] \tag{8.4b}$$

$$= \sum_{n=1}^\infty \frac{1}{n} \frac{e^{\beta n/2} + e^{-\beta n/2}}{[e^{\beta n/2} - e^{-\beta n/2}]^3} .$$

Our task is therefore to obtain a relationship between $g(N)$ and d_N for large N, which must hold if (8.3) and (8.4) are simultaneously true. This can be done by an elaborate analytic procedure based on the classic work of Hardy and Ramanujan.[11] However, it is more instructive, to obtain the answer by a thermodynamical analogy.

Observe that $F(\beta)$ defined by (8.2) is a partition function where β plays the role of inverse temperature, $\beta \to 1/T$, and $\triangle$ is the analog of the energy operator in the usual theory. This is as it should be — we have repeatedly emphasized that the dilation generator replaces the Hamiltonian in our method.

We may use this analogy to develop a thermodynamics based on the partition function $F(\beta)$. This thermodynamics is manifestly covariant. With the thermodynamical technique an expression for $g(N)$ is easily obtained.

We define the "free energy" by

$$A = -\frac{1}{\beta} \log F(\beta). \tag{8.5}$$

A is related to the "entropy" S and to the "internal energy" U by the well known equations

$$S = \beta^2 \frac{\partial}{\partial \beta} A \tag{8.6}$$

$$U = A + \frac{1}{\beta} S = -\frac{\partial}{\partial \beta} \cdot \log F(\beta). \tag{8.7}$$

Since we seek $g(N)$ only for large N, we are effectively in the high temperature limit $T \to \infty$ or $\beta \to 0$. In this region the "entropy" should be identified with $\log g(N)$ and the "internal energy" with the dilation eigenvalue d_N. Furthermore, the asymptotic form for $F(\beta)$ as $\beta \to 0$ is easy to obtain from (8.4b)

$$\log F(\beta) \longrightarrow \frac{\pi^4}{45\beta^3}. \tag{8.8}$$

Hence, we get from (8.5), (8.6), (8.7) and (8.8)

$$A \longrightarrow -\frac{\pi^4}{45\beta^4}, \tag{8.9}$$

$$S \longrightarrow \frac{\pi^4}{45\beta^3}, \tag{8.10}$$

$$U \longrightarrow \frac{\pi^4}{15\beta^4}. \tag{8.11}$$

Eliminating the temperature $1/\beta$ between (8.10) and (8.11) gives a relationship between the entropy and the internal energy, valid in the high temperature limit

$$S = cU^{3/4}$$

$$c = \frac{4\pi}{3(15)^{1/4}}. \tag{8.12}$$

With our interpretation for S and U, we finally derive

$$g(N) \xrightarrow[N \to \infty]{} \exp cd_N^{3/4} . \tag{8.13}$$

The same answer emerges, after much effort, when the computation is performed analytically,[4] which verifies the consistency and correctness of our thermodynamics.

The degeneracy, seen to be of the exponential variety, is reminiscent of the Hagedorn expression. To make contact with the latter formula, let us first recompute the degeneracy in δ dimensions, generalizing the 4-dimensional derivation given above. The answer, simply obtained by the same method, is

$$g_\delta(N) \xrightarrow[N \to \infty]{} \exp c_\delta d_N^{1-1/\delta}$$

$$c_\delta = \frac{\delta}{\delta - 1} \left[2(\delta - 1) \sum_{n=1}^{\infty} \frac{1}{n^\delta} \right]^{1/\delta} . \tag{8.14}$$

In particular, for 2-dimensions

$$g_2(N) \xrightarrow[N \to \infty]{} \exp \frac{2\pi}{\sqrt{3}} d_N^{1/2} . \tag{8.15}$$

Let us now postulate, that for reasons that are by no means apparent, there exists a "Law of Nature" relating physical mass to scale-dimensionality in this 2-dimensional space[12],

$$d_N = \frac{3\kappa}{4\pi^2} m^2 . \tag{8.16}$$

With this postulate, we find that the degeneracy of levels is $\exp \kappa m$, which is just the Hagedorn formula.

Since the derivation of the Hagedorn degeneracy is an open challenge to theoreticians, it is useful to elaborate on the possible circumstances that would lead to the new "Law of Nature" (8.16). Imagine that the correct, as yet unknown, physical theory is a field theory defined on some 2-dimensional, Euclidean parameter space. This parameter space is not the physical 4-dimensional space-time, though perhaps it is related to it. On the 2-dimensional space, dilations can be defined; they are generated by D. In addition, since this hypothetical theory is to describe the 4-dimensional physical world, there must exist a momentum operator, generating physical 4-dimensional translations. Associated with the momentum operator will be the physical mass operator M^2. The postulate (8.16) is then equivalent to the statement that the expression for M^2 in terms of the fundamental fields is proportional to the formula for D.

The observation that 2-dimensional models are in some way relevant to particle theory of the 4-dimensional physical world seems to be a very important though not understood feature of Nature. Historically, the relevance of a 2-dimensional transverse space was noted first in the analysis of multi-particle production in the "pionization" region.[13] A consistent, satisfactory description of single particle production in this region is obtained by cutting off large, transverse momenta and using simply the longitudinal, 2-dimensional phase space factor dE/E. More recently, the success of the parton model in interpreting many of the important features of high energy reactions has called attention to 2-dimensional field theories as a means of describing the parton distribution functions.[12] Indeed it was in this context that the relation (8.16) was first suggested. Finally, recent investigations of dual resonance models make essential use of the general formalism of 2-dimensional field theories.[14]

This collection of facts leads me to exhibit further applications of our techniques to 2-dimensional field theories.

IX. TWO-DIMENSIONAL FIELD THEORIES, THE VIRASORO ALGEBRA

Field theories in 2 dimensions have been first introduced as an interesting theoretical laboratory where explicitly solvable models can be studied. More recently, however, the interest in such models has become much less academic. We apply our formalism to 2-dimensional field theories and demonstrate that our Lorentz-invariant quantization rules in *Euclidean* space lead directly to the operator structure present in dual models.

It has been shown that the higher moments of the energy-momentum tensor, in a large class of 2-dimensional field theories, obey essentially the Virasoro algebra.[14,15] However, with a conventional approach, some difficulties persist in the interpretation of various integrals. With our formalism, we can obtain a clear, unambiguous derivation of the Virasoro algebra, without the difficulties previously encountered.

Another point of contact between the present approach and dual models is seen in the structure of the R ordered product appearing in our Euclidean perturbation theory, Eq. (7.5). This is identical to that of the operator form for the integrals in the n-point Veneziano formulas.

I shall begin by considering a theory in δ dimensions with a traceless, symmetric energy-momentum tensor (for the moment the metric is immaterial) and I construct the current

$$J_f^\mu(x) = \theta^{\mu\nu}(x) f_\nu(x)\,, \tag{9.1}$$

where $f^\mu(x)$ is a function of x. Special cases of (9.1) are well known. These are the currents of the conformal group

$$
\begin{array}{lll}
\text{Translation} & \text{current:} & f^\mu(x) = a^\mu \\
\text{Lorentz} & \text{current:} & f^\mu(x) = \omega^{\mu\nu} x_\nu,\ \omega^{\mu\nu} = -\omega^{\nu\mu} \\
\text{Dilation} & \text{current:} & f^\mu(x) = a x^\mu \\
\text{Conformal} & \text{current:} & f^\mu(x) = 2x^\mu a_\alpha x^\alpha - a^\mu x^2\,.
\end{array} \tag{9.2}
$$

Of course, all these currents are conserved when the energy-momentum tensor is traceless, symmetric and conserved. Are there any other forms for f^μ, such that $\partial_\mu J_f^\mu(x) = 0$? To answer this question, we differentiate (9.1)

$$
\begin{aligned}
\partial_\mu J_f^\mu &= \theta^{\mu\nu} \partial_\nu f_\mu \\
&= \frac{1}{2}\theta^{\mu\nu}\left[\partial_\mu f_\nu + \partial_\nu f_\mu - \frac{2}{\delta} g_{\mu\nu}\partial_\alpha f^\alpha\right].
\end{aligned}
$$

Clearly, the current will be conserved when

$$\partial_\mu f_\nu(x) + \partial_\nu f_\mu(x) - \frac{2}{\delta} g_{\mu\nu}\partial_\alpha f^\alpha(x) = 0\,. \tag{9.3}$$

Eq. (9.3) for $\delta > 2$ is solved only by the forms (9.2).[16] However for $\delta = 2$ a wider class of solutions exists. Thus if we wish to have the large symmetry, which leads to the conserved currents J_f^μ, with f^μ satisfying (9.3), we are again led to 2-dimensional theories. Since, as will emerge presently, this symmetry is related to the Virasoro algebra,[14] I shall now confine myself to 2-dimensional models.

Examples of field theories in 2 dimensions with a traceless, symmetric and conserved energy-momentum tensor are the following:

I. Free Boson Field

$$
\begin{aligned}
\mathcal{L} &= \frac{1}{2}\partial_\mu \phi \partial^\mu \phi \\
\theta^{\mu\nu} &= \partial^\mu \phi \partial^\nu \phi - \frac{1}{2}g^{\mu\nu}\partial^\alpha \phi \partial_\alpha \phi\,.
\end{aligned} \tag{9.4}
$$

II. Free Fermion Field

$$\mathcal{L} = \frac{i}{2}\bar{\psi}\gamma^\mu \overleftrightarrow{\partial}_\mu \psi$$

$$\theta^{\mu\nu} = \frac{i}{4}\{\bar{\psi}\gamma^\mu \overleftrightarrow{\partial}^\nu \psi + \bar{\psi}\gamma^\nu \overleftrightarrow{\partial}^\mu \psi\}.$$

(9.5)

III. Gradient Coupling Model

$$\mathcal{L} = \frac{1}{2}\partial_\mu \phi \partial^\mu \phi + \frac{i}{2}\bar{\psi}\gamma^\mu \overleftrightarrow{\partial}_\mu \psi - \lambda\bar{\psi}\gamma^\mu \psi \partial_\mu \phi$$

$$\theta^{\mu\nu} = \partial^\nu \phi \partial^\nu \phi + \frac{i}{4}\{\bar{\psi}\gamma^\mu \overleftrightarrow{\partial}^\nu \psi + \bar{\psi}\gamma^\nu \overleftrightarrow{\partial}^\mu \psi\}$$

$$- \frac{\lambda}{2}\{\bar{\psi}\gamma^\mu \psi \partial^\nu \phi + \bar{\psi}\gamma^\nu \psi \partial^\mu \phi\} - \frac{1}{2}g^{\mu\nu}\partial_\alpha \phi \partial^\alpha \phi.$$

(9.6)

IV. Thirring Model

$$\mathcal{L} = \frac{i}{2}\bar{\psi}\gamma^\mu \overleftrightarrow{\partial}_\mu \psi - \lambda\bar{\psi}\gamma^\mu \psi\bar{\psi}\gamma_\mu \psi$$

$$\theta^{\mu\nu} = \frac{i}{4}\{\bar{\psi}\gamma^\mu \overleftrightarrow{\partial}^\nu \psi + \bar{\psi}\gamma^\nu \overleftrightarrow{\partial}^\mu \psi\} - \lambda g^{\mu\nu}\bar{\psi}\gamma^\alpha \psi\bar{\psi}\gamma_\alpha \psi.$$

(9.7)

V. The "Covariant String"

$$\mathcal{L} = \left\{\frac{1}{2}[\varepsilon_{\mu\nu}\partial^\mu \phi^A \partial^\nu \phi^B][\varepsilon_{\alpha\beta}\partial^\alpha \phi^A \partial^\beta \phi^B]\right\}^{1/2}$$

$$= \{\det \partial_\mu \phi^A \partial_\nu \phi^A\}^{1/2}.$$

(9.8)

Here $\varepsilon^{\mu\nu}$ is the totally antisymmetric tensor, $\varepsilon^{01} = 1$. The indices A, B describe some additional degree of freedom and are summed over. (In applications to the dual model, A, B refer to physical space-time[17] — one may of course, allow models I–IV to possess such further degrees of freedom also.) In this model the energy-momentum tensor vanishes identically.

For all the above theories the currents J_f^μ will be conserved, provided f^μ satisfies (9.3). I shall discuss first the situation in *Minkowski* space, and shall demonstrate that a consistent algebraic description of this symmetry is not possible. In *Euclidean* space, when the theory is quantized according to our method, no difficulties arise.

To find the solutions to (9.3) take the ++ component and the −− component of that equation. (The +− component vanishes identically.) In Minkowski space we thus find

$$\partial_+ f_+ = 0, \quad \partial_- f_- = 0.$$

(9.9a)

The integrals clearly are (recall that $f_\pm = f^\mp$)

$$f^+ = f^+(x^+)$$
$$f^- = f^-(x^-).$$

(9.9b)

The reason for the existence of all these conserved currents is that 2-dimensional theories with a traceless, symmetric and conserved energy-momentum tensor are invariant under the coordinate transformation

$$\delta_f x^\mu = f^\mu(x),$$

(9.10a)

where f^μ is of the form (9.9) and the field is transformed as follows

$$\delta_f \phi(x) = f^\mu(x)\partial_\mu \phi(x) + \frac{\partial_\mu f_\alpha(x)}{2}\left[g^{\alpha\mu}d - \varepsilon^{\alpha\mu}\Sigma\right]\phi(x).$$

(9.10b)

Here d is the scale dimension of the field ϕ and $\varepsilon^{\alpha\mu}\Sigma$ is the spin matrix associated with ϕ. (In 2 dimensions we may set $\Sigma^{\alpha\mu} = \varepsilon^{\alpha\mu}\Sigma$.) The symmetry can be seen directly in models I–IV above. For the string model, example V above, the currents J_f^μ vanish identically since $\theta^{\mu\nu}$ is zero. The model is invariant under the transformation (9.10), even if f^μ does *not* satisfy (9.9). This is a gauge symmetry which does not lead to any conserved currents.[17] In the following, we consider only theories with non-vanishing $\theta^{\mu\nu}$.

The composition law for the transformation (9.10) is

$$\delta_g \delta_f x^\mu = f^\mu(x) + g^\mu(x) + f^\nu(x)\partial_\nu g^\mu(x)$$
$$\delta_f \delta_g x^\mu = g^\mu(x) + f^\mu(x) + g^\nu(x)\partial_\nu f^\mu(x).$$

(9.11a)

It is easy to see that if f^μ and g^μ satisfy (9.9), so does

$$h^\mu = g^\nu \partial_\nu f^\mu - f^\nu \partial_\nu g^\mu$$

(9.11b)

and the transformations form a group. By choosing

$$f: \quad \begin{aligned} f^+ &= 0 \\ f^- &= f(x^-), \end{aligned}$$

(9.12a)

$$f: \quad \begin{aligned} f^+ &= f(x^+) \\ f^- &= 0 \end{aligned}$$

(9.12b)

we recognize the fact that one is dealing with a direct product of two groups of transformation, and it suffices to consider only one of the factors. Henceforth, we make the choice (9.12a).

The charges can be constructed as follows

$$Q_f = \int_{-\infty}^{\infty} dx^1 \theta^0{}_\mu(x) f^\mu(x),$$

$$= \int_{-\infty}^{\infty} dx^1 \theta^{0+}(x) f(x^-) \tag{9.13a}$$

$$= \int_{-\infty}^{\infty} \frac{dx^1}{\sqrt{2}} [\theta^{++}(x) + \theta^{-+}(x)] f(x^-).$$

$\theta^{+-}(x)$ vanishes since the stress tensor is traceless. Also from the conservation of the stress tensor, $\partial_\mu \theta^{\mu+}(x) = 0 = \partial_+ \theta^{++}(x)$, we learn that θ^{++} depends only on x^-. Hence the charge is

$$Q_f = \int_{-\infty}^{\infty} dx^- \theta^{++}(x) f(x^-). \tag{9.13b}$$

From the composition law (9.11) one would expect that the charges satisfy the commutation relations

$$i[Q_f, Q_g] = Q_h$$
$$h = f'g - g'f. \tag{9.14}$$

In order to verify (9.14), consider

$$\int_{-\infty}^{\infty} dx^- dy^- f(x^-) g(y^-) [\theta^{++}(x), \theta^{++}(y)]$$

$$= \int_{-\infty}^{\infty} dx^- dy^- f(x^-) g(y^-) [\theta^{++}(x), \theta^{++}(y)]|_{x^+=y^+}, \tag{9.15}$$

where the equality is true simply because θ^{++} does not depend on x^+. The form of the local commutator of two energy momentum tensors which will insure the validity of (9.15) is

$$[\theta^{++}(x), \theta^{++}(y)]|_{x^+=y^+} = i[\theta^{++}(x) + \theta^{++}(y)]\partial_- \delta(x^- - y^-) \tag{9.16}$$

and canonical evaluation in simple models yields this result.

However, from general principles one can show that the commutator of the stress tensor with itself cannot be of the form given in (9.16). That the term proportional to the first derivative of the delta function is as indicated insures Poincaré invariance of the theory; indeed (9.16) is the analog of the Dirac-Schwinger commutator for the present problem. In addition, positivity and Lorentz covariance insure that a triple derivative is necessarily present. This is the analog of the usual Schwinger term in current commutators, and the failure of canonical manipulations to expose it is of course familiar.[18] The correct expression for the commutator which replaces (9.16) is

$$[\theta^{++}(x), \theta^{++}(y)]\Big|_{x^+ = y^+} = i[\theta^{++}(x) + \theta^{++}(y)]\partial_-\delta(x^- - y^-)$$

$$- \frac{ia}{2}\partial_-^3\delta(x^- - y^-) \qquad (9.17)$$

$$a > 0.$$

The commutator of the charges therefore is

$$i[Q_f, Q_g] = Q_h - \frac{a}{2}\int_{-\infty}^{\infty} dx^- f'''(x^-)g(x^-) \qquad (9.18)$$

and for arbitrary f and g the charges do not satisfy the "classical" algebra.

The additional c number poses also the problem of convergence of the integral. If $f(x)$ or $g(x)$ are powers of x, as they are for the Virasoro algebra, $\int_{-\infty}^{\infty} dx^- g(x^-)f'''(x^-)$ does not converge. Furthermore, one may question the convergence of the formula (9.13) for the charges, when f is a positive or negative power. Thus we conclude that the symmetry (9.10) and the associated algebra cannot be given an operator basis in a field theory defined on 2-dimensional Minkowski space.

The difficulties which have been encountered in giving a coherent field theoretic basis for the Virasoro algebra in a Minkowski space field theory do not exist if the field theory is quantized by our method, in Euclidean space. I shall now show this. The solution of Eq. (9.3) for f^μ in Euclidean space is

$$f_1(x) + if_2(x) = f(re^{i\theta})$$
$$f_1(x) - if_2(x) = g(re^{-i\theta})$$
$$x_1 = r\cos\theta, \qquad x_2 = r\sin\theta. \qquad (9.19)$$

Here f and g are arbitrary functions of the argument. As before we may set g to zero, and confine our attention to a non-vanishing f. The charge is given by an angular integral, [see (5.4)]

$$Q_f = i \int_0^{2\pi} d\theta x_\mu J_f^\mu(x) \, ,$$

$$= i \int_0^{2\pi} d\theta x_\mu \theta^{\mu\nu}(x) f_\nu(x) \tag{9.20}$$

$$= i \int_0^{2\pi} d\theta r e^{i\theta} f(re^{i\theta}) \theta(x) \, ,$$

where

$$\theta(x) \equiv \frac{1}{2} [\theta^{11}(x) - i\theta^{12}(x)] \, . \tag{9.21}$$

It is easy to show that θ is only a function of $x^1 + ix^2 = re^{i\theta}$, as a consequence of conservation, tracelessness and symmetry of $\theta^{\mu\nu}$. Hence, Q_f is also given by

$$Q_f = \int_c dz \, f(z)\theta(z) \tag{9.22}$$

and the integration contour is over the circle of radius r. Clearly, Q_f is independent of r, if $f(z)\theta(z)$ is analytic for $0 < |z| < \infty$.

To establish the algebraic properties of the charges, in our Euclidean space theory, we compute the $[Q_f, Q_g]$ commutator. Again what is needed is the $[\theta, \theta]$ commutator at equal r. The Euclidean space analog of (9.17) is

$$i[\theta(x), \theta(x')]|_{r=r'} = \frac{1}{r^2} \left\{ e^{-2i\theta}\theta(x) + e^{-2i\theta'}\theta(x') \right\} \partial_\theta \delta(\theta - \theta')$$

$$+ \frac{a}{2r^4} e^{-2i(\theta+\theta')} \{ \partial_\theta \delta(\theta - \theta') + \partial_\theta^3 \delta(\theta - \theta') \} \tag{9.23}$$

$$a > 0 \, .$$

Consequently, we find

$$i[Q_f, Q_g] = Q_h - \frac{ia}{2} \int_0^{2\pi} d\theta \, re^{i\theta} g(re^{i\theta}) f'''(re^{i\theta})$$

$$= Q_h - \frac{a}{2} \int_c dz \, g(z) f'''(z) \, . \tag{9.24}$$

When f and g are analytic for $0 < |z| < \infty$, the integral is independent of contour. In contrast to the Minkowski space formula (9.18), the present result (9.24) is free from divergences. Also the expression for the charge, (9.20) or (9.22), involves a finite range integral and no question about convergence need be raised. Thus the Euclidean quantization provides a consistent algebraic description of the large symmetry present in many 2-dimensional field theories. The only anomaly is the c number addition proportional to a.

The discrete Virasoro algebra is obtained by setting

$$f(z) = iz^{1-n}$$
$$g(z) = iz^{1-m} \, . \tag{9.25}$$

Then from (9.24) one gets

$$[Q_n, Q_m] = (m-n)Q_{n+m} - \pi a \delta_{n+m,0}(n^3 - n) \, . \tag{9.26}$$

Notice that the commutator anomaly is absent for n or $m = 0, \pm 1$. These cases correspond to a constant, linear or quadratic function f; *i.e.*, they give rise to the usual conformal group, Eq. (9.2). Thus the ordinary conformal algebra is realized in a conventional way, both for the Minkowski-space theory and for Euclidean-space theory. Only when $|n| > 1$ does the additional c number become relevant, and in that instance only the Euclidean method gives non-divergent results. The presence of a non-vanishing, c number anomaly shows that Q_n cannot annihilate the vacuum, for all n, and the symmetry is absent.

X. CONCLUSION

The aspects of this investigation that we find most interesting are the following. The development of a manifestly covariant quantization procedure which can replace the conventional non-covariant method provides a wealth of covariant canonical commutators. It will be interesting to examine current commutators in this context and to see whether previous successes of equal-time and light-like current algebra can be extended.

We have given an operator basis for Euclidean quantum field theory. The quantization procedure follows closely the method of solving partial differential equations of the Laplacian type in Euclidean space, where the initial value data is specified on a sphere, rather than on a plane. Our approach gives special prominence to the dilation operator. It makes contact with the explicitly conformally covariant formalism of Johnson and Adler.[19] Furthermore, it offers the intriguing possibility of studying anomalies of scale invariance as modifications of the dilation propagator.

A covariant thermodynamics has put us closer to the goal of deriving the Hagedorn spectrum from first principles. What still is missing is an *a priori* justification of the use of 2-dimensional field theory and of the identification of mass squared with the dilation eigenvalue. Nevertheless, our approach to field theory in *Euclidean* space offers an attractive alternative to the conventional field theoretic formulations of the dual resonance model.

REFERENCES

1. For a summary see V. d'Alfaro, S. Fubini, G. Furlan and C. Rossetti, *Currents in Hadron Physics* (North-Holland, Amsterdam, 1973).
2. For a summary see "Canonical Light-Cone Commutators and their Applications", reprinted in this volume on p. 309.
3. Our notational conventions in Minkowski space are $g^{00} = -g^{ii} = 1$; $x^{\pm} = \frac{1}{\sqrt{2}}(x^0 \pm x^3)$. In Euclidean space, the metric is positive and no distinction between upper and lower indices need be made.
4. S. Fubini, A. Hanson, and R. Jackiw, *Phys. Rev. D* **7**, 1732 (1973). In the initial stages of investigation, we collaborated with E. del Guidice. Light-cone quantization as well as the present method have been discussed by P. A. M. Dirac, *Rev. Mod. Phys.* **21**, 392 (1949).
5. J. Schwinger, *Proc. Nat. Acad. Sci.* **44**, 956 (1958); K. Symanzik, in *Mathematical Theory of Elementary Particles*, R. Goodman and I. Segal, eds. (MIT Press, Cambridge, MA, 1966).
6. A related problem is that translations are hard to define. It will be seen that our method diagonalizes D. Since D and P^{μ} satisfy the commutation relation $i[D, P^{\mu}] = P^{\mu}$, either D is not Hermitian or P^{μ} does not exist. In this connection see W. W. MacDowell and R. Roskies, *J. Math. Phys.* **13**, 1585 (1972).
7. It is at this stage that we part company with previous approaches to Euclidean field theory, Ref. 5. In these investigations the non-interacting Euclidean field does not satisfy the free equation (3.3). Also, for us the fields are non-commuting operators, while previously they were taken to commute.
8. The properties of the harmonics are as follows

$$Y_{\ell nm}(\alpha) = Y_{\ell nm}(\theta, \phi, \psi)$$

$$= N_{\ell nm} e^{im\phi} (\sin \theta)^n C_{\ell-n}^{n+1}(\cos \theta)(\sin \psi)^m C_{n-m}^{m+\frac{1}{2}}(\cos \psi).$$

The $C_{\nu}^{\lambda}(z)$ are Gegenbauer polynomials and $N_{\ell nm}$ is a normalization factor defined by

$$(N_{\ell nm})^{-2} = 2\pi E_0(\ell, n) E_1(n, m)$$

$$E_k(\ell, n) = \frac{\pi 2^{k-2n} \Gamma(\ell + n - k + 2)}{(2\ell + 2 - k)(\ell - n)! [\Gamma(n + 1 - k/2)]^2}.$$

The range of the indices is

$$\ell = 0, 1, \ldots, \infty$$
$$n = 0, 1, \ldots, \ell$$
$$m = -n, -n+1, \ldots, n \,.$$

The harmonics are complete

$$\sum_{\ell n m} Y^*_{\ell n m}(\alpha) Y_{\ell n m}(\alpha') = \delta^3(\alpha - \alpha')$$

$$= \frac{1}{\sin^2 \theta \sin \psi} \delta(\theta - \theta') \delta(\psi - \psi') \delta(\phi - \phi')$$

and orthogonal

$$\int_0^\pi d\theta \, \sin^2 \theta \int_0^\pi d\psi \, \sin \psi \int_0^{2\pi} d\phi Y^*_{\ell n m}(\alpha) Y_{\ell' n' m'}(\alpha') = \delta_{\ell \ell'} \delta_{n n'} \delta_{m m'} \,.$$

They satisfy the addition theorem

$$\sum_{n=0}^{\ell} \sum_{m=-n}^{n} Y^*_{\ell n m}(\alpha) Y_{\ell n m}(\alpha') = \frac{\ell + 1}{2\pi^2} C_\ell^1(\alpha \cdot \alpha') \,.$$

The harmonic is an eigenfunction of the angular derivative defined in (4.3)

$$L^2 Y_{\ell n m} = \ell(\ell + 2) Y_{\ell n m} \,.$$

9. A. di Sessa, *J. Math. Phys.* **15**, 1892 (1974).
10. C. Callan, S. Coleman and R. Jackiw, *Ann. Phys.* (NY) **59**, 42 (1970).
11. G. H. Hardy and S. Ramanujan, *Proc. London Math. Soc.* (2) **17**, 75 (1918).
12. E. del Guidice, P. di Vecchia, S. Fubini and R. Musto, *Nuovo Cim.* **12A**, 813 (1972).
13. D. Amati, S. Fubini and A. Stanghellini, *Nuovo Cim.* **26** 896 (1962); S. Fubini, *Lectures at the 1963 Scottish Universities Summer School*, R. G. Moorhouse, ed. (Oliver and Boyd, Edinburgh, UK, 1966).
14. For a summary see M. A. Virasoro, in *Duality and Symmetry in Hadron Physics*, E. Gotsman, ed. (Weizmann Science Press, Jerusalem, 1971).
15. S. Ferrara, R. Gatto and A. F. Grillo, *Nuovo Cim.* **12A**, 959 (1972).
16. L. Bianchi, *Lezioni di Geometria Differenziale* (Spoetti, Pisa, 1902), p. 375.
17. Y. Nambu, Lectures at the Copenhagen Summer Symposium (1970); O. Hara, *Prog. Theor. Phys.* (Kyoto) **46**, 1549 (1971); T. Goto, *Prog. Theor. Phys.* (Kyoto) **46**, 1560 (1971); L. N. Chang and F. Mansouri, *Phys. Rev. D* **5**, 2535 (1972); F. Mansouri and Y. Nambu, *Phys. Lett.* **39B**, 375 (1972); P. Goddard, J. Goldstone, C. Rebbi and C. B. Thorn, *Nucl. Phys.* **B56**, 109 (1973).
18. The analogous anomaly for commutators in 4-dimensions was given by D. Boulware and S. Deser, *J. Math. Phys.* **8**, 1468 (1967).
19. K. Johnson, unpublished; S. L. Adler, *Phys. Rev. D* **6**, 3445 (1972).

Paper IV.3

Dirac invented procedures for quantizing constrained, singular dynamical systems, like gauge theories. Subsequent developments created a methodology of great complexity. However, a simplification is also possible, which provides an alternative to Dirac's approach, and was described at a workshop devoted to "Constraint Theory and Quantization Methods" in Montepulciano (June 1993).

(CONSTRAINED) QUANTIZATION
WITHOUT TEARS

(Constraint Theory and Quantization Methods, F. Colomo, L. Lusanna and G. Marmo, eds. World Scientific, Singapore, 1994)

To accomplish conventional and elementary quantization of a dynamical system, one is instructed to: begin with a Lagrangian, eliminate velocities in favor of momenta by a Legendre transform that determines the Hamiltonian, postulate canonical brackets among coordinates and momenta and finally, define dynamics by commutation with the Hamiltonian. But this procedure may fail for several reasons: it may not be possible to solve for the velocities in terms of the momenta, or it may be that the Hamiltonian equations do not reproduce the desired dynamical equations. In such cases, one is dealing with so-called "singular" Lagrangians and "constrained" dynamics. Almost half a century ago, Dirac developed his method for handling this situation[1] and since that time, the subject has defined an area of specialization in mathematical physics, as is put into evidence by a recent monograph[2] and by this series of workshops.[3]

While Dirac's approach and subsequent developments can cope with most models of interest, my colleague Ludwig Faddeev and I realized that in many instances, Dirac's method is unnecessarily cumbersome and can be stream-lined and simplified. We have advertised[4] an alternative approach, based on Darboux's theorem, wherein one arrives at the desired results — formulas for brackets and for the Hamiltonian — without following Dirac step by step.

Very specifically, two aspects of the Dirac procedure are avoided. First, when it happens that the Lagrangian L depends linearly on the velocity $\dot{\xi}^i$ for one of the dynamical variables ξ^i, or even is independent of $\dot{\xi}^i$, the attempt to define the canonical momentum $\Pi_i = \frac{\partial L}{\partial \dot{\xi}^i}$, and to eliminate $\dot{\xi}_i$ in favor of Π_i obviously fails. In the Dirac procedure, one nevertheless defines a canonical momentum and views the $\dot{\xi}^i$-independent expression $\frac{\partial L}{\partial \dot{\xi}^i}$ as a constraint on Π_i. In our method, such constraints are never introduced. Second, in the Dirac procedure, constraints are classified and distinguished as first class or second class, primary or secondary. This distinction is not made in our method; all constraints are held to the same standard.

It is therefore clear that our approach eliminates useless paperwork, and here I shall give a description with the hope that this audience of specialists will appreciate the economy of our proposal and will further adopt and disseminate it.

I shall use notation appropriate to a mechanical system, with coordinates labeled by $\{i, j, \dots\}$ taking values in a set of integers up to N, and a summation convention for repeated indices. Field theoretic generalization is obvious: the discrete quantities $\{i, j, \dots\}$ become continuous spatial variables. Time dependence of dynamical variables is not explicitly indicated since all quantities are defined at the same time, but time-differentiation is denoted by an over-dot. Although the language of quantum mechanics is used, with $\hbar$ scaled to unity, ("commutation," *etc.*) ordering issues are not addressed — so more properly speaking, we are describing a classical Hamiltonian reduction of dynamics. Grassmann variables are not considered, since that complication is a straight-forward generalization. Finally, total time derivative contributions to Lagrangians are omitted whenever convenient.

Our starting point is a first-order Lagrangian formulation for the dynamics of interest; *i.e.*, we assume that the Lagrangian is at most linear in time deriva-tives. This is to be contrasted with the usual approach, where the starting point is a second-order Lagrangian, quadratic in time-derivatives, and a first-order Lagrangian is viewed as "singular" or "constrained." In fact, just because dynamics is described by first-order differential equations, it does not mean that there are constraints, and this is a point we insist upon and we view the conventional position to be inappropriate.

Indeed there are many familiar and elementary dynamical systems that are first-order, without there being any constraints: Lagrangians for the Schrödinger equation and the Dirac equation are first-order in time derivatives; in light-cone quantization, where $x^+ \equiv \frac{1}{\sqrt{2}}(t + x)$ is the evolution coordinate, dynamics is first-order in this "time"; the most compact descrip-tion of chiral bosons in two space-time dimensions is first-order in time.[5] It is clear that characterizing any of these systems as "singular" or "constrained" reflects awkward mathematics rather than physical fact.

Moreover, a conventional second order Lagrangian can be converted to first-order form by precisely the same Legendre transform used to pass from a Lagrangian to a Hamiltonian. The point is that the formula

$$H = \frac{\partial L}{\partial \dot{q}} \dot{q} - L, \tag{1}$$

$$p \equiv \frac{\partial L}{\partial \dot{q}}, \tag{2}$$

may also be read in the opposite direction,

$$L(p, q) = p \dot{q} - H(p, q) \tag{3}$$

and it is straightforward to verify that Euler-Lagrange equations for the first-order Lagrangian $L(p, q)$ coincide with the Hamiltonian equations based on $H(p, q)$. Thus given a conventional Hamiltonian description of dynamics, we can always construct a first-order Lagrangian whose *configuration space* coincides with the Hamiltonian *phase space.*

We begin therefore with a general first-order Lagrangian

$$L = a_i(\xi)\dot{\xi}^i - V(\xi).\tag{4}$$

Note that a_i has the character of a vector potential (connection) for an Abelian gauge theory, in that modifying $a_i(\xi)$ by a total derivative $a_i \to a_i + \frac{\partial}{\partial \xi^i}\theta$ does not affect dynamics, since the Lagrangian changes by a total time-derivative. Observe further that when a Hamiltonian is defined by the usual Legendre transform, velocities are absent from the combination $\frac{\partial L}{\partial \dot{\xi}^i}\dot{\xi}^i - L$, since L is first-order in $\dot{\xi}^i$, and V may be identified with the Hamiltonian

$$H = \frac{\partial L}{\partial \dot{\xi}^i}\dot{\xi}^i - L = V.\tag{5}$$

Thus the Lagrangian in (4) may be presented as

$$L = a_i(\xi)\dot{\xi}^i - H(\xi)\tag{6}$$

and the first term on the right side defines the "canonical one-form" $a_i(\xi)\,d\xi^i \equiv a(\xi)$.

To introduce our method in its simplest realization, we begin with a special case, which in fact will be shown to be quite representative: instead of dealing with a general $a_i(\xi)$, we take it to be linear in ξ^i

$$a_i(\xi) = \frac{1}{2}\xi^j\omega_{ji}.\tag{7}$$

The constant matrix ω_{ij} is antisymmetric, since any symmetric part merely contributes an irrelevant total time-derivative to L and can be dropped. The Euler-Lagrange equation that follows from (6) and (7) is

$$\omega_{ij}\dot{\xi}^j = \frac{\partial}{\partial \xi^i}H(\xi).\tag{8}$$

The development now goes to two cases. The first case holds when the antisymmetric matrix ω_{ij} possesses an inverse, denoted by ω^{ij}, in which case ω_{ij} must be even-dimensional, *i.e.*, the range N of $\{i, j, \dots\}$ is $2n = N$. It follows from (7) that ξ^i satisfies the evolution equation

$$\dot{\xi}^i = \omega^{ij}\frac{\partial}{\partial \xi^j}H(\xi)\tag{9}$$

and *there are no constraints*. Constraints are present only in the second case, when ω_{ij} has no inverse, and as a consequence, possesses N' zero modes z_a^i, $a = 1, \ldots, N'$. The system is then constrained by N' equations in which no time-derivatives appear

$$z_a^i \frac{\partial}{\partial \xi^i} H(\xi) = 0. \tag{10}$$

On the space orthogonal to that spanned by the $\{z_a\}$, ω_{ij} possesses an even-dimensional $(= 2n)$ inverse, so in this case $N = 2n + N'$.

For the moment, we shall assume that ω_{ij} *does* possess an inverse and that there are no constraints. The second, constrained case will be dealt with later.

With the linear form for $a_i(\xi)$ and in the absence of constraints, all dynamical equations are contained in (9). Brackets are defined so as to reproduce (9) by commutation with the Hamiltonian

$$\dot{\xi}^i = \omega^{ij} \frac{\partial}{\partial \xi^j} H(\xi) = i[H(\xi), \xi^i]$$

$$= i[\xi^j, \xi^i] \frac{\partial}{\partial \xi^j} H(\xi).$$

This implies that we should take

$$[\xi^i, \xi^j] = i\omega^{ij} \tag{11a}$$

or for general functions of ξ,

$$[A(\xi), B(\xi)] = i \frac{\partial A(\xi)}{\partial \xi^i} \omega^{ij} \frac{\partial B(\xi)}{\partial \xi^j}. \tag{11b}$$

It is reassuring to verify that a conventional dynamical model, when presented in the form (3), is a special case of the present theory with ξ^i comprising the two-component quantity $\binom{p}{q}$ and ω_{ij} the antisymmetric 2×2 matrix ε_{ij}, $\varepsilon_{12} = 1$. Eq. (11b) then implies $[q, p] = i$.

Next, let us turn to the more general case with $a_i(\xi)$ an arbitrary function of ξ^i, not depending explicitly on time. The Euler-Lagrange equation for (6) is

$$f_{ij}(\xi) \dot{\xi}^j = \frac{\partial}{\partial \xi^i} H(\xi), \tag{12}$$

where

$$f_{ij}(\xi) = \frac{\partial}{\partial \xi^i} a_j(\xi) - \frac{\partial}{\partial \xi^j} a_i(\xi). \tag{13}$$

f_{ij} behaves as a gauge invariant (Abelian) field strength (curvature) constructed from the gauge variant potential (connection). It is called the "symplectic two-form," $\frac{1}{2}f_{ij}(\xi)d\xi^i d\xi^j = f(\xi)$; evidently it is exact: $f = da$, and therefore closed: $df = 0$. In the non-singular, unconstrained situation the anti-symmetric $N \times N$ matrix f_{ij} has the matrix inverse f^{ij}, hence $N = 2n$, and (12) implies

$$\dot{\xi}^i = f^{ij}\frac{\partial}{\partial \xi^j}H(\xi)\,. \tag{14}$$

This evolution equation follows upon commutation with H provided the basic bracket is taken as

$$[\xi^i, \xi^j] = i f^{ij}(\xi)\,. \tag{15}$$

The Bianchi identity satisfied by f_{ij} ensures that (15) obeys the Jacobi identity.

The result (15) and its special case (11b) can also be derived by an alternative, physically motivated argument. Consider a massive particle, in any number of dimensions, moving in an external electromagnetic field, described by the vector potential $a_i(\xi)$ and scalar potential $V(\xi)$. The Lagrangian and Hamiltonian are expressions familiar from the theory of the Lorentz force,

$$L = \frac{1}{2}m\dot{\xi}^i\dot{\xi}^i + a_i(\xi)\dot{\xi}^i - V(\xi)\,, \tag{16a}$$

$$H = \frac{1}{2m}(p_i - a_i(\xi))^2 + V(\xi)\,, \tag{16b}$$

with p_i conjugate to ξ^i. It is seen that (4), (5) and (6) correspond to the $m \to 0$ limit of (16a) and (16b). Owing to the $O(m^{-1})$ kinetic term in (16b), the limit of vanishing mass can only be taken if $p_i - a_i(\xi) = m\dot{\xi}^i$ is constrained to vanish. Adopting for the moment the Dirac procedure, we recognize that vanishing of $m\dot{\xi}^i$ is a second class constraint, since the constraints do not commute,

$$[m\dot{\xi}^i, m\dot{\xi}^j] = [p_i - a_i(\xi), p_j - a_j(\xi)]$$

$$= i f_{ij}(\xi) \neq 0 \tag{17}$$

and a computation of the Dirac bracket $[\xi^i, \xi^j]$ regains (15).

In this way, we see that what one would find by following Dirac is also gotten by our method, but we arrive at the goal much more quickly. Also, the above discussion gives a physical setting for Lagrangians of the form (6): when dealing with a charged particle in an external magnetic field, in the strong field limit the Lorentz force term — the canonical one-form — dominates the kinetic term, which therefore may be dropped in first approximation. One is then left with quantum mechanical motion where the spatial coordinates fail to commute by terms of order of the inverse of the magnetic field. More specifically, with constant magnetic field B along the z-axis, energy levels of motion confined to the x-y plane form the well-known Landau bands. For strong fields, only the lowest band is relevant, and further effects of the additional potential $V(x, y)$ are approximately described by the "Peierls Substitution".[6] This states that the low-lying energy eigenvalues are

$$E = \frac{B}{2m} + \varepsilon_n \,, \tag{18}$$

where $\frac{B}{2m}$ is the energy of the lowest Landau level in the absence of V, while the ε_n are eigenvalues of the operator $V(x, y)$ (properly ordered!) with

$$i[x, y] = \frac{1}{B} \,. \tag{19}$$

Clearly, the present considerations about quantizing first-order Lagrangians give a new derivation[7] of this ancient result from condensed matter physics.[6] [One may also verify (18) by forming mH from (16b) and computing ε_n perturbatively in m.[8]]

While the development starting with arbitrary $a_i(\xi)$ and unconstrained dynamics appears more general than that based on the linear, special case (7), the latter in fact includes the former. This is because by using Darboux's theorem, one can show that an arbitrary vector potential [one-form $a_i d\xi^i$] whose associated field strength [two-form $d(a_i d\xi^i) = \frac{1}{2} f_{ij} d\xi^i d\xi^j$] is non-singular, in the sense that the matrix f_{ij} possesses an inverse, can be mapped by a coordinate transformation onto (7) with ω_{ij} non-singular. Thus apart from a gauge term, one can always present $a_i(\xi)$ as

$$a_i(\xi) = \frac{1}{2} Q^k(\xi) \omega_{k\ell} \frac{\partial Q^\ell(\xi)}{\partial \xi^i} \tag{20a}$$

correspondingly $f_{ij}(\xi)$ as

$$f_{ij}(\xi) = \frac{\partial Q^k(\xi)}{\partial \xi^i} \omega_{k\ell} \frac{\partial Q^\ell(\xi)}{\partial \xi^j} \tag{20b}$$

and in terms of new coordinates Q^i the curvature is ω_{ij} — a constant and non-singular matrix. Moreover, by a straightforward modification of the Gram-Schmidt argument a basis can be constructed such that the antisymmetric

$N \times N$ matrix ω_{ij} takes the block-off-diagonal form

$$\omega_{ij} = \begin{pmatrix} 0 & I \\ -I & 0 \end{pmatrix}_{ij}, \tag{21}$$

where I is the n-dimensional unit matrix ($N = 2n$). [With these procedures, one can also handle the case when a_i is explicitly time-dependent — a transformation to constant ω_{ij} can still be made.] In the Appendix, we present Darboux's theorem adopted for the present application, and we explicitly construct the coordinate transformation $Q^i(\xi)$. The coordinates in which the curvature two-form becomes (21) are of course, the canonical coordinates and they can be renamed p_i, q^i, $i = 1, \ldots, n$.

We conclude the discussion of non-singular, first-order dynamics by recording the functional integral for the quantum theory. The action of (4) obviously is

$$I = \int a_i(\xi) d\xi^i - \int H(\xi) dt \tag{22}$$

and the path integral involves, as usual, the phase exponential of the action. The measure however is non-minimal; the correct prescription is

$$Z = \int \prod_i \mathcal{D}\xi^i \, \det^{1/2} f_{jk} \, \exp i I. \tag{23}$$

The $\det^{1/2} f_{ij}$ factor can be derived in a variety of ways: One may use Darboux's theorem to map the problem onto one with constant canonical curvature (21), where the measure is just the Liouville measure $\prod_{i=1}^{2n} \mathcal{D}\xi^i = \prod_{i=1}^{n} \mathcal{D}p_i \, \mathcal{D}q^i$, and the Jacobian of the transformation is seen from (20b) to be $\det^{1/2} f_{ij}$. Alternatively, one may refer to our derivation based on Dirac's second class constraints, Eqs. (16), (17) and recall that the functional integral in the presence of second class constraints involves the square root of the constraints' bracket.[9] By either argument, one arrives at (23), which also exhibits the essential nature of the requirement that f_{ij} be a non-singular matrix.

We now turn to the second, more complicated case, where there are constraints because f_{ij} is singular. It is evident from the Appendix that the Darboux construction may still be carried out for the non-singular projection of f_{ij}, which is devoid of the zero-modes (10). This results in the Lagrangian

$$L = \frac{1}{2} \xi^i \omega_{ij} \dot{\xi}^j - H(\xi, z). \tag{24}$$

Here ω_{ij} may still be taken in the canonical form (21), but now in the Hamiltonian, there are N' additional coordinates denoted by z_a, $a = 1, \ldots, N'$, arising from the N' zero modes of f_{ij} and leading to N' constraint equations

$$\frac{\partial}{\partial z_a} H(\xi, z) = 0 \,. \tag{25}$$

This is the form that (10) takes in the canonical coordinates achieved by Darboux's theorem. The constrained nature of the z_a variables is evident: they do not occur in the canonical one-form $\frac{1}{2}\xi^i \omega_{ij} \dot{\xi}^j \, dt$ and there is no time-development for them.

In the next step, we examine the constraint equations (25) and recognize that for the z_a occurring *non-linearly* in $H(\xi, z)$ one can solve (25) for the z_a. [More precisely, this needs $\det \frac{\partial^2 H(\xi, z)}{\partial z_a \partial z_b} \neq 0$.] On the other hand, when $H(\xi, z)$ contains a constrained z_a variable *linearly*, Eq. (25) does not permit an evaluation of the corresponding z_a, because (25) in that case is a relation among the ξ^i, with z_a absent from the equation. Therefore using (25), we evaluate as many z_a's as possible, in terms of ξ^i's and other z_a's, and leave for further analysis the linearly occurring z_a's. Note that this step does not affect the canonical one-form in the Lagrangian.

Upon evaluation and elimination of as many z_a's as possible, we are left with a Lagrangian in the form

$$L = \frac{1}{2}\xi^i \omega_{ij} \dot{\xi}^j - H(\xi) - \lambda_k \Phi^k(\xi) \,, \tag{26}$$

where the last term arises from the remaining, linearly occurring z_a's, now renamed λ_k, and the only true constraints in the model are the Φ^k, which enter multiplied by Lagrange multipliers λ_k. To incorporate the constraints, it is not necessary to classify them into first class or second class. Rather we solve them, by satisfying the equations

$$\Phi^k(\xi) = 0 \tag{27}$$

which evidently give relations among the ξ^i — evaluating some in terms of others. This procedure obviously eliminates the last term in (26) and it reduces the number of ξ^i's below the $2n$ that are present in (26); also it replaces the diagonal canonical one-form by the expression $\bar{a}_i(\xi) d\xi^i$, where i ranges over the reduced set, and $\bar{a}_i$ is a non-linear function obtained by inserting the solutions to (27) into (26).

The Darboux procedure must now be repeated: the new canonical one-form $\bar{a}_i(\xi) d\xi^i$, which could be singular, is brought again to diagonal form, possibly leading to constraint equations, which must be solved. Eventually, one hopes that the iterations terminate and one is left with a completely reduced, unconstrained and canonical system.

Of course, there may be technical obstacles to carrying out the above steps: solving the constraints may prove too difficult, constructing the Darboux transformation to canonical coordinates may not be possible. One can then revert to the Dirac method, with its first and second class constraints, and corresponding modifications of brackets, subsidiary conditions on states, and non-minimal measure factors in functional integrals.

I conclude my presentation by exhibiting our method in action for electromagnetism coupled to matter, which for simplicity I take to be Dirac fields ψ, since their Lagrangian is already first-order. Also I include a gauge non-invariant mass term for the photon, to illustrate various examples of constraints. The electromagnetic Lagrangian in first-order form reads

$$L = \int d\mathbf{r} \left\{ -\mathbf{E} \cdot \dot{\mathbf{A}}^i + i\psi^* \dot{\psi} - \frac{1}{2}(\mathbf{E}^2 + \mathbf{B}^2 + \mu^2 \mathbf{A}^2) \right\},$$

$$- H_M((\nabla - i\mathbf{A})\psi), \tag{28}$$

$$- \int d\mathbf{r} \left\{ A_0(\rho - \nabla \cdot \mathbf{E}) - \frac{\mu^2}{2} A_0^2 \right\}.$$

Here $\mathbf{A}$ is the vector potential with $\mathbf{B} \equiv \nabla \times \mathbf{A}$, A_0 the scalar potential that is absent from the symplectic term, μ the photon mass. The matter Hamiltonian is not specified beyond an indication that coupling to $\mathbf{A}$ is through the covariant derivative, while $\rho = \psi^* \psi$. The Lagrangian is in the form (24); when μ is non-zero the constrained variable A_0 enters quadratically and

$$\frac{\delta H}{\delta A_0(\mathbf{r})} = 0 \tag{29}$$

leads to the evaluation of A_0

$$A_0 = \frac{1}{\mu^2}(\rho - \nabla \cdot \mathbf{E}) \tag{30}$$

so that the unconstrained Lagrangian becomes

$$L = \int d\mathbf{r} \left\{ -\mathbf{E} \cdot \dot{\mathbf{A}} + i\psi^* \dot{\psi} - \frac{1}{2}\left(\mathbf{E}^2 + \mathbf{B}^2 + \mu^2 \mathbf{A}^2 + \frac{1}{\mu^2}(\rho - \nabla \cdot \mathbf{E})^2 \right) \right\}$$
$$- H_M((\nabla - i\mathbf{A})\psi). \tag{31}$$

The canonical pairs are identified as $(-\mathbf{E}, \mathbf{A})$ and $(i\psi^*, \psi)$. In the absence of a photon mass, the Lagrangian (28) is of the form (26), with one Lagrange multiplier $\lambda = A_0$. Eq. (29) then leads to the Gauss law constraint

$$\mathbf{\nabla} \cdot \mathbf{E} = \rho. \tag{32}$$

To solve the constraint, we decompose both $\mathbf{E}$ and $\mathbf{A}$ into transverse and longitudinal parts,

$$\mathbf{E} = \mathbf{E}_T + \frac{\mathbf{\nabla}}{\sqrt{-\nabla^2}} E, \tag{33}$$

$$\mathbf{A} = \mathbf{A}_T + \frac{\mathbf{\nabla}}{\sqrt{-\nabla^2}} A, \tag{34}$$

$$\mathbf{\nabla} \cdot \mathbf{E}_T = \mathbf{\nabla} \cdot \mathbf{A}_T = 0$$

and (32) implies $E = \frac{-1}{\sqrt{-\nabla^2}}\rho$. Inserting this into (28) at $\mu^2 = 0$, we are left with

$$L = \int d\mathbf{r} \left\{ -\mathbf{E}_T \cdot \dot{\mathbf{A}}_T + \rho \frac{1}{\sqrt{-\nabla^2}} \dot{A} + i\psi^* \dot{\psi} - \frac{1}{2} \left(\mathbf{E}_T^2 + \mathbf{B}^2 - \rho \frac{1}{\nabla^2} \rho \right) \right\}$$

$$- H_M \left(\left(\mathbf{\nabla} - i\mathbf{A}_T - i\frac{\mathbf{\nabla}}{\sqrt{-\nabla^2}} A \right) \psi \right). \tag{35a}$$

While the constraint has been eliminated, the canonical one-form in (35a) is not diagonal. The Darboux transformation that is now performed replaces ψ by $(\exp i\frac{1}{\sqrt{-\nabla^2}}A)\psi$. This has the effect of canceling $\rho \frac{1}{\sqrt{-\nabla^2}}\dot{A}$ against a contribution coming from $i\psi^*\dot{\psi}$ and eliminating A from the Hamiltonian (since $\mathbf{B} = \mathbf{\nabla} \times \mathbf{A}_T$). We are thus left with the Coulomb-gauge Lagrangian

$$L = \int d\mathbf{r} \left\{ -\mathbf{E}_T \cdot \dot{\mathbf{A}}_T + i\psi^* \dot{\psi} - \frac{1}{2} \left(\mathbf{E}_T^2 + \mathbf{B}^2 - \rho \frac{1}{\nabla^2} \rho \right) \right\} - H_M((\mathbf{\nabla} - i\mathbf{A}_T)\psi) \tag{35b}$$

without ever selecting the Coulomb gauge! The canonical pairs are $(-\mathbf{E}_T, \mathbf{A}_T)$ and $(i\psi^*, \psi)$.

We recall that the Dirac approach would introduce a canonical momentum Π_0 conjugate to A_0 and constrained to vanish. The constraints (30) or (32) would then emerge as secondary constraints, which must hold so that $[H, \Pi_0]$ vanish. Finally, a distinction would be made between the $\mu \neq 0$ and $\mu = 0$ theories: in the former the constraint is second class, in the latter it is first class.[9] None of these considerations are necessary for successful quantization. Our method also quantizes very efficiently Chern-Simons theories, with or without a conventional kinetic term for the gauge field[10] [indeed the phase space reductive limit of taking the kinetic term to zero, as in (16), (17) above, can be clearly described[11]] as well as gravity theories in first-order form, be they the Einstein model[12] or the recently discussed gravitational gauge theories in lower dimensions.[13]

Finally, we record a first-order Lagrangian for Maxwell theory with external, conserved sources $(\rho, \mathbf{j})$, $\dot{\rho} + \nabla \cdot \mathbf{j} = 0$, which depends only on field strengths $(\mathbf{E}, \mathbf{B})$ (rather than potentials) and is self-dual in the absence of sources[14]

$$L = \int d\mathbf{r} d\mathbf{r}' \left(\dot{E}^i(\mathbf{r}) + j^i(\mathbf{r}) \right) \omega_{ij}(\mathbf{r} - \mathbf{r}') B^j(\mathbf{r}')$$

$$- \frac{1}{2} \int d\mathbf{r} (\mathbf{E}^2 + \mathbf{B}^2) - \int d\mathbf{r} \left(\lambda_1 (\rho - \nabla \cdot \mathbf{E}) + \lambda_2 \nabla \cdot \mathbf{B} \right), \quad (36)$$

$$\omega_{ij}(\mathbf{r}) \equiv \varepsilon^{ijk} \frac{\partial_k}{\nabla^2} \delta(\mathbf{r}) = \frac{1}{4\pi} \varepsilon^{ijk} \frac{r^k}{r^3}. \quad (37)$$

Varying the $\mathbf{E}$ and $\mathbf{B}$ fields as well as the two Lagrange multipliers $\lambda_{1,2}$ gives the eight Maxwell equations. The duality transformation $\mathbf{E} \to \mathbf{B}$, $\mathbf{B} \to -\mathbf{E}$, supplemented by $\lambda_1 \to -\lambda_2$, $\lambda_2 \to \lambda_1$ changes the Lagrangian by a total time derivative, when there are no sources. The canonical one-form is spatially non-local, owing to the presence of ω_{ij}, which has the inverse

$$\omega^{ij}(\mathbf{r}) = -\varepsilon^{ijk} \partial_k \delta(\mathbf{r}) \quad (38)$$

when restricted to transverse fields — these are the only unconstrained degrees of freedom in (36). It then follows that the non-vanishing commutator is the familiar formula

$$[E_T^i(\mathbf{r}), B_T^j(\mathbf{r}')] = -i\varepsilon^{ijk} \partial_k \delta(\mathbf{r} - \mathbf{r}'). \quad (39)$$

This self-dual presentation of electrodynamics is similar to formulations of self-dual fields on a line[5] and on a plane.[10]

APPENDIX: DARBOUX'S THEOREM

We give a constructive derivation of Darboux's Theorem. Specifically, we show that subject to regularity requirements stated below, any vector potential (connection one-form) $a_i(\xi)$ may be presented, apart from a gauge transformation, as

$$a_i(\xi) = \frac{1}{2} Q^m(\xi) \omega_{mn} \frac{\partial Q^n(\xi)}{\partial \xi^i} \quad (A.1)$$

and correspondingly the field strength $f_{ij}(\xi)$ (curvature two-form) as

$$f_{ij}(\xi) = \frac{\partial Q^m(\xi)}{\partial \xi^i} \omega_{mn} \frac{\partial Q^n(\xi)}{\partial \xi^j} \tag{A.2}$$

with ω_{mn} constant and antisymmetric. The proof also gives a procedure for finding $Q^m(\xi)$. It is then evident that a coordinate transformation from ξ to Q renders f_{ij} constant and a further adjustment of the basis puts ω_{ij} in the canonical form (21).

We consider a continuously evolving transformation $Q^m(\xi; \tau)$, to be specified later, with the property that at $\tau = 0$, it is the identity transformation

$$Q^m(\xi; 0) = \xi^m \tag{A.3a}$$

and at $\tau = 1$, it arrives at the desired $Q^m(\xi)$, (which will be explicitly constructed)

$$Q^m(\xi; 1) = Q^m(\xi). \tag{A.3b}$$

$Q^m(\xi; \tau)$ is generated by $v^m(\xi; \tau)$, in the sense that

$$\frac{\partial}{\partial \tau} Q^m(\xi; \tau) = v^m(Q(\xi; \tau); \tau). \tag{A.4}$$

Note, that v^m depends explicitly on τ. Also we need to define the transform by $Q^m(\xi; \tau)$ of quantities relevant to the argument: connection one-form, curvature two-form *etc.* The definition is standard: the transform, denoted by T_Q, acts by

$$T_Q a_i(\xi) = a_m(Q) \frac{\partial Q^m}{\partial \xi^i}, \tag{A.5a}$$

$$T_Q f_{ij}(\xi) = f_{mn}(Q) \frac{\partial Q^m}{\partial \xi^i} \frac{\partial Q^n}{\partial \xi^j}. \tag{A.5b}$$

To give the construction, we consider the given $a_i(\xi)$ to be embedded in a one-parameter family $a_i(\xi; \tau)$, such that at $\tau = 0$ we have $a_i(\xi)$ and at $\tau = 1$ we have $\frac{1}{2} \xi^m \omega_{mi}$, where ω_{mi} is constant and antisymmetric

$$a_i(\xi; 0) = a_i(\xi), \tag{A.6a}$$

$$a_i(\xi; 1) = \frac{1}{2} \xi^m \omega_{mi}. \tag{A.6b}$$

It is then true that

$$\frac{d}{d\tau} [T_Q a_i(\xi; \tau)] = T_Q \left[L_v a_i(\xi; \tau) + \frac{\partial}{\partial \tau} a_i(\xi; \tau) \right], \tag{A.7}$$

where L_v is the Lie derivative, with respect to the vector v^m that generates the transformation, see (A.4). Eq. (A.7) is straightforwardly verified by differentiating with respect to τ, and recalling that both the transformation and a_i are τ-dependent. Next we use the identity[15]

$$L_v a_i = v^n f_{ni} + \partial_i(v^n a_n) \tag{A.8}$$

and observe that when the generator is set equal to

$$v^n(\xi;\tau) = -f^{ni}(\xi;\tau)\frac{\partial}{\partial\tau}a_i(\xi;\tau). \tag{A.9}$$

Eq. (A.7) leaves

$$\frac{d}{d\tau}(T_Q a_i) = T_Q(\partial_i(v^n a_n)). \tag{A.10}$$

Thus $\frac{d}{d\tau}(T_Q a_i)$ is a gauge transformation, so that $T_Q a_i$ at $\tau = 0$, i.e., $a_i(\xi)$, differs from its value at $\tau = 1$, i.e., $\frac{1}{2}Q^m(\xi)\omega_{mn}\frac{\partial Q^n(\xi)}{\partial\xi'}$, by a gauge transformation. This is the desired result, and moreover $Q^m(\xi;\tau)$ and $Q^m(\xi) \equiv Q^m(\xi;1)$ are here explicitly constructed from the algebraic definition (A.9) for v^n [once an interpolating $a_i(\xi;\tau)$ is chosen], and integration of (A.4) (the latter task need not be easy).

Clearly (A.9) requires that $f_{ij}(\xi;\tau)$ possesses the inverse $f^{ij}(\xi;\tau)$; hence both the starting and ending forms, $f_{ij}(\xi)$ and ω_{ij}, must be non-singular. Also $f_{ij}(\xi;\tau)$ must remain non-singular for all intermediate τ. In fact this is not a restrictive requirement, because one may always choose ω_{ij} to be the value of $f_{ij}(\xi)$ at some point $\xi = \xi_0$, and then by change of basis transform ω_{ij} to any desired form.

This description of Darboux's theorem was prepared with the assistance of B. Zwiebach, whom I thank.

REFERENCES

1. P. A. M. Dirac, *Canad. J. Math.* **2**, 129 (1950); *Phys. Rev.* **114**, 924 (1959); *Lectures on Quantum Mechanics* (Yeshiva University, New York, NY, 1964).
2. M. Henneaux and C. Teitelboim, *Quantization of Gauge Systems* (Princeton University Press, Princeton, NJ, 1992).
3. *Constraint's Theory and Relativistic Dynamics*, G. Longhi and L. Lusanna, eds. (World Scientific, Singapore, 1987).
4. L. Faddeev and R. Jackiw, *Phys. Rev. Lett.* **60**, 1692 (1988).
5. R. Floreanini and R. Jackiw, *Phys. Rev. Lett.* **59**, 1873 (1987).
6. R. Peierls, *Z. Phys.* **80**, 763 (1933).

7. G. Dunne and R. Jackiw, in *Common Trends in Condensed Matter and High Energy Physics*, L. Alvarez-Gaumé, A. Devoto, S. Fubini and C. Trugenberger, eds. (*Nucl. Phys.* **B**, Proc. Suppl.) 33C (1993).

8. M. Berry, private communication.

9. P. Senjanovic, *Ann. Phys.* (NY) **100**, 227 (1976).

10. S. Deser and R. Jackiw, *Phys. Lett.* **139B**, 371 (1984); G. Dunne, R. Jackiw and C. Trugenberger, *Ann. Phys.* (NY) **194**, 197 (1989).

11. G. Dunne, R. Jackiw and C. Trugenberger, *Phys. Rev. D* **41**, 661 (1990).

12. R. Arnowitt, S. Deser and C. Misner, *Phys. Rev.* **116**, 1322 (1959); L. Faddeev, *Usp. Fiz. Nank.* **136**, 435 (1982) [English translation: *Sov. Phys. Usp.* **25**, 130 (1982)].

13. see "Gauge Theories for Gravity on a Line," reprinted in this volume on p. 197.

14. For other self-dual formulation of electromagnetism, see S. Deser, *J. Phys.* **A15**, 1053 (1982); J. Schwarz and A. Sen, *Nucl. Phys.* **B411**, 35 (1994).

15. R. Jackiw, *Phys. Rev. Lett.* **41**, 1635 (1978).

The Schrödinger representation provides the most elementary realization for quantum mechanics. Its use in atomic, molecular and other many-body physics has led to the development of many approximation techniques and much physical intuition. Quantum field theory on the other hand is usually presented in the Lorentz covariant Heisenberg picture or by functional integrals. Since much of my research in quantum field theory was informed by my student experience with quantum mechanics, I thought it useful to analyze the complexities of the former with the "physically" accessible methods of the latter. The work was described at schools in Montréal (June 1988) and Campo de Jordao (January 1989).

ANALYSIS ON INFINITE-DIMENSIONAL MANIFOLDS — SCHRÖDINGER REPRESENTATION FOR QUANTIZED FIELDS

Field Theory and Particle Physics, O. Éboli, M. Gomes and A. Santoro, eds. *(World Scientific, Singapore, 1990)*

I. INTRODUCTION, OVERVIEW, BOSONIC THEORIES

My lectures concern an unfamiliar approach to quantum field theory that closely parallels a similar technique in quantum mechanics and is called the *field theoretic Schrödinger representation*. The essence of the method is the following: Dynamical quantities are expressed in terms of fixed-time canonical variables. $\Phi(\mathbf{x})$ and $\Pi(\mathbf{x})$ for a bosonic field theory, which upon quantization satisfy canonical commutation relations. The quantum field theory involves states $|\Psi\rangle$ that are realized in the Schrödinger representation as functionals, $|\Psi\rangle \to \Psi(\varphi)$, of a time-independent, c-number field $\varphi(\mathbf{x})$. The operator $\Phi(\mathbf{x})$ acts on these states by multiplication, $\Phi(\mathbf{x})|\Psi\rangle \to \varphi(\mathbf{x})\Psi(\varphi)$, while the canonical momentum operator $\Pi(\mathbf{x})$ acts by functional differentiation, $\Pi(\mathbf{x})|\Psi\rangle \to \frac{1}{i}\frac{\delta}{\delta\varphi(\mathbf{x})}\Psi(\varphi)$.

Clearly, we are taking over into field theory the most elementary and naive approach to quantum mechanics. The motive for doing this is the desire to use the physical/mathematical intuition derived from ordinary [many-body] quantum mechanics to advance understanding and to develop exact or approximate results about quantum field theory. After all, one can view a field $\Phi(\mathbf{x})$ as a collection of mechanical variables Q_i $(i = 1, \ldots, N)$ for N degrees of freedom, in the limit that N becomes continuously infinite: $Q \to \Phi$, $i \to \mathbf{x}$.

However, as we shall see, things are not so simple, because of quantum field theoretic divergences. Indeed, the field theoretic Schrödinger representation is not new, but it has not been as widely used as the Green's function method for Fock space matrix elements of Heisenberg picture operators. The reason is that isolating and renormalizing perturbative divergences is effected more conveniently in the Lorentz covariant Green's function formalism than in the non-covariant Schrödinger representation. At a time when infinities posed an ill-defined obstacle to extracting unambiguous results from quantum field theory, Lorentz covariance was an indispensable guide for finding one's way. However, by now we have become accustomed, for better or worse, to dealing with field theoretic infinities, and their occurrence is not so threatening. Indeed, recent results establish renormalizability of the Schrödinger representation, for static[1] as well as time-dependent[2] problems.

Although the formalism that I wish to acquaint you with is truly elementary and should be immediately comprehensible to anyone familiar with quantum mechanics, it has been my experience that many in an audience schooled to think in terms of Fock spaces, Green's functions *etc.* are frequently puzzled by what I am trying to explain. I shall therefore begin by briefly reviewing a familiar quantum mechanical story, but in an order inverse to text-book presentations.

Consider the harmonic oscillator with a quadratic Hamiltonian,

$$H = \frac{1}{2}(P^2 + \omega^2 Q^2).$$

(1.1)

One way to unravel the dynamics is to introduce annihilation and creation operators

$$a = \frac{1}{\sqrt{2}}(\omega^{1/2}Q + i\omega^{-1/2}P),$$

(1.2a)

$$a^\dagger = \frac{1}{\sqrt{2}}(\omega^{1/2}Q - i\omega^{-1/2}P)$$

(1.2b)

from which the dynamical variables may be reconstructed

$$Q = \frac{1}{\sqrt{2\omega}}(a + a^\dagger),$$

(1.3a)

$$P = \frac{1}{i}\sqrt{\frac{\omega}{2}}(a - a^\dagger).$$

(1.3b)

As a consequence of the canonical commutator obeyed by P and Q, these annihilation and creation operators satisfy the important commutation relation

$$[a, a^\dagger] = 1$$

(1.4)

and the Hamiltonian is expressed in terms of them

$$H = \omega a^\dagger a + \frac{1}{2}\omega.$$

(1.5)

[All operators are evaluated at a common, fixed time and the commutator (1.4) is an "equal-time" commutator. The time argument is suppressed.] Next, it is asserted that there exists a normalized vacuum state $|0\rangle$, annihilated by a, which is also the ground state of the Hamiltonian

$$a|0\rangle = 0,$$

(1.6)

$$H|0\rangle = E_0|0\rangle, \quad E_0 = \frac{1}{2}\omega$$

(1.7)

while the excited states are obtained by repeated application of $a^\dagger$ to $|0\rangle$

$$|n\rangle = \frac{1}{\sqrt{n!}}(a^\dagger)^n|0,\rangle$$

(1.8)

$$H|n\rangle = E_n|n\rangle, \quad E_n = n\omega + \frac{1}{2}\omega.$$

(1.9)

These eigenstates are a basis for arbitrary states $|\psi\rangle$ in the model

$$|\psi\rangle = \sum_n c_n |n\rangle . \tag{1.10}$$

This elementary and familiar story is repeated in the conventional approach to field theory. For a free bosonic, spinless field with mass m, we take a quadratic Hamiltonian

$$H = \frac{1}{2} \int_{\mathbf{x}} \left(\Pi^2(\mathbf{x}) + \Phi(\mathbf{x})(-\nabla^2 + m^2)\Phi(\mathbf{x}) \right) . \tag{1.11a}$$

Here and below, $\int_{\mathbf{x}}$ is the integral over space which may be d-dimensional, $\int_{\mathbf{x}} \equiv \int d^d\mathbf{x}$; $\int_p$ is the corresponding Fourier space integral, $\int_p \equiv \int \frac{d^d p}{(2\pi)^d}$. A self-evident matrix/kernel notation will be used: for example (1.11a) is also written as

$$H = \frac{1}{2} \int (\Pi^2 + \Phi\omega^2\Phi) , \tag{1.11b}$$

where ω^2 is the kernel [operator]

$$\omega^2(\mathbf{x}, \mathbf{x}') = (-\nabla^2 + m^2)\delta(\mathbf{x} - \mathbf{x}') \tag{1.12a}$$

and a double integration is understood in the last term of (1.11b). The identity kernel I in position space is $\delta(\mathbf{x} - \mathbf{x}')$ and in momentum space $(2\pi)^d\delta(\mathbf{p} - \mathbf{p}')$. The kernel ω^2 is not diagonal in position space, but because of translation invariance it is diagonal in momentum space

$$\omega^2(\mathbf{p}, \mathbf{p}') \equiv \int_{\mathbf{x}, \mathbf{x}'} e^{i\mathbf{p}\cdot\mathbf{x}} \omega^2(\mathbf{x}, \mathbf{x}') e^{-i\mathbf{p}\cdot\mathbf{x}'}$$
$$= (p^2 + m^2)(2\pi)^d\delta(\mathbf{p} - \mathbf{p}') . \tag{1.12b}$$

Hence

$$\omega(\mathbf{p}, \mathbf{p}') = \omega(p)(2\pi)^d\delta(\mathbf{p} - \mathbf{p}')$$
$$\omega(p) \equiv \sqrt{p^2 + m^2} . \tag{1.12c}$$

As in the quantum mechanical example, the canonical variables, which now are the fields $\Phi(\mathbf{x})$, and $\Pi(\mathbf{x})$, are replaced by annihilation and creation operators, compare with (1.2) – (1.4)

$$a(\mathbf{p}) = \frac{1}{\sqrt{2}} \int_{\mathbf{x}} e^{-i\mathbf{p}\cdot\mathbf{x}} \left[\omega^{1/2}(p)\Phi(\mathbf{x}) + i\,\omega^{-1/2}(p)\Pi(\mathbf{x}) \right] , \tag{1.13a}$$

$$a^\dagger(\mathbf{p}) = \frac{1}{\sqrt{2}} \int_{\mathbf{x}} e^{i\mathbf{p}\cdot\mathbf{x}} \left[\omega^{1/2}(p)\Phi(\mathbf{x}) - i\,\omega^{-1/2}(p)\Pi(\mathbf{x}) \right] \tag{1.13b}$$

from which the fields can be reconstructed

$$\Phi(\mathbf{x}) = \int_{\mathbf{p}} \frac{1}{\sqrt{2\omega(p)}} \left[a(\mathbf{p})e^{i\,\mathbf{p}\cdot\mathbf{x}} + a^\dagger(\mathbf{p})e^{-i\,\mathbf{p}\cdot\mathbf{x}} \right] , \qquad (1.14a)$$

$$\Pi(\mathbf{x}) = \int_{\mathbf{p}} \frac{1}{i} \sqrt{\frac{\omega(p)}{2}} \left[a(\mathbf{p})e^{i\,\mathbf{p}\cdot\mathbf{x}} - a^\dagger(\mathbf{p})e^{-i\,\mathbf{p}\cdot\mathbf{x}} \right] . \qquad (1.14b)$$

The field theoretical version of (1.4) is

$$[a(\mathbf{p}), a^\dagger(\mathbf{p}')] = (2\pi)^d \delta(\mathbf{p} - \mathbf{p}') \qquad (1.15)$$

and the Hamiltonian in these variables reads

$$H = \frac{1}{2} \int_{\mathbf{p},\mathbf{p}'} \left[a^\dagger(\mathbf{p})\omega(\mathbf{p},\mathbf{p}')a(\mathbf{p}') + a(\mathbf{p})\omega(\mathbf{p},\mathbf{p}')a^\dagger(\mathbf{p}') \right]$$

$$= \int_{\mathbf{p}} \omega(p)a^\dagger(\mathbf{p})a(\mathbf{p}) + \frac{1}{2} \int_{\mathbf{p}} \omega(\mathbf{p},\mathbf{p}) . \qquad (1.16)$$

The last term can be written as $\frac{1}{2}\,\mathrm{tr}\,\omega$, where the trace of the kernel ω is taken in the functional sense. Because of translation invariance, which insures the presence of a delta function in $\omega(\mathbf{p},\mathbf{p}') = \omega(p)\int_{\mathbf{x}} e^{i\mathbf{x}\cdot(\mathbf{p}-\mathbf{p}')}$, the diagonal part possesses an infrared divergence proportional to the volume V of space: $\omega(\mathbf{p},\mathbf{p}) = \omega(p)\int_{\mathbf{x}} = \omega(p)V$. Furthermore, the trace is also ultraviolet divergent, owing to the growth of $\omega(p)$ with p: $\mathrm{tr}\,\omega = V\int_{\mathbf{p}} \omega(p)$.

This first and simplest quantum field theoretic infinity is usually eliminated by normal ordering: a procedure that ties the formalism to the specific Hamiltonian (1.11a), (1.16). On the contrary, for our purposes it is better to keep this contribution. We can justify manipulating with an undefined quantity by observing that nothing in the development thus far depends on the specific form of ω in (1.11b), and we could take a regulated expression such that $\frac{1}{2}\,\mathrm{tr}\,\omega$ exists. In fact, we shall ignore infrared divergences [for example, by pretending that space has finite volume], thus we may even take translation invariant formulas for ω. Of course, Poincaré invariance of the theory is lost if ω differs from (1.12), but should be regained in final results, when regulators are removed.

Continuing now with the usual field theoretic development, it is asserted, as in the harmonic oscillator, that there exists a normalized vacuum state, annihilated by $a(\mathbf{p})$, which is also the ground state of the Hamiltonian (1.11), (1.16)

$$a(\mathbf{p})|0\rangle = 0 , \qquad (1.17)$$

$$H|0\rangle = E_0|0\rangle, \qquad E_0 = \frac{1}{2}\operatorname{tr}\omega \qquad (1.18)$$

while the excited states are obtained by repeated application of $a^\dagger(\mathbf{p})$ to $|0\rangle$

$$|\mathbf{p}_1, \mathbf{p}_2, \ldots, \mathbf{p}_n\rangle = \frac{1}{\sqrt{n!}} a^\dagger(\mathbf{p}_1) a^\dagger(\mathbf{p}_2) \ldots a^\dagger(\mathbf{p}_n)|0\rangle,$$

$$H|\mathbf{p}_1, \mathbf{p}_2, \ldots, \mathbf{p}_n\rangle = E_n(\mathbf{p}_1, \mathbf{p}_2, \ldots, \mathbf{p}_n)|\mathbf{p}_1, \mathbf{p}_2, \ldots, \mathbf{p}_n\rangle, \qquad (1.19)$$

$$E_n(\mathbf{p}_1, \mathbf{p}_2, \ldots, \mathbf{p}_n) = \sum_i^n \omega(p_i) + \frac{1}{2}\operatorname{tr}\omega. \qquad (1.20)$$

These eigenstates, the so-called *Fock states*, provide a basis for all other states in the theory. They comprise the *Fock space* of the Hamiltonian (1.16) with an inner product that derives from the fact that the Fock basis states are orthogonal. It is evident that the field theoretic quantities consist of an infinite superposition of the corresponding harmonic oscillator formulas.

But there is another, more elementary approach to quantum mechanics, in which a Hamiltonian is not at the center of the development. Rather one begins by representing arbitrary states by functions of [half] the canonical variables, say q and an inner product is defined on this space of functions

$$|\psi\rangle \longrightarrow \psi(q), \qquad (1.21)$$

$$\langle\psi_1|\psi_2\rangle \longrightarrow \int dq \psi_1^*(q)\psi_2(q). \qquad (1.22)$$

Operators $\mathcal{O}$ of the theory are represented by kernels

$$\mathcal{O}|\psi\rangle \longrightarrow \int d\tilde{q}\,\mathcal{O}(q, \tilde{q})\psi(\tilde{q}). \qquad (1.23)$$

For the position operator Q, a diagonal kernel is used

$$Q \longrightarrow q\delta(q - \tilde{q}) \qquad (1.24a)$$

while the canonical commutation relation is satisfied by representing the momentum operator as

$$P \longrightarrow \frac{1}{i}\frac{\partial}{\partial q}\delta(q - \tilde{q}). \qquad (1.24b)$$

Thus, Q acts by multiplication on functions of q, and P by differentiation. In this way, the action of any operator constructed from P and Q is determined

$$\mathcal{O}(P, Q)|\psi\rangle \longrightarrow \mathcal{O}\left(\frac{1}{i}\frac{\partial}{\partial q}, q\right)\psi(q). \qquad (1.25)$$

In this *Schrödinger representation*, the eigenstates of a Hamiltonian are found by solving differential equations *eg.*, for the harmonic oscillator

$$\left(-\frac{1}{2}\frac{\partial^2}{\partial q^2} + \omega^2 q^2\right)\psi_n(q) = E_n\psi_n(q).$$ (1.26)

The ground state is represented by a Gaussian

$$|0\rangle \longrightarrow \psi_0(q) = \left(\frac{\omega}{\pi}\right)^{1/4} e^{-\frac{1}{2}q\omega q},$$ (1.27a)

$$E_0 = \frac{1}{2}\omega.$$ (1.27b)

The annihilation and creation operators are also expressed in this formalism, with the former annihilating $\psi_0(q)$

$$a|0\rangle \longrightarrow \frac{1}{\sqrt{2}}\left(\omega^{1/2}q + \omega^{-1/2}\frac{\partial}{\partial q}\right)\psi_0(q) = 0$$ (1.28)

and the latter creating the excited states, which are represented by polynomials in q — the Hermite polynomials — multiplying $\psi_0(q)$. They may be found by repeatedly solving the differential equation (1.26), or by representing the formal expressions (1.8)

$$|1\rangle = a^\dagger|0\rangle \longrightarrow \psi_1(q) = \frac{1}{\sqrt{2}}\left(\omega^{1/2}q - \omega^{-1/2}\frac{\partial}{\partial q}\right)\psi_0(q),$$

$$= \sqrt{2}\omega^{1/2}q\psi_0(q),$$ (1.29a)

$$H\psi_1(q) = E_1\psi_1(q), \quad E_1 = \omega + \frac{1}{2}\omega,$$

$$|2\rangle = \frac{1}{\sqrt{2}}a^\dagger|1\rangle \rightarrow \psi_2(q),$$

$$= \frac{1}{2}\left(\omega^{1/2}q - \omega^{-1/2}\frac{\partial}{\partial q}\right)\psi_1(q) = \frac{1}{\sqrt{2}}(2q\omega q - 1)\psi_0(q),$$ (1.29b)

$$H\psi_2(q) = E_2\psi_2(q), \quad E_2 = 2\omega + \frac{1}{2}\omega,$$

etc.

The set of oscillator eigenfunctions $\{\psi_n(q)\}$ is complete; it can serve as a basis for our function space of states $\{\psi\}$. Or one can take other bases, *eg.*, a set oscillator eigenfunctions with a different frequency — one is expandable in terms of the other.

It is now clear that the above can be taken over *almost* in its entirety to field theory. As mentioned already, for a bosonic theory, states are represented by functionals of a time-independent, c-number field $\varphi(\mathbf{x})$

$$|\Psi\rangle \longrightarrow \Psi(\varphi).$$

(1.30)

An inner product is defined by functional integration

$$\langle\Psi_1|\Psi_2\rangle \longrightarrow \int \mathcal{D}\varphi \Psi_1^*(\varphi)\Psi_2(\varphi).$$

(1.31)

Operators of the theory are realized by functional kernels

$$\mathcal{O}|\Psi\rangle \longrightarrow \int \mathcal{D}\tilde\varphi \mathcal{O}(\varphi,\tilde\varphi)\Psi(\tilde\varphi).$$

(1.32)

Specifically, the kernels for the canonical variables

$$\Phi(\mathbf{x}) \longrightarrow \varphi(\mathbf{x})\delta(\varphi-\tilde\varphi),$$

(1.33a)

$$\Pi(\mathbf{x}) \longrightarrow \frac{1}{i}\frac{\delta}{\delta\varphi(x)}\delta(\varphi-\tilde\varphi)$$

(1.33b)

make use of a functional delta function and functional differentiation. The action of any operator constructed from these variables is then determined

$$\mathcal{O}(\Pi,\Phi)|\Psi\rangle \longrightarrow \mathcal{O}\Big(\frac{1}{i}\frac{\partial}{\partial\varphi},\varphi\Big)\Psi(\varphi).$$

(1.34)

Dynamical calculations require positing a Hamiltonian $H(\Pi,\Phi)$, and the dynamical equation is the functional Schrödinger equation for time-dependent functionals

$$i\frac{\partial}{\partial t}\Psi(\varphi;t) = H\Big(\frac{1}{i}\frac{\delta}{\delta\varphi},\varphi\Big)\Psi(\varphi;t).$$

(1.35)

Of course, for time-independent Hamiltonians a separation of variables is possible, and one is led to a functional eigenvalue problem

$$\Psi(\varphi;t) = e^{-iEt}\Psi_E(\varphi)$$

$$H\Big(\frac{1}{i}\frac{\delta}{\delta\varphi},\varphi\Big)\Psi_E(\varphi) = E\Psi_E(\varphi)$$

(1.36)

which can be solved for a quadratic Hamiltonian

$$H = \frac{1}{2}\int (\Pi^2 + \Phi h\Phi).$$

(1.37)

Here h is a symmetric real kernel, not necessarily of the free-field form in (1.11) – (1.12): we call it the *first quantized Hamiltonian*. The Fock ground state $\Psi_0(\varphi)$ that solves

$$\frac{1}{2}\int\Big(-\frac{\delta^2}{\delta\varphi\delta\varphi} + \varphi h\varphi\Big)\Psi_0(\varphi) = E_0\Psi_0(\varphi)$$

(1.38)

is a Gaussian functional,

$$\Psi_0(\varphi) = \det{}^{1/4}\left(\frac{\omega}{\pi}\right)e^{-\frac{1}{2}\int \varphi\omega\varphi} \tag{1.39a}$$

with covariance ω determined by

$$\omega^2 = h \tag{1.39b}$$

and eigenvalue

$$E_0 = \frac{1}{2}\operatorname{tr}\omega. \tag{1.39c}$$

Ψ_0 is annihilated by the [position space] annihilation operator

$$a = \frac{1}{\sqrt{2}}\int\left(\omega^{1/2}\varphi + \omega^{-1/2}\frac{\delta}{\delta\varphi}\right) \tag{1.40a}$$

which together with its conjugate creation operator

$$a^\dagger = \frac{1}{\sqrt{2}}\int\left(\omega^{1/2}\varphi - \omega^{-1/2}\frac{\delta}{\delta\varphi}\right) \tag{1.40b}$$

satisfy the expected commutation relation

$$[a(\mathbf{x}), a^\dagger(\mathbf{y})] = \delta(\mathbf{x} - \mathbf{y}). \tag{1.41}$$

Higher excited states are obtained by first determining the eigenfunctions and eigenvalues of the real and symmetric kernel ω

$$\int_\mathbf{y}\omega(\mathbf{x}, \mathbf{y})f_\varepsilon(\mathbf{y}) = \varepsilon f_\varepsilon(\mathbf{x}). \tag{1.42}$$

The one-particle state is then constructed by the action of $\int_\mathbf{x} f_\varepsilon(\mathbf{x})a^\dagger(\mathbf{x})$ on Ψ_0; the two-particle state by the action of $\frac{1}{\sqrt{2}}\int_{\mathbf{x},\mathbf{y}} f_{\varepsilon'}(\mathbf{x})f_\varepsilon(\mathbf{y})a^\dagger(\mathbf{x})a^\dagger(\mathbf{y})$ etc. In this way we find

$$\Psi_1(\varphi) = \frac{1}{\sqrt{2}}\int_{\mathbf{x},\mathbf{y}} f_\varepsilon(\mathbf{x})\left(\omega^{1/2}(\mathbf{x},\mathbf{y})\varphi(\mathbf{y}) - \omega^{-1/2}(\mathbf{x},\mathbf{y})\frac{\delta}{\delta\varphi(\mathbf{y})}\right)\Psi_0(\varphi),$$

$$= \sqrt{2\varepsilon}\int f_\varepsilon\varphi\Psi_0(\varphi), \tag{1.43}$$

$$E_1 = \varepsilon + \frac{1}{2}\operatorname{tr}\omega,$$

$$\Psi_2(\varphi) = \frac{1}{2} \int_{\mathbf{x},\mathbf{y}} f_{\varepsilon'}(\mathbf{x}) \Big(\omega^{1/2}(\mathbf{x},\mathbf{y})\varphi(\mathbf{y}) - \omega^{-1/2}(\mathbf{x},\mathbf{y})\frac{\delta}{\delta\varphi(\mathbf{y})} \Big) \Psi_1(\varphi) ,$$

$$= \frac{1}{\sqrt{2}} \Big[2\sqrt{\varepsilon\varepsilon'} \int f_{\varepsilon'}\varphi \int f_{\varepsilon}\varphi - \int f_{\varepsilon'}f_{\varepsilon} \Big] \Psi_0(\varphi) , \qquad (1.44)$$

$$E_2 = \varepsilon + \varepsilon' + \frac{1}{2}\operatorname{tr}\omega ,$$

etc.

The set of all Fock functionals $\{\Psi_n(\varphi)\}$ comprises a basis for the *Fock functional space*. As we shall see presently, this is not the full functional space.

In the Poincaré invariant model of (1.12), $\omega = \sqrt{-\nabla^2 + m^2}$, $f_\varepsilon(\mathbf{x}) = \varepsilon^{i\mathbf{p}\cdot\mathbf{x}}$ and $\varepsilon = \omega(p) = \sqrt{p^2 + m^2}$. Then of course, the ground state energy (1.39c) is infinite, but this divergence need not disturb us, because it is unambiguously isolated in the formalism. Also the normalization factor

$$\det^{1/4}\Big(\frac{\omega}{\pi}\Big) = \exp\frac{1}{4}\operatorname{tr}\ln\frac{\omega}{\pi} = \exp\frac{V}{4}\int_{\mathbf{p}}\ln\frac{\sqrt{p^2 + m^2}}{\pi} \qquad (1.45)$$

is infrared infinite because of the volume factor V, and ultraviolet infinite because of the divergence in the momentum representation. This infinity also disappears for matrix elements of operators between states in the Fock space, because the normalization is precisely so chosen that it cancels. For example, the expectations of operator bilinears are well-defined

$$\langle 0|\Phi(\mathbf{x})\Phi(\mathbf{y})|0\rangle = \int \mathcal{D}\varphi \Psi_0^*(\varphi)\varphi(\mathbf{x})\varphi(\mathbf{y})\Psi_0(\varphi) ,$$

$$= \frac{1}{2}\omega^{-1}(\mathbf{x},\mathbf{y}) = \int_{\mathbf{p}} e^{-i\mathbf{p}\cdot(\mathbf{x}-\mathbf{y})}\frac{1}{2\sqrt{p^2 + m^2}} , \qquad (1.46\text{a})$$

$$\langle 0|\Pi(\mathbf{x})\Pi(\mathbf{y})|0\rangle = \int \mathcal{D}\varphi \Big(\frac{1}{i}\frac{\delta}{\delta\varphi(\mathbf{x})}\Psi_0(\varphi)\Big)^* \Big(\frac{1}{i}\frac{\delta}{\delta\varphi(\mathbf{y}))}\Psi_0(\varphi)\Big) ,$$

$$= \frac{1}{2}\omega(\mathbf{x},\mathbf{y}) = \int_{\mathbf{p}} e^{-i\mathbf{p}\cdot(\mathbf{x}-\mathbf{y})}\frac{1}{2}\sqrt{p^2 + m^2} , \qquad (1.46\text{b})$$

$$\langle 0|\Phi(\mathbf{x})\Pi(\mathbf{y})|0\rangle = \int \mathcal{D}\varphi \Psi_0^*(\varphi)\varphi(\mathbf{x})\frac{1}{i}\frac{\delta}{\delta\varphi(\mathbf{y})}\Psi_0(\varphi) ,$$

$$= \frac{i}{2}\delta(\mathbf{x}-\mathbf{y}) \qquad (1.46\text{c})$$

and coincide with the conventional free-field correlation functions.

Notice also that the equation (1.39b), which determines the covariance ω of the ground state Gaussian from the first quantized Hamiltonian h, leaves a sign undetermined. For example, for translationally invariant quantities, (1.39b) is solved by $\omega(p) = (\text{sign})\sqrt{h(p)}$, where the choice of sign could even vary with $\mathbf{p}$. The positive sign is selected by the requirement that the wave functional damp for large field strength φ, rather than diverge, as would be the case when the sign is negative. This localization in field space also assures a positive energy spectrum in the quantized field theory.

The analogy to the quantum mechanical oscillator states (1.26) – (1.29) is clear — the formulas are the same, except for an infinite replication, arising from the infinite number of degrees of freedom in a field theory. But precisely because of this infinity, the analogy is not complete. The set of all Fock functionals $\{\Psi_n(\varphi)\}$, which by definition forms a basis for the Fock space of the Hamiltonian (1.37), is inadequate for expanding every functional one might construct in our functional space. For example, consider two Fock vacua, with translationally invariant covariance ω_1 and ω_2. These are ground states of two Hamiltonians H_1 and H_2, whose first quantized form involve $h_1 = \omega_1^2$ and $h_2 = \omega_2^2$. As I mentioned already, in the quantum mechanical case, the ground state of each Hamiltonian can be expanded in terms of the energy eigenstates of the other. However, in the field theory, the overlap between the two Fock vacua is

$$\int \mathcal{D}\varphi \Psi_{\omega_1,0}^*(\varphi)\Psi_{\omega_2,0}(\varphi) = e^{-N}\,, \tag{1.47a}$$

$$N = \frac{V}{2}\int_{\mathbf{p}} \ln \frac{1}{2}\left(\sqrt{\frac{\omega_1(p)}{\omega_2(p)}} + \sqrt{\frac{\omega_2(p)}{\omega_1(p)}}\right)\,. \tag{1.47b}$$

Even ignoring the [infrared] infinity associated with infinite volume V, N/V will still diverge in the ultraviolet, unless ω_1 and ω_2 approach each other rapidly at large p. Since the integrand in (1.47b) is positive, the divergence sets e^{-N} to zero, and the overlap (1.47a) vanishes. Then also the overlap between all higher Fock functionals built on the respective two Fock vacua will possess the factor (1.47), and the two Fock bases, as well as the two Fock spaces built on these bases are mutually orthogonal. In other words, our functional space contains within it inequivalent Fock spaces.

This difference between ordinary quantum mechanics and quantum field theory in the Schrödinger representation is a consequence of the fact that the space on which the quantum mechanics is defined is a finite dimensional manifold, accommodating a finite number of quantum mechanical degrees of freedom, while the field theory, with its infinite degrees of freedom, uses an infinite dimensional manifold. The inequivalent Fock spaces provide inequivalent representations of the canonical commutation relations — a situation that does not arise in quantum mechanics. Moreover, as I shall later explain, representations of transformation groups on inequivalent Fock spaces are in general inequivalent. Specifically, answers can depend on the covariance that enters in the definition of the Fock vacuum. On the contrary, when we represent transformations by operator kernels $\mathcal{O}(\varphi, \tilde{\varphi})$ as in (1.32), a unique representation is constructed, for which a specific Fock space need not be pre-selected. This is one of the advantages of the Schrödinger representation formalism.

For further discussion, it is useful to generalize the concept of Fock vacuum, freeing it from its role as the ground state of some Hamiltonian. We shall call any Gaussian wave functional a "Fock vacuum" and characterize it by its covariance $\Omega(\mathbf{x}, \mathbf{y})$, which is symmetric in $\mathbf{x} \leftrightarrow \mathbf{y}$. However, we shall allow Ω to be complex, but with positive definite real part

$$
|\Omega\rangle \to \Psi_\Omega(\varphi) = \det^{1/4}\left(\frac{\Omega_R}{\pi}\right) \exp{-\frac{1}{2}\int \varphi\Omega\varphi}
$$
$$
\Omega(\mathbf{x}, \mathbf{y}) = \Omega(\mathbf{y}, \mathbf{x}), \quad \Omega = \Omega_R + i\Omega_I .
$$

(1.48)

Also Ω can be time-dependent in which case the Fock vacuum $\Psi_\Omega \ (\varphi; t)$ is a time-dependent functional. The reason for the nomenclature, *i.e.*, for calling the above Gaussian a "Fock vacuum", is that it is annihilated by an operator $A(\mathbf{x})$ that is linear in Φ and Π

$$
A = \frac{1}{\sqrt{2}}\int \Omega_R^{-1/2}(\Omega\Phi + i\,\Pi) \to \frac{1}{\sqrt{2}}\int \Omega_R^{-1/2}\left(\Omega\varphi + \frac{\delta}{\delta\varphi}\right) .
$$

(1.49a)

The expected commutation relations between A and its conjugate $A^\dagger$

$$
A^\dagger = \frac{1}{\sqrt{2}}\int \Omega_R^{-1/2}(\Omega^*\Phi - i\,\Pi) \to \frac{1}{\sqrt{2}}\int \Omega_R^{-1/2}\left(\Omega^*\varphi - \frac{\delta}{\delta\varphi}\right)
$$

(1.49b)

are obeyed

$$
[A(\mathbf{x}), A^\dagger(\mathbf{y})] = \delta(\mathbf{x} - \mathbf{y}) .
$$

(1.50)

Higher basis states are obtained by repeated action of $A^\dagger$ on Ψ_Ω. Of course when $\Omega_I = 0$ and Ω_R is static, the above reverts to the usual Fock basis for a Hamiltonian with first quantized form Ω^2. But the generalization will be useful when we discuss time-dependent systems, for which the concept of energy eigenstates cannot be used.

With (1.48) the expectations of Φ and Π vanish. Bilinears become

$$\langle \Omega | \Phi(\mathbf{x})\Phi(\mathbf{y}) | \Omega \rangle = \frac{1}{2}\Omega_R^{-1}(\mathbf{x}, \mathbf{y}), \tag{1.51a}$$

$$\langle \Omega | \Pi(\mathbf{x})\Pi(\mathbf{y}) | \Omega \rangle = \frac{1}{2}\Omega_R^{-1}(\mathbf{x}, \mathbf{y}) + \frac{1}{2}(\Omega_I \Omega_R^{-1} \Omega_I)(\mathbf{x}, \mathbf{y}), \tag{1.51b}$$

$$\langle \Omega | \Phi(\mathbf{x})\Pi(\mathbf{y}) | \Omega \rangle = \frac{i}{2}\delta(\mathbf{x} - \mathbf{y}) - \frac{1}{2}(\Omega_R^{-1}\Omega_I)(\mathbf{x}, \mathbf{y}). \tag{1.51c}$$

[It is possible to consider a more general Gaussian that contains a term linear in φ in the exponential. Then $\langle \Phi \rangle$ and $\langle \Pi \rangle$ no longer vanish and bilinears contain a disconnected part. For the sake of simplicity, I shall not develop this generalization.]

Let me remark that there exists an approach within which one can view the wave functions of quantum mechanics and the wave functionals of quantum field theory, and also the kernels that implement the operators, as matrix elements. In quantum mechanics, one defines continuously normalized eigenstates of the position operator: $Q|q\rangle = q|q\rangle$, $\langle q|\tilde{q}\rangle = \delta(q - \tilde{q})$. Then the wave function $\psi(q)$ is the overlap between an abstract state $|\psi\rangle$ and $\langle q|$: $\psi(q) = \langle q|\psi\rangle$. Also the kernel $\mathcal{O}(q, \tilde{q})$ corresponding to the operator $\mathcal{O}$ is the matrix element $\langle q|\mathcal{O}|\tilde{q}\rangle$. Analogously in a bosonic field theory, we can define eigenstates $|\varphi\rangle$ of the quantum field operator, $\Phi(\mathbf{x})|\varphi\rangle = \varphi(\mathbf{x})|\varphi\rangle$, with functional delta function normalization $\langle \varphi|\tilde{\varphi}\rangle = \delta(\varphi - \tilde{\varphi})$. Then we can interpret our wave functionals as overlaps, $\Psi(\varphi) = \langle \varphi|\Psi\rangle$, and the operator kernels as matrix elements, $\mathcal{O}(\varphi, \tilde{\varphi}) = \langle \varphi|\mathcal{O}|\tilde{\varphi}\rangle$. However, we do not dwell on this formalism, because it does not generalize to fermion fields.

Finally, it should be appreciated that the functional Schrödinger representation is the natural setting for a third quantization if that step should indeed be taken, as has been occasionally suggested and is again mentioned these days. To third quantize, one would promote the wave functionals $\Psi(\varphi)$ to operators and postulate commutation relations between Ψ and Ψ^* that involve the functional delta function.

In my next lecture, I shall describe how gauge theories appear within the Schrödinger representation, and how their various topological intricacies are readily established. The third lecture concerns the representation of transformation groups. I demonstrate that the functional method allows exhibiting unambiguously non-trivial features, like extensions in the algebra of infinitesimal transformations and cocycles in the finite transformations. This is done without normal ordering or choosing any specific Fock vacuum. Then in the fourth lecture, I review various variational approximations that are suggested for the field theory by their well-known quantum mechanical antecedents. This is done for static and time-dependent problems, for pure and mixed states. Thus I shall develop the subject of quantum fields in and out of thermal equilibrium. The last lecture is devoted to a reprise of the material in the first lecture, but now for Fermi fields.

II. APPLICATION TO GAUGE THEORIES

We discuss gauge theories, mostly without matter interactions.[3] The vector potential A_μ, which for the non-Abelian theory may be presented as an element of the gauge group's Lie Algebra

$$A_\mu = A_\mu^a T_a$$

$$T_a^\dagger = -T_a, \quad [T_a, T_b] = f_{abc} T_c, \quad \text{tr } T_a T_b = -\frac{1}{2}\delta_{ab} \tag{2.1}$$

is the basic Lagrangian variable in the Lagrange density

$$\mathcal{L} = \mathcal{L}_{\text{YM}} \equiv -\frac{1}{4} F^{\mu\nu a} F_{a\mu\nu} = \frac{1}{2} \text{tr } F^{\mu\nu} F_{\mu\nu} \tag{2.2}$$

$$F_{\mu\nu}^a = \partial_\mu A_\nu^a - \partial_\nu A_\mu^a + f_{abc} A_\mu^b A_\nu^c$$

$$F_{\mu\nu} = F_{\mu\nu}^a T_a = \partial_\mu A_\nu - \partial_\nu A_\mu + [A_\mu, A_\nu]. \tag{2.3}$$

The coupling strength has been scaled from the potential, so that it does not appear in the field equations. The theory is invariant against local gauge transformations, effected by an element g of the gauge group

$$A_\mu \to A_\mu^g \equiv g^{-1} A_\mu g + g^{-1} \partial_\mu g, \tag{2.4a}$$

$$F_{\mu\nu} \to F_{\mu\nu}^g = g^{-1} F_{\mu\nu} g. \tag{2.4b}$$

The field equation follows from (2.2) by varying A_μ

$$\partial_\mu F^{\mu\nu} + [A_\mu, F^{\mu\nu}] \equiv D_\mu F^{\mu\nu} = 0. \tag{2.5}$$

Also from its definition, $F_{\mu\nu}$ satisfies the Bianchi identity

$$D_\alpha F_{\beta\gamma} + D_\beta F_{\gamma\alpha} + D_\gamma F_{\alpha\beta} = 0. \tag{2.6}$$

A canonical formulation is achieved by adopting the Weyl gauge $(A_0 = 0)$. The Hamiltonian is

$$H = \frac{1}{2} \int_{\mathbf{x}} \left(\mathcal{E}_a^2(\mathbf{x}) + \frac{1}{2} F_{ij}^a(\mathbf{x}) F_{ij}^a(\mathbf{x}) \right)$$

$$\mathcal{E}^i = F_{0i} = \dot{A}_i, \tag{2.7}$$

where the overdot signifies time-differentiation. The canonical momentum identified from (2.2) [with $A_0 = 0$]

$$\Pi_a^i \equiv \frac{\partial \mathcal{L}_{\text{YM}}}{\partial \dot{A}_a^i} = -\mathcal{E}_a^i.$$ (2.8)

Hence, the non-vanishing fixed-time canonical commutation relation

$$[\mathcal{E}_a^i(\mathbf{x}), A_b^j(\mathbf{y})] = i\delta^{ij}\delta_{ab}\delta(\mathbf{x} - \mathbf{y})$$ (2.9)

indicates that in a Schrödinger representation $\mathcal{E}_a^i(\mathbf{x})$ is replaced by $i\frac{\delta}{\delta A_a^i(\mathbf{x})}$ acting on functionals of A_a^i, $\Psi(A)$.

Since $A_0 = 0$, the time component of the field equation (2.5) — the Gauss law — cannot be obtained, neither by varying the Lagrangian, nor as a Hamiltonian equation because it is seen from (2.5) that its time component is a fixed time constraint between canonical variables

$$G \equiv D_i \mathcal{E}^i = 0.$$ (2.10)

On the other hand, it is recognized that the theory in the Weyl gauge admits as symmetry transformations time-independent gauge transformations g, whose infinitesimal form $(g = I + \theta)$ reads

$$\delta A^i = -D_i\theta.$$ (2.11)

Furthermore, one verifies that G_a is conserved, and generates (2.11)

$$i\left[H, \int_{\mathbf{x}} \theta^a(\mathbf{x})G_a(\mathbf{x})\right] = 0,$$ (2.12a)

$$i\left[\int_{\mathbf{y}} \theta^b(\mathbf{y})G_b(\mathbf{y}), A_a^i(\mathbf{x})\right] = \delta A_a^i(\mathbf{x}).$$ (2.12b)

Since the G_a commute with the Hamiltonian, and satisfy an algebra that follows the group Lie algebra

$$i\left[G_a(\mathbf{x}), G_b(\mathbf{y})\right] = f_{abc}G_c(\mathbf{x})\delta(\mathbf{x} - \mathbf{y})$$ (2.13)

one can regain the Gauss law (2.10) in the quantum theory by imposing it as a condition on physical states [rather than as an operator equation]

$$G|\Psi\rangle = 0.$$ (2.14a)

Therefore, in the Schrödinger representation Gauss's law implies a functional differential equation that must be satisfied by functional physical states

$$G|\Psi\rangle = 0 \Longrightarrow \left(\partial_i \frac{\delta}{\delta A_a^i(\mathbf{x})} - f_{abc}A_b^i(\mathbf{x})\frac{\delta}{\delta A_c^i(\mathbf{x})}\right)\Psi(A) = 0.$$ (2.14b)

It is clear that Eq. (2.14) demands that $\Psi(A)$ is invariant against gauge transformations that can be obtained by an iterative build up of an infinitesimal gauge transformation. This is an additional condition on energy eigenstates, which also satisfy the functional Schrödinger equation

$$\int_{\mathbf{x}} \left(-\frac{1}{2} \frac{\delta^2}{\delta A_a^i(\mathbf{x})\delta A_a^i(\mathbf{x})} + \frac{1}{4} F_{ij}^a(\mathbf{x}) F_{ij}^a(\mathbf{x}) \right) \Psi_E(A) = E\Psi_E(A) \,. \qquad (2.15)$$

Next, we discuss further properties of various gauge theories in various dimensions.

A. Abelian Theory — Quantum Electrodynamics (QED)

In the Abelian gauge theory of free electrodynamics, Eq. (2.14) reduces to

$$\partial_i \frac{\delta}{\delta A^i(\mathbf{x})} \Psi(A) = 0 \,. \qquad (2.16)$$

The general solution is an arbitrary functional of transverse components of A^i; $\Psi(A)$ does not depend on the longitudinal components of A^i. The transverse components of course are gauge invariant: only the longitudinal component changes under Abelian gauge transformations

$$A^i \to A^i - \partial_i \theta \,. \qquad (2.17)$$

Hence, functionals that satisfy the Abelian Gauss law are gauge invariant

$$\partial_i \frac{\delta}{\delta A^i} \Psi(A) = 0 \Longrightarrow \Psi(A - \partial\theta) = \Psi(A) \,. \qquad (2.18)$$

The Schrödinger Hamiltonian is quadratic

$$H_{\text{EM}} = \frac{1}{2} \int \left(-\frac{\delta^2}{\delta A^i \delta A^i} + A^i h_{ij} A^j \right)$$
$$h_{ij} = -\nabla^2 \delta_{ij} + \partial_i \partial_j \qquad (2.19)$$

with ground state solution

$$\Psi_0(A) \propto e^{-\frac{1}{2} \int A^i \omega_{ij} A^j}$$

$$\omega_{ij}(\mathbf{x}, \mathbf{y}) = (-\nabla^2 \delta_{ij} + \partial_i \partial_j) \int_{\mathbf{p}} e^{-i\,\mathbf{p}\cdot(\mathbf{x}-\mathbf{y})} \frac{1}{|\mathbf{p}|} \qquad (2.20a)$$

which may also be written in manifestly gauge invariant form

$$\Psi_0(A) \propto e^{-\frac{1}{4} \int F^{ij} \frac{1}{\sqrt{-\nabla^2}} F^{ij}} \,. \qquad (2.20b)$$

Thus the ground state of (2.19) automatically satisfies the Gauss law (2.16) — it is unnecessary to impose it additionally because the vacuum is unique — there is no spontaneous breaking of the gauge symmetry. [These formulas make sense in any spatial dimension greater than 1.]

I leave it as an exercise to construct excited states and to analyze another instructive, solvable problem: Abelian electrodynamics in the presence of an external, static and classical charge density.

B. Yang-Mills in Four-Dimensional Space-Time

The question arises whether physical Yang-Mills wave functionals that satisfy (2.14) are invariant against *arbitrary* finite gauge transformations g, not just against those that are built up from iterating infinitesimal gauge transformations, $g = I + \theta$. Of course, since the Hamiltonian is invariant against all gauge transformations, a physical state can respond to a finite gauge transformation at most by a phase.

To answer this question, consider the following functional which is defined in three-space

$$W(A) = \int \omega(A),$$

$$\omega(A) = -\frac{1}{16\pi^2}\varepsilon^{ijk} \operatorname{tr}\left(F_{ij}A_k - \frac{2}{3}A_iA_jA_k\right),$$

$$= -\frac{1}{4\pi^2}\varepsilon^{ijk}\left(\frac{1}{2}\partial_iA_jA_k + \frac{1}{3}A_iA_jA_k\right).$$

(2.21)

This is called the Chern-Simons term; it has the property that

$$\frac{\delta W(A)}{\delta A_i^a} = \frac{1}{16\pi^2}\varepsilon^{ijk}F_{jk}^a$$

(2.22)

and, as a consequence of the Bianchi identity (2.6), satisfies the Gauss law constraint

$$\left(D_i\frac{\delta}{\delta A^i}\right)_a W(A) = 0.$$

(2.23)

On the other hand, explicit calculation shows that $W(A)$ is not invariant under an arbitrary gauge transformation

$$W(A^g) - W(A) = \frac{1}{24\pi^2}\int \varepsilon^{ijk} \operatorname{tr} g^{-1}\partial_i g g^{-1}\partial_j g g^{-1}\partial_k g.$$

(2.24)

The right-hand is an integer that measures the *winding number* of the gauge transformation g. Gauge transformations that are built up by iterating an infinitesimal gauge transformation $(g = I + \theta)$ have zero winding number, but it is easy to construct gauge functions g with non-zero, integer winding number n. We conclude that the most general physical state in four-dimensional Yang-Mills theory is of the form

$$\Psi(A) = e^{-i\theta W(A)}\Psi'(A)\,, \qquad (2.25)$$

where $\Psi'(A)$ is invariant against all gauge transformations, and thereby also satisfies Gauss's law. The response of a physical state to a gauge transformation is therefore,

$$\Psi(A^g) = e^{-in\theta}\Psi(A)\,. \qquad (2.26)$$

This is the origin of the famous vacuum angle, here established without any "instanton" approximation.[3]

Another interesting calculation that has been performed for 4-dimensional Yang-Mills theory exhibits the most general explicit solution of Gauss's law. In the Abelian case, this is elementary: as stated already, a functional satisfying the Abelian Gauss law is an arbitrary functional depending only on the gauge invariant, transverse components of the Abelian vector potential. In the non-Abelian theory a much more complicated answer emerges, which is known explicitly.[4]

C. Yang-Mills Theory in Three-Dimensional Space Time

For planar gauge theories, in three-dimensional space-time, there is no vacuum angle. However, a related phenomenon arises from the possibility of adding the Chern-Simons term (2.21), now defined in three-dimensional space-time, to the action of the Lagrangian (2.2)

$$\int \mathcal{L} = \int \mathcal{L}_{\text{YM}} + 8\pi^2 \mu W(A)\,. \qquad (2.27)$$

The "coupling strength" μ has dimension of mass, and the numerical factor is for later convenience. The field equation that follows from varying (2.27) is

$$D_\mu F^{\mu\nu} + \mu \frac{1}{2}\varepsilon^{\nu\alpha\beta}F_{\alpha\beta} = 0 \qquad (2.28)$$

and analysis of the quadratic portion — or equivalently, of the Abelian theory — shows that the excitations are massive, this despite the gauge-invariant nature of the field equation. This is an example of a topologically massive gauge theory.[3]

The Hamiltonian for (2.27) is again (2.7) — the mass term is invisible when the energy is expressed in terms of $\mathcal{E}_a^i$ and F_{ij}^a. However, the connection between the electric field and the canonical momentum is modified, owing to the derivative interaction provided by the Chern-Simons term. In the Weyl gauge, we find

$$\Pi^i = -\mathcal{E}^i + \frac{1}{2}\mu\varepsilon^{ij}A^j \,. \tag{2.29}$$

Again, the time component of the field equation (2.28)

$$D_i\mathcal{E}^i + \frac{\mu}{2}\varepsilon^{ij}F_{ij} = 0 \tag{2.30a}$$

is imposed as a constraint on states. In canonical variables the left-hand side of (2.30a) reads

$$-(D_i\Pi^i)_a + \frac{\mu}{2}\varepsilon^{ij}\partial_i A_j^a \equiv -G_a \,. \tag{2.30b}$$

G_a commutes with the Hamiltonian, continues to satisfy the Lie algebra (2.13) and generates infinitesimal time-independent gauge transformations, whose iteration exhausts all static gauge transformations in the plane — in contrast to four-dimensional Yang-Mills theories considered in sub-Section B, in three-space-time dimensions static gauge transformations have zero winding number. Requiring that G_a annihilate physical states leads to the equation

$$\left(D_i\frac{\delta}{\delta A^i}\right)_a \Psi(A) - i\frac{\mu}{2}\varepsilon^{ij}\partial_i A_j^a \Psi(A) = 0 \tag{2.31a}$$

or upon iteration

$$e^{i\int\theta^a G_a}\Psi(A) = \Psi(A) \,. \tag{2.31b}$$

The left-hand side of (2.31b) is explicitly found to be

$$e^{i\int\theta^a G_a}\Psi(A) = e^{i8\pi^2\mu\omega(A;g)}\Psi(A^g) \,. \tag{2.32}$$

Thus in the presence of the Chern-Simons term, Gauss's law does not require physical states to be gauge invariant — rather they must satisfy

$$\Psi(A^g) = e^{-i8\pi^2\mu\omega(A;g)}\Psi(A) \,. \tag{2.33}$$

The phase $\omega(A;g)$ is as follows. The Chern-Simons density $\omega(A)$, (2.21), is not gauge invariant. Rather it satisfies

$$\omega(A^g) - \omega(A) = \frac{1}{8\pi^2}\varepsilon^{\alpha\beta\gamma}\,\mathrm{tr}\,\partial_\alpha(\partial_\beta g g^{-1}A_\gamma)\,,$$

$$+ \frac{1}{24\pi^2}\varepsilon^{\alpha\beta\gamma}\,\mathrm{tr}\,g^{-1}\partial_\alpha g g^{-1}\partial_\beta g g^{-1}\partial_\gamma g\,, \tag{2.34}$$

$$= \partial_\alpha\omega^\alpha(A;g)\,.$$

The fact $\omega(A^g) - \omega(A)$ is a total divergence is manifest for one of the terms in (2.34), for the other it is locally, but not globally true. The quantity $\omega(A; g)$ in the phase (2.32) and (2.33) is obtained by taking the time component of ω^α at fixed time $(t = 0)$ and integrating over the plane

$$\omega(A; g) = \int_{\mathbf{x}} \omega^0(A; g). \tag{2.35}$$

However, because $\omega^\alpha(A; g)$ is defined only locally, not globally $\omega(A; g)$ is ambiguous by integers, *i.e.*, it is defined only modulo an integer. Thus, in order that formula (2.33) have a unique well-defined meaning, *i.e.*, in order that the ambiguity in the phase not produce an ambiguity when exponentiated, we must have

$$\mu = \frac{n}{4\pi} \tag{2.36}$$

i.e., Gauss's law is only integrable when the topological mass is quantized. Other values of the mass parameter do not lead to a consistent quantum field theory. This situation is the field theoretic analog of the familiar quantum mechanical phenomenon that magnetic monopole strength must be quantized for the quantum theory to be consistent.

Another way of describing the situation in Eq. (2.32) is to observe that the unitary operator $U(g) \equiv e^{i \int \theta^a G_a}$ that effects the finite gauge transformation g on the dynamical variable A_a^i

$$U(g) A U^\dagger(g) = A^g \tag{2.37}$$

acts on functionals of the dynamical variable, $\Psi(A)$, not merely by transforming the argument, but also by multiplying the wave functional by a phase. The composition law for these unitary operators follow the composition law of the group

$$U(g_1) U(g_2) = U(g_{1 \circ 2})$$
$$g_1 g_2 = g_{1 \circ 2}. \tag{2.38}$$

This is because the Lie algebra of the generators follows the Lie algebra of the group. As a consequence of (2.38), the phase $\omega(A; g)$ must satisfy a consistency condition

$$\omega(A^{g_1}; g_2) - \omega(A; g_{1 \circ 2}) + \omega(A; g_1) = 0 \quad \text{mod integer}. \tag{2.39}$$

That our phase does indeed satisfy (2.39) follows from (2.34) and (2.35).

Any quantity that depends on one group element [here g] as well as [possibly] on a point in the manifold on which the group acts [here A] and satisfies (2.39) is called a *1-cocycle*.

We see therefore that the gauge group of Yang-Mills theory with a Chern-Simons term is realized with 1-cocycle. A 1-cocycle always arises in quantum theory when representing a symmetry transformation that does not leave a Lagrangian invariant but changes it by a total time derivative. In the present instance, the contribution to the gauge theory Lagrangian of the Chern-Simons term is

$$L_{\rm CS} = 8\pi^2 \mu \int_{\bf x} \omega(A) \,. \tag{2.40}$$

Under a gauge transformation, this changes by

$$
\begin{aligned}
L_{\rm CS} &\to 8\pi^2 \mu \int_{\bf x} \omega(A^g) \\
&= L_{\rm CS} + 8\pi^2 \mu \int_{\bf x} (\omega(A^g) - \omega(A)) \\
&= L_{\rm CS} + 8\pi^2 \mu \int_{\bf x} \partial_\alpha \omega^\alpha(A; g) \\
&= L_{\rm CS} + \frac{d}{dt} 8\pi^2 \mu \int_{\bf x} \omega^0(A; g) \\
&= L_{\rm CS} + \frac{d}{dt} 8\pi^2 \mu \omega(A; g) \,.
\end{aligned}
\tag{2.41}
$$

We have used (2.34), the definition (2.35), and we recognize that the differentiated term in the last equality of (2.41) is precisely the 1-cocycle that occurs in (2.32) and (2.33).

Another, more elementary example of a 1-cocycle is found in quantum mechanics. The free Lagrangian $L = \frac{1}{2} m \dot{Q}^2$ admits the Galilean boost $Q \to Q + vt$ as an Abelian symmetry group of transformations, generated by $G = Pt - mQ$. Here v is the parameter of the transformation specifying the group element. The operator $U(v) = e^{ivG}$ implements the finite transformation, $U(v) Q U^\dagger(v) = Q + vt$, but acts on functions of q by $U(v)\psi(q) = e^{-i\omega(q;v)}\psi(q+vt)$, where the 1-cocycle $\omega(q; v) = mqv + \frac{1}{2}mtv^2$ also arises in the change of the Lagrangian $L \to \frac{1}{2}m(\dot{Q}+v)^2 = L + m\dot{Q}v + \frac{1}{2}mv^2 = L + \frac{d}{dt}\omega(Q; v)$.

D. QED in Two-Dimensional Space Time

We begin by considering free electrodynamics, which in one spatial dimension is governed by the Hamiltonian

$$H = \frac{1}{2} \int \mathcal{E}^2 \tag{2.42}$$

and Gauss's law is

$$\frac{d}{dx}\mathcal{E}(x) = 0 \tag{2.43}$$

in the quantum field theory (2.14), (2.16) and (2.43) require that physical states obey

$$\frac{d}{dx}\frac{\delta}{\delta A(x)}\Psi(A) = 0. \tag{2.44}$$

The most general solution of this is $\Psi(A) = f(\int A)$, *i.e.*, Ψ is an arbitrary function [not functional] of $\int_x A(x)$. Next the energy eigenstates are obtained by solving

$$E\Psi(A) = Ef(\int A) = H\Psi(A)$$

$$= -\frac{1}{2}\int_x \frac{\delta^2}{\delta A(x)\delta A(x)}f(\int A) = -\frac{L}{2}f''(\int A), \tag{2.45}$$

where L is the length of space and the double dash signifies double differentials with respect to the argument $\int_x A(x)$ of the function f. The eigenstates solving (2.45) are

$$\Psi_E(A) = e^{-i\mathcal{E}_0 \int_x A(x)} \tag{2.46}$$

with eigenvalue $E = \frac{1}{2}L\mathcal{E}_0^2 = \frac{1}{2}\int_x \mathcal{E}_0^2$. Also

$$\mathcal{E}(x)\Psi_E(A) = i\frac{\delta}{\delta A(x)}\Psi_E(A) = \mathcal{E}_0\Psi_E(A) \tag{2.47}$$

so $\mathcal{E}_0$ is a background electric field and E is the energy carried by this field.

We may perform a gauge transformation on our states $A(x) \to A(x) - \frac{d}{dx}\Lambda(x)$, $\int_x A(x) \to \int_x A(x) - \triangle\Lambda$. We define a winding number for a gauge transformation by $\triangle\Lambda = 2\pi n$, so that $e^{-i\triangle\Lambda} = 1$, and we see that just as in 4-dimensional space-time, there is an angle in the response of the wave functional to a gauge transformation

$$\Psi_E(A) \to e^{-in\theta}\Psi_E(A). \tag{2.48}$$

Here the θ angle is now identified with the background electric field, $\theta = 2\pi\mathcal{E}_0$. We conclude that pure electrodynamics in 2 space-time dimensions is trivial — there are no propagating excitations — save for inequivalent vacuum sectors labeled by θ, or equivalently by a background field $\mathcal{E}_0$.

The θ angle may be cast into the Lagrangian and removed from the states by adding to the conventional $\mathcal{L}_{EM} = -\frac{1}{4}F^{\mu\nu}F_{\mu\nu} = \frac{1}{2}\mathcal{E}^2$ the total time derivative $\dot{A} = -\mathcal{E}$

$$\mathcal{L} = \frac{1}{2}\mathcal{E}^2 - \frac{\theta}{2\pi}\mathcal{E}. \tag{2.49}$$

The Hamiltonian is unaffected, but the relation between $\mathcal{E}$ and the canonical momentum changes

$$\Pi = \frac{\partial \mathcal{L}_\theta}{\partial \dot{A}} = \dot{A} + \frac{\theta}{2\pi} \, . \tag{2.50}$$

The Gauss law constraint (2.43), (2.44) remains unaltered, but the Schrödinger operator now involves $\frac{1}{2} \int \left(i \frac{\delta}{\delta A} + \frac{\theta}{2\pi} \right)^2$. The eigenfunctions now acquire relative to (2.46) an additional phase, $e^{i\frac{\theta}{2\pi} \int_x A(x)}$, cancelling the phase that is present there, and thus become gauge invariant [constants].

Let us now include fermions. [Because the Schrödinger picture for fermions is not developed until later, I shall present the argument formally.] The Lagrangian (2.49) becomes enlarged to

$$\mathcal{L} = -\frac{1}{4} F^{\mu\nu} F_{\mu\nu} - \frac{\theta}{4\pi} \varepsilon^{\mu\nu} F_{\mu\nu} + \bar{\psi}(i \slashed{\partial} - \slashed{A} - m)\psi \, . \tag{2.51}$$

In the massless case $m = 0$ [Schwinger model], the Fermi field ψ may be redefined by a chiral transformation $\psi \to e^{\alpha\gamma_5}\psi$. $\gamma_5^\dagger = -\gamma_5$, which however does not leave the Lagrangian invariant because of the chiral anomaly. After such a field redefinition, which cannot affect physical content, $\mathcal{L}$ acquires an additional term proportional to $\varepsilon^{\mu\nu} F_{\mu\nu}$ that alters the θ parameter by an arbitrary amount, showing that θ has no physical relevance for $m = 0$. However, for massive fermions, $m \neq 0$, a chiral redefinition is no longer useful since the mass term is not invariant, and the θ parameter retains physical significance.[5]

All this is an amusing mimicry of the situation in 4-dimensional QCD: in particular $\varepsilon^{\mu\nu} F_{\mu\nu}$ is the 2-dimensional analog of the 4-dimensional Pontryagin density $^*F^{\mu\nu} F_{\mu\nu}$.[3]

E. Confinement in Non-Abelian Gauge Theories

I have presented several successful analyses of aspects of gauge theories in the Schrödinger representation. These analyses are essentially kinematical — except for the 2-dimensional models, dynamics is too intricate to unravel. However, the gauge symmetry, as it manifests itself in the Gauss law constraint, leads to a tractable problem.

There is another feature of non-Abelian theories that is generally believed to be kinematical in origin, arising from the non-Abelian symmetry. I have in mind the phenomenon of color confinement, and it is natural to ask whether this too can be understood in the Schrödinger representation. In fact, no complete proof of color confinement has been given within this [or any other] formalism, but R. Feynman in one of his last investigations gave qualitative arguments to this end. I now summarize his ideas.[6]

The question to be answered may be formulated in the following way: is there a finite gap in the energy spectrum of the Yang-Mills Hamiltonian above the ground state vacuum, so that the first excited state is a massive excitation — presumably a glueball — or can the gap be made arbitrarily small so that the excitation is a massless gluon. Of course, the latter situation holds in electrodynamics, so we must pinpoint a relevant difference between the Abelian and non-Abelian gauge theories.

Let us first examine quantum mechanics and inquire how energy eigenfunctions $\psi_E(q)$ for a system with an energy gap differ from those where there is no gap. It appears that the absence of a gap is correlated with the fact that the dynamical variable q can become arbitrarily large: compare free motion in an unbounded region [no gap — continuous spectrum] with either free motion in a compact domain. *eg.*, a circle, or bound motion like that of the harmonic oscillator [gap exists — spectrum is discrete]. In the former case, q can get large and the wave function remains appreciable: in the latter, it cannot because on a compact domain there are no "large" values of q, alternatively for bound motion the wave function rapidly vanishes at large q.

Turning now to field theory, we observe that in electrodynamics a low energy photon, whose energy level is arbitrarily close to the vacuum energy corresponds to a long wavelength vector potential that extends over large regions of space. The question is: can such "large" configurations exist in a non-Abelian theory?

Next, we recall that in a quantum gauge field theory the configuration space is not the space of *all* vector potentials, rather — as a consequence of Gauss's law — it is the space of vector potentials, with gauge copies identified. In the Abelian theory, this means that the true space is that of transverse vector potentials which are otherwise unrestricted and can take on the low wavelength profile, thus bringing the energy arbitrarily close to the ground state. Feynman argues, and presents some rough calculations to support his argument, that in a non-Abelian gauge theory such "large" configurations are gauge equivalent to "small" configurations, so that the energy gap cannot be closed.

In other words, according to Feynman, the dynamical variable cannot become large — not because the [functional] space is closed and compact [it is not] but because "large" configurations are either gauge equivalent to small ones, or damped by the magnetic potential energy.

A related, more mathematical development is due to I. Singer. Observe that our functional Schrödinger equation involves a Laplacian — to be sure an infinite-dimensional Laplacian on the infinite-dimensional manifold of functions. Singer has shown that it is possible to define and estimate the curvature of such manifolds, and he finds that the space of all vector potentials modulo non-Abelian gauge transformations has positive curvature.[7]

Of course, properties of the [infinite-dimensional] manifold and its curvature cannot be the whole story, because the Hamiltonian without the magnetic "potential" $\frac{1}{4}F_{ij}^a F_{ij}^a$ still involves a Laplacian on the same infinite-dimensional manifold, but no gap is evident in this unphysical, but solvable example. Evidently, magnetic damping is also crucial.

Completing the above arguments would obviously be most gratifying. What is needed is a better understanding of the mathematics on infinite-dimensional spaces.

F. Measure of Integration

I conclude this lecture by briefly mentioning an effect of the Gauss law constraint on the functional measure of the integration in the inner product (1.31). Because $\left(D_i \frac{\delta}{\delta A^i}\right)_a \Psi(A) = 0$, $\Psi(A)$ does not depend on the gauge degrees of freedom in A_a^i. Consequently, physical states are not normalizable with respect to [functional] integration over *all* components of A_a^i, because $\Psi(A)$ does not depend on some of them. For example, QED wave functionals [in space-time greater than two] do not depend on the logitudinal component of A^i, while in two-dimensional space-time, the wave functionals do not depend on the local form of $A(x)$, merely on $\int_x A(x)$. Clearly, an infinity arises if we integrate over variables on which $\Psi(A)$ does not depend.

The problem and the cure is analogous to many-body quantum mechanics with translation invariance, in the center-of-mass frame (vanishing total momentum). These wave functions do not depend on the center-of-mass coordinate, and one avoids infinite norm by integrating only over relative coordinates.

Similarly, in quantized gauge theories, gauge functionals must not be integrated over the gauge components of vector potentials. In QED for example, the integration is over transverse vector potentials only.

However, it may be inconvenient to integrate over some but not all components of the potentials. Then one may alternatively integrate over *all* components provided one inserts a [functional] delta function of the gauge degrees of freedom. In QED this means inserting $\delta(\partial_i A^i)$: in non-Abelian gauge theories, where the gauge degrees of freedom are less manifest, the Faddeev-Popov procedure does the job.

I trust that the above seven examples have persuaded you that the functional Schrödinger representation is the most efficient way of arriving at those results about quantized gauge field theories.

III. REPRESENTING TRANSFORMATION GROUPS

We now turn to another kinematical topic that can be usefully analyzed in the Schrödinger representation: the implementation of a group of symmetry transformation in quantum theory.

A. Representing Transformation Groups in Quantum Theory

A transformation group with elements $\{g_i\}$ may be specified abstractly by giving the composition law

$$g_1 g_2 = g_{1 \circ 2} \, . \tag{3.1}$$

When dealing with Lie groups, the group elements are parametrized by generators θ of infinitesimal transformations

$$g = e^{\theta} \tag{3.2}$$

and the abstract Lie group is described by its abstract Lie algebra

$$[\theta_1, \theta_2] = \theta_{1 \circ 2} \, . \tag{3.3}$$

In quantum theory, these structures are frequently represented in a non-trivial, unstraightforward way. To give a representation, one identifies the quantum variables, *eg.*, P and Q for one-dimensional quantum mechanics, and seeks a unitary operator, U, $U^{\dagger} = U^{-1}$, which implements the transformation

$$Q \underset{g}{\to} Q^g = U(g) Q U^{\dagger}(g) \, , \tag{3.4a}$$

$$P \underset{g}{\to} P^g = U(g) P U^{\dagger}(g) \, . \tag{3.4b}$$

One may then ask how this operator acts on states, *i.e.*, in the Schrödinger representation, on wave functions $\psi(q)$ depending on q [or on p].

The simplest rule is that U transforms the argument of the function according to (3.4)

$$U(g)\psi(q) = \psi(q^g) \, . \tag{3.5}$$

By use of the group composition law (3.1), it follows that (3.5) implies the same composition law (3.1) for the unitary representation operators

$$U(g_1)U(g_2) = U(g_{1\circ2}) \tag{3.6}$$

and also it further follows from (3.5) or (3.6) that the composition law is associative

$$(U(g_1)U(g_2))\,U(g_3) = U(g_1)\,(U(g_2)U(g_3))\,. \tag{3.7}$$

The above, simplest story is realized in quantum mechanics, for example by the Abelian group of spatial translations and the non-Abelian group of spatial rotations.

However, this simplest story may be elaborated. We have already seen one elaboration, when (3.5) is relaxed by allowing the action of the unitary operator on wave functions not only to transform the argument, but also to multiply the wave function by a phase. Instead of (3.5), we have

$$U(g)\psi(q) = e^{-2\pi i\omega_1(q;g)}\psi(q^g)\,. \tag{3.8}$$

However, the composition law (3.6) is still retained, and it implies the 1-cocycle condition on ω_1

$$\omega_1(q^{g_1}; g_2) - \omega_1(q; g_{1\circ2}) + \omega_1(q; g_1) = 0 \quad \text{mod integer}\,. \tag{3.9}$$

As I showed, the above happens for Galileo boosts in non-relativistic quantum mechanics and for gauge transformations in three-dimensional topologically massive gauge theories with Chern-Simons term.

The next generalization within representation theory consists of abandoning (3.6) as well, by allowing a phase to intervene in the composition law of the unitary operators. Instead of (3.6) we can have

$$U(g_1)U(g_2) = e^{-2\pi i\omega_2(q;g_1,g_2)}U(g_{1\circ2})\,. \tag{3.10}$$

Nevertheless, associativity is retained and (3.7) implies

$$\omega_2(q^{g_1}; g_2, g_3) - \omega_2(q; g_{1\circ2}, g_3) + \omega_2(q; g_1, g_{2\circ3})$$
$$- \omega_2(q; g_1, g_2) = 0 \quad \text{mod integer}\,. \tag{3.11}$$

When a quantity depends on two group elements and possibly on q — the point on the manifold where the group acts — and also satisfies (3.11), it is called a *2-cocycle*.

Representations that make use of 2-cocycles are called *projective* or *ray* representations and they occur in quantum mechanics, for example in the representation of the full Galileo group of transformations [coordinate translations and boosts]

$$Q \to Q + vt + q_0 \equiv Q + a, \qquad (3.12a)$$

$$P \to P + mv \equiv P + b. \qquad (3.12b)$$

Transformations of this Abelian group of translations on phase space are implemented by the operator

$$U(a,b) = e^{i(aP - bQ)}, \qquad (3.13)$$

$$U(a,b)QU^\dagger(a,b) = Q + a, \qquad (3.14a)$$

$$U(a,b)PU^\dagger(a,b) = P + b \qquad (3.14b)$$

which composes according (3.10), with a 2-cocycle

$$2\pi\omega_2 = \frac{1}{2}(a_1 b_2 - a_2 b_1). \qquad (3.15)$$

[In this example, the group element is specified by (a,b). Also, ω_2 does not depend on the transformed point q.]

     For continuous Lie groups of transformations, the discussion may be carried out in infinitesimal terms. The infinitesimal action of the transformation is represented by generators G_i and $U(g) = e^{i\theta^i G_i}$, where the θ^i are infinitesimal parameters. G_i generates the infinitesimal transformation on dynamical variables: the infinitesimal version of (3.4) reads

$$Q \to Q^g = Q + \delta Q + \dots$$
$$\delta Q = i \left[\theta^i G_i, Q \right], \qquad (3.16a)$$

$$P \to P^g = P + \delta P$$
$$\delta P = i \left[\theta^i G_i, P \right]. \qquad (3.16b)$$

As is well-known, when the operators implementing finite transformations compose faithfully according to the group composition law (3.1) *i.e.*, when (3.6) holds, the generators follow the Lie algebra of the group (3.3)

$$[\theta_1^i G_i, \theta_2^j G_j] = i\theta_{1 \circ 2}^i G_i \qquad (3.17)$$

while the requirement of associativity (3.7) translates into the statement that the Jacobi identity holds. On the other hand, when the composition is projective with a 2-cocycle as in (3.10), then the algebra of generators differs from the group algebra by an *extension* $\delta\omega$, which is the infinitesimal portion of the 2-cocycle

$$[\theta_1^i G_i, \theta_2^j G_j] = iG_i \theta_{1 \circ 2}^i + i\theta_1^i \theta_2^j \delta\omega_{ij}. \qquad (3.18)$$

The 2-cocycle condition (3.11), in infinitesimal form, assures that (3.18) is consistent with the Jacobi identity. When the 2-cocycle does not depend on the transformed point, the extension is called a *central* extension. For a central extension, $\delta\omega$ is a *c*-number.

Returning to the example of translations on phase space, we recognize that the Heisenberg algebra

$$[Q, P] = i \tag{3.19}$$

represents the abstract Lie algebra of generators for that Abelian group, but the representation is not faithful, since the generators do not commute; rather there is a central extension in the commutator (3.19), coming from the 2-cocycle (3.15) in the quantum mechanical composition law (3.10). Thus we see that a projective representation is the very essence of quantum theory, a fact stressed by Weyl and Bargmann already half a century ago.[8]

B. Problems with Infinities in Quantum Field Theory

From the above discussion, we conclude that in general we must expect transformation groups to be realized projectively in a quantum theory. The surprise about the discoveries of recent times has been that in quantum *field* theory, which is beset by infinities, the infinities must be carefully controlled to expose the projective nature of a representation. Indeed, initial calculations were formal: they treated infinities cavalierly and overlooked the projective phases. Since typically the calculations concerned commutators of generators, what was missed was the extension, and when eventually found, it was called an *anomaly*, even though now we recognize it to be a natural and slight modification of the simplest representation theory.

In quantum field theory, extensions arise in various Lie algebras. First, there is the extension which is present in the canonical commutator between field momentum $\Pi(\mathbf{x})$ and field variable $\Phi(\mathbf{x})$

$$[\Phi(\mathbf{x}), \Pi(\mathbf{y})] = i\delta(\mathbf{x} - \mathbf{y}) . \tag{3.20}$$

This continuum field analog of the Heisenberg commutator (3.19) is not an "anomaly": indeed, it provides the starting point for canonical quantization. Anomalous extensions arise in the algebra of generators of local gauge transformations in any even-dimensional space-time, and in the algebra of the two-dimensional conformal group.

Two dimensions is especially interesting because one can prove the *necessary* occurrence of an extension from general principles of relativistic and unitary quantum field theory. The argument begins with Schwinger's well-known proof that the equal-time commutator between time and space components of a symmetry current, as well as between the time-time (energy density) and the time-space (momentum density) components of the energy momentum tensor, contains non-canonical terms, as a consequence of Lorentz invariance and positivity of the inner product. For chiral fermions in two dimensions, the space component of the current is proportional to the time component, so one may use Schwinger's argument to infer the existence of an anomalous commutator between the time components of currents, but these are just the generators of local gauge transformations on the Fermi fields. Also, two-dimensional conformal transformations are generated by components of the energy momentum tensor; hence once again the necessary anomalous commutator provides an extension. In higher dimensions, Schwinger's argument still assures that the commutator between temporal and spatial components of tensors contains extra terms — but the spatial components are no longer related to generators. Hence there is no necessity for an extension, and indeed one can construct models that are carefully adjusted so that the generator algebra is faithful and without extension. Such constructions are important for unified gauge theories, for particle phenomenology and for string theories.

Anomalies have many further ramifications, and their study has become a vast subject. One important consequence is that anomalous commutators between transformation generators may affect other commutators and in particular a generator may fail to commute with the Hamiltonian when the extension is taken into account. In this way, symmetries of a classical theory may disappear in a quantum theory, without any explicit symmetry-breaking interactions or spontaneous symmetry-breaking instabilities. This *quantum mechanical* symmetry violation has important phenomenological consequences [low energy dynamics of mesons, anomaly cancellation constraints]. The study of anomalies has also seeded a rebirth of an interaction between physicists and mathematicians, which is accelerating with impetus from string research.[3]

Here I shall discuss how field commutator extensions are found in the Schrödinger representation. As we shall see, there are certain advantages to this approach over the conventional Fock space method.

In the most frequent situation, the generators of the relevant symmetry transformation are first constructed formally; typically they are polynomials in the canonical variables, Π and Φ, and they generate the infinitesimal transformation on Π and Φ by commutation. Moreover, their commutators when evaluated formally follow the Lie algebra of the group

$$[\theta_1^i Q_i, \theta_2^j Q_j] = i\,\theta_{1\circ2}^i Q_i\,.$$

(3.21)

[I shall now use Q, rather than G, to describe the generator.] Equation (3.21) holds in classical field theory, with Poisson bracketing. It appears to hold in quantum theory with canonical commutation; but this is misleading since Q is an ill-defined quantum mechanical operator containing products of operators at the same point.

To arrive at well-defined generators, a three-step procedure is adopted. First, the formal expression for Q is regulated in some fashion so that no ill-defined products occur: $Q \to Q^R$. [Henceforth, we suppress the label i in the generators.] Second, the singular portions of Q^R, which are ill-defined in the absence of the regularization, are isolated and removed. For the simple models that we consider, a c-number subtraction q^R suffices. Finally, third, the regulators are removed from the subtracted expression, leaving well-defined generators, which we denote by $: Q :$, even though the colons do not necessarily signify normal ordering

$$: Q : \equiv \lim_R (Q^R - q^R)\,.$$

(3.22)

The generators $: Q :$, well-defined in the above manner, continue to generate the infinitesimal transformations on the canonical variables. However, non-linear relations like (3.21) can be modified. From (3.21) and (3.22) one gets

$$[: Q_1 :, : Q_2 :] = i : Q_{1\circ2} : + i \lim_R q_{1\circ2}^R\,.$$

(3.23)

If the limit of $q_{1\circ2}^R$ is non-zero, the quantum field theoretic realization of the Lie algebra acquires an extension, not seen in the classical theory — this is the origin of an anomaly.

It still remains to decide how the regularizing subtraction q^R should be determined. In the conventional approach, a Fock vacuum is chosen, q^R is the expectation of Q^R in that state and $: Q :$ is *normal ordered* with respect to that vacuum. While this procedure may produce well-defined results, they depend on the covariance of the vacuum, specifically the extension in the Lie algebra can depend on the vacuum.[9] Furthermore, normal ordering with respect to the inequivalent vacua that define inequivalent Fock spaces can produce inequivalent representations of the transformation group.

This must be viewed as a disadvantage of the conventional procedure. Choice of vacuum is a dynamical issue, while representation of a transformation group is kinematical. It should be possible within field theory to represent transformations without reference to a dynamical Hamiltonian, just as in ordinary quantum mechanics. Moreover, even when dynamics is specified by a definite Hamiltonian, there need not be a unique or natural choice for the vacuum, as for example when the Hamiltonian is time-dependent.

The Schrödinger representation allows determining the subtraction intrinsically, without normal ordering and without preselecting any Fock space.[10] We begin by considering the Schrödinger representation of a regulated generator. It is a functional kernel

$$Q^R(\Pi, \Phi) \to Q^R\Big(\frac{1}{i}\frac{\delta}{\delta\varphi}, \varphi\Big)\delta(\varphi - \tilde{\varphi})$$ (3.24)

involving a functional delta function. Because of the functional delta function this expression does not yield useful information about the singularities of Q^R when the regulators are removed. But we may also consider the functional representation kernel for the operator that implements the finite [regulated] transformation on functional states

$$e^{-i\tau Q^R} \to U_\tau^R(\varphi, \tilde{\varphi}),$$ (3.25)

$$e^{-i\tau Q^R}|\Psi\rangle \to \int \mathcal{D}\tilde{\varphi} U_\tau^R(\varphi, \tilde{\varphi})\Psi(\tilde{\varphi}).$$ (3.26)

Evidently, U_τ^R satisfies a functional Schrödinger-like equation

$$i\frac{\partial}{\partial\tau}U_\tau^R(\varphi, \tilde{\varphi}) = Q^R\Big(\frac{1}{i}\frac{\delta}{\delta\varphi}, \varphi\Big)U_\tau^R(\varphi, \tilde{\varphi})$$ (3.27)

with initial condition

$$U_{\tau=0}^R(\varphi, \tilde{\varphi}) = \delta(\varphi, \tilde{\varphi}).$$ (3.28)

Because the regulators are present, everything is well-defined and a unique solution may be found. $U_\tau^R(\varphi, \tilde{\varphi})$ is a functional of φ and $\tilde{\varphi}$ [rather than a functional distribution like the representation (3.24) for Q^R] with specific dependence on φ and $\tilde{\varphi}$ as well as on the regulators. So its behavior when regulators are removed may be explicitly studied.

Generically, U_τ^R becomes singular as the regulators are removed, but in the simple models that we consider, the interesting infinities are confined to a φ-independent phase $e^{-i\tau q^R}$. Thus $e^{i\tau q^R}U_\tau^R(\varphi, \tilde{\varphi})$ possesses a well-defined limit. Since

$$e^{-i\tau(Q^R-q^R)} \to e^{i\tau q^R}U_\tau^R(\varphi, \tilde{\varphi})$$ (3.29)

the regularizing subtraction for the generator Q^R is determined to be q^R. In our Schrödinger representation approach, q^R is determined intrinsically without referring to any Fock space or pre-selecting any vacuum. Note finally that when q^R survives in the limit that the regulators are removed, the above construction yields a projective representation for the transformation group. Even when there is no cocycle in the composition law for the regulated transformation kernels, multiplied by the renormalizing phase factor, introduces one. Hence, the cocycle is explicitly determined by our procedure, as is its infinitesimal part, which gives the extension for the Lie algebra.

C. Two-Dimensional Conformal Transformation

Quantum field theoretic representation for conformal transformations on two-dimensional space-time illustrates well our program.[10] Conformal transformations in two dimensions form a doubly infinite transformation group, whereby $x \pm t$ are taken into arbitrary, and in general, different functions of $x \pm t$. At fixed time, the infinitesimal transformation law for the coordinate x

$$\delta_f x = -f(x) \tag{3.30}$$

obeys a Lie algebra given by the Lie bracket,

$$[\delta_f, \delta_g]x = -\delta_{(f,g)}x \tag{3.31a}$$

$$(f,g) = fg' - gf'. \tag{3.31b}$$

[The dash signifies differentiation with respect to the argument x.] The infinitesimal transformation law for a general field $\Theta(x)$ may be taken as

$$\delta_f \Theta = f\Theta' + df'\Theta + \frac{\alpha}{2\pi}f^{(d+1)}. \tag{3.32}$$

Here d is a numerical constant, called the *scale dimension*, and the inhomogeneous term, involving an arbitrary constant α and $d+1$ derivatives, may be present for $d = 0, 1, 2$. One verifies that the composition law for two transformations (3.32) with functions f and g, follows the defining composition law (3.31).

We obtain first a representation in terms of a quantized boson field, for the case $d = 1$, $\alpha = 0$. To this end, we consider a field operator $\chi(x)$ that satisfies the [equal time] commutation relation consistent with scale dimensionality 1,

$$[\chi(x), \chi(y)] = i\delta'(x-y) \equiv k(x,y) = \int \frac{dp}{2\pi}e^{-ip(x-y)}p. \tag{3.33}$$

[One may think of χ as $\frac{1}{\sqrt{2}}(\Pi + \Phi')$ where Π and Φ are canonically conjugate, but this is not necessary.] The formal generator of the transformation

$$Q_f = \frac{1}{2}\int dx\, \chi(x)f(x)\,\chi(x) \tag{3.34}$$

transforms the field operator χ as

$$\delta_f \chi = i\,[Q_f, \chi] = (f\chi)' \qquad (3.35)$$

which coincides with (3.32) for $d = 1$ and $\alpha = 0$. The generators follow the Lie algebra (3.31)

$$[Q_f, Q_g] = iQ_{(f,g)} \qquad (3.36)$$

when the commutator is evaluated formally, without care about the product of χ with itself at the same point.

To regulate the generator, we promote f to a bilocal function $F(x, y)$ and define,

$$Q_f^R \equiv Q_F = \frac{1}{2} \int \chi(x) F(x, y) \chi(y) \qquad (3.37)$$

while removing the regulator consists of passing to the local limit

$$F(x, y) \rightarrow \frac{1}{2} \left(f(x) + f(y) \right) \delta(x - y)\,. \qquad (3.38)$$

$F(x, y)$ is taken to be real and symmetric in (x, y) and sufficiently well-behaved near $x \approx y$ to permit all formal manipulations. Commutators of the regulated generators may be readily evaluated since there are no singularities associated with operators at coincident points

$$[Q_F, Q_G] = iQ_{(F,G)}$$
$$i(F, G) \equiv FkG - GkF\,. \qquad (3.39)$$

Of course Q_F no longer generates conformal transformations, rather general linear canonical transformations.

In the Schrödinger representation for the above quantities, we represent χ by the kernel

$$\chi(x) \rightarrow \frac{1}{\sqrt{2}} \left(\frac{1}{i} \frac{\delta}{\delta\varphi(x)} + \varphi'(x) \right) \delta(\varphi - \tilde{\varphi}) \qquad (3.40)$$

so that (3.33) is regained. As a consequence, the regulated transformation kernel

$$U_\tau^R(\varphi, \tilde{\varphi}) \equiv U(\varphi, \tilde{\varphi}; \tau F) \qquad (3.41)$$

satisfies the equation,

$$i\frac{\partial}{\partial\tau} U(\varphi, \tilde{\varphi}; \tau F) = \frac{1}{4} \int \left(\frac{1}{i} \frac{\delta}{\delta\varphi} + \varphi' \right) F \left(\frac{1}{i} \frac{\delta}{\delta\varphi} + \varphi' \right) U(\varphi, \tilde{\varphi}; \tau F) \qquad (3.42)$$

which is the differential expression of the composition law

$$\int \mathcal{D}\tilde{\varphi}\, U(\varphi_1, \tilde{\varphi}; F)U(\tilde{\varphi}, \varphi_2; G) = U(\varphi_1, \varphi_2; F \circ G)$$

$$F \circ G = F + G + \frac{1}{2}(F, G) + \dots . \tag{3.43}$$

The solution is a Gaussian, times a normalization factor N_F

$$U(\varphi_1, \varphi_2; F) = N_F \exp - \int \varphi_1 k\, \varphi_2 \exp \frac{i}{2} \int (\varphi_1 - \varphi_2) K_F(\varphi_1 - \varphi_2), \tag{3.44}$$

$$N_F = \det^{-1/2} F^{1/2}\left(\frac{2\pi i}{\mathcal{F}} \sin \frac{\mathcal{F}}{2}\right) F^{1/2}, \tag{3.45}$$

$$K_F = F^{-1/2}\left(\mathcal{F} \operatorname{ctn} \frac{\mathcal{F}}{2}\right) F^{-1/2}, \tag{3.46}$$

$$\mathcal{F} \equiv F^{1/2} k\, F^{1/2}. \tag{3.47}$$

Constant factors in N_F have been fixed by the initial condition

$$U(\varphi_1, \varphi_2; 0) = \delta(\varphi_1 - \varphi_2). \tag{3.48}$$

The representation is clearly unitary. That the composition law (3.43) is indeed satisfied may be verified with the help of trigonometric identities when F if proportional to G; the general case is checked by expanding $F \circ G$.

In the local limit, K_F attains a well-defined expression

$$K_F(x, y) \to K_f(x, y) = \frac{1}{f(x)}\left\{ \int \frac{d\lambda}{2\pi}\left(\lambda \operatorname{ctn} \frac{1}{2}\lambda\right) \exp\left(-i\lambda \int_y^x \frac{dz}{f(z)}\right)\right\} \frac{1}{f(y)}$$

$$= -i\pi \frac{1}{f(x)} P \csc^2\left\{\pi \int_y^x \frac{dz}{f(z)}\right\} \frac{1}{f(y)}. \tag{3.49}$$

[P means principle value.] The normalization constant, N_F however, diverges. The divergence resides in an unimportant constant factor Z which may be removed by redefining the measure of functional integration and in a phase e^{-iq_F}, which is determined in imaginary τ [$F \to -iF$]; this continuation, rather than $F \to iF$, is appropriate when the energy spectrum is bounded below

$$q_F = \frac{1}{4} \operatorname{tr} F\omega, \tag{3.50}$$

$$\omega(x, y) \equiv |k|(x, y) = \int \frac{dp}{2\pi} e^{-ip(x-y)}|p| = -P\frac{1}{\pi(x-y)^2}. \tag{3.51}$$

[We use the notation $|\ldots|$ on a kernel to represent the absolute value kernel, defined through its spectral representation by taking the absolute value of the eigenvalues, as in (3.33) and (3.51).] It follows that

$$Z^{-1}e^{-i(Q_F-q_F)} \rightarrow Z^{-1}e^{iq_F}U(\varphi_1,\varphi_2;F) \qquad (3.52)$$

possesses a well-defined local limit, and we are led to define the renormalized generator by

$$:Q_f: \equiv \lim_{F\to f}\left(Q_F - \frac{1}{4}\operatorname{tr}F\omega\right). \qquad (3.53)$$

Notice that the renormalizing subtraction $q^R \equiv q_F$ has been determined without choosing any vacuum state. Since q_F is a numerical quantity, independent of φ, the subtraction is a c-number which does not change (3.35). But the non-linear commutator (3.36) is modified as in (3.23), and the Lie algebra of the renormalized generators acquires a central extension

$$[:Q_f:,:Q_g:] = i:Q_{(f,g)}: -\frac{i}{48\pi}c\int(fg''' - gf''') \qquad (3.54)$$

$$c = 1.$$

Evidently, the composition law (3.43) acquires a 2-cocycle

$$2\pi\omega_2(F,G) = -\frac{1}{4}\operatorname{tr}(F\circ G - F - G)\,\omega. \qquad (3.55)$$

With bilocal F and G this is a "trivial" cocycle — it can be removed by redefining phases, indeed it arose because an additional phase was inserted in the redefinition of U. But in the local limit, *i.e.*, when the regularization is removed and the bilocal functions become local, ω_2 is non-trivial — no phase redefinition eliminates it.

Of course, the subtraction is ambiguous up to terms that are finite in the local limit; they are obviously "trivial" in the sense that they may be adjusted at will by a finite redefinition of the generators. But the result for the non-trivial part of the extension — not removable by redefining generators — is unique and is not specific to the representation (3.40) for χ, which may be generalized to

$$\chi \rightarrow \frac{i}{\sqrt{2}}\left(\alpha\frac{\delta}{\delta\varphi} + \beta k\varphi\right)\delta(\varphi - \bar{\varphi}) \qquad (3.56a)$$

provided the kernels α and β satisfy

$$\frac{1}{2}(\alpha k\beta^T + \beta k\alpha^T) = k \qquad (3.56b)$$

as required by (3.33). One may verify that the representation kernel which arises from the more general formula (3.56) possesses a different Gaussian φ_1, φ_2, dependence and the infinite constant Z is modified. But the F dependent infinity (3.50) is unaffected by the generalization (3.56), so the center in (3.54) remains the same.

Our functional transformation kernel allows computing how states transform under conformal transformations

$$\Psi(\varphi) \xrightarrow[F]{} \Psi_F(\varphi) = e^{iq_F} \int \mathcal{D}\bar\varphi \, U(\varphi, \bar\varphi; F)\Psi(\bar\varphi) \,. \tag{3.57}$$

In particular, for a Gaussian with covariance Ω, the transformed state is again Gaussian with transformed covariance

$$\Omega_F = \Omega - (\Omega + k)(\Omega - iK_F)^{-1}(\Omega - k) \,. \tag{3.58a}$$

Also the transformed state acquires an additional phase θ_F

$$\theta_F = q_F + \mathrm{Im} \left[\ln N_F - \frac{1}{2} \, \mathrm{tr} \, \ln(\Omega - iK_F) \right] \,. \tag{3.58b}$$

The Fock vacuum of a massless theory possesses $\Omega = \omega$, which is invariant under the $SO(2,1)$ subgroup conformal transformation generated by $f(x) = (1, x, x^2)$. More generally, Ω is a representation for the conformal algebra without center; the latter resides in the representation provided the phase θ_F.

It is possible to obtain inequivalent representations of the conformal Lie algebra with centers $c > 1$ in (3.54). For these, we employ the transformation law (3.32) with $d = 1$, but $\alpha \neq 0$. The inhomogeneous transformation law can be realized on our field χ by "improving" the generator (3.34) with the addition $\frac{\alpha}{\sqrt{2\pi}} \int f'\chi$. The modified generators

$$Q_f^\alpha = \frac{1}{2} \int f \chi^2 + \frac{\alpha}{\sqrt{2\pi}} \int f'\chi \tag{3.59}$$

effect the transformation

$$i\,[Q_f^\alpha, \chi] = (f\chi)' + \frac{\alpha}{2\pi} f''(x) \tag{3.60}$$

and they possess an extension already on the classical level: when their commutator is evaluated with Poisson brackets one finds $c = 12\alpha^2$. In the quantum theory, after proper definition of the bilinear, the quantum extension adds the additional 1, so that $: Q_f^\alpha :$ satisfies

$$[: Q_f^\alpha :, : Q_g^\alpha :] = i : Q_{(f,g)}^\alpha : - \frac{i}{48\pi}(1 + 12\alpha^2) \int (fg''' - gf''') \,. \tag{3.61}$$

The transformation kernel corresponding to the improved generator is the previous expression, (3.44), appropriate to $\alpha = 0$, times the additional factor

$$\exp -i\frac{\alpha}{\sqrt{2\pi}} \int (\ln f)' \left(\varphi_1 - \varphi_2 - \frac{\alpha}{\sqrt{16\pi}} f' \right).$$

Fields that transform with $d = 2$ and inhomogeneously are provided by $: \chi^2 :$. This is seen from (3.61) with $\alpha = 0$, which upon functional differentiation with respect to the arbitrary function $g(x)$ yields

$$i\left[: Q_f :, : \chi^2 : \right] = f \; : \chi^2 :' +2f : \chi^2 : -\frac{1}{12\pi} f'''. \tag{3.62}$$

Here, unlike (3.59), the inhomogeneous term is a quantum effect, not present classically.

Finally, fields with $d = 0$ are constructed from our χ by

$$\theta(x) = \frac{1}{2} \int dy\, \varepsilon(x - y)\chi(y). \tag{3.63}$$

When χ is transformed inhomogeneously as in (3.60), θ transforms as in (3.32) with $d = 0$. Note that by virtue of (3.33) θ satisfies commutation relations appropriate to $d = 0$

$$[\theta(x), \theta(y)] = -\frac{i}{2}\varepsilon(x - y). \tag{3.64}$$

Although our presentation has been purposefully without reference to dynamics, let me conclude this topic by recalling that the conformal group is a symmetry group for the free, massless two-dimensional boson theory and for the Liouville theory [exponential interaction]. The action is invariant when the field transforms with $d = 0$, and in the free case any α is allowed, while in the Liouville model α must be the coupling constant. Then χ is identified with $\frac{1}{\sqrt{2}}(\Pi + \Phi')$ and the generator is constructed from the energy-momentum tensor, appropriately improved for $\alpha \neq 0$.

An even simpler conformally invariant theory is that of a self-dual field governed by a non-local Lagrangian[11]

$$L = \frac{1}{4} \int dx\, dy\, \chi(t, x)\, \varepsilon(x - y)\, \dot{\chi}(t, y) - \frac{1}{2} \int dx\, \chi^2(t, x). \tag{3.65}$$

[The dot signifies time differentiation.] The equation of motion satisfied by the field χ is the self-dual one

$$\dot{\chi} = \chi' \tag{3.66}$$

and the Lagrangian changes by the total derivative $\frac{d}{dt}\left(-\frac{\alpha}{\sqrt{8\pi}} f'\chi\right)$ when the field is transformed as

$$\delta \chi(t,x) = (f(t+x)\chi(t,x))' + \frac{\alpha}{\sqrt{2\pi}} f''(t+x) \qquad (3.67)$$

so that the conserved generator is Q_f^α, and $\frac{1}{2}\chi^2$ is the Hamiltonian density. Canonical quantization of this first-order theory leads to the commutator (3.33) for χ. Thus our realization of the conformal group coincides with the fixed-time ($t = 0$) quantization of the above model.

Observe also that θ, introduced in (3.63), is a self-dual field, when χ is. The dynamics for θ may be deduced from (3.63) and (3.66). The Lagrangian is local

$$L = \frac{1}{2} \int \left(\theta'\dot\theta - (\theta')^2 \right). \qquad (3.68)$$

The non-locality, hidden in (3.63), appears only in the commutator (3.64).

Are there models with $c < 1$? It is known that unitarity restricts values of c below 1 to a discrete series.[12] The first permitted is $c = \frac{1}{2}$ and this is realized for free Majorana fermions, as will be discussed below. Dirac fermions, even when interacting as in the Thirring model, possess $c = 1$. For $\frac{1}{2} < c < 1$, there are known statistical models at discrete values of c, but we have not found a field theoretic projective representation for these cases.

The conformal group in two-dimensions is represented efficiently and unambiguously in the Schrödinger functional formalism. It may also be represented when the generators are normal ordered with respect to a Fock vacuum, and the two agree when the covariance of the vacuum satisfies $\omega(p) \xrightarrow[p\to\infty]{} |p|$. Thus, ultimately it is a matter of choice which method is used, although the intrinsic approach seems preferable to me since a unique answer emerges without preselecting a specific ground state of some quadratic Hamiltonian.

D. Field Theory in de Sitter Space

I now present an analysis of a problem where the Hamiltonian is quadratic, but does not select a unique Fock space. The example concerns field theory in de Sitter space. Because the Hamiltonian is time-dependent, stationary energy eigenstates do not exist, and a conventional Fock space cannot be associated to this quadratic Hamiltonian. Therefore, transformation generators cannot be renormalized by conventional normal ordering methods. Only with the development of our Schrödinger picture approach has it been possible to resolve a long-standing puzzle of how conformal symmetries are realized field theoretically in de Sitter space.[13]

In $1+1$-dimensional de Sitter space, the line element is

$$ds^2 = dt^2 - e^{2ht}dx^2 \tag{3.69}$$

providing a metric for a particular open slicing of the general space, defined by the surface of a hyperboloid of revolution in $2+1$ dimensions. The space-time possesses constant curvature $2h^2$. It is convenient to pass to conformal coordinates, by defining a "conformal time"

$$\tilde{t} = \frac{e^{-ht}}{h}, \tag{3.70a}$$

$$ds^2 = \frac{1}{\tilde{t}^2 h^2}(d\tilde{t}^{\,2} - dx^2). \tag{3.70b}$$

Both metrics are time-dependent. [There does exist a choice of coordinates in which the metric is time-independent, but then it is singular in space, so we do not use it.] We shall work with (3.70) and we suppress the tilde.

The Killing equation

$$f_{\alpha;\beta} + f_{\beta;\alpha} = 0 \tag{3.71}$$

for the infinitesimal isometries f^μ of de Sitter space-time admits three linearly independent solutions. The simplest corresponds to spatial translations

$$f_1^\mu = (0, 1). \tag{3.72}$$

The other two — a dilatation and a restricted spatial conformal transformation [translation in inverted space] —

$$f_2^\mu = (t, x), \tag{3.73}$$

$$f_3^\mu = \left[tx, \frac{1}{2}(t^2 + x^2)\right] \tag{3.74}$$

together with (3.72) close on the $SO(2,1)$ de Sitter Lie algebra

$$[f_1, f_2]^\mu = f_1^\mu, \quad [f_1, f_3]^\mu = f_2^\mu, \quad [f_2, f_3]^\mu = f_3^\mu$$
$$[f_i, f_j]^\mu = f_i^\alpha \partial_\alpha f_j^\mu - f_j^\alpha \partial_\alpha f_i^\mu. \tag{3.75}$$

A scalar field Φ with mass m, without self-interaction but in de Sitter space, is governed by the Lagrange density

$$\mathcal{L} = \sqrt{-g}\,\frac{1}{2}\,[g^{\mu\nu}\partial_\mu\Phi\partial_\nu\Phi - m^2\phi^2] \tag{3.76}$$

and the covariantly conserved energy-momentum tensor is

$$T_{\mu\nu} = \partial_\mu\Phi\partial_\nu\Phi - \frac{1}{2}g_{\mu\nu}(g^{\alpha\beta}\partial_\alpha\Phi\partial_\beta\Phi - m^2\Phi^2). \tag{3.77}$$

Time-independent generators of symmetry transformations are constructed from the energy-momentum tensor projected on the Killing vectors

$$Q_i = \int dx \sqrt{-g}\, T^0{}_\nu f_i^\nu \,. \tag{3.78a}$$

In this way, we are led to three generators, which formally are given by

$$Q_i = \int dx \left\{ \frac{1}{2} \left(\Pi^2 + \Phi'^2 + \frac{m^2}{h^2 t^2} \Phi^2 \right) f_i^0 + \Pi \Phi' f_i^1 \right\}, \tag{3.78b}$$

where the canonical momentum Π is defined by $\Pi = \frac{\delta \mathcal{L}}{\delta \dot{\Phi}} = \dot{\Phi}$. Observe that Q_1 is just the total momentum. The $SO(2,1)$ invariance group generated by (3.78) — the de Sitter group — is the analog in de Sitter space of the flat space-time Poincaré group. [Note the two have the same number of generators.]

The first question that arises in analyzing this model concerns what are the interesting states. Since the Hamiltonian

$$\begin{aligned}
H &= \int dx \left(\Pi \dot{\Phi} - \mathcal{L} \right) = \int dx \sqrt{-g}\, T^0_0 \\
&= \frac{1}{2} \int dx \left(\Pi^2 + \Phi'^2 + \frac{m^2}{h^2 t^2} \Phi^2 \right)
\end{aligned} \tag{3.79}$$

is time-dependent owing to the mass term, it makes no sense to define the vacuum as the lowest energy eigenvalue: the eigenvalues are time-dependent. Equivalently, the time-dependent [functional] Schrödinger equation

$$i\frac{\partial}{\partial t} \Psi(\varphi; t) = \frac{1}{2} \int dx \left(-\frac{\delta^2}{\delta \varphi(x) \delta \varphi(x)} + \varphi'^2(x) + \frac{m^2}{h^2 t^2} \varphi^2(x) \right) \Psi(\varphi; t) \tag{3.80}$$

does not separate in time.

A plausible definition for the Fock vacuum is that it is a Gaussian solution to (3.80). Making the Gaussian *Ansatz*

$$\Psi_\Omega(\varphi; t) = N e^{-\frac{1}{2} \int \varphi \Omega \varphi} \tag{3.81}$$

we find that the covariance $\Omega(x, y; t)$ solves

$$-i\frac{\partial}{\partial t} \Omega = -\Omega^2 + k^2 + \frac{m^2}{h^2 t^2} \tag{3.82}$$

and the normalization factor $N(t)$ satisfies

$$i\frac{\partial}{\partial t} \ln N = \frac{1}{2} \operatorname{tr} \Omega \,. \tag{3.83}$$

But even with the Gaussian *Ansatz*, the state is not uniquely determined because the first order differential equation (3.82) for Ω needs an initial condition for unique solution. [Constants in N may be fixed by requiring that the state be normalized to unity.] If the space-time were asymptotically flat, one could prescribe conventional boundary conditions by requiring that the asymptotic isometries of Poincaré invariance be implemented in the vacuum state, *i.e.*, asymptotically it should be annihilated by the generators of the Poincaré group. However, our de Sitter space does not possess a flat limit. One might look for physical principles to remove the arbitrariness. De Sitter space may be relevant to the extreme conditions of the very early universe. Unfortunately, the initial conditions there are certainly not known.

Faced with this impasse, the following strategy is employed. Rather than fixing Ω by specifying initial conditions, we attempt to get a unique formula by requiring the state be invariant against the isometries of the background metric, in this case $SO(2, 1)$. This is a plausible strategy: in flat space, one may also solve the time-dependent Schrödinger equation with a Gaussian *Ansatz*. When Poincaré invariance is demanded one arrives at the unique Fock vacuum with covariance $\omega(p) = \sqrt{p^2 + m^2}$.

Thus we are led to seek Gaussian solutions of (3.82) that are also $SO(2, 1)$ invariant. Translation invariance is easily implemented by choosing a translation invariant covariance

$$\Omega(x, y; t) = \int \frac{dp}{2\pi} e^{ip(x-y)} \Omega(|p|; t) \qquad (3.84)$$

i.e., Ω depends only on the kernel $|k|$; in the momentum representation Ω is diagonal. With this choice, the total momentum — the translation generator — $Q_1 = i \int \Phi k \Pi = i \int dx \varphi'(x) \frac{\delta}{\delta \varphi(x)}$ annihilates the state. The solutions to (3.82) and (3.83) are now determined as

$$\Omega = -i \frac{\partial}{\partial t} \ln D \qquad (3.85)$$

$$N = \det^{-1/2} D \qquad (3.86)$$

$$D = \left(\frac{1 - r^2}{r}\right)^{1/2} \left(\frac{\pi}{|k|\Theta'_\nu(t\,|k|)}\right)^{1/2} \cos\left[\Theta_\nu\left(t\,|k|\right) - \alpha\left(|k|\right)\right]. \qquad (3.87)$$

Here, Θ_ν is the real phase of $e^{\frac{i\nu\pi}{2}}$ times the Hankel function of order $\nu \equiv \sqrt{\frac{1}{4} - \frac{m^2}{\hbar^2}}$

$$e^{\frac{i\nu\pi}{2}} H_\nu^{(1)} = M_\nu e^{i\Theta_\nu}$$

$$M_\nu^2(x)\Theta'_\nu(x) = \frac{2}{\pi x}. \qquad (3.88)$$

[The reason for introducing the factor $e^{\frac{i\nu\pi}{2}}$, which is absent from the conventional formulas for phase and modulus of Hankel functions, is that our Θ_ν remains real as ν becomes imaginary.] The single integration constant α is complex, and at this stage of the argument may still depend on $|k|$. Its imaginary part determines $r \equiv \tanh(\operatorname{Im}\alpha)$. Also, Ω is complex, and may be separated into real and imaginary parts

$$\Omega = \Omega_R + i\Omega_I. \tag{3.89}$$

Various multiplicative constants are adjusted so that $\int \mathcal{D}\varphi \Psi^* \Psi = 1$.

More explicitly, the formulas read

$$N = \det^{-1/2}\left\{\left(\frac{\pi^2 t}{2}\right)^{1/2} H_\nu^{(1)}(t\,|k|)\right\}$$

$$\times \det^{-1/2}\left\{\frac{r^{-1/2}\big(\Theta_\nu(t\,|k|) - \theta\big) + ir^{1/2}\sin\big(\Theta_\nu(t\,|k|) - \theta\big)}{\cos\big(\Theta_\nu(t\,|k|) - \theta\big) + i\sin\big(\Theta_\nu(t\,|k|) - \theta\big)}\right\}. \tag{3.90}$$

A t-independent phase has been dropped, and $\theta \equiv \operatorname{Re}\alpha$

$$\Omega_R = |k|\frac{r\Theta_\nu'(t\,|k|)}{\cos^2\big(\Theta_\nu(t\,|k|) - \theta\big) + r^2\sin^2\big(\Theta_\nu(t\,|k|) - \theta\big)}, \tag{3.91a}$$

$$\Omega_I = |k|\left(\frac{1}{2}\frac{\Theta_\nu''(|k|)}{\Theta_\nu'(t\,|k|)} + \Theta_\nu'(t\,|k|)\frac{(1 - r^2)\tan\big(\Theta_\nu(t\,|k|) - \theta\big)}{1 + r^2\tan^2\big(\Theta_\nu(t\,|k|) - \theta\big)}\right). \tag{3.91b}$$

Owing to translation invariance, N diverges: D and Ω are diagonal in the momentum representation, proportional to $(2\pi)\delta(p - p')$, times a function whose form is as in (3.87) and (3.91) with the kernel $|k|$ replaced by the variable $|p|$. Hence, the determinant, which involves a trace, acquires $(2\pi)\delta(0)$ — the length of space, L. This infrared divergence may be avoided by regulating the translationally invariant kernel $k(x,y) \equiv i\delta'(x - y)$ in some translationally non-invariant way. We shall not take the time here to exercise this care, which would lengthen the presentation without altering results. The careful treatment of this point has been given in the literature.[13]

We have arrived at formulas which contain one arbitrary complex function α of $|k|$, or two real functions, r and θ. To specify the state further, we consider the remaining two $SO(2,1)$ generators in (3.78): the dilatation $i = 2$ (3.73), and the restricted spatial conformal transformation $i = 3$, (3.74). Invariance of our Gaussian will be achieved if these also annihilate the state, but first they must be well-defined by regulating and renormalizing them. This being a two-dimensional theory without self-interactions, a subtraction should suffice, but now we see that the conventional Fock space method does not provide a way of determining the subtraction. We cannot subtract the vacuum expectation value of Q_i because we do not have a unique vacuum, but a whole α-dependent family of candidate vacua. Without knowledge of the subtraction, the most we can require is that off-diagonal states not be created by the application of Q_i to the Fock vacuum that we seek.

In the translation invariant situation, the conditions

$$Q_i \Psi_\Omega \propto \Psi_\Omega \qquad (3.92)$$

for the remaining two isometries, $i = 2$ and 3, imply one equation on Ω

$$t \frac{\partial}{\partial t} \frac{\Omega(|p|;t)}{p} = p \frac{\partial}{\partial p} \frac{\Omega(|p|;t)}{p} \qquad (3.93)$$

which is satisfied provided that α is constant, as we henceforth require. This leaves the state depending on one complex parameter, α. Moreover, one may check that the restricted conformal generator Q_3 annihilates our translation invariant state, because its action involves $\int dx\, f_3^0(x)$, which vanishes by parity since $f_3^0(x) = tx$. [Recall that restricted conformal transformations are translations in an inverted coordinate system.]

In the absence of further information, this is as far as one can go and the situation has been summarized in the older literature by the statement that there is a one parameter (α) family of de Sitter "invariant" vacua.[14] But our more careful analysis shows that at best the "invariance" is a phase invariance, which is all that (3.92) insures.

With the intrinsic subtraction method available in the Schrödinger representation, one may go further. By requiring that the subtraction q_2 for the dilatation generator Q_2 be such that the kernel corresponding to $e^{-i(Q_2 - q_2)}$ be well-defined, q_2 may be uniquely determined, up to finite terms. One finds, by the method explained earlier

$$q_2 = \frac{t}{2} \operatorname{tr} |k| \Theta_\nu' (t\,|k|) \qquad (3.94)$$

and the renormalized dilatation generator : Q_2 : $= Q_2 - q_2$ acting on our candidate vacuum states gives

$$: Q_2 : |\Omega\rangle = (r-1)\frac{L}{t} \int_0^\infty \frac{dk}{2\pi} k\Theta_\nu'(k) \frac{1 - (r+1)\sin^2(\Theta_\nu(k) - \theta)}{1 + (r^2 - 1)\sin^2(\Theta_\nu(k) - \theta)} |\Omega\rangle . \qquad (3.95)$$

The eigenvalue is not only infrared divergent owing to L, but also ultraviolet divergent owing to the k integral. Moreover, it is time-dependent. Thus, the one parameter of de Sitter "invariant" vacua is seen to be only phase-invariant, with infinite phase, except for the special case $r = 1$, where the eigenvalue vanishes and the state is truly invariant. The remaining θ dependence is an unavoidable phase arbitrariness — the arbitrary origin of time.

It may appear puzzling that two generators of the non-Abelian $SO(2,1)$ group annihilate all vacua, but a third does not. The resolution lies in the infinite eigenvalue in (3.95). The situation is analogous to the Poincaré group in flat space-time, where the Lorentz generator when commuted with the momentum gives the Hamiltonian. However, the former two annihilate the ground state, which is Lorentz and translation invariant, while the Hamiltonian possesses an infinite eigenvalue — the zero point energy. Physically, what is being said is that one cannot translate or boost an infinitely heavy object. Similar remarks apply to our theory in de Sitter space.

It should be appreciated that we cannot redefine the generators so that some other vacuum, with $r \neq 1$, becomes invariant at the expense of phase-invariance of $r = 1$ vacuum. The point is that only finite redefinition is permitted at this stage, but the eigenvalue in (3.95) is infinite and it cannot be removed.

The $r = 1$ vacuum, known as the *Bunch-Davies vacuum*, has been previously preferred: it is the one naturally arising in a Euclidean formulation, it allows a sensible definition of energy, it is relevant to the inflationary program. We see that also it is the unique, completely de Sitter invariant state.

IV. DYNAMICAL (VARIATIONAL) CALCULATIONS

Thus far our investigations have been kinematical. We now turn to dynamical problems. Of course dynamical calculations in quantum field theory are much more difficult than in quantum mechanics, and even in the latter, simpler theory exact results are rare. There one resorts to quantum mechanical approximation methods: the ones that have been developed and utilized in quantum mechanics are much more extensive than those available for quantum field theory. The purpose of formulating quantum field theory in a representation similar to the quantum mechanical one is to permit adopting the techniques of the latter to the former.

Specifically, we shall speak about variational methods. The general strategy is the following. Dynamical equations are formulated as variational problems; relevant equations are obtained as the condition that some quantity be stationary against arbitrary variations. This is then implemented approximately, by choosing the quantities to be varied to have a specific form, parameterized by unknown parameters. The approximation consists of varying these parameters rather than performing arbitrary variations, *i.e.*, the arbitrary variation is replaced by a parametric variation in the Rayleigh-Ritz manner. In this way, one arrives at approximate but self-consistent equations which capture some of the non-linearities of the exact, intractable problem.

A. Variational Methods for Pure States

Variational principles can be time-independent — these are widely known: also there are less well-known time-dependent variational principles. Both occur already in classical mechanics, where static solutions stationarize the Hamiltonian [energy] as a function of p and q, as is seen from the Hamiltonian equations

$$0 = \dot{q} = \frac{\partial H(p,q)}{\partial p}, \qquad 0 = -\dot{p} = \frac{\partial H(p,q)}{\partial q}. \tag{4.1}$$

The full time-dependent Newtonian equations are derived by Hamilton's variational principle, which requires stationarizing the classical action I_{cl} — the time integral of the Lagrangian L, and a functional of $q(t)$

$$I_{\mathrm{cl}} = \int dt\, L$$

$$\frac{\delta L_{\mathrm{cl}}(q)}{\delta q(t)} = 0. \tag{4.2}$$

Notice that the static variations in (4.1) do not require boundary conditions: on the contrary the time-dependent ones of (4.2) must vanish at the endpoints of the time integral that defines the action.

Both static and time-dependent variation principles of classical physics have their quantum analogs. The former translates into the requirement that expectation values of the Hamiltonian be stationary; this yields the time-independent Schrödinger equation. The latter's analog is Dirac's little-known time-dependent variational principle, which results in the time-dependent Schrödinger equation.

For the static variation, we vary $\langle \Psi | H | \Psi \rangle$ subject to the constraint $\langle \Psi | \Psi \rangle = 1$, where $|\Psi\rangle$ is time-independent, obtaining the time-independent Schrödinger equation

$$H\,|\Psi\rangle = E\,|\Psi\rangle \tag{4.3}$$

as the stationarity condition. For the time-dependent variations, following Dirac, we consider the action-like quantity

$$I = \int dt\, \langle \Psi; t\, |\, i\partial_t - H\, |\Psi; t \rangle, \tag{4.4a}$$

where $|\Psi; t\rangle$ is a time-dependent wave function. This function is varied; demanding that I be stationary gives the time-dependent Schrödinger equation

$$i\partial_t |\Psi; t\rangle = H |\Psi; t\rangle. \tag{4.4b}$$

Clearly (4.4) implies (4.3) if $|\Psi; t\rangle$ is chosen to be $e^{-iEt}|\Psi\rangle$.

All this can be done for field theory in the Schrödinger representation. With an arbitrary polynomial Hamiltonian, we take a time-dependent Fock vacuum as in (1.48), with arbitrary covariance Ω, which is viewed as the variational parameter. We evaluate the Gaussian functional integral to form $\langle\Psi|H|\Psi\rangle$ or I, vary the covariance and obtain variational equations for the covariance. Rather than giving details now, let me pass to a more general problem, which encompasses (4.3) and (4.4).[15]

B. Variational Methods for Mixed States

The time-dependent Schrödinger equation governs the time evolution of a quantum system, when the initial state is a pure state, described by a definite wave function [or wave functional in the quantum field theory]. However, frequently one is interested in collective phenomena, where initially the system is in a mixed state, described by a density matrix. In quantum mechanics, this is

$$\rho(q_1, q_2) = \sum_n p_n \psi_n(q_1)\psi_n^*(q_2), \tag{4.5a}$$

$$\text{tr}\,\rho = \int dq\,\rho(q, q) = 1. \tag{4.5b}$$

Here $\{\psi_n\}$ is a complete set of wave functions and p_n is the probability that the system is in the state n. Average values of physical quantities described by operators $\mathcal{O}$, which in turn are represented by kernels $\mathcal{O}(q_1, q_2)$, are given by a trace over the density matrix

$$\langle\mathcal{O}\rangle = \text{tr}\,\rho\,\mathcal{O} = \int dq\,d\tilde{q}\,\mathcal{O}(q, \tilde{q})\rho(\tilde{q}, q). \tag{4.6}$$

The time development of these averages is determined, once the time dependence of the density matrix is known.

The above Schrödinger picture development can obviously be given in field theory as well. The density matrix becomes a functional of φ_1 and φ_2

$$\rho(\varphi_1, \varphi_2) = \sum_n p_n \Psi_n(\varphi_1)\Psi_n^*(\varphi_2) \tag{4.7}$$

and integrals like in (4.5) and (4.6) become functional.

Next, we come to the question of how to determine the time dependence of ρ. When our system is in equilibrium with its environment at some temperature $T \equiv \frac{1}{k\beta}$ [k is Boltzmann's constant] the complete set of wave functions in (4.5a) or wave functionals in (4.7) comprises the energy eigenstates with eigenvalues E_n and p_n is Boltzmann distributed

$$p_n = e^{-\beta E_n} \Big/ \sum_{n'} e^{-\beta E_{n'}} . \tag{4.8}$$

The density matrix corresponds to the canonical ensemble, and its time evolution is trivial: ρ remains constant in time because the p_n's are constant and the time dependence in energy eigenfunctions is a phase that disappears from $\Psi\Psi^*$. For systems out of equilibrium, both the probabilities and wave functions are time-dependent. Discovering this time dependence becomes enormously difficult, owing to the complicated interactions between the system and its environment, and the problem passes into the wide subject of non-equilibrium statistical mechanics.

There is however, a set of simplifying assumptions that still retains physical interest and yet allows successful analysis of a non-equilibrium situation. We suppose that the dynamical variables of the system [the coordinates of quantum mechanics or the fields of quantum field theory] are governed by a Hamiltonian that does not refer to dynamics of the environment. However, we further suppose that the effect of the environment is still felt in our Hamiltonian in that its parameters are time-varying, for example the coefficient of a quadratic term can go from positive [stable] to negative [unstable], or there can be a time-dependent background metric, like field theory in de Sitter space. Moreover, we assume that the occupation probabilities p_n are constant in time. Since the entropy is given by

$$S = -k \sum_n p_n \ln p_n , \tag{4.9}$$

the last assumption means that entropy does not change.

With these assumptions one can derive an equation for the time evolution of the density matrix. Simply by differentiating ρ with respect to time, and using the time-dependent Schrödinger equation for the wave functions, we arrive at the Liouville-von Neumann equation, which holds both in quantum mechanics and quantum field theory

$$\frac{d\rho}{dt} = i\,[\,\rho, H\,] . \tag{4.10}$$

[Note the sign is opposite to a Heisenberg equation of motion.] We contemplate using this equation in the following situations. A specific Hamiltonian of interest is selected, with time varying parameters. The time dependence is present during the interval $t_i < t < t_f$. For times earlier than t_i, the Hamiltonian is assumed constant, and the initial data for the Liouville equation will always be specified in this static regime, where we shall take ρ to be given by $\rho_i(\beta_i)$ — the initial Hamiltonian's Boltzmann distributed density matrix at some initial β_i, or an approximation thereto. The solution to (4.10) is then examined in the late period. $t > t_f$, where the Hamiltonian is again static but perhaps with different parameters. We wish to determine whether at late times ρ is static or not, and if static, whether it is given by a Boltzmann distribution but perhaps at some other temperature. [In some examples, $t_{i,f}$ may be $\mp\infty$.]

Terminology that we shall use: a time-independent density matrix is said to describe a system in *equilibrium*; is its form corresponds to a Boltzmann distribution, we say that the system is in *thermal equilibrium*, but non-thermal equilibria are also possible. When the density matrix is time-dependent, we say that the system is *out-of-equilibrium*.

Therefore, we are considering the problem of a system in thermal equilibrium, which becomes disturbed by the environment so that Hamiltonian parameters change. We wish to know whether there is a return to equilibrium, in particular to thermal equilibrium after the disturbance ceases, and also we wish to follow the behavior of various interesting quantities through the disturbance. Our methods also allow considering an arbitrary initial distribution, not necessarily in thermal equilibrium, and we can calculate the time evolution in this more general situation.

The processes that we describe are isentropic but energy non-conserving. In the language of statistical mechanics, we are speaking of *adiabatic* processes — closely related to but not identical with the *quantum adiabatic theorem*. Although we are led to this sub-class of non-equilibrium phenomena by the desire to have a closed system of equations, the physics that is described is not without interest; precisely such isentropic, energy non-conserving processes are thought to have taken place in the early universe.

Our goal is now settled: obtain solutions to the Liouville equation (4.10) with a definite time-dependent Hamiltonian H and canonical distribution density matrix as initial condition. But the method for achieving this goal must still be developed because (4.10) cannot be integrated directly, except for linear [non-interacting] problems described by a quadratic Hamiltonian. To this end, we use a variational principle, first stated in the many body context by Balian and Vénéroni, which yields, under arbitrary variation, the Liouville equation. An approximate application of this principle with a restricted variational *Ansatz*, in the Rayleigh-Ritz manner, leads to approximate but tractable equations for the density matrix that may be integrated.[16]

Moreover, the equations that we derive may be specialized to produce a variational approximation to the static eigenvalue equation (4.3), or to the time-dependent Schrödinger equation (4.4) for a pure state.

The variational principle that yields Liouville's equation is stated as follows. Consider

$$I = - \int_{t_i}^{t_f} dt \, \text{tr} \, \rho \left(\frac{d}{dt} \Lambda + i \left[H, \Lambda \right] \right) - \text{tr} \left(\rho \Lambda \right) \Big|_{t=t_i} . \tag{4.11}$$

Here, Λ, ρ and H are time-dependent kernels, they are functions of two coordinates in quantum mechanics and functionals of two fields in quantum field theory; the trace is over these variables. Λ is a Lagrange multiplier which will disappear from the formalism. Both Λ and ρ are varied in (4.11); also an initial condition is set on ρ and a final condition on Λ: at initial times [where the Hamiltonian is time-independent] $\rho \big|_{t=t_i}$ is the canonical density matrix; at final times, it is convenient to take Λ to be the identity kernel. Because of the fixed initial and final condition, the corresponding variations vanish. Therefore, varying Λ in (4.11) gives the Liouville equation for ρ, while varying ρ gives the Liouville condition for Λ. However, the latter is simply solved by $\Lambda = I$, in view of the final condition $\Lambda \big|_{t=t_f} = I$. In this way, the Liouville equation (4.10) is derived variationally.

C. Variational Ansatz for the Density Matrix

An approximate implementation of the variational principle for the Liouville equation consists of choosing trial forms for ρ and Λ depending on time-dependent parameters, evaluating I, and varying these parameters to obtain equations for them.

The trial form for ρ that we take is a Gaussian; this is motivated by (1) the fact that for quadratic Hamiltonians, the exact density matrix is a Gaussian; (2) with a Gaussian all the traces — functional integrations — may be explicitly evaluated. Thus we adopt the formula

$$\rho(\varphi_1, \varphi_2) = N^2 \exp - \frac{1}{2} \int \left\{ \varphi_1 \left(\frac{G^{-1}}{2} - 2i\Pi \right) \varphi_1 \right.$$

$$\left. + \varphi_2 \left(\frac{G^{-1}}{2} + 2i\,\Pi \right) \varphi_2 - \varphi_1 G^{-1/2} \xi G^{-1/2} \varphi_2 \right\} . \tag{4.12}$$

Here N^2 is a real normalization constant

$$N = \det{}^{-1/4} 2\pi G^{1/2} (1 - \xi_R)^{-1} G^{1/2} . \tag{4.13}$$

The three, possibly time-dependent kernels, G, Π and ξ are functions of x and y; the first two being real and symmetric, while ξ has a symmetric real part ξ_R and an antisymmetric imaginary part $G^{1/2} \xi_I G^{1/2}$. The significance of these quantities is brought out by evaluating averages of field bilinears with the above density matrix

$$\langle \Phi(\mathbf{x})\Phi(\mathbf{y}) \rangle = \left(G^{1/2}(1 - \xi_R)^{-1} G^{1/2} \right) (\mathbf{x}, \mathbf{y}) , \tag{4.14a}$$

$$\langle \Pi(\mathbf{x})\Pi(\mathbf{y}) \rangle = \frac{1}{4} \left(G^{1/2}(1 + \xi_R)^{-1} G^{-1/2} \right) (\mathbf{x}, \mathbf{y})$$
$$+ 4 \left((\Pi + \xi_I) G^{1/2}(1 - \xi_R)^{-1} G^{1/2}(\Pi - \xi_I) \right) (\mathbf{x}, \mathbf{y}) , \tag{4.14b}$$

$$\langle \Phi(\mathbf{x})\Pi(\mathbf{y}) \rangle = \frac{i}{2}\delta(\mathbf{x} - \mathbf{y}) + 2 \left(G^{1/2}(1 - \xi_R)^{-1} G^{1/2}(\Pi - \xi_I) \right) (\mathbf{x}, \mathbf{y}) . \tag{4.14c}$$

Moreover, when ξ vanishes, the density matrix (4.12) describes a pure state, compare (1.48)

$$\rho(\varphi_1, \varphi_2)|_{\xi=0} = \Psi(\varphi_1) \Psi^*(\varphi_2) , \tag{4.15a}$$

$$\Psi(\varphi) = N \, e^{-\frac{1}{2} \int \varphi \Omega \varphi} , \tag{4.15b}$$

$$\Omega = \frac{G^{-1}}{2} - 2i \, \Pi . \tag{4.15c}$$

Hence, we call ξ the "degree of mixing"; obviously it is related to the entropy. Since entropy is conserved in our theory, we expect ξ to be time-independent.

Also we need a trial form for Λ. Because ρ involves four variational parameters, we need to obtain four equations for them. This is achieved by taking a four-parameter trial form for Λ. A useful choice is the operator formula

$$\Lambda = \Lambda^{(I)} + \int_{\mathbf{x},\mathbf{y}} \Lambda^{(\Pi\Pi)}(\mathbf{x},\mathbf{y}) \Pi(\mathbf{x})\Pi(\mathbf{y})$$
$$+ \int_{\mathbf{x},\mathbf{y}} \Lambda^{(\Phi\Pi)}(\mathbf{x},\mathbf{y}) \left(\Phi(\mathbf{x})\Pi(\mathbf{y}) + \Pi(\mathbf{y})\Phi(\mathbf{x}) \right) , \tag{4.16a}$$
$$+ \int_{\mathbf{x},\mathbf{y}} \Lambda^{(\Phi\Phi)}(\mathbf{x},\mathbf{y}) \Phi(\mathbf{x})\Phi(\mathbf{y})$$

which corresponds to the kernel

$$\Lambda(\varphi_1, \varphi_2) = \Lambda^{(I)}\delta(\varphi_1 - \varphi_2) - \int_{\mathbf{x},\mathbf{y}} \Lambda^{(\Pi\Pi)}(\mathbf{x}, \mathbf{y}) \frac{\delta^2}{\delta\varphi_1(\mathbf{x})\delta\varphi_1(\mathbf{y})}\delta(\varphi_1 - \varphi_2)$$

$$+ i \int_{\mathbf{x},\mathbf{y}} \Lambda^{(\Phi\Pi)}(\mathbf{x}, \mathbf{y})\,(\varphi_1(\mathbf{x}) + \varphi_2(\mathbf{x}))\, \frac{\partial}{\partial\varphi_1(\mathbf{y})}\delta(\varphi_1 - \varphi_2)\,, \quad (4.16b)$$

$$+ \int_{\mathbf{x},\mathbf{y}} \Lambda^{(\Phi\Phi)}(\mathbf{x}, \mathbf{y})\varphi_1(\mathbf{x})\varphi_1(\mathbf{y})\delta(\varphi_1 - \varphi_2)\,.$$

Finally, select a Hamiltonian. Several interesting models have been analyzed: free field theory with a varying mass term [here the Gaussian *Ansatz* is exact]; the same model with a quartic self-interaction; either of the previous two in a Robertson–Walker, time-varying background. Since all these examples possess space-translation invariant dynamics, we may take all the variational parameters to be translation invariant, and diagonal in momentum space. Upon evaluating (4.11) with (4.13), (4.16) and a definite Hamiltonian, we vary the $\Lambda^{(i)}$ to find the approximate equations determining the density matrix. [Also by varying the parameters in ρ, we obtain equations for $\Lambda^{(i)}$, but these need not be analyzed.]

To exemplify the procedure we record here, the equations that arise when a scalar field in flat space-time possesses a quartic self-interaction. The Hamiltonian is

$$H = \frac{1}{2}\int\left(\Pi^2 + \Phi h\Phi + \frac{\lambda}{3}\Phi^4\right). \quad (4.17)$$

The quadratic term, h, varies in time during the interval $t_i < t < t_f$. The variational equations for the parameters G and Π, in momentum space, where they are diagonal, read

$$\dot{G}(\mathbf{p}) = 4\Pi(\mathbf{p})G(\mathbf{p})\,, \quad (4.18a)$$

$$\dot{\Pi}(\mathbf{p}) = \frac{1 - \xi^2(\mathbf{p})}{8G^2(\mathbf{p})} - 2\Pi^2(\mathbf{p}) - \frac{1}{2}\left(h(\mathbf{p}) + 2\lambda\int_{\mathbf{k}}\frac{G(\mathbf{k})}{1 - \xi(\mathbf{k})}\right). \quad (4.18b)$$

Also the equation for the normalization constant is solved by the anticipated formula (4.13), and as expected ξ is time-independent.

Equation (4.18) needs to be renormalized owing to a possible ultraviolet divergence in the $\mathbf{k}$ integral. This can be done, and (4.18) poses a well-defined system of equations, once a form for $h(\mathbf{p})$ and its time-dependence is fixed, and an initial condition for G, Π and ξ is selected.[15]

One may also specialize (4.18) to the variational equations for a pure state by setting ξ — the degree of mixing — to zero. In that case (4.18) reduce to the field-theoretic time-dependent Hartree-Fock equations, previously derived from an approximate application of Dirac's time-dependent variational principle (4.4).[17] Finally, with a time-independent h, the static version of (4.18) corresponds to the mean-field approximation for quantum field theory.

Here is not the place, nor do we have the time, to discuss detailed properties of the solutions to (4.18). Let me nevertheless summarize what one finds. For generic time-dependence in h, the system does not return to equilibrium for late times, $t > t_f$, where h becomes static again. However, the departures from equilibrium decrease with increasing coupling strength λ, also the departures from equilibrium becomes arbitrarily small when the passage in time from $h(t_i)$ to $h(t_f)$ is arbitrary slow. This is in keeping with ones *a priori* physical expectations. Moreover, with very specific time-profiles for $h(t)$, equilibrium can be regained at $t > t_f$ — the density matrix becomes time-independent. This unexpected phenomenon can arise even in the absence of interactions and the mathematical analysis is curiously analogous to the analysis of reflectionless potentials.[15]

The study of quantum fields out of thermal equilibrium is a subject to which particle theorists are turning these days, because of the probable occurrence of this phenomenon in the early universe. The functional Schrödinger formalism, together with the approximate variational principle for determining the density matrix provides a useful framework for analyzing this problem.

V. FERMION SYSTEMS

A. Functional Space for Fermions

A description of a Schrödinger picture for fermionic quantum fields, analogous to that for bosonic fields is now presented. The challenge here is of course that fermionic quantum fields do not commute with themselves, and to keep things as simple as possible I shall examine the problem in its most elementary setting: a massless Weyl-Majorana field in two-dimensional space-time.[18] Such a field is described by a real one-component spinor, defined on a line, and satisfies the fixed-time anticommutation relations, which may be viewed as an infinite-dimensional Clifford algebra

$$\{\psi(x), \psi(y)\} = \delta(x - y) \equiv I(x, y) \tag{5.1}$$

$$\psi^\dagger = \psi \tag{5.2}$$

We wish to define a functional space to represent states in this quantum field theory, determine an action of the operator $\psi(x)$ on this space, give a procedure for taking adjoins, and define an inner product. All this we want to do without specifying a Hamiltonian with the usual positive and negative first-quantized energy eigenmodes, which play a central role in the conventional approach to this problem.

To begin, our functional space consists of functionals $\Psi(u)$ of anti-commuting Grassmann fields $u(x)$ at fixed time

$$\{u(x), u(y)\} = 0. \tag{5.3}$$

We seek to represent the [infinite] Clifford algebra (5.1) in terms of the Grassmann variables (5.3). To this end, we represent the action of $\psi(x)$ by

$$|\Psi\rangle \longrightarrow \Psi(u)$$

$$\psi(x)|\Psi\rangle \longrightarrow \frac{1}{\sqrt{2}}\left(u(x) + \frac{\delta}{\delta u(x)}\right)\Psi(u). \tag{5.4}$$

To understand how the dual state should be represented, let us first analyze the problem on a space $\{x\}$ consisting of two points $i = 1, 2$ on which two fermion operators $\psi(i)$ are defined, satisfying the two-dimensional Clifford algebra

$$\{\psi(i), \psi(j)\} = \delta_{ij}. \tag{5.5}$$

A specific state $|\Psi_f\rangle$ is represented by a functional of $u(i)$ that can be expanded in a four-dimensional basis

$$|\Psi_f\rangle \to \Psi_f(u) = f_0 + f_1 u(1) + f_2 u(2) + f_{12} u(1) u(2), \tag{5.6}$$

where the f's are numbers. The inner product with another state $|\Psi_g\rangle$ is defined in the natural way

$$\langle\Psi_g|\Psi_f\rangle = g_0^* f_0 + g_1^* f_1 + g_2^* f_2 + g_{12}^* f_{12} = \langle\Psi_f|\Psi_g\rangle^*. \tag{5.7}$$

This can be expressed as

$$\langle\Psi_g|\Psi_f\rangle = \int d^2 u \, \Psi_g^*(u)\Psi_f(u) \tag{5.8}$$

provided $\langle\Psi_g|$ is represented by

$$\langle\Psi_g| \to \Psi_g^*(u) = g_{12}^* + g_2^* u(1) - g_1^* u(2) + g_0^* u(1)u(2). \tag{5.9}$$

Equation (5.7) follows from (5.6), (5.8) and (5.9) since only the following

Grassmann integral is non-vanishing

$$\int d^2u\, u(1)u(2) = 1 . \tag{5.10}$$

The duality transformation that takes (5.6) into (5,9) is that of differential forms in two dimensions: there are four basis forms: the zero-form, two one-forms dx^μ, $\mu = 1,2$ and one area two-form $\frac{1}{2}\varepsilon_{\mu\nu}dx^\mu dx^\nu$. These are the analogs of the four basis elements in (5.6), while the duals of these forms are the basis elements in dual formula (5.9).

Our dualization procedure can be formulated analytically by the Berezin integral. Introduce auxiliary variables $\bar{u}(i)$ and an auxiliary dual functional $\bar{\Psi}_g(\bar{u})$ defined by the natural formula

$$\bar{\Psi}_g(\bar{u}) = g_0^* + g_1^*\bar{u}(1) + g_2^*\bar{u}(2) + g_{12}^*\bar{u}(2)\bar{u}(1) . \tag{5.11}$$

Then our dual (5.9) is given by

$$\Psi_g^*(u) = \int d^2\bar{u}\, e^{\left(\bar{u}(1)u(1)+\bar{u}(2)u(2)\right)} \bar{\Psi}_g(\bar{u}) \tag{5.12}$$

where the only non-vanishing Grassmann integral is

$$\int d^2\bar{u}\, \bar{u}(2)\bar{u}(1) = 1 . \tag{5.13}$$

With these rules the adjoint of $u(i)$ is $\frac{\delta}{\delta u(i)}$ and $\frac{1}{\sqrt{2}}\left(u(i) + \frac{\delta}{\delta u(i)}\right)$ is Hermitian.

Note that our representation is reducible: we require four elements to represent (5.5), hence the representation is four-dimensional. But any two Pauli matrices also reproduce (5.5) two-dimensionally.

This example points to the continuum field theoretic generalization. From the functional $\Psi(u)$, construct $\bar{\Psi}(\bar{u})$ by complex conjugating scalars, reversing the order of Grassmann variables and replacing $u(x)$ by $\bar{u}(x)$. Then the dual of $\Psi(u)$ is given by

$$\Psi^*(u) = \int \mathcal{D}\bar{u}\, e^{\int_x \bar{u}(x)u(x)} \bar{\Psi}(\bar{u}) \tag{5.14}$$

and $\frac{1}{\sqrt{2}}\left(u(x) + \frac{\delta}{\delta u(x)}\right)$ is Hermitian, as it must be if it is to represent the Hermitian operator $\psi(x)$.

For a Gaussian state, which we also call a Fock vacuum in analogy with the bosonic case,

$$|\Omega\rangle \rightarrow \Psi_\Omega(u) = \det^{-1/4}\Omega \exp -\frac{1}{2}\int u\Omega u \tag{5.15a}$$

the dual is

$$\langle\Omega| \to \Psi_\Omega^*(u) = \det^{-1/4}(\Omega^{\dagger-1}) \exp \frac{1}{2} \int u\Omega^{\dagger-1}u. \qquad (5.15b)$$

Here Ω is an antisymmetric kernel

$$\Omega(x,y) = -\Omega(y,x). \qquad (5.16)$$

B. Two-Dimensional Conformal Transformations

Two-dimensional conformal transformations $\delta_f x = -f(x)$, discussed in Section III, act also on fermion Majorana fields and can be represented by these variables. The formal generator

$$Q_f = \frac{i}{4} \int dx \left[\psi(x)f(x)\psi'(x) - \psi'(x)f(x)\psi(x) \right] \qquad (5.17)$$

gives the field transformation law (3.32) with $d = 1/2$ and $\alpha = 0$,

$$\delta_f \psi = i\,[Q_f, \psi] = f\psi' + \frac{1}{2}f'\psi. \qquad (5.18)$$

Q_f formally satisfies the algebra (3.31), (3.36), but suffers from singularities owing to the coincident-point operator product. For a well-defined regularized generator, we take

$$Q_F = \frac{1}{2} \int \psi F \psi. \qquad (5.19)$$

This generates canonical transformations, but (5.18) is regained when the antisymmetric Hermitian kernel $F(x,y)$ tends to

$$F(x,y) \to \frac{i}{2}\,[f(x) + f(y)]\,\delta'(x-y) = \frac{1}{2}\,[f(x) + f(y)]\,k(x,y). \qquad (5.20)$$

[Notice that this is like (3.38) with the interchange $k \leftrightarrow \delta$; see also below.] Q_F satisfies

$$[Q_F, Q_G] = i\,Q_{(F,G)}$$
$$(F,G) = -i\,[F,G]. \qquad (5.21)$$

The representation and intrinsic renormalization of these quantities is an important application of the fermionic Schrödinger picture formalism.

The kernel $U(u_1, u_2; F)$ that represents the finite transformation satisfies the differential equation

$$i\frac{\partial}{\partial \tau} U(u_1, u_2; \tau F) = \frac{1}{4} \int dx dy \left(u_1(x) + \frac{\delta}{\delta u_1(x)} \right) F(x, y)$$

$$\times \left(u_1(y) + \frac{\delta}{\delta u_1(y)} \right) U(u_1, u_2; \tau F) \tag{5.22}$$

with an initial condition at $\tau = 0$.

$$U(u_1, u_2; 0) = \delta(u_1 - u_2). \tag{5.23}$$

The solution is Gaussian

$$U(u_1, u_2; F) = N_F \exp - \int u_1 u_2 \exp \frac{i}{2} \int (u_1 - u_2) K_F(u_1 - u_2), \tag{5.24}$$

$$N_F = \det{}^{1/2} i \sin \frac{F}{2}, \tag{5.25}$$

$$K_F = \operatorname{ctn} \frac{F}{2}. \tag{5.26}$$

Constants are adjusted in N_F so that (5.23) holds. That U is indeed the correct transformation kernel, satisfying the composition law

$$\int \mathcal{D} u\, U(u_1, u; F) U(u, u_2; G) = U(u_1, u_2; F \circ G) \tag{5.27}$$

$$F \circ G = F + G + \frac{1}{2}(F, G) + \cdots$$

can be verified explicitly when F is proportional to G; the more general case can be checked by expanding $F \circ G$.

The definition of the inner product on our space determines the form of the adjoint kernel

$$U^\dagger(u_1, u_2; F) = N_F^* \exp - \int u_1 u_2 \exp -\frac{i}{2} \int (u_1 - u_2) K_F^\dagger(u_1 - u_2). \tag{5.28}$$

Since F and K_F are Hermitian, the above is just $U(u_1, u_2; -F)$; hence, the representation is unitary.

The kernel (5.24) should be compared to the corresponding bosonic one (3.44). We appreciate that the first exponentials coincide once the commutator of the bosonic fields, k is replaced by the anti-commutator of the fermionic fields, $I \equiv \delta$. The analogy of the remaining formulas is brought out, if we similarly replace k in $\mathcal{F}$ by I, i.e., replace $\mathcal{F}$ by F. Of course, differences in the Jacobian factor between fermions and bosons have to be taken into account.

The local limit when the regulator is removed can be evaluated. K_F attains a well-defined expression

$$K_F(x,y) \to K_f(x,y) = \frac{1}{\sqrt{f(x)}}\left\{ \int \frac{d\lambda}{2\pi}\left(\text{ctn}\frac{\lambda}{2}\right)\exp\left(-i\lambda \int_y^x \frac{dz}{f(x)}\right)\right\}\frac{1}{\sqrt{f(y)}}$$

$$= \frac{1}{\sqrt{f(x)}}\left\{ P\,\text{ctn}\pi \int_y^x \frac{dz}{f(z)}\right\}\frac{1}{\sqrt{f(y)}}. \tag{5.29}$$

The normalization factor N_F diverges. The analysis, as in the bosonic case, is performed for imaginary $\tau\,[F \to -iF]$, where N_F becomes $e^{\frac{1}{2}\,\text{tr}\,\ln\sinh\frac{1}{2}|F|}$ and the antisymmetric kernel F is replaced by the absolute value kernel. The result, after returning to real τ, is that apart from an infinite constant Z, $e^{iq_F}N_F$ attains a limit, provided

$$q_F = -\frac{1}{4}\,\text{tr}\,F\frac{k}{|k|}. \tag{5.30}$$

To renormalize, we absorb the constant divergence Z in the functional integration measure, and remove the divergent phase. Thus $Z^{-1}e^{iq_F}U(u_1,u_2;F)$ possesses a finite limit, but the composition law (5.27) acquires a trivial cocycle

$$2\pi\omega_2(F,G) = \frac{1}{4}\,\text{tr}\,(F\circ G - F - G)\frac{k}{|k|} \tag{5.31}$$

which becomes non-trivial in the local limit. This implies that the renormalized charges

$$:Q_f: = \lim\left(Q_F + \frac{1}{4}\,\text{tr}\,F\frac{k}{|k|}\right) \tag{5.32}$$

satisfy (3.54) with $c=1/2$, in agreement with a general theorem.[12]

As in the bosonic case, this central extension is not sensitive to the way that the field operator is represented. One can verify that the divergent phase (5.30) is unaffected by the following generalization of (5.4)

$$\psi = \frac{1}{\sqrt{2}}\left(\alpha u + \alpha^*\frac{\delta}{\delta u}\right), \tag{5.33a}$$

$$\frac{1}{2}(\alpha^*\alpha^T + \alpha\alpha^\dagger) = I. \tag{5.33b}$$

[The last condition on the kernel α is required by (5.1).] But also, as in the bosonic case, the opposite sign for the center results if the analysis of divergences is carried out after continuation to imaginary τ in the opposite sense from ours: $F \to iF$ rather than $F \to -iF$, the latter being the conventional one for theories with energy spectra bounded below but not above.

The action of the transformation kernel on a generic state $\Psi(u)$ is given by

$$\Psi(u) \underset{F}{\longrightarrow} \Psi_F(u) = e^{iq_F} \int \mathcal{D}\tilde{u}\, U(u, \tilde{u}; F)\Psi(\tilde{u}) \,. \qquad (5.34)$$

In particular, the transform of a properly normalized Gaussian, with covariance Ω, is again a Gaussian with transformed covariance

$$\Omega_F = \Omega + (I - \Omega)(\Omega + iK_F)^{-1}(I + \Omega) \qquad (5.35a)$$

and an additional phase θ_F

$$\theta_F = q_F + \text{Im}\left[\ln N_F + \frac{1}{2}\, \text{tr } \ln(\Omega + iK_F) \right] . \qquad (5.35b)$$

As in the bosonic case [compare (3.58)], Ω is a representation for the conformal algebra without center; the latter resides in the representation provided by the phase θ_F.

The fermionic representation of the conformal group has been constructed without reference to a Hamiltonian for the fermions, but it is useful to remark that the dynamics of a free, massless Majorana-Weyl field or Dirac field is conformally invariant. Indeed, bosonization of the Dirac model produces the self-dual theory governed by (3.65).[11]

C. Fock-Space Dynamics in the Schrödinger Representation

We now examine dynamical second quantized fermion theories in our formalism. This we do in order to explore further properties of the fermionic Schrödinger picture and also to give explicit realization to the Dirac sea phenomenon in fermionic field theories.

Let us begin by recalling the algebraic structures associated with the field theoretic, Majorana-Weyl Hamiltonian

$$H = \frac{1}{2} \int \psi h \psi \,. \qquad (5.36)$$

The "first quantized" Hamiltonian h is antisymmetric and imaginary

$$h(x, y) = -h^*(x, y) \qquad (5.37)$$

with a complete, orthonormal set of "first quantized" eigenmodes

$$h f_\varepsilon = \varepsilon f_\varepsilon \,, \qquad (5.38)$$

$$\int dx\, f_{\varepsilon'}^*(x) f_\varepsilon(x) = \delta(\varepsilon - \varepsilon') \,, \qquad (5.39a)$$

$$\int d\varepsilon f_\varepsilon^*(x) f_\varepsilon(y) = \delta(x-y)\,. \tag{5.39b}$$

[Whenever appropriate, summation over discrete eigenvalues and replacement of the δ-function by a Kronecker delta is understood.] According to (5.37), the eigenvalues are paired in sign

$$h f_\varepsilon^* = -\varepsilon f_\varepsilon^* \tag{5.40}$$

and we assume that there are no isolated vanishing eigenvalues. The field operator may be expanded in the first quantized modes

$$\psi(x) = \int d\varepsilon a_\varepsilon f_\varepsilon(x)\,, \tag{5.41}$$

$$a_\varepsilon = a_{-\varepsilon}^\dagger = \int dx f_\varepsilon^*(x)\psi(x)\,. \tag{5.42}$$

The anti-commutator of the a_ε's is dictated by (5.1)

$$\{a_\varepsilon, a_{-\varepsilon'}\} = \{a_\varepsilon, a_{\varepsilon'}^\dagger\} = \delta(\varepsilon - \varepsilon')\,. \tag{5.43}$$

The mode operator a_ε is a shift operator for the second quantized Hamiltonian

$$[a_\varepsilon, H] = \varepsilon a_\varepsilon\,. \tag{5.44}$$

Hence, the spectrum of H is unbounded from above and below, unless a_ε annihilates states. Of course, all this is familiar, and is realized for example for the Poincaré invariant model

$$h(x,y) = i\,\delta'(x-y)\,. \tag{5.45}$$

We now seek eigenstates of H within our Grassmann functional space. For the Fock vacuum we chose a Gaussian

$$|\Omega\rangle = \det^{-1/4}\Omega \exp\frac{1}{2}\int u\Omega u\,, \tag{5.46}$$

where Ω is antisymmetric. The eigenvalue equation

$$H|\Omega\rangle = \frac{1}{4}\int\left(u+\frac{\delta}{\delta u}\right)h\left(u+\frac{\delta}{\delta u}\right)|\Omega\rangle = E_V|\Omega\rangle \tag{5.47}$$

requires that

$$(I-\Omega)\,h\,(I+\Omega) = 0 \tag{5.48}$$

and the vacuum energy is

$$E_V = \frac{1}{4}\operatorname{tr} h\Omega\,. \tag{5.49}$$

Excited states are polynomials in ψ operating on $|\Omega\rangle$, which in our formalism become polynomials in $\frac{1}{2}(I + \Omega)u \equiv u_+$ multiplying the Fock vacuum.

We now show that Ω is not determined uniquely by (5.48). Take Ω to be simultaneously diagonalized with h, so (5.48) requires $\Omega^2 = I$; in the ε-representation, h is diagonal and we have

$$\Omega(\varepsilon, \varepsilon') = \Omega(\varepsilon)\delta(\varepsilon - \varepsilon'), \tag{5.50a}$$

$$\Omega(\varepsilon) = -\Omega(-\varepsilon), \tag{5.50b}$$

$$\Omega(\varepsilon) = \pm 1, \tag{5.50c}$$

where the variation in sign can occur for any matrix element. In other words, there is an infinity of solutions for Ω depending on the different ways one assigns signature.

To understand further the form of Ω, and to select one sign signature from the infinity of possibilities, we compute the effect of a_ε

$$a_\varepsilon|\Omega\rangle = \int f_\varepsilon^* \psi |\Omega\rangle = \frac{1}{\sqrt{2}} \int f_\varepsilon^* \left(u + \frac{\delta}{\delta u}\right)|\Omega\rangle = \frac{1}{\sqrt{2}}[1 + \Omega(\varepsilon)] \int f_\varepsilon^* u|\Omega\rangle. \tag{5.51}$$

It is seen that a_ε annihilates $|\Omega\rangle$ whenever $\Omega(\varepsilon)$ is -1. Thus choosing Ω is equivalent to choosing the prescription for filling the Dirac sea to define a field theoretic vacuum. When $\Omega(\varepsilon) = -1$ for positive ε and $+1$ for negative ε, *i.e.*,

$$\Omega(\varepsilon) = \begin{cases} -1 & \text{for } \varepsilon > 0 \\ 1 & \text{for } \varepsilon < 0 \end{cases} \tag{5.52}$$

a_ε annihilates $|\Omega\rangle$ for $\varepsilon > 0$ but not for $\varepsilon < 0$. This is the conventional choice of filled negative energy sea, and corresponds to

$$E_V = -\frac{L}{2} \int_0^\infty d\varepsilon\, \varepsilon, \tag{5.53}$$

where L is the length of space.

More generally, $\frac{1}{2}[1 + \Omega(\varepsilon)]$ is the filling factor, vanishing for empty states. Choices other than (5.52) for Ω are also possible, corresponding to other filling prescriptions, but they would be unconventional, and would in general define inequivalent theories. Note that the overlap between two Gaussian vacua is proportional to

$$\langle \Omega_1 | \Omega_2 \rangle \propto \det^{1/2}(\Omega_1 + \Omega_2). \tag{5.54}$$

Since Ω_1 and Ω_2 can differ only in the sign of one or more eigenvalues, $\Omega_1 + \Omega_2$ has a zero eigenvalue. Whether the overlap vanishes, depends on the weight of the zero in $\Omega_1 + \Omega_2$. If different signs are associated to a discrete mode in

Ω_1 and Ω_2, then the determinant certainly vanishes; for continuum modes, a sufficiently infinite number of modes must be differently filled between Ω_1 and Ω_2 for the determinant to vanish. When the overlap of two vacua vanishes, so will the overlap between corresponding excited states, and the Fock spaces — the different theories built with different Ω's — are inequivalent.

As a final check of our formalism, we compute the equal-time correlation functions. Following the rules we have put forward, it is easy to show that

$$\rho(x,y) \equiv \langle\Omega|\psi(x)\psi(y)|\Omega\rangle = \frac{1}{2}(I - \Omega)(x,y). \tag{5.55}$$

For the Hamiltonian (5.36), (5.45), and the choice (5.52) for the vacuum, this is $\int \frac{dp}{2\pi} e^{-ip(x-y)}\theta(p)$, which is the conventional result.

In summary, let us contrast our fermionic Schrödinger picture with the previously described bosonic one. In both cases, the functional space contains inequivalent Fock spaces. Choosing a specific quadratic Hamiltonian can select a specific Fock space for bosons, but not for fermions. In the former case, there is no sign ambiguity for the Gaussian covariance Ω because we require convergence of a Gaussian normalization integral, hence $\text{Re}\,\Omega > 0$; in the latter, the integral is Grassmannian, all integrals converge, and the sign of Ω is not fixed. Stated differently, a particle state is localized in the bosonic functional space, while there is no concept of localization in the Grassmann space. A unique fermionic Fock space requires prescribing a filling factor.

Finally, we note that when the first quantized Hamiltonian h possesses an isolated zero mode, Ω is undetermined in that channel. This is the origin in the present formalism of vacuum degeneracy associated with charge fractionalization.[19]

RERERENCES

1. K. Symanzik, *Nucl. Phys.* **B190** [FS3], 1 (1983); M. Lüscher, *Nucl. Phys.* **B254**, 52 (1985).

2. F. Cooper and E. Mottola, *Phys. Rev. D* **36**, 3114 (1987); S.-Y. Pi and M. Samiullah, *Phys. Rev. D* **36**, 3128 (1987).

3. For a comprehensive discussion of gauge theories, their topological properties and the associated anomalies, see S. Treiman, R. Jackiw, B. Zumimo and E. Witten, *Current Algebra and Anomalies* (Princeton University Press, Princeton, NJ/World Scientific, Singapore, 1985).

4. J. Goldstone and R. Jackiw, *Phys. Lett.* **B74**, 81 (1978); A. Izergin, V. Korepin, M. Semenov–Tian–Shansky and L. Faddeev. *Teor. Mat. Fiz.* **38**, 3 (1979) [English translation: *Theor. Math. Phys.* **38**, 1 (1979)].

5. S. Coleman, R. Jackiw and L. Susskind, *Ann. Phys.* (NY) **93**, 267 (1975); S. Coleman, *Ann. Phys.* (NY) **101**, 239 (1976).

6. R. Feynman, *Nucl. Phys.* **B188**, 479 (1981).

7. I. Singer, *Astérisque*, 323 (1985).

8. More discussion of 1- and 2-cocycles and generalization to 3-cocycles, which signal failure of associativity, may be found in Ref. 3.

9. A. Niemi and G. Semenoff, *Phys. Lett.* **B176**, 108 (1986); H. Neuberger, A. Niemi and G. Semenoff, *Phys. Lett.* **B181**, 244 (1986); R. Floreanini and R. Jackiw, *Phys. Rev. D* **37**, 2206 (1988).

10. The following development is based on the research papers by R. Floreanini and R. Jackiw, *Phys. Lett.* **B175**, 428 (1986), *Ann. Phys.* (NY) **178**, 227 (1987), *Phys. Rev. Lett.* **59**, 1873 (1987), and Ref. 9. For further reviews see R. Jackiw, in *Super Field Theories*, H. Lee, V. Elias, G. Kunstatter, R. Mann and K. Viswanathan, eds. (Plenum, New York, NY, 1987), and in *Conformal Field Theory, Anomalies and Superstrings*, B. Baaquie, C. Chew, C. Oh and K. K. Phua, eds. (World Scientific, Singapore, 1988).

11. R. Floreanini and R. Jackiw, *Phys. Rev. Lett.* **59**, 873 (1987); M. Bernstein and J. Sonnenschein, *Phys. Rev. Lett.* **60**, 1772 (1988); M. Henneaux and C. Teitelboim, *Phys. Lett.* **B206**, 650 (1988).

12. D. Friedan and S. Shenker, in *Unified String Theories*, M. Green and D. Gross, eds. (World Scientific, Singapore, 1986).

13. I follow R. Floreanni, C. Hill and R. Jackiw, *Ann. Phys.* (NY) **175**, 345 (1987); see also Ref. 10.

14. For a survey of the older literature, see N. Birell and P. Davies, *Quantum Fields in Curved Space* (Cambridge University Press, Cambridge, UK, 1982).

15. O. Èboli, R. Jackiw and S.-Y. Pi, *Phys. Rev. D* **37**, 3557 (1988).

16. R. Balian and M. Vénéroni, *Ann. Phys.* (NY) **164**, 334 (1985).

17. F. Cooper, S.-Y. Pi and P. Stancioff, *Phys. Rev. D* **34**, 3831 (1986).

18. The presentation follows Floreanni and Jackiw, Ref. 9; see also Ref. 10.

19. For a review see A. Niemi and G. Semenoff, *Phys. Rep.* **135**, 99 (1986).

Section V

SOLITONS, INSTANTONS AND SEMI-CLASSICAL QUANTUM FIELD THEORY

Tunneling, coherent states, dynamical modification of spin and other quantum numbers, quantization of coupling constant parameters, all are phenomena familiarly seen in quantum mechanics, but unanticipated — expect by Skyrme — in quantum field theory, because theorists would not stray from covariant perturbation theory when seeking to expose the physical content of a model. Although perturbative calculation could not be carried far and therefore structures could not be probed deeply, that conservative method seemed to open the only safe path around the contradictions lurking in field theoretical infinities. But in the 1970's, we became emboldened to follow Skyrme into non-perturbative analyses, and thereby discovered novel physical effects within otherwise conventional models.

Paper V.1

The commemoration of a famous, post World-War II conference in June 1983 was striking in that the original event was highlighted by discussion of experimental discoveries, while its reprise was fueled by theoretical speculations — a change of emphasis in physics that continues to evolve.

NON-PERTURBATIVE AND TOPOLOGICAL METHODS IN QUANTUM FIELD THEORY

Shelter Island II, R. Jackiw, N. Khuri, S. Weinberg and E. Witten, eds. (MIT Press, Cambridge, MA, 1985)

449

As we heard yesterday, renormalized perturbation theory, whose development in the West begin immediately after the first Shelter Island Conference with Bethe's Lamb shift calculation, gave us in the intervening third of a century the principal tool for extracting physical content from a quantum field theory. Aside from various resummation techniques of the Bethe-Salpeter variety, little effort was expended on the study of non-perturbative effects, even though there can be no doubt that these are necessarily present, since a quantum field theory is, after all, a quantum mechanical system, to be sure one with an infinite number of degrees of freedom, but nevertheless capable of giving rise to familiar quantal phenomena like collective macroscopic excitations, or tunneling — to name two, about which I shall have more to say. Just as the Born series for a quantum mechanical potential will not directly illuminate such processes, so also the Dyson-Feynman-Schwinger perturbation expansion has nothing to say about them in field theory. Of course, some individuals made isolated suggestions that there might exist interesting field theoretical structures inaccessible to conventional perturbative analysis — the most important and prescient are Skyrme's conjectures[1] — but the subject did not engender significant research. One can mention two reasons for this lack of interest by the physics community. First, the infinities of field theory had only recently been tamed by the renormalization procedure, which is defined perturbatively. Consequently, there was much uncertainty whether one could obtain unambiguous results outside the perturbative framework, or whether such calculations were hopelessly contaminated by the infinities. Second, there were doubts about the relevance of field theoretical dynamics to elementary particle physics, so that close study of any particular field theory was judged only mathematically but not physically interesting.

In the last decade, these attitudes have evolved. For better or worse, we have grown accustomed to the infinities and are less intimidated by them, although Dirac and Schwinger counsel against this complacency. Also local field dynamics has vigorously re-established itself as the only viable theoretical framework for elementary-particle physics; first through the successes of current algebra, then through light-cone and quark/parton analyses of deep inelastic processes, and finally through the contemporary synthesis in quantum chromodynamics and quantum flavor dynamics. While perturbation theory remains the preeminent calculational method, now we have also perfected alternative approximate analyses, which have exposed a wealth of non-perturbative effects in quantum field theory. I shall briefly summarize what we have learned, describe the impact on practical physical questions, and then towards the end,

I shall indicate some new research directions in this area, of the kind mentioned by Singer at the begining of his talk yesterday.

It has been established that a quantum field theory will in general give rise to particle states and also to processes beyond those that are seen in the Born series. The novel states do not arise by quantizing small fluctuations of a local field present in the Lagrangian, nor are they conventional bound states; rather they occur, when a symmetry is spontaneously broken, as collective, coherent excitations of all the elementary quanta. These are the celebrated soliton states, and are present in any number of dimensions: the kink in one, the Abrikosov-Landau-Ginsburg [Nielsen-Olesen] vortex in two, the 't Hooft-Polyakov monopole in three. While solitons in classical field theories have a long history, it is only recently that we have learned how to quantize them and how to perform approximate yet accurate quantum mechanical calculations.[2] The procedure is an extension of techniques familiar from broken symmetry studies. One finds a finite energy solution to the classical field equations — the soliton — shifts the quantum field by the classical solution, and quantizes the shifted field. In the broken symmetry case, the classical background is constant and is interpreted in the quantum theory as an approximation to the quantum field's vacuum expectation value; while in the soliton case the background is inhomogeneous and gives an approximation to the field form-factor in the soliton state. There are technical problems: one needs to preserve translation invariance; infrared divergences from zero Goldstone modes must be cured; quantum numbers that characterize the soliton states must be identified. All these problem have been solved,[2] and the following picture emerges: for weak coupling, g, the solitons are heavy with mass of order g^{-2} times the characteristic mass scale of the theory. Though more massive than the elementary excitations, the solitons are nevertheless stable, owing to an infinite energy barrier that separates them from ordinary particles. Their stability is further assured by a conserved charge — kink number, vortex number, monopole number — which however does not directly arise by Noether's theorem from a Lagrangian symmetry, but rather characterizes the topological, *i.e.*, large distance, properties of the field configuration. The interactions between solitons are strong, $O(g^{-1})$, hence difficult to calculate. Those between solitons and ordinary particles are $O(g^0)$; of course, conventional particles interact with strength $O(g)$. Therefore, there are three interaction scales. Furthermore, the soliton has both stable and metastable excited states, which are formed either by binding with ordinary particles, or by excitation of the soliton's internal degrees of freedom; for example, the charged dyon is an excited state of the monopole.[3] A remarkable property of the solitons is that they may possess unexpected spin and may obey unexpected statistics: although arising in a bosonic theory with integer-spin fields, they can be fermions with half-integer-spin. Further novel effects are seen when a Dirac fermion is coupled to a soliton: the most dramatic of these is the emergence of states with fractional fermion number.[4]

Turning now to processes, rather than states, the principal result here establishes the occurrence of tunneling in a Yang-Mills theory between energy-degenerate states, in close analogy to the tunneling that occurs in a periodic potential, giving rise to band formation.[5] In general, one does not expect such tunneling to happen in a quantum field theory, owing to infinite energy barriers between degenerate states — indeed, it is the absence of tunneling that permits spontaneous symmetry breaking. However, in a non-Abelian gauge field theory, there are paths in the gauge potential configuration space which avoid the infinite energy barriers and along which tunneling can occur. Semi-classical analysis of tunneling in potential theory is familiarly carried out by identifying these paths as solutions to the classical dynamical equations in imaginary time; similar solutions in Yang-Mills theory are the Belavin, Polyakov, Schwartz, Tyupkin instantons, and they have been used to give a semi-classical description of tunneling in Yang-Mills theory.

All this research has involved particle physicists in a novel activity: solving partial differential non-linear equations, and this has put us in contact with mathematicians who, as it happens, had come to similar equations on their own. Aside from the satisfaction of participating in a confluence of separate streams of research, we have benefited by learning and using some of the mathematician's techniques, principally topological analyses of field theory. With these, it has been possible to go beyond the non-perturbative but approximate calculations, that I have been describing so far, and to establish results rigorously.

One very surprising fact that has been found in this way about the Yang-Mills quantum theory is that it is characterized by a hidden parameter, the vacuum angle θ.[5] This comes about in the following way. Yang-Mills theory, like Maxwell theory, is gauge invariant, but its non-Abelian gauge transformations, unlike the electromagnetic ones, cannot all be gotten by iterating an infinitesimal gauge transformation. Physical results must be gauge invariant, yet quantum mechanical states need not be; according to Winger, a change of phase is permitted. Of course, Gauss' law requires states to be invariant against infinitesimal gauge transformations, and iterating these produces a finite gauge transformation, which by construction is continuously deformable to the identity. All electromagnetic gauge transformations can be reached in this way, but not all the non-Abelian ones. Thus we learn that under "large" gauge transformations — as those that cannot be continuously deformed to the identity are called, to distinguish them from the "small" ones which can be so deformed — physical states of a Yang-Mills theory are only invariant up to a phase, and this phase is the hidden θ parameter of the model. The detailed classification of possible gauge transformations to which I have alluded makes use of the topological fact that the group of all integers is equivalent to Π_3 of the gauge group, *i.e.*, the mappings of the 3-sphere into the gauge group can be classified by integers. It should be emphasized that the existence of the angle does not rely on instantons; it is an exact quantum mechanical statement, while instantons provide an approximate method for calculating the consequence of the angle. Again, the analogy with a periodic potential is a useful one: Bloch-Floquet theory makes the exact statement that the wave function acquires a phase when its argument is shifted by the potential's period. These shifts are analogs of large gauge transformations. The consequent tunneling and band formation can be analyzed in the tight-binding approximation and this is analogous to instanton calculations.

Another important topological lesson has taught us that the axial vector anomaly,[6] which these days is central to many issues in gauge theories, is not a curious byproduct of field theoretical infinities, but rather a consequence of the topologically non-trivial possibilities that are present in gauge theories, as described by the Pontryagin index.

One further topological technique has proved useful in physics. Frequently, we need to analyze the spectrum of a linear differential operator, for example to determine quantum fluctuations in some background. Zero eigenvalue modes, when they occur, are the most crucial ones for physical application. While one may solve the differential equations explicitly, to determine these important modes, we have now learned that there exist *a priori* procedures, called "index theorems", which classify the zero modes. The Atiyah-Singer index theorem, first used in the physics of instantons, is an important example[7] while its various extensions appear in other applications. I shall not here have time to discuss the mathematical or physical import of these results, but Singer's talk yesterday and Witten's tomorrow address the point adequately.

I now must assess the impact on practical physics that these theoretical discoveries have had. The oldest results — those associated with the axial vector anomaly — seem to be in good shape: anomaly controlled low energy processes like $\pi^0 \to 2\gamma$ proceed at the calculated rate[8]; our prediction that quarks and leptons must balance in number, so there be no anomalous obstruction to renormalizability, keeps pace with the discovery of new leptons and quarks[9]; the symmetry that is anomalously violated — chiral $U(1)$ — does not occur in Nature.[10] However, further, more recent results pose problems: the vacuum angle is P and T non-invariant; moreover, its magnitude is not calculable owing to the axial anomaly. While one is intrigued by this unexpected source of CP violation, the stringent bounds on the neutron's dipole moment demand θ to be effectively zero, yet there is no principle guaranteeing this — we are facing a puzzle not unlike that of the cosmological constant in gravity theory.

Among solitons, the monopole is the most important for particle physicists, since it should arise in those theories that unify the strong interactions with the electro-weak. But we all are disappointed that after fifty years of looking, we still have not found one — maybe I should say that we haven't found more than one.[11] This now poses a problem — the "unified" theory demands a monopole, but it is not seen. There are more problems: It has been suggested that monopoles catalyze proton decay[12] — but we hear from Goldhaber that the proton seems stable; more recently, it was startlingly alleged that monopoles in unified theories can obstruct color gauge invariance[13] — but colored states are not supposed to exist. Is there something wrong with the predictions, or are we working with the wrong model, or will experiments eventually come into line? At present, we do not know — so the monopole soliton remains obviously important, perhaps only in a negative way, and it is unclear what its ultimate application will be.

Thus far, particle physics has not absorbed all the non-perturbative results that have been uncovered; especially the more exotic ones like boson-fermion conversion or fractional fermion charge seem far from phenomenological application. But in condensed matter physics, the last of these — fractional charge formation — has found practical application and is used to explain experimental data. Unlike particle physicists, who hope that one Hamiltonian explains all phenomena, condensed matter people have many Hamiltonians, each one describing some different material, and among this great variety one can find realization of all quantum mechanical possibilities. Certainly, there are systems with spontaneous symmetry breaking — the effect was first established within condensed matter physics. Certainly, there are solitons — these are the domain walls between different symmetry states of the system. Of course, there are fermions moving in the field of the soliton — the electrons of the conduction band. In one such system — polyacetylene — these three facts come together to produce observed phenomena which has been interpreted as soliton-induced charge fractionization.[14] This story is very beautiful, because it can be presented very generally without reference to details, utilizing only the quantum mechanical properties of fermions interacting with topological solitons. Also the discussion may be given from many points of view: chemistry, condensed matter physics, particle field theory, mathematics. This vividly demonstrates how different disciplines have come, each in their own way, to the same fact about Nature.

Let me sketch briefly a very simple pictorial description of the fractionization phenomenon in polyacetylene. That material is a one-dimensional polymer with double degenerate ground states. [The degeneracy arises for reasons that are called the "Peirles instability" but need not concern us here.[14]] Two distinct patterns for bonding electrons to the carbon atoms, which comprise the polyacetylene chain, characterize the two ground states; call them A and B. They are illustrated in the first two lines of the Figure.

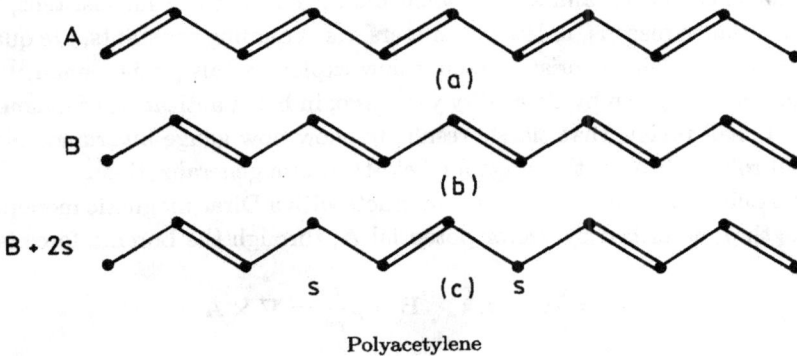

Polyacetylene

Figure (a): vacuum A
Figure (b): vacuum B
Figure (c): vacuum B with two solitons

A kink or soliton is a defect in the regular single bond – double bond alteration pattern, and state B with two solitons is depicted in the Figure's lowest line. Let us compare state B without solitons [middle line] to that with two solitons [last line]. They differ only in the region between solitons and the one without solitons carries five links in the interval, while the two-soliton state has four links. Inserting two solitons produces a defect of one link. If we now imagine separating the two solitons, so each acts independently of the other, the quantum numbers of the missing link must be equally shared by each soliton, *i.e.*, the link number fractionizes into halves. This is the essence of fermion fractionization, whose further consequences have been experimentally observed.[15] I must stress that the fractionization is without quantum fluctuation; it concerns eigenvalues and not merely expectation values.[16]

Let me now describe a new direction and new results of topological investigations in field theory. This current topic concerns one of the oldest questions of modern physics — quantization of physical quantities. The first such quantization, the ordinary quantum mechanical one, quantizes dynamical attributes of a system, like energy, angular momentum, and it involves Planck's constant $\hbar$. In recent times, we have seen another type of quantization, arising classically for topological reasons — the quantization of soliton number or instanton number. This quantization makes no reference to $\hbar$; finiteness of the classical energy or action requires it. However, there is a third mechanism for quantization which arises from the conjunction of topological and quantum mechanical principles. This happens when one is dealing with a system which produces consistent classical dynamics, but when quantized becomes inconsistent, for gauge invariance reasons, unless parameters, *i.e.*, coupling constants, are quantized. Although field theorists are only now exploring this phenomenon,[17] its first example was given by Dirac fifty years ago, in his quantization of magnetic charge. I shall review that classic result, to show how gauge invariance plays a central role and to set the stage for field-theoretic generalizations.

A particle of mass m and charge e interacts with a Dirac magnetic monopole of strength g, described by vector potential $\mathbf{A}$, through the Lorentz force law,

$$m\ddot{\mathbf{r}} = \frac{e}{c}\dot{\mathbf{r}} \times \mathbf{B}, \qquad \mathbf{B} = g\frac{\mathbf{r}}{r^3} = \boldsymbol{\nabla} \times \mathbf{A}$$

which follows from the Lagrangian

$$L = \frac{1}{2}m\dot{\mathbf{r}}^2 + \frac{e}{c}\dot{\mathbf{r}} \cdot \mathbf{A}(\mathbf{r}).$$

The equation of motion is obviously gauge invariant; however, the Lagrangian is not. Under a gauge transformation $\mathbf{A} \to \mathbf{A} + \boldsymbol{\nabla}\Theta$, L changes by a total derivative: $L \to L + \frac{e}{c}\dot{\mathbf{r}} \cdot \boldsymbol{\nabla}\Theta = L + \frac{d}{dt}\left(\frac{e}{c}\Theta\right)$, so the action $I = \int dt\, L$ changes by end-point contributions. But classical mechanics is entirely determined by the dynamical equation of motion, which is gauge invariant, hence obviously consistent; classical mechanics does not care about the action! Quantum mechanics, on the other hand, uses the exponential of the action $\exp \frac{i}{\hbar} I$, which must be gauge invariant for consistency. Hence, any gauge change of the action must be an integral multiple of $2\pi\hbar$. One can show that for a magnetic monopole, the end-point terms in the gauge transformed action give $4\pi \frac{eg}{c}$; therefore, gauge invariance requires the Dirac quantization condition on the parameters of the problem: $\frac{eg}{2} = \frac{\hbar}{2} n$. This derivation relies on the action, which is appropriate to a functional formulation of quantum mechanics like Feynman's, and we see that the monopole is described by a class of multivalued actions labeled by integers. One may also give a Hamiltonian argument. Recall that the angular momentum in a charged particle-monopole system has a radial component of magnitude $\frac{eg}{c}$. Therefore, conventional quantum mechanics of angular momentum gives the quantization condition by requiring that amount to be $\hbar$ times an integer or half-integer. The mathematical background to the Dirac quantization condition is the fact that $\Pi_1\big(U(1)\big)$ is equivalent to the group of integers, *i.e.*, the map of the unit circle into the gauge group, here $U(1)$, is classified by integers.

The first field theoretical example which requires quantized parameters is a three-dimensional gauge theory. In there dimensions, it is possible to have a term in the Lagrangian which explicitly gives a mass to the gauge potential; nevertheless, the equation of motion is gauge-covariant.[18] The Lagrangian, written in matrix notation for the gauge fields,

$$L = \frac{1}{2g^2} \operatorname{tr} F^{\mu\nu} F_{\mu\nu} - \frac{m}{2g^2} \varepsilon^{\mu\nu\alpha} \operatorname{tr} \left(F_{\mu\nu} A_\alpha - \frac{2}{3} A_\mu A_\nu A_\alpha \right)$$

implies the following equation of motion

$$\mathcal{D}_\mu F^{\mu\nu} + \frac{m}{2} \varepsilon^{\nu\alpha\beta} F_{\alpha\beta} = 0 \,.$$

Here g is a coupling with dimension of $[\text{mass}]^{1/2}$ and $\mathcal{D}$ is the covariant derivative. [Upon examining the linear or Abelian theory one learns that the excitations indeed carry mass m.[19]] Under a gauge transformation $A_\mu \rightarrow U^{-1} A_\mu U + U^{-1} \partial_\mu U$, the action changes according to

$$I = \int d^3x \mathcal{L} \longrightarrow I + m \frac{8\pi^2}{g^2} W(U)$$

$$W(U) = \frac{1}{24\pi^2} \int d^3x \varepsilon^{\alpha\beta\gamma} \operatorname{tr} [\partial_\alpha U U^{-1} \partial_\beta U U^{-1} \partial_\gamma U U^{-1}] \,.$$

Again because Π_3 (gauge group) is the group of all integers, $W(U)$ is an integer, and gauge invariance sets a quantization condition $\frac{4\pi m}{g^2} = \hbar n$.[20] Thus, three-dimensional gauge theories are defined by multivalued gauge invariant actions, analogous to the magnetic monopole example. One can also give a Hamiltonian derivation. In a Hamiltonian formulation, gauge invariance is achieved by requiring that physical states satisfy Gauss' law, which in our massive theory reads

$$\left(\mathcal{D}_i F^{i0} + \frac{m}{2}\varepsilon^{ij} F_{ij} \right) | \text{state} \rangle = 0.$$

But this requirement cannot be met unless m is quantized.[21]

A second example, now in four dimensions, has been given by Witten,[22] who considers an $SU(2)$ gauge theory with N doublets of Weyl fermions in the fundamental representation. Because $SU(2)$ is anomaly free, this appears to be a consistent theory for any N. Yet again gauge invariance sets a limitation in the quantum theory: N must be even. The argument, in a functional integral formulation, proceeds as follows. We begin with the quantum generating functional,

$$\mathcal{Z} = \int d\psi d\bar{\psi} dA_\mu \exp \frac{i}{\hbar} \int d^4 x \mathcal{L}$$

$$\mathcal{L} = \frac{1}{2g^2} \operatorname{tr} F^{\mu\nu} F_{\mu\nu} + i\hbar \sum_{n=1}^{N} \bar{\psi}_n \gamma^\mu (\partial_\mu + A_\mu) \psi_n$$

and integrate out the Weyl fermions, leaving an effective non-local action for gauge field dynamics,

$$I = \int d^4 x \frac{1}{2g^2} \operatorname{tr} F^{\mu\nu} F_{\mu\nu} - i\hbar N \ln \operatorname{Det}^{\frac{1}{2}} \left[\gamma^\mu (\partial_\mu + A_\mu) \right]$$

whose gauge transformation properties must still be investigated. [The square root on the determinant occurs because one is integrating four-component Weyl fermions.] Witten finds, as a consequence of $\Pi_4 \big(SU(2) \big)$ being the group of two integers $(0, 1)$ under modulo 2 addition, that a gauge transformation changes the effective action by a discrete amount: $I \to I + \hbar N\pi$. Hence the restriction: N must be even. The Hamiltonian version of this argument, given by Goldstone,[23] is especially intriguing since it presents the result as a consequence of the triangle anomaly, which plays no role in the above Lagrangian discussion. The statement that the $SU(2)$ gauge theory is anomaly-free means that the chiral $SU(2)$ currents, to which the $SU(2)$ gauge fields couple, are conserved. However, the $U(1)$ chiral fermion number current possesses an anomalous divergence and as usual, the conserved fermion number operator F is not gauge invariant, but changes under a gauge transformation by an integral multiple of N.[24] In the quantum theory, there are various transformations one can perform on a physical state, for example rotations by 2π, whose effect is to multiply the state by $(-1)^F$. Clearly if F is ambiguous by N, N must be even for consistency.

One further, four dimensional example has also been given by Witten.[25] This concerns a non-linear σ-model, which summarizes the low energy theorems of $SU(3) \times SU(3)$ current algebra, or, as one would say today, describes the low energy dynamics of QCD. The most obvious and simplest form for the action is

$$I_\sigma^0 = -\frac{F_\pi^2}{16} \int d^4x \operatorname{tr} \partial_\mu U \partial^\mu U^{-1} \,,$$

where U is an $SU(3)$ matrix and F_π is fitted by experiment to 190 MeV. However, when gauged, I_σ^0 does not reproduce the axial vector anomalies. Wess and Zumino[26] showed how to modify I_σ^0 to include anomalies: they add a further term I_{WZ}, [whose form is compicated]. An arbitrary coefficient c multiplies the Wess-Zumino addition; however, to reproduce the precise numerical value of the anomaly, as calculated from quark triangle graphs, the coefficient must be N_c — the number of colors. Nevertheless, for the moment we leave the constant unspecified. Once again, the transformation properties of the action $I_\sigma = I_\sigma^0 + cI_{\mathrm{WZ}}$ reflect a topological fact: $\Pi_5\big(SU(3)\big)$ is the group of all integers, and as a consequence, consistency of the quantized theory requires the constant c to be an integer, just as the microscopic quark theory predicts. We do not learn the value of the integer, nevertheless, it is remarkable that a bosonic effective Lagrangian somehow knows about the underlying fermionic structure![27]

Finally, let me comment on the physical import of these new results. I need not add anything more to all that has been said about magnetic monopoles. About the three dimensional gauge theory, you may be surprised to hear that it is not merely a theoretician's toy, but also is physically relevant, because physical four-dimensional theories, when studied at high temperature, are effectively described by theories in three dimensions.[28] Therefore, our three-dimensional gauge theory may provide a phenomenological Lagrangian for four-dimensional high temperature QCD, and the three-dimensional quantized mass term may arise from high temperature magnetic screening due to the

Pontryagin density in QCD, which is responsible for topological effects in four dimensions. While we have no proof of this conjecture, let us point out that both the Pontryagin density and our mass term are P and T non-invariant. Moreover, there is a mathematical connection between them, through the formula

$$\text{Pontryagin density} = -\frac{1}{16\pi^2} \text{tr} \, {}^*F^{\mu\nu} F_{\mu\nu} = \partial_\mu X^\mu$$

$$X^\mu = -\frac{1}{16\pi^2} \varepsilon^{\mu\alpha\beta\gamma} \text{tr} \left(F_{\alpha\beta} A_\gamma - \frac{2}{3} A_\alpha A_\beta A_\gamma \right).$$

The topological current X^μ involves the same combination of gauge fields as our mass term does; indeed the latter's contribution to the action is just the three-dimensional integral of one component of X^μ. This relationship between Pontryagin density and topological current has a well established place in mathematics under the name of "Chern-Simons secondary characteristic class."[29]

Witten's restriction that the number of $SU(2)$ Weyl fermion doublets in the fundamental representation must be even is surely fascinating, but it does not advance phenomenology since no one is considering models with such fermion content.

Finally, we come to the Wess-Zumino term, with its coefficient set necessarily to an integer; we take it to be three — the number of colors. It has recently been shown that by taking the Skyrme non-linear σ model,[1] [in the Skyrme action, I_σ^0 is supplemented by higher derivative interactions that give rise to solitons][30] together with the anomaly-producing Wess-Zumino term

$$I = I_{\text{Skyrme}} + cI_{\text{WZ}} = I_\sigma^0 + \alpha \int d^4x \, \text{tr} \left\{ [\partial_\mu U U^{-1}, \partial_\nu U U^{-1}]^2 \right\} + cI_{\text{WZ}}$$

$$c = N_c = 3$$

the solitons of the resulting theory — which may still be viewed as a description of low-energy QCD — are fermions with quantum numbers of the low-lying baryons.[31] I spoke of Skyrme at the beginning — he suggested years ago that baryons might be solitons — and his suggestion has now come to fruition in a beautiful synthesis of several topological and non-perturbative ideas: axial vector anomaly, solitons and their unexpected quantum numbers, quantized Lagrangian parameters.

This is where the subject stands today. While the experimental questions that it has produced remain unsettled, I hope you find, as I do, that it possesses the elegance and generality appropriate to descriptions of fundamental phenomena in Nature.

REFERENCES

1. *Selected Papers, with Commentary of Tony Hilton Royle Skyrme*, G. E. Brown, ed. (World Scientific, Singapore, 1994).
2. For reviews see R. Jackiw, *Rev. Mod. Phys.* **49**, 681 (1977) and R. Rajaraman, *Solitons and Instantons* (North-Holland, Amsterdam, 1982).
3. R. Jackiw in *Gauge Theories and Modern Field Theory*, R. Arnowitt and P. Nath, eds. (MIT Press, Cambridge, MA, 1976).
4. For a review see "Fermion Fractionization", reprinted in this volume on p. 79.
5. For reviews see R. Jackiw, *Rev. Mod. Phys.* **52**, 661 (1980) and Rajaraman, Ref. 2.
6. For a review see S. Treiman, R. Jackiw, B. Zumino and E. Witten, *Current Algebra and Anomalies* (Princeton University Press/World Scientific, Princeton NJ/Singapore, 1985).
7. For reviews see R. Jackiw, C. Nohl and C. Rebbi in *Particles and Fields*, D. Boal and A. Kamal, eds. (Plenum Press, New York, NY, 1978); T. Eguchi, P. Gilkey and A. Hanson, *Phys. Rep.* **66**, 213 (1980).
8. J. Steinberger, *Phys. Rev.* **76**, 1180 (1949); H. Fukuda and Y. Miyamoto, *Prog. Theor. Phys.* (Kyoto) **4**, 347 (1949); J. Schwinger, *Phys. Rev.* **82**, 664 (1951); J. Bell and R. Jackiw, *Nuovo Cim.* **60A**, 47 (1969); S. Adler, *Phys. Rev.* **177**, 2426 (1969); S. Glashow, R. Jackiw and S. Shei, *Phys. Rev.* **187**, 1416 (1969).
9. D. Gross and R. Jackiw, *Phys. Rev.* D **6**, 477 (1972); C. Bouchiat, J. Iliopoulos and P. Meyer, *Phys. Lett.* **38B**, 519 (1972).
10. This anomalous non-conservation is caused by the tunneling mentioned above. G. 't Hooft, *Phys. Rev. Lett.* **37**, 8 (1976) and *Phys. Rev.* D **14**, 3432 (1976); R. Jackiw and C. Rebbi, *Phys. Rev. Lett.* **37**, 172 (1976); C. Callan, R. Dashen and D. Gross, *Phys. Lett.* **B63**, 334 (1976).
11. B. Cabrera, *Phys. Rev. Lett.* **48**, 1378 (1982).
12. V. Rubakov, *Zh. Eksp. Teor. Fiz., Pis'ma Redakt.* **33**, 658 (1981) [English translation: *JETP Lett.* **33**, 644 (1981)] and *Nucl. Phys.* **B203**, 311 (1982); C. Callan, *Phys. Rev.* D **D25**, 2141 (1982) and **D26**, 2058 (1982); V. Rubakov and M. Sevebryakov, *Nucl. Phys.* **B218**, 240 (1983).
13. P. Nelson and A. Manohar, *Phys. Rev. Lett.* **50**, 943 (1983); A. Balachandran, G. Marmo, N. Mukunda, J. Nilsson, E. Sudarshan and F. Zaccaria, *Phys. Rev. Lett.* **50**, 1553 (1983).
14. W. Su, J. Schrieffer and A. Heeger, *Phys. Rev. Lett.* **42**, 1698 (1979) and *Phys. Rev. B* **22**, 2099 (1980). For a comparison of this condensed matter approach to fermion fractionization with the earlier field theoretic analysis [R. Jackiw and C. Rebbi, *Phys. Rev.* D **13**, 3398 (1976)] see R. Jackiw and J. Schrieffer, *Nucl. Phys.* **B190** [**FS3**], 253 (1981).
15. For a summary of experiments see A. Heeger, *Comments Solid State Physics* **10**, 53 (1981).

16. S. Kivelson and J. Schrieffer, *Phys. Rev.* B **25**, 6447 (1982); R. Rajaraman and J. Bell, *Phys. Lett.* **116B**, 151 (1982); J. Bell and R. Rajaraman, *Nucl. Phys.* **B220** [**FS8**], 1 (1983); Y. Frishman and B. Horovitz, *Phys. Rev.* B **27**, 2565 (1983); R. Jackiw, A. Kerman, I. Klebanov and A. Semenoff, *Nucl. Phys.* **B225** [**FS9**], 233 (1983).

17. S. Deser, R. Jackiw and S. Templeton, *Phys. Rev. Lett.* **48**, 975 (1982) and *Ann. Phys.* (NY) **140**, 372 (1982).

18. R. Jackiw and S. Templeton, *Phys. Rev.* D **23**, 2291 (1981); J. Schonfeld, *Nucl. Phys.* **B185**, 157 (1981).

19. These models exhibit gauge invariant masses for gauge fields, without the intervention of Higgs fields. Thus they supplement the two-dimensional Schwinger model [massless spinor QED in two dimensions] as examples of this phenomenologically important effect. It is noteworthy that in both cases the mass arises for topological reasons: the Schwinger model uses the Pontryagin class through its axial vector anomaly; the three-dimensional models rely on the Chern-Simons characteristic — as explained below. For a review of topological mass generation see R. Jackiw, in *Asymptotic Realms of Physics*, A. Guth, K. Huang and R. Jaffe, eds. (MIT Press, Cambridge, MA, 1983). Also three-dimensional gravity allows construction of a similar topological mass term; see Ref. 17.

20. Ref. 17. For a review see R. Jackiw in *Springer Lecture Notes in Physics* **181**, 157 (1983).

21. J. Goldstone and E. Witten (1982) (unpublished). Their argument is given in the review of Ref. 20.

22. E. Witten, *Phys. Lett.* **117B**, 324 (1982).

23. J. Goldstone (1982) (unpublished).

24. I. Gerstein and R. Jackiw, *Phys. Rev.* **181**, 1955 (1969); Jackiw and Rebbi, Ref. 10.

25. E. Witten, *Nucl. Phys.* **B223**, 422 (1983).

26. J. Wess and B. Zumino, *Phys. Lett.* **37B**, 95 (1971).

27. All this is paralleled in two dimensions: two-dimensional QCD also has axial vector anomalies, hence an effective non-linear σ-model must have a Wess-Zumino type term to reproduce them. Its coefficient will again be quantized.

28. For a review see D. Gross, R. Pisarski and L. Yaffe, *Rev. Mod. Phys.* **53**, 43 (1981).

29. S. Chern, *Complex Manifolds without Potential Theory*, 2nd ed. (Springer-Verlag, Berlin, 1979).

30. See also L. Faddeev, *Lett. Math. Phys.* **1**, 289 (1976).

31. A. Balachandran, V. Nair, S. Rajeev and A. Stern, *Phys. Rev. Lett.* **49**, 1124 (1982); E. Witten, *Nucl. Phys.* **B223**, 433 (1983); Phenomenological application of these ideas is being attempted; see M. Rho, A. Goldhaber and G. Brown, *Phys. Rev. Lett.* **51**, 747 (1983) and G. Adkins, C. Nappi and E. Witten, *Nucl. Phys.* **B228**, 552 (1983).

Paper V.2

The most recent examples of soliton phenomena in quantum field theory make use of many themes in modern mathematical physics: Chern-Simons action, non-linear Schrödinger equation, Liouville/Toda equations. A review of this material was presented in collaboration with So Young Pi in Japan and Korea (July-August 1991).

SELF-DUAL CHERN-SIMONS SOLITONS

Progr. of Theor. Phys. (Kyoto), Suppl. **107**, *1(1992);*
Recent Developments in Field Theory, J. E. Kim, ed.
(Min Eum Sa, Seoul, 1992)

I. INTRODUCTION

These lectures are devoted to the two-year old topic of Chern-Simons solitons. Solitons have interested particle theorists since the mid-1970's[1]; Chern-Simons dynamics, since the early 1980's.[2] Now in the decade of the 1990's, we are combining the two subjects.

The Chern-Simons theories that we consider exist in three-dimensional space-time. They possess pedagogical interest — teaching us [we hope] useful lessons that may be relevant to the physical four-dimensional world. Also they are mathematically fascinating, giving rise to effects and structures that invite further study. Finally, they can have a physical role in descriptions of actual, laboratory processes that are confined to a spatial plane. In a very precise sense, Chern-Simons models are field theoretical relatives of Landau's results on charged particles in an external magnetic field and of the Aharonov-Bohm [Ehrenberg-Siday] effect. These days, the quantum Hall effect and high-T_c superconductivity — two planar phenomena — are analyzed in terms of Chern-Simons dynamics.

In our lectures, we shall describe the structures that are supported by Chern-Simons interactions, specifically the soliton solutions to the appropriate equations of motion.

Although we shall work with Chern-Simons theories in three-dimensional space-time, to set the stage for our calculations we start from a different place in mathematical physics — the (1+1)-dimensional non-linear Schrödinger equation, given by

$$i\frac{\partial}{\partial t}\psi(t,x) = -\frac{1}{2}\frac{\partial^2}{\partial x^2}\psi(t,x) - g\psi^*(t,x)\psi(t,x)\psi(t,x). \qquad (1.1)$$

This equation enjoys two honorable and important roles.

First, we can take ψ to be a classical, c-number, complex function of its arguments, and (1.1) is a non-linear partial differential equation that determines ψ. In this context (1.1) is one of those famous non-linear partial differential equations that is integrable and possesses a complete set of soliton solutions. Understanding the complete integrability and soliton structure of the non-linear Schrödinger equation was an important milestone along the road that now has led to the elaborate theory of completely integrable systems.

In the second, alternative view, ψ is a quantum field operator, satisfying an equal-time commutation relation with its Hermitian conjugate,

$$\left[\psi(t,x),\psi^\dagger(t,y)\right] = \delta(x-y) \qquad (1.2)$$

and (1.1) is the Heisenberg equation of motion for the operator [$\psi^\dagger$ replaces ψ^* and $\hbar$ is scaled to unity]. This then is a non-relativistic quantum field theory and because the number operator N is conserved

$$N = \int dx\, \psi^\dagger(t,x)\psi(t,x) \qquad (1.3)$$

$$\frac{d}{dt}N = 0 \qquad (1.4)$$

the quantum Hilbert space may be spanned by energy and number eigenstates $|E, N\rangle$. In a well-known way, it follows that in the N-sector, one is dealing with a N-body quantum-mechanical [linear] Schrödinger problem where the potential [coded in the non-linearity of (1.1)] is the pairwise δ-function interaction. Specifically, by defining

$$e^{-iEt}u_E(x_1,\ldots,x_n) = \langle 0|\psi(t,x_1)\ldots\psi(t,x_n)|E,N\rangle, \qquad (1.5)$$

where $|0\rangle$ is the no-particle, zero-energy state, one finds that the wave function u_E satisfies

$$\left(-\frac{1}{2}\sum_i \frac{\partial^2}{\partial x_i^2} - g\sum_{ij}\delta(x_i - x_j)\right)u_E(x_1,\ldots,x_N) = Eu_E(x_1,\ldots,x_N). \qquad (1.6)$$

This many-body problem is also completely solvable and the energy spectrum and S-matrix are known.

A final remarkable fact about the model is that the same quantum mechanical spectrum is obtained by quantizing the classical soliton solutions of the first approach, using the methods of soliton quantization. Thus the above two aspects of the non-linear Schrödinger equation are two sides of the same coin.

In view of the remarkable properties exhibited by the non-linear Schrödinger equation in (1+1)-dimensional space-time, it is natural to try extending the model to higher dimensionality, for example to (2+1)-dimensions, by positing, instead of (1.1)

$$i\partial_t\psi(t,\mathbf{r}) = -\frac{1}{2}\nabla^2\psi(t,\mathbf{r}) - g\psi^*(t,\mathbf{r})\psi(t,\mathbf{r})\psi(t,\mathbf{r}). \qquad (1.7)$$

The single coordinate x has been replaced by the two-vector $\mathbf{r}$ and the second derivative with respect to x becomes the two-dimensional Laplacian. Unfortunately, nothing particularly interesting has been found for this planar generalization of (1.1).

The new suggestion that we make, and which does produce an interesting dynamical model, consists of introducing a gauge potential $A^\mu(t, \mathbf{r}) = \left(A^0(t, \mathbf{r}), \mathbf{A}(t, \mathbf{r})\right)$ by promoting all derivatives in (1.7) to gauge covariant derivatives and thereby allowing ψ to interact with A^μ in a gauge-invariant fashion. [When writing expressions involving gauge fields, we shall frequently use Lorentz covariant notation, even though matter dynamics is non-relativistic. We use the Minkowski metric, with signature $(1, -1, -1)$ and set c to unity.]

II. NON-RELATIVISTIC ABELIAN MODEL

The equation that we propose to study is[3]

$$iD_t\psi = -\frac{1}{2}\mathbf{D}^2\psi - g\psi^*\psi\psi \,, \tag{2.1}$$

where the ordinary derivatives of (1.7) are replaced by gauge covariant derivatives

$$D_\mu \equiv \partial_\mu + iA_\mu$$
$$D_t = \partial_t + iA^0 \tag{2.2}$$
$$\mathbf{D} = \boldsymbol{\nabla} - i\mathbf{A} \,.$$

Equation (2.1) is also an evolution equation

$$i\partial_t\psi(t, \mathbf{r}) = -\frac{1}{2}\left(\boldsymbol{\nabla} - i\mathbf{A}(t, \mathbf{r})\right)^2\psi(t, \mathbf{r}) + A^0(t, \mathbf{r})\psi(t, \mathbf{r})$$
$$- g\psi^*(t, \mathbf{r})\psi(t, \mathbf{r})\psi(t, \mathbf{r}) \tag{2.3}$$

but to close the system we still need an equation specifying the time-evolution of A^μ. Owing to the gauge invariant coupling of A^μ to ψ, we expect that the equation determining A^μ should also be gauge-invariant, *i.e.*, involve only the field strength $F_{\mu\nu}$,

$$F_{\mu\nu} = \partial_\mu A_\nu - \partial_\nu A_\mu \,. \tag{2.4}$$

The field components, in non-relativistic notation are,

$$\mathbf{B} = \boldsymbol{\nabla} \times \mathbf{A} \,, \tag{2.5a}$$

$$\mathbf{E} = -\boldsymbol{\nabla}A^0 - \partial_t\mathbf{A} \,. \tag{2.5b}$$

The natural and obvious equation for gauge field dynamics is the Maxwell equation, relating a derivative of $F^{\mu\nu}$ to the conserved matter current j^μ

$$j^\mu = \begin{cases} j^0 = \rho = \psi^*\psi \,, \\ \mathbf{j} = \mathrm{Im}\,\psi^*\mathbf{D}\psi \end{cases} \tag{2.6}$$

$$\partial_\mu F^{\mu\nu} = j^\nu \,. \tag{2.7a}$$

However, as is by now well-known, in three-dimensional space-time, there exists another term that may be added to (2.7a): the modified gauge field equation of topologically massive electrodynamics could be adopted[2]

$$\partial_\mu F^{\mu\nu} + \frac{1}{2}\kappa\varepsilon^{\nu\alpha\beta}F_{\alpha\beta} = j^\nu \, . \tag{2.7b}$$

Here κ has dimensions of mass and measure the strength of the modification. It is therefore also true that at low energies [large distances] the modification dominates the Maxwell term, so if one is considering a low energy phenomenological model, it makes sense to drop the Maxwell term altogether, giving rise to a field-current identity,[4] which we in fact adopt as the gauge field equation, complementing the gauged, non-linear Schrödinger equation

$$\frac{1}{2}\varepsilon^{\mu\alpha\beta}F_{\alpha\beta} = \frac{1}{\kappa}j^\mu \, . \tag{2.7c}$$

This is of course the equation of the Abelian Chern-Simons gauge theory; in components it reads

$$B = -\frac{1}{\kappa}\rho \, , \tag{2.8}$$

$$E^i = \frac{1}{\kappa}\varepsilon^{ij}j^j \, . \tag{2.9}$$

Equation (2.8) — the time-component of (2.7c) — is the Gauss law of Chern-Simons dynamics. It shows that $\mathbf{A}$ is determined by the matter density, and upon integrating (2.8) over all space one arrives at the important consequence that

$$\Phi = \int d^2\mathbf{r}\, B = -\frac{1}{\kappa}\int d^2\mathbf{r}\,\rho \equiv -\frac{1}{\kappa}N \, . \tag{2.10}$$

Therefore, in this theory excitations carrying charge $Q = N$ also possess magnetic flux Φ. Indeed all that is interesting about the present model is encapsulated in (2.10), which in fact also holds in the topologically massive theory (2.7b): the space integral of the time component in that equation still gives (2.10). Thus, the adoption of (2.7c), (2.8) and (2.9) may be viewed as the promotion of the spatially global formula (2.10) to the spatially local relation (2.8).

The equations (2.1), (2.3), (2.7c), (2.8) and (2.9) are the Euler-Lagrange equations of the Lagrange density

$$\mathcal{L} = \frac{\kappa}{4}\varepsilon^{\alpha\beta\gamma}A_\alpha F_{\beta\gamma} + i\psi^* D_t\psi - \frac{1}{2}|\mathbf{D}\psi|^2 + \frac{g}{2}(\psi^*\psi)^2 \, . \tag{2.11}$$

The first term is the Chern-Simons density. It is not gauge invariant, even though the equations that follow from varying it are. [In the quantized theory, which relies on the action not merely on the equations of motion, the issue of gauge invariance has to be re-examined.[2] One finds that in the Abelian theory there are no quantum-mechanical constraints on κ, regardless of base-space topology — contrary statements appear occasionally, but they are wrong.]

The dynamical equations may also be given a Hamiltonian formulation, involving solely the matter variables. The Hamiltonian H is

$$H = \frac{1}{2} \int d^2 \mathbf{r} \left(|\mathbf{D}\psi|^2 - g(\psi^*\psi)^2 \right), \tag{2.12}$$

where the vector potential $\mathbf{A}$ occurring in $\mathbf{D}\psi$ is not an independent variable, but is expressed in terms of $\rho = \psi^*\psi$ by

$$\mathbf{A}(t, \mathbf{r}) = \frac{1}{\kappa} \int d^2 \mathbf{r}' \mathbf{G}(\mathbf{r} - \mathbf{r}') \rho(t, \mathbf{r}'), \tag{2.13}$$

where $\mathbf{G}$ is the Greens function

$$G^i(\mathbf{r}) = \frac{1}{2\pi} \epsilon^{ij} \partial_j \ln r = \frac{\epsilon^{ij}\hat{r}^j}{2\pi r} \tag{2.14a}$$

satisfying

$$\mathbf{\nabla} \times \mathbf{G}(\mathbf{r}) = -\delta^2(\mathbf{r}) \tag{2.14b}$$

so that $\mathbf{A}$ solves (2.5a), (2.8). Equation (2.3) then is given as

$$i\partial_t \psi(t, \mathbf{r}) = \frac{\delta H}{\delta \psi^*(t, \mathbf{r})}. \tag{2.15}$$

Note that A^0 does not appear in H; A^0 arises in (2.15) when varying with respect to ψ^* the vector potential $\mathbf{A}$, which is present in the spatial covariant derivative occurring in H and depends on ψ^* through (2.13). A^0 enters (2.15) as

$$A^0(t, \mathbf{r}) = \frac{1}{\kappa} \int d^2 \mathbf{r}' \mathbf{G}(\mathbf{r} - \mathbf{r}') \cdot \mathbf{j}(t, \mathbf{r}) \tag{2.16}$$

which is easily shown to satisfy (2.5b), (2.9).

Note that the Hamiltonian formulation, with gauge fields expressed in terms of matter variables, achieves a fixing of the gauge freedom

$$A_\mu \to A_\mu + \partial_\mu \omega, \tag{2.17a}$$

$$\psi \to e^{-i\omega}\psi \tag{2.17b}$$

present in the formulation based on the Lagrange density (2.11). This is because the potentials are uniquely prescribed by the integral representations (2.13) and (2.16). Alternatively, one may state the gauge choice as the Coulomb condition, $\nabla \cdot \mathbf{A} = 0$, supplemented by large distance fall-off requirements on the differential equations satisfied by A^μ; the integral representations imply

$$\lim_{r \to \infty} r A^i(t, \mathbf{r}) = \frac{1}{2\pi\kappa} \varepsilon^{ij} \hat{r}^j N \,, \qquad (2.18a)$$

$$\lim_{r \to \infty} A^0(t, \mathbf{r}) = 0 \,. \qquad (2.18b)$$

The gauge fixed Lagrangian becomes

$$L = \int d^2 \mathbf{r} \, i\psi^* \partial_t \psi - H \qquad (2.19a)$$

and this defines the gauge-fixed action I

$$I = \int dt \, d^2 \mathbf{r} \, i\psi^* \partial_t \psi - \int dt \, H = \int dt \, L \,. \qquad (2.19b)$$

While issues of gauge invariance, the form of the action, *etc.* are largely irrelevant in the classical theory, which is fully determined by the equations of motion, these matters play a significant role in the quantum theory, and will concern us below.

Just as the non-linear Schrödinger equation on a line, so also our gauged, planar non-linear Schrödinger equation may be considered to be a classical field theory, or alternatively a non-relativistic quantum field theory. In the former context, we shall explore soliton solutions. In the latter, we again observe number conservation

$$\frac{d}{dt} N = 0 \qquad (2.20)$$

and recognize that we are dealing with the second-quantized N-body anyon problem, where the anyons possess, in addition to their statistical [Chern-Simons] interaction, a further two-body δ-function attraction of strength g, coming from the $(\psi^*\psi)^2$ non-linearity.

In what follows, we are principally concerned with the classical theory. We shall choose a particular value of g, which allows further explicit analysis.

III. TRANSFORMATIONS ON THE MODEL

Before presenting classical solutions to our dynamical equations, we record various useful transformations that can be performed. Some of these are symmetries of the equations; others map our system into other interesting dynamical models. Familiarity with these transformations assists in understanding the properties of the solutions.

The Chern-Simons contribution to the action $\int d^3 x \mathcal{L}$ of (2.11) is a topological invariant, hence invariant against all coordinate transformations, which we present in space-time notation

$$x^\mu \to X^\mu , \tag{3.1}$$

$$A_\mu \to \tilde{A}_\mu , \tag{3.2a}$$

$$\tilde{A}_\mu(x) = A_\nu(X) \frac{\partial X^\nu}{\partial x^\mu} . \tag{3.2b}$$

Owing to the field-current identity (2.7c), the matter current must then transform as a contravariant density,

$$\tilde{j}^\mu(x) = \frac{1}{\triangle} j^\nu(X) \frac{\partial x^\mu}{\partial X^\nu} \tag{3.3}$$

with $\triangle \equiv \det (\partial x^\mu / \partial X^\nu)$.

But of course, the action for the non-relativistic matter dynamics is not invariant under this general coordinate transformation. Nevertheless, it is interesting to consider a subclass of transformations for which the time coordinate is transformed into a function only of time and the spatial coordinate is rescaled by a time-dependent factor

$$x^\mu = \begin{pmatrix} t \\ \mathbf{r} \end{pmatrix} \longrightarrow \begin{pmatrix} T \\ \mathbf{R} \end{pmatrix} = \begin{pmatrix} T(t) \\ \sqrt{\dot{T}(t)}\,\mathbf{r} \end{pmatrix} , \tag{3.4}$$

$$\psi \to \tilde{\psi} , \tag{3.5a}$$

$$\tilde{\psi}(t, \mathbf{r}) = \sqrt{\dot{T}}\, e^{-ir^2 \ddot{T}/4\dot{T}} \psi(T, \mathbf{R}) . \tag{3.5b}$$

[The vector potential continues to transform as in (3.2).]

For arbitrary $T(t)$, (3.4) and (3.5) is not a symmetry transformation; rather, our dynamical system is mapped into one with an external harmonic force, with time-dependent frequency given by[5]

$$\omega^2(t) = \frac{\dddot{T}}{2\dot{T}} - \frac{3}{4} \left(\frac{\ddot{T}}{\dot{T}} \right)^2 = -\sqrt{\dot{T}} \frac{d^2}{dt^2} \frac{1}{\sqrt{\dot{T}}} . \tag{3.6}$$

This feature will be exploited below. However, for three particular forms of $T(t)$, the matter action remains invariant, and the transformations comprise the $SO(2,1)$ group.

A. Symmetry Transformations

The three invariances comprise

$$T(t) = t + a \qquad \text{time translation}, \tag{3.7}$$

$$T(t) = at \qquad \text{dilation}, \tag{3.8}$$

$$T(t) = \frac{t}{1 - at} \qquad \text{conformal transformation}. \tag{3.9}$$

With these, $\tilde{\psi}$ of (3.5b) is a solution to our equations provided ψ solves them. The time translations are of course generated by the Hamiltonian (2.12). The generator of dilations D is

$$D = tH - \frac{1}{2} \int d^2\mathbf{r}\, \mathbf{r} \cdot \mathbf{j}(t, \mathbf{r}) \tag{3.10}$$

while the conformal transformation is generated by K

$$K = -t^2 H + 2tD + \frac{1}{2} \int d^2\mathbf{r}\, r^2 \rho(t, \mathbf{r}). \tag{3.11}$$

All three generators are constants of motion:[6] $\frac{d}{dt}H = \frac{d}{dt}D = \frac{d}{dt}K = 0$.

Additionally there are the conventional, expected symmetries: spatial variables may be translated, rotated or boosted: $t \to t$, $\mathbf{r} \to \mathbf{R}$

$$\mathbf{R} = \mathbf{r} + \mathbf{a} \qquad \text{space translation}, \tag{3.12}$$

$$R^i = \mathcal{R}^{ij}(\varphi)R^j \qquad \text{rotation}, \tag{3.13}$$

$$\mathbf{R} = \mathbf{r} - \mathbf{v}t \qquad \text{boost}. \tag{3.14}$$

Here $\mathcal{R}^{ij}(\varphi)$ is the two-dimensional matrix implementing a rotation by angle φ: $\mathcal{R}^{ij}(\varphi) = \delta^{ij}\cos\varphi - \epsilon^{ij}\sin\varphi$, while $\mathbf{v}$ is the velocity of the boost. The transformed field takes the obvious form for translations and rotations

$$\tilde{\psi}(t, \mathbf{r}) = \psi(t, \mathbf{r} + \mathbf{a}), \tag{3.15}$$

$$\tilde{\psi}(t, \mathbf{r}) = \psi(t, \mathcal{R}\mathbf{r}) \tag{3.16}$$

while boosts require the familiar 1-cocycle

$$\tilde{\psi}(t, \mathbf{r}) = e^{i\left(\mathbf{r} \cdot \mathbf{v} - \frac{v^2}{2} t\right)} \psi(t, \mathbf{r} - \mathbf{v}t). \tag{3.17}$$

The time-independent generators of these symmetry transformations are, respectively, the momentum $\mathbf{P}$,

$$\mathbf{P} = \int d^2 \mathbf{r} \, \mathbf{j}(t, \mathbf{r}) \tag{3.18}$$

the angular momentum J,

$$J = \int d^2 \mathbf{r} \, \mathbf{r} \times \mathbf{j}(t, \mathbf{r}) \tag{3.19}$$

and the Galilean boost $\mathbf{B}$,

$$\mathbf{B} = t\mathbf{P} - \int d^2 \mathbf{r} \, \mathbf{r} \rho(t, \mathbf{r}) \tag{3.20}$$

with $\frac{d\mathbf{P}}{dt} = \frac{dJ}{dt} = \frac{d\mathbf{B}}{dt} = 0$. Finally, phase invariance insures number conservation (2.20).

The constants of motion/symmetry generators can be given in terms of a non-relativistic energy-momentum tensor. We define the energy density $\mathcal{H}$

$$\mathcal{H} = T^{00} = \frac{1}{2}\left(|\mathbf{D}\psi|^2 - g\rho^2\right) \tag{3.21}$$

which together with the energy flux T^{i0}

$$T^{i0} = -\frac{1}{2}\left[(D_t\psi)^* D_i\psi + (D_i\psi)^*(D_t\psi)\right] \tag{3.22}$$

satisfies the continuity equation,

$$\partial_t T^{00} + \partial_i T^{i0} = 0 \tag{3.23}$$

thus ensuring that energy

$$E = H = \int d^2 \mathbf{r} \, T^{00} \tag{3.24}$$

is time-independent. The momentum density $\mathcal{P}$

$$\mathcal{P}^i = T^{0i} = j^i \tag{3.25}$$

satisfies a continuity equation with the momentum flux — the stress tensor T^{ij} —

$$T^{ij} = \frac{1}{2}\left((D_i\psi)^*(D_j\psi) + (D_j\psi)^*(D_i\psi) - \delta^{ij}(D_k\psi)^*(D_k\psi)\right)$$
$$+ \frac{1}{4}(\delta^{ij}\nabla^2 - 2\partial_i\partial_j)\rho + \delta^{ij}\mathcal{H}, \tag{3.26}$$

$$\partial_t T^{0i} + \partial_j T^{ji} = 0 \tag{3.27}$$

which renders the momentum

$$P^i = \int d^2\mathbf{r}\, T^{0i} \tag{3.28}$$

conserved. Also of course, there is the current continuity equation

$$\partial_t \rho + \nabla \cdot \mathbf{j} = 0 \tag{3.29}$$

by virtue of which N in (2.20) is time-independent.

Note that the energy-momentum tensor is not symmetric: energy flux does not equal momentum density $T^{i0} \neq T^{0i}$; the theory is not Lorentz invariant. However, the stress-tensor is symmetric $T^{ij} = T^{ji}$; this is a consequence of rotational invariance, and the conservation of angular momentum

$$J = \int d^2\mathbf{r}\, \varepsilon^{ij} r^i T^{0j} \tag{3.30}$$

follows from the continuity equation (3.27) and the symmetry of T^{ij}.

The current continuity equation (3.29), together with (3.25) and (3.28) insure the conservation of the boost generator (3.20).

Finally, the dilation and conformal generators may be presented as

$$D = \int d^2\mathbf{r}\left[t T^{00} - \frac{1}{2}r^i T^{0i}\right], \tag{3.31}$$

$$K = \int d^2\mathbf{r}\left[t^2 T^{00} - t r^i T^{0i} + \frac{1}{2}r^2 \rho\right]. \tag{3.32}$$

Their time-independence follows from the previously stated continuity equations and also from the fact that the trace of the stress tensor is twice the energy density

$$\sum_{i=1}^{2} T^{ii} = 2T^{00}. \tag{3.33}$$

[Recall that in a relativistic theory the trace relation for $T^{\mu\nu}$ that encodes conformal invariance is without the factor 2.]

That the above conformal invariance is respected in classical field theory by the $(\psi^*\psi)^2$ interaction is easily understood from the quantum viewpoint. As mentioned already, this field theoretic contact term is the second-quantized form of a $\delta^2(\mathbf{r})$ interaction. But on the plane, $\delta^2(\mathbf{r})$ scales as $1/r^2$, *i.e.*, in the same way as the kinetic Laplacian operator. Hence, the interaction is scale-free. However, this consideration also puts into evidence a quantum mechanical complication. The $\delta^2(\mathbf{r})$ potential in quantum mechanics requires renormalization or self-adjoint extension, which in fact spoils the scale invariance — there occurs anomalous scale symmetry breaking.[7] Correspondingly in the quantum field theoretic formulation, the coupling strength of the $(\psi^*\psi)^2$ term needs renormalization; the scale symmetry and the trace relation (3.33) acquire anomalies and must be analyzed by renormalization group-type arguments. This is especially striking when it is recalled that the *relativistic* $(\psi^*\psi)^2$ interaction is super-renormalizable in three-dimensional space-time; it requires no coupling constant renormalization and has been "constructed" by axiomatists. Consequently, the pathologies that beset the non-relativistic theory, both in the field theoretic and the quantum mechanical formulations, must arise as the non-relativistic limits is taken.[8]

Since here we are concerned with the classical theory, these problems do not arise, but it is important to keep them in mind for an eventual quantum mechanical application.

B. Other Transformations

We have remarked that for general $T(t)$ in (3.4) and (3.5) our problem is mapped onto one with an external harmonic force, where the frequency is the time-dependent quantity (3.6). This observation allows constructing solutions $\tilde\psi$ to the external field problem from solutions ψ of the original problem.[5]

Within this approach, it is useful to view the dynamics as involving solely matter variables, with the potentials given by (2.13) and (2.16). It then follows that the transformation rules for A_μ (3.2b) and J^μ (3.3) are consequences of the transformation law (3.5) for ψ and the formulas (2.6), (2.13) and (2.16) for these quantities in terms of the fields.

If we take $T(t) = \frac{1}{\omega}\tan\omega t$, the harmonic frequency (3.6) is constant: $\omega(t) = \omega$. Therefore from (3.5) we know that if $\psi(t,\mathbf{r})$ solves the equations in the absence of external forces, then

$$\psi_\omega(t, \mathbf{r}) = \frac{1}{\cos \omega t} e^{-i\frac{1}{2}\omega r^2 \tan \omega t} \psi(T, \mathbf{R})$$

$$T = \frac{1}{\omega} \tan \omega t, \qquad \mathbf{R} = \frac{\mathbf{r}}{\cos \omega t}$$

(3.34)

solves the problem in the presence of an additional harmonic force with frequency ω [The subscript ω indicates quantities in the external harmonic field.]

$$i\partial_t \psi_\omega = -\frac{1}{2}\mathbf{D}^2\psi_\omega + \left(A_\omega^0 - g\rho_\omega + \frac{1}{2}\omega^2 r^2\right)\psi_\omega,$$

(3.35)

where $\mathbf{A}_\omega$, from which $\mathbf{D}$ is constructed, and A_ω^0 are given by (2.13) and (2.16), with density of charge ρ_ω and current $\mathbf{j}_\omega$ constructed from ψ_ω by (2.6). The conserved energy and angular momentum now read

$$H_\omega = \frac{1}{2}\int d^2\mathbf{r}\left(|\mathbf{D}\psi_\omega|^2 - g\rho_\omega^2 + \omega^2 r^2 \rho_\omega\right),$$

(3.36)

$$J_\omega = \int d^2\mathbf{r}\,\mathbf{r} \times \mathbf{j}_\omega,$$

(3.37)

where ρ_ω and $\mathbf{j}_\omega$ are constructed from ψ_ω according to (2.6), or equivalently are related to ρ and $\mathbf{j}$ by (3.3). Substituting formula (3.34) for ψ_ω gives

$$E_\omega = E + \omega^2 K,$$

(3.38)

$$J_\omega = J.$$

(3.39)

The angular momentum is unchanged, while the energy acquires a contribution proportional to the value in the force-free problem of the conformal generator K, (3.11).[9]

The above ideas and methods can be extended to the construction of a solution in the presence of an external, constant magnetic field from a solution without the magnetic field.[5] The equation of motion in the presence of the constant magnetic field $\mathcal{B}$, described in the redial gauge by

$$\mathcal{A}^i = -\frac{1}{2}\varepsilon^{ij} r^j \mathcal{B}$$

(3.40)

is

$$i\partial_t \psi_{\mathcal{B}} = -\frac{1}{2}\mathbf{D}_{\mathcal{B}}^2\psi_{\mathcal{B}} + (A_{\mathcal{B}}^0 - g\rho_{\mathcal{B}})\psi_{\mathcal{B}}$$

$$\mathbf{D}_{\mathcal{B}} = \nabla - i\mathbf{A}_{\mathcal{B}} - i\mathcal{A},$$

(3.41)

where A_B^0 and $\mathbf{A}_B$ are still given by (2.13) and (2.16), with density of charge ρ_B and current $\mathbf{j}_B$, which are constructed from ψ_B by formulas like (2.6), except that the covariant derivative in the definition of $\mathbf{j}_B$ involves the external magnetic field. [The subscript B indicates quantities in the presence of the external magnetic field.] One may verify that the solution to (3.41) is constructed from the solution $\psi(t, \mathbf{r})$ in the absence of B by

$$\psi_B(t,\mathbf{r}) = \frac{1}{\cos\frac{B}{2}t} e^{-i\frac{B}{4}r^2\tan\frac{B}{2}t} e^{i(N/4\pi\kappa)Bt}\psi(T,\mathbf{R})$$

$$T = \frac{2}{B}\tan\frac{B}{2}t, \qquad R^i = \frac{1}{\cos\frac{B}{2}t}\mathcal{R}^{ij}\Big(\frac{B}{2}t\Big)r^j . \tag{3.42}$$

The transformation exhibited in (3.42) is a time-dependent dilation [like in (3.34)] supplemented by a time-dependent rotation by angle $\frac{B}{2}t$ [a generalization of (3.13) and (3.16)] and finally, a gauge transformation with the gauge function $-\frac{B}{4\pi\kappa}Nt$ [like in (2.17)].

The gauge transformation is needed to assure that the gauge potential A_B^0 satisfy (2.18), or what is equivalent, that the equation of motion (3.41) follow from the matter Hamiltonian H_B

$$H_B = \frac{1}{2}\int d^2\,\mathbf{r}\big(|\mathbf{D}_B\psi_B|^2 - g\rho_B^2\big) . \tag{3.43}$$

As a consequence of the gauge transformation, the vector potential transformation formula (3.2) is modified,

$$A_{\mu B}(x) = A_\nu(X)\frac{\partial X^\nu}{\partial x^\mu} - \partial_\mu\Big(\frac{B}{4\pi\kappa}Nt\Big) \tag{3.44}$$

but the transformation law (3.3) for the gauge invariant current remains unchanged.

In addition to the energy, the momentum $\mathbf{P}_B$ and angular momentum J_B are conserved. Their expressions, in the presence of the external magnetic field, are

$$P_B^i = \int d^2\,\mathbf{r}j_B^i - B\int d^2\,\mathbf{r}\epsilon^{ij}r^j\rho_B , \tag{3.45}$$

$$J_B = \int d^2\,r\,\mathbf{r}\times\mathbf{j}_B + \frac{1}{2}B\int d^2\,\mathbf{r}\,r^2\rho_B . \tag{3.46}$$

On (3.42) these constants of motion are expressed in terms of the $\mathcal{B} = 0$ constants, and one finds

$$E_{\mathcal{B}} = E + \frac{\mathcal{B}^2}{4}K - \frac{\mathcal{B}}{2}J, \qquad (3.47)$$

$$P_{\mathcal{B}}^i = P^i + \frac{\mathcal{B}}{2}\epsilon^{ij}B^j, \qquad (3.48)$$

$$J_{\mathcal{B}} = J. \qquad (3.49)$$

The energy acquires contributions from the conformal (K) and angular momentum (J) generators; the momentum is supplemented by a term involving the boost generator $(\mathbf{B})$; while the angular momentum remains unchanged.

The above transformations can be used to generate soliton solutions in the presence of external fields from those in their absence.[10]

We conclude this discussion by remarking that the symmetry transformations of the force-free problem, which lead to eight conserved generators $(H, D, K, \mathbf{P}, J, \mathbf{B})$ and are discussed in sub-Section A above, can be "imported" to the external harmonic and magnetic field problems by the transformations (3.34) and (3.42). In this way, one finds conserved generators/constants of motion that are additional to those in (3.36), (3.37) and (3.43), (3.45), (3.46).[11]

IV. STATIC SOLUTIONS

A. General Properties

We consider static solutions to the model described in Section II; *viz.* we seek solutions to

$$-\frac{1}{2}(\boldsymbol{\nabla} - i\mathbf{A})^2\psi + A^0\psi - g\rho\psi = 0, \qquad (4.1)$$

$$\boldsymbol{\nabla} \times \mathbf{A} = -\frac{1}{\kappa}\rho, \qquad (4.2)$$

$$\boldsymbol{\nabla} \times A^0 = \frac{1}{\kappa}\mathbf{j}. \qquad (4.3)$$

These are the time-independent version of (2.1), (2.3), (2.5), (2.8) and (2.9).

Before proceeding, let us observe the following important consequences of the symmetries in our problem. Galileo invariance insures that the boost generator $\mathbf{B}$ in (3.20) is time-independent. But for static solutions $\int d^2\mathbf{r}\,\mathbf{r}\,\rho$ is also time-independent. It therefore follows that momentum $\mathbf{P} = \int d^2\mathbf{r}\,\mathbf{j}$

vanishes on static solutions — a natural and not unexpected fact. An additional, remarkable vanishing theorem follows from conformal invariance, which requires that K in (3.11) be time-independent. Since static solutions lead to time-independent $\int d^2 \mathbf{r} \, r^2 \rho$, it must be that H and D also vanish, which from (3.10) further implies that $\int d^2 \mathbf{r} \, \mathbf{r} \cdot \mathbf{j} = 0$. Thus static solutions carry zero energy and momentum.[12] [It should be stated that implicitly we are making various regularity assumptions, some of which will be made explicitly later.]

In order to explore the implication of vanishing energy [for static solutions] we make use of the following identity

$$|\mathbf{D}\psi|^2 = |(D_x - i\varepsilon(\kappa)D_y)\psi|^2 - \varepsilon(\kappa)(B\rho + \nabla \times \mathbf{j}), \quad \varepsilon(\kappa) = \text{sign}\,\kappa \quad (4.4)$$

to present the Hamiltonian (2.12) as

$$H = \frac{1}{2}\int d^2 \mathbf{r} \left\{ |(D_x - i\varepsilon(\kappa)D_y)\psi|^2 - \left(g - \frac{1}{|\kappa|}\right)\rho^2 \right\}, \quad (4.5)$$

where $\int d^2 \mathbf{r} \, \nabla \times \mathbf{j}$ has been dropped with the hypothesis that $\mathbf{j}$ is sufficiently well-behaved. Since the first term in the integrand of (4.5) is non-negative and so is ρ^2, a non-trivial ($\psi \neq 0$) zero-energy configuration is possible only for $g \geq \frac{1}{|\kappa|}$, since otherwise the energy is positive for non-zero ψ.

Henceforth, we take $g = \frac{1}{|\kappa|}$ so that the field non-linearity [δ-function interaction] is attractive and with strength correlated to the Chern-Simons coupling, κ. Also, without loss of generality κ may be taken positive, so the Hamiltonian becomes

$$H = \frac{1}{2}\int d^2 \mathbf{r} \, |(D_x - iD_y)\psi|^2 . \quad (4.6)$$

Since time-independent solutions stationarize H and moreover H must also then vanish, we see that static solutions require a self-duality condition,

$$(D_x - iD_y)\psi = 0, \quad (4.7a)$$

$$\mathbf{D}\,\psi = i\,\mathbf{D} \times \psi \quad (4.7b)$$

and of course, the vector potential $\mathbf{A}$ in the above covariant derivatives also satisfies

$$\nabla \times \mathbf{A} = -\frac{1}{\kappa}(\psi^*\psi). \quad (4.8)$$

Thus the system of first-order equations (4.7), (4.8) is completely equivalent [with regularity assumptions] to the second-order (4.1), (4.2), (4.3) at $g = \frac{1}{\kappa}$.

That the first-order equations imply the second-order ones is easily verified; that the implication also goes in the other direction is a consequence of conformal symmetry.

Note that self-duality (4.7) has the consequence that

$$\mathbf{j} = \frac{1}{2}\boldsymbol{\nabla} \times \rho. \tag{4.9}$$

Hence according to (4.3)

$$A^0 = \frac{1}{2\kappa}\rho \tag{4.10}$$

and also it is obvious from (4.9) that $\int d^2\,\mathbf{r}\,\mathbf{j}$ and $\int d^2\,\mathbf{r}\,\mathbf{r}\cdot\mathbf{j}$ do both indeed vanish, as was earlier established by a general argument. The angular momentum (3.19) is expressed in terms of N, which need not be an integer

$$J = \int d^2\,\mathbf{r}\,\mathbf{r} \times \left(\frac{1}{2}\boldsymbol{\nabla} \times \rho\right) = \int d^2\,\mathbf{r}\,\rho = N\,. \tag{4.11}$$

This is related to flux Φ by the Chern-Simons constraint

$$\Phi = \int d^2\,\mathbf{r}\,B = -\frac{1}{\kappa}\int d^2\,\mathbf{r}\,\rho = -\frac{1}{\kappa}N\,. \tag{4.12}$$

It will be seen that Φ is quantized. The remaining, non-vanishing symmetry generators/constants of motion are

$$\mathbf{B} = -\int d^2\,\mathbf{r}\,\mathbf{r}\,\rho \equiv -N\langle\mathbf{r}\rangle\,, \tag{4.13}$$

and

$$K = \frac{1}{2}\int d^2\,\mathbf{r}\,r^2\,\rho \equiv \frac{1}{2}N\langle r^2\rangle\,. \tag{4.14}$$

The values of $\langle\mathbf{r}\rangle$ and $\langle r^2\rangle$ depend on the form of the solution.

Finally, we discuss the particular choice $g = 1/\kappa$ for the non-linearity. To be sure, this choice is made to facilitate the subsequent self-dual formulation, which leads to explicit solution, see below. Nevertheless, the choice has several "natural" features that are worth noting.

Observe that the Pauli equation

$$\boldsymbol{\sigma} \cdot \left(\frac{1}{i}\boldsymbol{\nabla} - \mathbf{A}\right)\chi = \varepsilon\chi \tag{4.15}$$

for a two-spinor χ, and in two dimensions involves two Pauli matrices. Upon iteration one gets

$$\left\{\boldsymbol{\sigma}\cdot\left(\frac{1}{i}\boldsymbol{\nabla}-\mathbf{A}\right)\right\}^2\chi = \left\{-(\boldsymbol{\nabla}-iA)^2 - \sigma^3 B\right\}\chi = \varepsilon^2\chi. \qquad (4.16a)$$

One may choose the χ spinor to be an eigenstate of σ^3, $\sigma^3\chi_\pm = \pm\chi_\pm$, and also owing to the Chern-Simons constraint, $B = -\frac{1}{\kappa}\chi_\pm^*\chi_\pm$. Hence (4.16a) becomes

$$\left\{-\frac{1}{2}(\boldsymbol{\nabla}-iA)^2 \pm \frac{1}{2\kappa}\chi_\pm^*\chi_\pm\right\}\chi_\pm = \frac{\varepsilon^2}{2}\chi_\pm. \qquad (4.16b)$$

With the lower sign, the left-hand side coincides with (2.3) and (4.1) at $g = 1/\kappa$ and A^0 evaluated from (4.10). Thus we see that the non-linearity may be alternatively viewed as a magnetic moment interaction with a magnetic field that is given self-consistently by the Chern-Simons equations. Moreover, when $g = 1/\kappa$ the magnetic moment is the "minimal" one in a spinorial formulation, and therefore there is a "hidden" supersymmetry.[3]

When we later discuss a relativistic soliton model, we shall demonstrate that $g = 1/\kappa$ corresponds to the non-relativistic limit of that theory, which also possesses supersymmetric features.

B. Self-Dual Solutions

In fact the system (4.7) and (4.8) may be solved completely and explicitly. Equation (4.7) implies that

$$A^x - iA^y = i(\partial_x - i\partial_y)\ln\psi. \qquad (4.17)$$

When this form for $\mathbf{A}$ is inserted in (4.8) one finds that $\rho = \psi^*\psi$ satisfies the Liouville equation

$$\nabla^2\ln\rho = -\frac{2}{\kappa}\rho. \qquad (4.18)$$

Since

$$\psi = \rho^{1/2}e^{i\omega} \qquad (4.19)$$

we see that there is no equation determining the phase ω; rather it is fixed by regularity requirements: At the zeroes of ψ, equivalently at the zeroes of ρ, $\ln\psi$ and $\ln\rho$ may be singular. Demanding that $\mathbf{A}$, determined in (4.17) by the holomorphic derivative of $\ln\psi$, be non-singular fixes the phase of ψ, which is then further constrained by the requirement that ψ be single-valued near its zeroes. This is illustrated below in detail. It is also understood that (4.18) holds away from the zeroes of ρ, and the non-singular ρ is determined near its zeroes by continuity.

The general solution to (4.18) is known. Let us consider first in (r, θ) variables the radially symmetric, θ-independent solutions, which can be constructed by quadrature from the second-order r-differential equation that is implied by (4.18). One finds the most general regular, positive and θ-independent ρ to be[3]

$$\rho(\mathbf{r}) = \frac{4\kappa\mathcal{N}^2}{r^2} \left[\left(\frac{r_0}{r} \right)^{\mathcal{N}} + \left(\frac{r}{r_0} \right)^{\mathcal{N}} \right]^{-2} \tag{4.20}$$

where r_0 and $\mathcal{N}$ are two positive constants of integration. Clearly, r_0 is a scale reflecting the scale invariance of the dynamics. To fix $\mathcal{N}$, let us observe that regularity at the origin, where $\rho \underset{r \to 0}{\sim} r^{2\mathcal{N}-2}$, and at infinity, where $\rho \underset{r \to \infty}{\sim} r^{-2\mathcal{N}-2}$, require $\mathcal{N} \geq 1$. Moreover, since $\rho^{1/2}$ has a zero at the origin of order $\mathcal{N} - 1$, singularities in $\mathbf{A}$ are avoided if the phase of ψ is chosen to be $\omega = (1 - \mathcal{N})\theta$. Finally, for single-valued ψ, $\mathcal{N}$ must be an integer. In this way, we conclude that[3]

$$\psi(\mathbf{r}) = \frac{2\sqrt{\kappa}\mathcal{N}}{r} \left(\left(\frac{r_0}{r} \right)^{\mathcal{N}} + \left(\frac{r}{r_0} \right)^{\mathcal{N}} \right)^{-1} e^{i(1-\mathcal{N})\theta}, \quad \mathcal{N} = 1, 2, \ldots . \tag{4.21}$$

This is interpreted as an $\mathcal{N}$-soliton/vortex solution, where the solitons are superimposed as the origin, and each has the same scale r_0. Note that with this solution

$$\Phi = -\frac{1}{\kappa} \int d^2\,\mathbf{r}\,\rho(\mathbf{r}) = -2\pi(2\mathcal{N}) . \tag{4.22}$$

In units of 2π, which is the natural unit of flux with our conventions, our solution carries an even number, $2\mathcal{N}$, flux units. Also we get

$$J = N = -\kappa\Phi = 2\pi\kappa(2\mathcal{N}) \tag{4.23}$$

which is not an integer for arbitrary κ. Owing to the rotational symmetry, the Galileo boost generator vanishes

$$\mathbf{B} = -N\langle \mathbf{r} \rangle = 0 . \tag{4.24}$$

The conformal charge is

$$K = \frac{1}{2}N\langle r^2 \rangle = \frac{1}{2}Nr_0^2 \left(\frac{\pi/\mathcal{N}}{\sin \pi/\mathcal{N}} \right) . \tag{4.25}$$

This diverges for $\mathcal{N} = 1$, and tends to $\frac{1}{2}Nr_0^2$ for $\mathcal{N} \to \infty$.

The general solution to the Liouville equation, producing a physically acceptable ρ, involves a holomorphic function $f(z)$, $z = r\,e^{i\theta}$, and its complex conjugate

$$\rho(\mathbf{r}) = \frac{4\kappa\,|f'(z)|^2}{\left(1 + |f(z)|^2\right)^2}.$$ (4.26)

The previous solution, with $\mathcal{N}$ solitons superimposed at the origin, corresponds to $f(z) = c_0 z^{-\mathcal{N}}$. Separated solitons are obtained by taking[3]

$$f(z) = \sum_{n=1}^{\mathcal{N}} \frac{c_n}{z - z_n} \qquad z_n = x_n + iy_n.$$ (4.27)

This describes $\mathcal{N}$ solitons at $2\mathcal{N}$ real positions z_n and $2\mathcal{N}$ real scales and phases coded in the complex numbers c_n. Thus the solution depends on $4\mathcal{N}$ real parameters. That this is the most general solution is confirmed by index theory.[13]

It is easy to evaluate the flux carried by the general solution (4.26), (4.27); it is given by asymptotic data and coincides with (4.22)

$$\Phi = -2\pi(2\mathcal{N}).$$ (4.28)

Angular momentum as well as number follow as before

$$J = N = 2\pi\kappa(2\mathcal{N}).$$ (4.29)

The boost generator is given by the center-of-mass, which is now non-vanishing

$$\mathbf{B} = -4\pi\kappa \sum_{n=1}^{\mathcal{N}} \mathbf{r}_n.$$ (4.30)

The conformal generator cannot be in general evaluated explicitly. But it should be remarked that since $f(z) \xrightarrow[z\to\infty]{} \sum_{\kappa=1}^{\mathcal{N}} c_n/z$, the large distance behavior of the generic $\mathcal{N}$-soliton solution is the same as that of a single soliton, unless $\sum_{\kappa=1}^{\mathcal{N}} c_n = 0$. Consequently, in the generic case K diverges.

The flux in our static solution is quantized by even integers $(2\mathcal{N})$, as in (4.22) and (4.28) A solution involving an arbitrary integer is available, but it describes a soliton/vortex condensate. To achieve this one takes $f(z)$ in (4.26) to be doubly periodic, producing a periodic charge density. Flux must now be computed over a unit cell in space. With a particular choice for f, ρ possesses a single zero in the unit cell, leading to a single unit of flux.[14]

Finally, let us remark that it has been occasionally stated that the Chern-Simons interaction has no dynamical effect save modifying spin and statistics in the quantum theory. This assertion appears to be based on the observation that the Green's function $\mathbf{G}$ occurring in (2.13) and given by (2.14) can be presented as a gradient of the azimuthal angle $\theta(\mathbf{r}) \equiv \tan^{-1} y/x$,

$$\mathbf{G}(\mathbf{r}) = -\frac{1}{2\pi} \boldsymbol{\nabla}\theta \tag{4.31}$$

so that

$$\mathbf{A}(t, \mathbf{r}) = -\frac{1}{2\pi\kappa} \int d^2\mathbf{r}' \boldsymbol{\nabla}\theta(\mathbf{r} - \mathbf{r}')\, \rho(t, \mathbf{r}') \,. \tag{4.32}$$

If one could interchange the $\mathbf{r}'$-integral with the $\mathbf{r}$-gradient, one would conclude that $\mathbf{A}$ is a [multi-valued] pure gauge, removable by a singular gauge transformation, which only affects phases of quantum mechanical wave functions, and has no other dynamical role. However, the integral *cannot* be exchanged with the derivative, because the integrand is multi-valued; $\mathbf{A}$ is *not* a pure gauge.[3] That this is indeed true is put into evidence by the existence of non-trivial soliton solutions: if the gauge field interaction is removable, the only dynamics would be provided by the scalar field self-interaction, but it is known that this cannot bind static solitons on the plane. [Consistency between (2.14b) and (4.31) is achieved by virtue of the unexpected formula $\boldsymbol{\nabla} \cdot \boldsymbol{\nabla} \times \theta = 2\pi\delta^2(\mathbf{r})$.]

V. TIME-DEPENDENT SOLUTIONS

A complete analysis of the time-dependent, gauged non-linear Schrödinger equation on the plane has not been achieved thus far, and it is not known whether the time-dependent system is as completely integrable in (2+1)-dimensional space-time, as is the static system on the two-dimensional plane. Our model is tantalizingly similar to the Davey–Stewartson equation, which is known to be integrable in (2+1) dimensions, for a range of its parameters.[15] But the similarity has not been elevated to an identity.

Some time-dependent solutions have been found. All of them are related to our previously described static solutions, and they arise both in the absence and in the presence of external [harmonic, magnetic] interactions. We describe these in turn.

A. Solutions in the Absence of External Fields

In the absence of external fields, the time-dependent solutions that have been constructed thus far are symmetry transforms of the static solutions. Both a Galilean boost [with velocity **v**] and a conformal boost [with parameter a] will introduce time-dependence in the static soliton/vortex profile, while preserving the validity of the equation of motion. The transformation can also be a combination of the two boosts [K and **B** commute in the Lie algebra]. Therefore the boosted function [described by subscript b] is[16]

$$\psi_b(t, \mathbf{r}) = \frac{1}{1 - at} e^{iv^2/2a} e^{-i\left(a/2(1-at)\right)(\mathbf{r} - \mathbf{v}/a)^2} \psi(\mathbf{R}), \qquad (5.1)$$

$$\mathbf{R} = \frac{\mathbf{r} - \mathbf{v}t}{1 - at} \qquad (5.2)$$

where $\psi(\mathbf{r})$ is one of our static solutions. The transformation laws for the potentials and current are as in (3.2) and (3.3), with $X^\mu = \left(\frac{t}{1-at}, \mathbf{R}\right)$.

The constants of motion on these boosted generators differ from their static forms [save for K and **B**, which remain unchanged since they commute]. The momentum and energy no longer vanish; rather one finds boosted expressions

$$\mathbf{P}_b = N\mathbf{v}_a, \qquad (5.3)$$

$$E_b = \frac{1}{2}Nv_a^2 + \frac{1}{2}Na^2\langle\triangle\mathbf{r}\rangle^2. \qquad (5.4)$$

The angular momentum is also boosted

$$J_b = \langle\mathbf{r}\rangle \times N\mathbf{v_a} + N. \qquad (5.5)$$

Here $\mathbf{v}_a \equiv \mathbf{v} - a\langle\mathbf{r}\rangle$, $(\triangle\mathbf{r})^2 \equiv \langle r^2\rangle - \langle\mathbf{r}\rangle^2$. With a pure Galileo boost ($a = 0$), we see that the soliton moves as a particle with mass $M = N$. [Note that the mass m in the Schrödinger equation has been scaled to unity.] A conformal boost ($a \neq 0$) shifts the velocity $\mathbf{v} \to \mathbf{v}_a$ and shows that there is internal structure, since $(\triangle\mathbf{r})^2 \neq 0$. Finally, the dilation generator reads

$$D_b = -\frac{1}{2}\langle\mathbf{r}\rangle \cdot \mathbf{P}_b + \frac{1}{2}Na(\triangle\mathbf{r})^2. \qquad (5.6)$$

Specific forms for the static solution are needed to evaluate $\langle\mathbf{r}\rangle$ and $\langle r^2\rangle$, defined by (4.13) and (4.14). For the rotationally symmetric profile, $\langle\mathbf{r}\rangle$ vanishes while $\langle r^2\rangle$ converges only when $\mathcal{N} > 1$; see (4.25). For the generic $\mathcal{N}$-vortex solution $\langle r^2 \rangle$ diverges, unless the previously mentioned sum rule holds. There is no physical answer to the question why infinite energy is required to boost conformally the generic vortex.

B. Solutions in the Presence of External Fields

Transforming according to (3.5) the solutions in the absence of external fields, both the static solutions and the above-described, time-dependent boosted ones, produces time-dependent solutions[10] in the presence of (i) a harmonic force[17] with frequency ω and/or (ii) a constant magnetic field[10] B. We describe these new solutions, but restrict the discussion to transforms of static solutions; transforms of boosted solutions are discussed in the literature,[10,17] but do not appear to provide new information.

In the presence of an external harmonic force with frequency ω,

$$\psi_\omega(t, \mathbf{r}) = \frac{1}{\cos \omega t} e^{-i\frac{1}{2}\omega r^2 \tan \omega t} \psi\left(\frac{\mathbf{r}}{\cos \omega t}\right) \tag{5.7}$$

is a periodic solution to (3.35), with period $2\pi/\omega$, where $\psi(\mathbf{r})$ is a static, zero-energy solution at $\omega = 0$, see (3.34). The energy of this solution, from (3.38), is

$$E_\omega = \frac{1}{2} N \omega^2 \langle r^2 \rangle \tag{5.8}$$

and may diverge, as explained before. According to (3.39) the angular momentum does not see the harmonic force

$$J_\omega = N. \tag{5.9}$$

The momentum which is *not* a constant of motion, is found to be

$$P_\omega(t) = \int d^2 \mathbf{r} \, \mathbf{j}_\omega(t, \mathbf{r}) = -N \langle \mathbf{r} \rangle \, \omega \sin \omega t. \tag{5.10}$$

Because (5.7) is periodic in time it may be quantized by the semi-classical [Bohr-Sommerfeld] method. To this end, we integrate the canonical one-form $\int d^2 \mathbf{r} \, i \psi_\omega^* \partial_t \psi_\omega$ over the periods $2\pi/\omega$, and equate this to $2\pi n$, where n is a "principal quantum number." Since $\int d^2 \mathbf{r} \, i \psi_\omega^* \partial_t \psi_\omega = L_\omega + H_\omega$ where L_ω and H_ω are the Lagrangian and Hamiltonian [see (2.19) and (3.36)], an energy quantization condition follows

$$E_\omega = \omega \left(n - \frac{1}{2\pi} \int_0^{2\pi/\omega} dt \, L_\omega \right). \tag{5.11}$$

When the Lagrangian is evaluated on a solution to (3.35) only the non-quadratic terms survive

$$L_\omega = \int d^2\mathbf{r} \left(A_\omega^0 - \frac{1}{2\kappa}\rho_\omega \right) \rho_\omega . \tag{5.12}$$

Substituting the solution (5.7) into (5.12) shows that L_ω vanishes on the solution. Thus the semi-classical quantization rule (5.11) becomes[18]

$$E_\omega = \omega n \tag{5.13}$$

which of course coincides with that for a point-particle in a harmonic potential on a plane.

A similar analysis can be performed for the problem with external magnetic field $\mathcal{B}$.[10,18] The solution to (3.41), which is transformed from the static $\mathcal{B} = 0$ solution, is given by (3.42) as

$$\psi_\mathcal{B}(t,\mathbf{r}) = \frac{1}{\cos\frac{\mathcal{B}}{2}t} e^{-i\frac{\mathcal{B}}{4}r^2 \tan\frac{\mathcal{B}}{2}t} e^{i(N/4\pi\kappa)\mathcal{B}t} \psi(\mathbf{R})$$

$$R^i = r^i - \tan\left(\frac{\mathcal{B}}{2}t\right)\varepsilon^{ij}r^j . \tag{5.14}$$

The kinematical characteristics of this solution are coded in the constants of motion. The energy, from (3.47), is

$$E_\mathcal{B} = \frac{1}{8}N\mathcal{B}^2\langle r^2 \rangle - \frac{\mathcal{B}}{2}N \tag{5.15}$$

and this may diverge. The angular momentum of (3.49) again does not see the external field

$$J_\mathcal{B} = N . \tag{5.16}$$

For the momentum one gets from (3.48)

$$P_\mathcal{B}^i = -\frac{1}{2}N\mathcal{B}\varepsilon^{ij}\langle r^j \rangle . \tag{5.17}$$

Note that

$$E_\mathcal{B} + \mathcal{B}J_\mathcal{B} = \frac{P_\mathcal{B}^2}{2N} + \frac{\mathcal{B}}{2}N + \frac{N}{8}\mathcal{B}^2(\triangle\mathbf{r})^2 . \tag{5.18}$$

This shows that apart from the internal energy, proportional to $(\triangle\mathbf{r})^2$, the kinematics of our vortex-soliton is that of a point particle, with mass $M = N$, spin N and unit g-factor, moving in the external magnetic field $\mathcal{B}$.

The semi-classical quantization condition may be imposed on $\psi_{\mathcal{B}}$.[18,19] Note that $\tan \frac{\mathcal{B}}{2}t$ has period $2\pi/\mathcal{B}$. The Jacobian factor in (5.14), *viz.* $1/\cos \frac{\mathcal{B}}{2}t$, really enters with absolute value, hence the period of $\psi_{\mathcal{B}}(t, \mathbf{r})$ is $2\pi/\mathcal{B}$, and the quantization condition, analogous to (5.11) is

$$E_{\mathcal{B}} = \mathcal{B}\left(n - \frac{1}{2\pi} \int_0^{2\pi/\mathcal{B}} dt\, L_{\mathcal{B}}\right), \qquad (5.19)$$

where the Lagrangian, evaluated on the solution reads

$$L_{\mathcal{B}} = \int d^2\mathbf{r} \left(A_{\mathcal{B}}^0 - \frac{1}{2\kappa}\rho_{\mathcal{B}}\right) \rho_{\mathcal{B}} \,. \qquad (5.20\text{a})$$

Evaluating $L_{\mathcal{B}}$ with the explicit solution (5.4) leaves,

$$L_{\mathcal{B}} = -\frac{\mathcal{B}N^2}{8\pi\kappa} \,. \qquad (5.20\text{b})$$

Therefore the quantization condition becomes

$$E_{\mathcal{B}} = \mathcal{B}\left(n + \frac{N^2}{8\pi\kappa}\right). \qquad (5.21)$$

Note that the divergence in the classical harmonic oscillator and magnetic energy, evaluated on solutions with slow large-distance fall off, for which have no physical interpretation, does not appear to affect the semi-classical quantization results (5.13) and (5.21).

Finally, let us remark that static solutions in the presence of a magnetic field have also been considered[10] — only numerical results are available.

VI. NON-ABELIAN GENERALIZATION

In a non-Abelian generalization of our gauged, planar, non-linear Schrödinger equation, the matter fields comprise a multiplet ψ_n, which transforms under non-Abelian local gauge transformations according to some definite representation U of the non-Abelian gauge group

$$\psi \to U^{-1}\psi \,. \qquad (6.1)$$

The gauge-covariant derivative D_i

$$D_\mu \psi = \partial_\mu \psi + A_\mu \psi \qquad (6.2)$$

acts also by multiplication with the vector potential, which is now an anti-Hermitian matrix in the group's Lie algebra, in the representation of the matter fields, and can be written components as

$$A_\mu = A_\mu^a T^a \,, \tag{6.3}$$

$$[T^a, T^b] = f_{abc} T^c \,, \tag{6.4a}$$

$$(T^a)^\dagger = -T^a \,. \tag{6.4b}$$

Thus the field equation with which we are concerned is

$$i\partial_t \psi = -\frac{1}{2} \mathbf{D}^2 \psi - iA_0 \psi + \frac{i}{\kappa} \rho\psi \,. \tag{6.5}$$

The last term in (6.5) involves the matter density; in component notation this reads

$$\rho^a = -i\psi^\dagger T^a \psi \tag{6.6a}$$

while its anti-Hermitian matrix version is

$$\rho = T^a \rho^a = -iT^a (\psi^\dagger T^a \psi) \,. \tag{6.6b}$$

The transformation law for the potentials A_μ, which accompanies (6.1) and leaves the equation of motion (6.5) unchanged, is of course a gauge transformation

$$A_\mu \to U^{-1} A_\mu U + U^{-1} \partial_\mu U \,. \tag{6.7}$$

The gauge-potentials present in (6.5) are determined by the matter variables through the non-Abelian Chern-Simons equation, involving the coupling constant κ:

$$\partial_x A_y - \partial_y A_x + [A_x, A_y] \equiv -B = \frac{1}{\kappa}\rho \,, \tag{6.8}$$

$$\partial_i A_0 - \partial_t A_i + [A_i, A_0] \equiv -E^i = -\frac{1}{\kappa}\epsilon^{ij} j^j \,. \tag{6.9}$$

Here j^i is the spatial current density,

$$j^i = -\frac{1}{2} T^a \big[\psi^\dagger T^a D_i \psi - (D_i \psi)^\dagger T^a \psi \big] \tag{6.10}$$

which together with the matter density (6.6) satisfies a covariant continuity equation, as a consequence of (6.5):

$$\partial_t \rho^a + f_{abc} A_0^b \rho^c + \partial_i j^{ia} + f_{abc} A_i^b j^{ic} = 0 \,. \tag{6.11}$$

Unlike the Abelian case, we cannot solve explicitly for the potentials in terms of the matter fields. Thus we must remain with the coupled set of equations (6.5), (6.8) and (6.9).

Since the gauge fields and currents belong to the adjoint representation, Eqs. (6.8) and (6.9) take the same form in any representation, not only in the one of the matter variables.

The strength of the non-linearity in (6.5) is correlated with the Chern-Simons coupling, as in the Abelian case, and as a consequence the static version of Eqs. (6.5), (6.8) and (6.9) is solved by ψ satisfying the self-dual equation,

$$(D_x - iD_y)\psi = 0, \tag{6.12a}$$

$$\mathbf{D}\psi = i\,\mathbf{D} \times \psi \tag{6.12b}$$

and the vector potential $\mathbf{A}$ in $\mathbf{D}$ obeying (6.8).

To prove this assertion, observe that (6.8) and (6.12) imply

$$\mathbf{D}^2\psi = i\,\mathbf{D}\cdot\mathbf{D} \times \psi = -iB\psi = \frac{i}{\kappa}\rho\psi. \tag{6.13}$$

Hence in the absence of time dependence, (6.5) becomes, with the help of the above,

$$0 = i\left(A_0 - \frac{1}{2\kappa}\rho\right)\psi$$

and is solved when

$$A_0 = \frac{1}{2\kappa}\rho. \tag{6.14}$$

It remains to establish that (6.9) also holds; for static fields it reads

$$\partial_i A_0 + [A_i, A_0] = -\frac{1}{\kappa}\epsilon^{ij}j^j.$$

But from (6.10) and (6.12b) it follows that

$$\epsilon^{ij}j^j = -\frac{1}{2}\left(\partial_i\rho + [A_i, \rho]\right) \tag{6.15}$$

which shows that (6.14) is indeed the correct form of the scalar potential.

Equation (6.5) can be derived from a Hamiltonian

$$H = \frac{1}{2}\int d^2\mathbf{r}\left((\mathbf{D}\psi)^\dagger \cdot (\mathbf{D}_i\psi) - \frac{1}{\kappa}\rho^a\,\rho^a\right). \tag{6.16}$$

H also is the conserved generator of time translations; moreover, the further transformations that are symmetries of the Abelian theory continue to be symmetries in the non-Abelian generalization. Therefore, as before, we conclude from the conformal symmetry that static solutions carry zero energy. Since it is true that

$$|\mathbf{D}\psi|^2 = |(D_x - iD_y)\psi|^2 + i\psi^\dagger B\psi + \boldsymbol{\nabla} \times \mathbf{V}$$

$$\mathbf{V} \equiv \frac{i}{2}\left[\psi^\dagger \mathbf{D}\psi - (\mathbf{D}\psi)^\dagger\psi\right] \qquad (6.17)$$

for sufficiently well-behaved fields such that the integral of $\mathbf{V}$ over a circle at infinity vanishes, the Hamiltonian (6.16) may be rewritten with the help of (6.6) and (6.8) as

$$H = \frac{1}{2}\int d^2\,\mathbf{r}\,|(D_x - iD_y)\psi|^2\,. \qquad (6.18)$$

This is non-negative and achieves its minimum, zero, when the self-dual equation (6.12) is satisfied.

Analogous to the Abelian case, the self-dual equation, which is a consequence of the special choice for the strength of the non-linearity, can be given a spinorial formulation.

Finally, we recall that in the quantized non-Abelian theory, gauge invariance enforces a quantization condition: $4\pi\kappa$ must be an integer.[2] But no such demand need be placed on the classical theory.

We saw that the $U(1)$ case can be analyzed and solved in complete generality once one has achieved the self-dual formulation. However, when the gauge group is non-Abelian we must make further specifications and *Ansätze* before constructing explicit solutions. For one thing, the representation of the gauge group that governs matter field transformations has to be fixed.

A. Adjoint Representation

In the adjoint representation the analysis can be presented most elegantly and carried the farthest.[20,21] Here the matter fields transform according to the same representation as the gauge fields; Lie-algebra representation matrices are constructed from the structure constants

$$(T^a)_{mn} = f_{man} \qquad (6.19)$$

and the number of components in the matter-field multiplet ψ_n coincides with the number of generators and gauge potentials A^a_μ. Then the equations take a compact form.

We define a matter field matrix Ψ by contracting the multiplet ψ with matrices representing the Lie algebra. Although any representation may be used, the most convenient of course is the defining one, where we denote the representation matrices by $T^a = T^a [T^a$ is $(2i)^{-1}$ times the Pauli matrices or the Gell-Mann matrices for $SU(2)$ and $SU(3)$, respectively.]

$$\Psi_{mn} = \psi_a (T^a)_{mn} \, . \tag{6.20}$$

It follows from (6.2), (6.4a), (6.19) and (6.20) that the covariant derivative acts by commutation. Moreover, upon defining holomorphic and anti-holomorphic derivatives as well as components by $\partial_\pm \equiv \partial_x \pm i\partial_y$, $A_\pm = -A_\mp^\dagger \equiv A_x \pm iA_y$, the self-dual equation (6.12) may be presented as

$$\partial_- \Psi + [A_-, \Psi] = 0 \, , \tag{6.21a}$$

$$\partial_+ \Psi^\dagger + [A_+, \Psi^\dagger] = 0 \, , \tag{6.21b}$$

where now the vector potentials also are matrices in the defining representation T^a.

The matter density reads

$$\rho = -iT^a(\psi_m^\dagger f_{man}\psi_n) = i\psi_m^\dagger [T^m, T^n]\,\psi_n = i\,[\Psi, \Psi^\dagger] \, . \tag{6.22}$$

The Chern-Simons equation (6.8) is presented in the defining representation

$$\partial_- A_+ - \partial_+ A_- + [A_-, A_+] = \frac{2}{\kappa}[\Psi^\dagger, \Psi] \, . \tag{6.23}$$

It is equations (6.21) and (6.23) that we wish to solve.

There now follows a remarkable fact that makes further development possible: Eqs. (6.21) and (6.23) combine into a zero-curvature condition for the connection/potential[21]

$$\mathcal{A}_+ = A_+ - \left(\frac{2}{\kappa}\right)^{1/2}\Psi \, , \tag{6.24a}$$

$$\mathcal{A}_- = A_- + \left(\frac{2}{\kappa}\right)^{1/2}\Psi^\dagger \, . \tag{6.24b}$$

While (6.21) and (6.23) imply that $\mathcal{A}_\pm$ carries zero curvature, the converse is not true: even if the zero curvature condition on $\mathcal{A}_\pm$ is satisfied, one must still solve (6.21). This is accomplished as follows. Since $\mathcal{A}$ has no curvature, we may set

$$\mathcal{A}_\pm = g^{-1}\partial_\pm g\,, \tag{6.25}$$

where g is an element of the group. Then (6.24) implies

$$A_+ = g^{-1}\partial_+ g + \sqrt{\frac{2}{\kappa}}\,\Psi\,, \tag{6.26a}$$

$$A_- = g^{-1}\partial_- g - \sqrt{\frac{2}{\kappa}}\,\Psi^\dagger\,. \tag{6.26b}$$

Substituting this into (6.21) gives an equation for Ψ which is solved by

$$\Psi = \sqrt{\frac{\kappa}{2}}\,g^{-1}\chi g \tag{6.27}$$

provided χ obeys

$$\partial_-\chi = [\chi^\dagger,\chi]\,, \tag{6.28a}$$

$$\partial_+\chi^\dagger = [\chi^\dagger,\chi]\,. \tag{6.28b}$$

In other words, solving our problem, defined by (6.21) and (6.23), involves choosing a gauge. There is a special choice such that the solution is

$$\overset{\circ}{\Psi} = \sqrt{\frac{\kappa}{2}}\,\chi\,, \tag{6.29}$$

$$\overset{\circ}{A}_+ = \chi\,, \tag{6.30a}$$

$$\overset{\circ}{A}_- = -\chi^\dagger \tag{6.30b}$$

with χ satisfying (6.28). The general solution is then a gauge transform of (6.29) and (6.30), *viz.* (6.26) and (6.27).

Thus it suffices to analyze (6.28), which also implies a zero-curvature condition. Define the current components

$$J_+ = 2\chi\,, \tag{6.31a}$$

$$J_- = -J_+^\dagger = -2\chi^\dagger\,. \tag{6.31b}$$

It then follows from (6.28) that J, viewed as a connection, carries no curvature

$$\partial_+ J_- - \partial_- J_+ + [J_+, J_-] = 0\,. \tag{6.32}$$

However, satisfying (6.32) by setting

$$J_\pm = h^{-1}\partial_\pm h \,, \tag{6.33}$$

where h is a group element, does not yet solve (6.28), which requires a further condition on h. Substituting (6.31) and (6.33) into (6.28) gives

$$\nabla^2 h = \frac{1}{2}\partial_+ h h^{-1}\partial_- h + \frac{1}{2}\partial_- h h^{-1}\partial_+ h \,. \tag{6.34}$$

This is recognized as the conservation equation for the current $J_i = h^{-1}\partial_i h$,

$$\partial_i J^i = 0 \tag{6.35}$$

and shows that (6.34) or (6.35) are the equations for the two-dimensional principal chiral model, governed by an Euclidean action

$$I = -\frac{1}{2}\int d^2\mathbf{r}\,\mathrm{tr}\,J_i J^i = \frac{1}{2}\int d^2\mathbf{r}\,\mathrm{tr}\,\partial_i h \partial^i h^{-1} \,. \tag{6.36}$$

Let us record one more zero-curvature condition, that *does* encapsulate all our equations. The current J is divergence-free, (6.35), and curl-free in the non-Abelian sense, (6.32). These two requirements are combined in the statement that the connection

$$\alpha_\pm(\lambda) = \frac{\lambda}{\lambda \mp i} J_\pm \tag{6.37}$$

carries no curvature for arbitrary λ.

A final reformulation is the following. Define

$$\chi = \partial_+\omega \,, \tag{6.38a}$$
$$\chi^\dagger = \partial_-\omega^\dagger \,. \tag{6.38b}$$

Equation (6.38) shows that $\omega^\dagger$ may differ from ω by a harmonic function: $\omega = \omega_R + i\omega_I$, $\omega_{R,I}^\dagger = \omega_{R,I}$, $\nabla^2\omega_I = 0$. We represent ω_I by $\eta + \eta^\dagger$ where η is a function only of $x + iy$, and then recognize that $\pm i\partial_\pm\omega_I = \pm i\partial_\pm(\eta + \eta^\dagger) = \partial_\pm(i\eta - i\eta^\dagger)$. Since $i\eta - i\eta^\dagger$ is Hermitian, we conclude that in fact ω in (6.38) can be chosen Hermitian, or

$$\omega = i\omega^a T^a \,, \tag{6.39}$$

where ω^a is real. [According to (6.31) this is equivalent to presenting the flat connection J in a transverse gauge.] Thus (6.28) may be combined into one equation for the Hermitian matrix ω,

$$\nabla^2 \omega = [\partial_- \omega, \partial_+ \omega] \qquad (6.40a)$$

or alternatively for the real quantity ω^a,

$$\nabla^2 \omega^a = i f_{abc} \partial_- \omega^b \partial_+ \omega^c = -f_{abc} \varepsilon^{ij} \partial_i \omega^b \partial_j \omega^c . \qquad (6.40b)$$

All these reformulations of our self-dual and Chern-Simons equations reflect the fact that they also arise as a two-dimensional reduction of the self-dual Yang-Mills equation in four-dimensional space. With four-vector potentials W_μ and field strengths $F_{\mu\nu} = \partial_\mu W_\nu - \partial_\nu W_\mu + [W_\mu, W_\nu]$, the self-dual equation is

$$F^{\mu\nu} = \frac{1}{2} \varepsilon^{\mu\nu\alpha\beta} F_{\alpha\beta} . \qquad (6.41)$$

With the *Ansätze* $\partial_3 W_\mu = 0 = \partial_4 W_\mu$, the identification

$$x = x^2 , \quad y = x^1 ; \quad A_x = W_2 , \quad A_y = W_1 ; \quad \Psi = \sqrt{\frac{\kappa}{2}} (W_3 - iW_4) \quad (6.42)$$

shows that (6.41) is equivalent to (6.21) and (6.23).[20] It is of course well-known that dimensionally reducing the self-dual Yang-Mills system gives rise to many interesting two-dimensional equations.[22]

While the above reformulations of the general problem lead to mathematically interesting equations, explicit solutions are more easily found by returning to the starting point: the self-dual (6.21) and the Chern-Simons (6.23) equations. In order to exhibit solutions we employ some standard Lie group notation, which we first review, so that our presentation is self-contained.

The group algebra generators are given in the Cartan-Weyl basis, with the commuting set, comprising the Cartan sub-algebra, denoted by $H^i = (H^i)^\dagger$, and the ladder generators denoted by $E^{\pm n} = (E^{\mp n})^\dagger$. The index i ranges over the rank r of the group while n ranges up to s such that $2s + r = d$, the dimension of the group. The non-trivial commutators are

$$[H^i, E^{\pm n}] = \pm v_n^i E^{\pm n} , \qquad (6.43)$$

$$[E^n, E^{-n}] = \sum_{i=1}^{r} v_n^i H^i , \qquad (6.44)$$

$$[E^n, E^{n'}] = N(n, n') E^{n+n'} \qquad n \neq -n' . \qquad (6.45)$$

The $v_n^i = -v_{-n}^i$ comprise s real "root vectors" [distinguished by $n = 1, \ldots, s$], with r components [labeled by $i = 1, \ldots, r$]. Of course only r of the s vectors are linearly independent and may be used as a basis for the r-dimensional vector space. The remaining $s - r$ root vectors are then linear combinations. $N(n, n')$ in (6.45) is non-zero if and only if $v_n^i + v_{n'}^i$ is a root vector. [To avoid possible confusion later, we shall *not* use a repeated index summation convention for the Lie algebra indices — all such summations are explicitly indicated.]

Specifically considering $SU(N)$, $d = N^2 - 1$, $r = N - 1$, $s = N(N-1)/2$. For $SU(2)$ there is only one member of the Cartan sub-algebra, conventionally taken as iT^3, while the ladder generators are $E^{\pm 1} = \frac{i}{\sqrt{2}}(T^1 \pm iT^2)$, with the single, one-component root "vector" $v = 1$. For $SU(3)$, the two commuting generators are iT^3 and iT^8; the ladder generators are $E^{\pm 1} = \frac{i}{\sqrt{2}}(T^1 \pm iT^2)$, $E^{\pm 2} = \frac{i}{\sqrt{2}}(T^6 \pm iT^7)$, $E^{\pm 3} = \frac{i}{\sqrt{2}}(T^4 \pm iT^5)$ with root vectors, respectively, $v_1^i = (1, 0)$, $v_2^i = (-1/2, \sqrt{3}/2)$, $v_3^i = (1/2, \sqrt{3}/2)$; evidently, $v_1^i + v_2^i = v_3^i$.

It is always possible to select an r-member subset of the ladder operators that satisfy

$$[E^n, E^{-n'}] = \delta_{nn'} \sum_{i=1}^{r} v_n^i H^i \,.$$

Upon introducing a new symbol, e^α, $\alpha = 1, \ldots, r$, to distinguish members of this set from the remaining $2s - r = d - 2r$ ladder operators and to allow a numerical proportionality factor c_α to be present [$e^\alpha = c_\alpha E^\alpha$], we have

$$[e^\alpha, e^{-\alpha'}] = \delta_{\alpha\alpha'} h^\alpha \,, \tag{6.46}$$

$$[h^\alpha, e^\beta] = K_{\beta\alpha} e^\beta \,, \tag{6.47}$$

where h^α is the sum of the H^i weighted by the components of the root vector associated with e^α

$$h^\alpha = |c_\alpha|^2 \sum_{i=1}^{r} v_\alpha^i H^i \,. \tag{6.48}$$

The matrix K in (6.47), determined from (6.43) and (6.48) to be

$$K_{\alpha\beta} = |c_\beta|^2 \sum_{i=1}^{r} v_\alpha^i v_\beta^i \tag{6.49}$$

is non-singular and is called the Cartan matrix. The proportionality factors c_α can be chosen so that K has integer entries. For $SU(N)$ all the c_α are the same and the Cartan matrix is symmetric. [These normalizations and conventions correspond to the Chevalley basis and the set $\{v_\alpha\}$ is just the set of simple roots.] Finally, we remark that having selected the set $\{e^\alpha\}$ one may identify a "maximal" ladder generator E^M that commutes with the $\{e^\alpha\}$.

For $SU(2)$ the set $\{e^\alpha\}$ consists of one element and can be chosen as $e^1 = \sqrt{2}E^1$, [or its Hermitian conjugate $\sqrt{2}E^{-1}$]. The Cartan matrix is the single number 2, while the maximal ladder generator is also e^1 [or, respectively, its Hermitian conjugate]. For $SU(3)$ the two members of the set may be chosen as $\{\sqrt{2}E^1, \sqrt{2}E^2\}$ or $\{\sqrt{2}E^1, \sqrt{2}E^{-3}\}$ or $\{\sqrt{2}E^2, \sqrt{2}E^{-3}\}$ [or their Hermitian conjugates]. The Cartan matrix always is $\begin{pmatrix} 2 & -1 \\ -1 & 2 \end{pmatrix}$, while the maximal ladder generator for the three alternatives is, respectively, E^3, E^{-2}, E^{-1} [or their Hermitian conjugates]. For $SU(N)$ the Cartan matrix is $(N-1) \times (N-1)$, with all diagonal entries equal to 2 and -1 entered above and below the diagonal:

$$
K_{\alpha\beta} = \begin{pmatrix}
2 & -1 & 0 & \cdots & 0 & 0 & 0 \\
-1 & 2 & -1 & \cdots & 0 & 0 & 0 \\
0 & -1 & 2 & \cdots & 0 & 0 & 0 \\
\vdots & \vdots & \vdots & & \vdots & \vdots & \vdots \\
0 & 0 & 0 & \cdots & 2 & -1 & 0 \\
0 & 0 & 0 & \cdots & -1 & 2 & -1 \\
0 & 0 & 0 & \cdots & 0 & -1 & 2
\end{pmatrix}.
$$

With these definitions and conventions in hand, we can now discuss solutions to (6.21) and (6.23). In the Abelian case, the complete and general solution was found because there the two equations — self-dual and Chern-Simons — combine into the Liouville equation. This generality does not appear attainable in the non-Abelian case. The best we can do is give various *Ansätze* for Ψ and A, from which familiar equations can be derived.

The most appealing results emerge with[20]

$$
\Psi = \sum_{\alpha=1}^{r} u_\alpha e^\alpha , \tag{6.50}
$$

$$
A_- = \sum_{\alpha=1}^{r} A_\alpha h^\alpha , \tag{6.51a}
$$

$$
A_+ = - \sum_{\alpha=1}^{r} A_\alpha^* h^\alpha . \tag{6.51b}
$$

Note that the matter density $i\,[\Psi, \Psi^\dagger]$ arising from (6.50) lies in the Cartan sub-algebra, as does the magnetic field of (6.51). A gauge covariant quantity, like ρ or B, can always be moved into the Cartan sub-algebra by a gauge transformation. Nevertheless, the above equations are not gauge equivalent to the most general possible structures: while some gauge freedom remains in choosing Ψ and $A_\pm$, (6.50) and (6.51) present a restrictive *Ansatz*.

The self-dual equation (6.21) requires that

$$\partial_- u_\alpha + u_\alpha \sum_{\beta=1}^r K_{\alpha\beta} A_\beta = 0, \tag{6.52}$$

where (6.47) has been used. This determines A_α

$$A_\alpha = -\sum_{\beta=1}^r K_{\alpha\beta}^{-1} \partial_- \ln u_\beta. \tag{6.53}$$

On the other hand, the Chern-Simons equation (6.23) implies, by virtue of (6.46),

$$\partial_- A_\alpha^* + \partial_+ A_\alpha = \frac{2}{\kappa} |u_\alpha|^2. \tag{6.54a}$$

When this is combined with (6.53), it follows that the matter density components $\rho_\alpha \equiv |u_\alpha|^2$ satisfy the Toda equation[20]

$$\nabla^2 \ln \rho_\alpha = -\frac{2}{\kappa} \sum_{\beta=1}^r K_{\alpha\beta} \rho_\beta. \tag{6.54b}$$

For $SU(2)$, this is just the Liouville equation for a single density, $\sigma = 2\rho_1$

$$\nabla^2 \ln \sigma = -\frac{2}{\kappa} \sigma \tag{6.55}$$

with the well-known solution

$$\sigma = \kappa \nabla^2 \ln\left(1 + |\varphi|^2\right), \tag{6.56}$$

where φ is a function of $z = x + iy$. For $SU(3)$, (6.54b) gives two equations,

$$\nabla^2 \ln \rho_1 = -\frac{2}{\kappa}(2\rho_1 - \rho_2)$$
$$\nabla^2 \ln \rho_2 = -\frac{2}{\kappa}(-\rho_1 + 2\rho_2) \tag{6.57}$$

that are solved by

$$\rho_1 = \frac{\kappa}{2}\nabla^2 \ln\left(1 + |\varphi_1|^2 + \frac{1}{4}|\varphi_1\varphi_2 + \Phi|^2\right)$$

$$\rho_2 = \frac{\kappa}{2}\nabla^2 \ln\left(1 + |\varphi_2|^2 + \frac{1}{4}|\varphi_1\varphi_2 - \Phi|^2\right),$$

(6.58a)

where $\varphi_{1,2}$ and Φ depend only on z with

$$\Phi' = \varphi_2\varphi_1' - \varphi_1\varphi_2'.$$

(6.58b)

The form of the solutions (6.56) and (6.58) shows that integrals of the matter density components, $Q_\alpha = \int d^2\mathbf{r}\,\rho_\alpha$, which also determine the flux components, are simply given by contributions from singularities of φ_α in the complex z-plane. By choosing these singularities appropriately one readily achieves finite flux.[21]

Upon recalling that the complete solution of the Abelian $U(1)$ problem also involves the Liouville equation, we recognize that the above $SU(2)$ special solution represents an embedding of an $U(1)$ in $SU(2)$. Similar embedding of $U(1)$ within $SU(N)$ $N > 2$ can be achieved in several ways. For example, one may take the special solution to the Toda equation where all the matter densities are proportional to each other: $\rho_\beta = \varepsilon_\beta\sigma$, and (6.54b) becomes the Liouville equation for σ when the ε_β are suitably chosen so that $\sum_{\beta=1}^{r} K_{\alpha\beta}\varepsilon_\beta = 1$ for all α. [For $SU(3)$, $\rho_1 = \rho_2 = \sigma$; for $SU(N)$, $\rho_\alpha = \frac{\alpha}{2}(N - \alpha)\sigma$.] Alternatively, one may take Ψ proportional to a single ladder operator, and the vector potential proportional to the corresponding element of the Cartan sub-algebra

$$\Psi = uE^n,$$

(6.59)

$$A_- = A\sum_{i=1}^{r} v_n^i H^i,$$

(6.60a)

$$A_+ = -A^*\sum_{i=1}^{r} v_n^i H^i.$$

(6.60b)

It is then easy to derive the Liouville equation for $\sigma = \left(\sum_{i=1}^{r} v_n^i v_n^i\right)|u|^2$.

More germane is the question whether generalizing the *Ansätze* (6.50), (6.51) still produces recognizable and interesting equations. This indeed is the case when the generalization involves the maximal ladder generator E^M. We generalize the matter field *Ansatz* to[21]

$$\Psi = \sum_{\alpha=1}^{r} u_\alpha e^\alpha + u_M E^{-M}$$

(6.61)

while the gauge potential remains in the Cartan sub-algebra

$$A_- = \sum_{\alpha=1}^{r} A_\alpha h^\alpha \,, \tag{6.62a}$$

$$A_+ = -\sum_{\alpha=1}^{r} A_\alpha^* h^\alpha \,. \tag{6.62b}$$

The self-dual equation (6.21) gives the previous (6.52) leading to the result (6.53), and another equation from the E^{-M} component,

$$\partial_- u_M - u_M \sum_{\alpha,i=1}^{r} v_M^i v_\alpha^i |c_\alpha|^2 A_\alpha = 0 \,, \tag{6.63a}$$

where (6.43) and (6.48) are used. By hypothesis the roots v_α^i form a basis, therefore v_M^i is expressible in terms of them, and for $SU(N)$ the expansion coefficients are unity: $v_M^i = \sum_{\alpha=1}^{r} v_\alpha^i$. Therefore with (6.49), (6.63a) becomes

$$\partial_- u_M - u_M \sum_{\alpha,\beta=1}^{r} K_{\beta\alpha} A_\alpha = 0 \tag{6.63b}$$

or as a consequence of (6.53)

$$\partial_- u_M + u_M \sum_{\alpha=1}^{r} \partial_- \ln u_\alpha = 0 \,. \tag{6.63c}$$

The solution involves F^*, an arbitrary function of $z^* = x - iy$, annihilated by ∂_-

$$u_M = F^* \bigg/ \prod_{\alpha=1}^{r} u_\alpha \,. \tag{6.63d}$$

The left-hand side of the Chern-Simons equation (6.23) is of course unmodified, because the gauge potential in (6.62) is the same as in (6.51). However, the right-hand side, *viz.* the matter density, changes

$$\frac{2}{\kappa}[\Psi^\dagger, \Psi] = \frac{2}{\kappa}\left[-\sum_{\alpha=1}^{r} \rho_\alpha \, h^\alpha + |u_M|^2 \sum_{i=1}^{r} v_M^i H^i\right]$$

$$= \frac{2}{\kappa}\sum_{\alpha=1}^{r}\left(-\rho_\alpha + \frac{|u_M|^2}{|c_\alpha|^2}\right) h^\alpha \tag{6.64}$$

$$= \frac{2}{\kappa}\sum_{\alpha=1}^{r}\left(-\rho_\alpha + |F|^2 \bigg/ \prod_{\beta=1}^{r} |\rho_\beta|^2\right) h^\alpha \,.$$

The sequence of equalities follows when v_M^i is expressed as a sum of the v_α^i, h^α is inserted from (6.48), and u_M is taken from (6.63d) with the constant $|c_\alpha|^2$, which is α-independent for $SU(N)$, absorbed in $|F|^2$. The Chern-Simons equation therefore reads

$$\partial_- A_\alpha^* + \partial_+ A_\alpha = -\frac{2}{\kappa}\left(\rho_\alpha - |F|^2 \Big/ \prod_{\beta=1}^r \rho_\beta\right). \tag{6.65a}$$

With A_α from (6.53) we find that the matter density components ρ_α satisfy

$$\nabla^2 \ln \rho_\alpha = -\frac{2}{\kappa}\sum_{\beta=1}^r K_{\alpha\beta}\,\rho_\beta + |F|^2 \sum_{\beta=1}^r K_{\alpha\beta} \Big/ \prod_{\beta=1}^r \rho_\beta. \tag{6.65b}$$

Note that for $SU(N)$, $\sum_{\beta=1}^r K_{\alpha\beta}$ vanishes for $\alpha \neq 1,\ r$, so the last term in (6.65b) contributes only to the equations for ρ_1 and ρ_r.

Equation (6.65) may also be presented[21] in the Toda form (6.54b) by using the $N \times N$ *affine* Cartan matrix $\tilde{K}$. For $SU(N)$ this is a symmetric, singular matrix obtained by adjoining to the $(N-1)\times(N-1)$ Cartan matrix a row and column which insures $\sum_{\beta=1}^N \tilde{K}_{\alpha\beta} = 0$ for all α; viz. the $SU(N)$ affine Cartan matrix is

$$\tilde{K}_{\alpha\beta} = \begin{pmatrix} & & & & & -1 \\ & & & & & 0 \\ & & & & & 0 \\ & & K_{\alpha\beta} & & & \vdots \\ & & & & & 0 \\ & & & & & -1 \\ -1 & 0 & 0 & \cdots & 0 & -1 & 2 \end{pmatrix}.$$

The affine Toda equation possesses N components,

$$\nabla^2 \ln \rho_\alpha = -\frac{2}{\kappa}\sum_{\beta=1}^N \tilde{K}_{\alpha\beta}\,\rho_\beta \tag{6.66}$$

but because $\tilde{K}_{\alpha\beta}$ is singular the equation for ρ_N is not independent of the others. Eliminating ρ_N from (6.66) gives (6.65b).

For $SU(2)$, (6.65b) or (6.66) becomes after suitable rescaling

$$\nabla^2 \ln \sigma = -\frac{2}{\kappa}\sigma + \frac{|F|^2}{\sigma} \tag{6.67a}$$

or

$$\nabla^2 \ln \tilde{\sigma} = -\sqrt{\frac{2}{\kappa}} |F| \left(\tilde{\sigma} - \frac{1}{\tilde{\sigma}} \right), \qquad (6.67\text{b})$$

where $\sigma = \sqrt{\frac{\kappa}{2}} |F| \tilde{\sigma}$. With constant F this is just the "sinh-Gordon" equation. For $SU(3)$, the affine Toda equation gives

$$\nabla^2 \ln \rho_1 = -\frac{2}{\kappa} (2\rho_1 - \rho_2) + \frac{|F|^2}{\rho_1 \rho_2}, \qquad (6.68\text{a})$$

$$\nabla^2 \ln \rho_2 = -\frac{2}{\kappa} (-\rho_1 + 2\rho_2) + \frac{|F|^2}{\rho_1 \rho_2}. \qquad (6.68\text{b})$$

The special case $\rho_1 = \rho_2 = \sigma$ is interesting in that the rescaling $\sigma = \left(\frac{\kappa}{2} |F|^2 \right)^{1/3} \tilde{\sigma}$ results in

$$\nabla^2 \ln \tilde{\sigma} = -\left(\frac{2}{\kappa} |F| \right)^{1/3} \left(\tilde{\sigma} - \frac{1}{\tilde{\sigma}^2} \right) \qquad (6.69)$$

which for constant F is the "Bullough–Dodd" equation.

When the arbitrary analytic function F is set to zero in (6.65b), we regain the Toda equation, with its known explicit solutions that lead to finite flux. However, for non-zero F, even when the equation is known to be integrable, as for example the sinh-Gordon and Bullough–Dodd equations, there do not seem to exist regular solutions with finite $\int d^2 \, \mathbf{r} \, \rho$.

For example, we consider (6.67b) with $F(z) = \sqrt{\frac{8}{\kappa}} (\mu z)^{-2}$. In terms of radial variables with the redefinitions $\mu\tau = \ln r$, $\mu\theta' = \theta$, $\tilde{\sigma} = e^\varphi$, that equation becomes

$$\left(\frac{\partial}{\partial \tau^2} + \frac{\partial^2}{\partial \theta'^2} \right) \varphi = -\sinh \varphi \qquad (6.70\text{a})$$

and radially symmetric [θ-independent] solutions satisfy

$$\frac{1}{2} \left(\frac{d}{d\tau} \varphi \right)^2 + \cosh \varphi = \varepsilon \qquad (6.70\text{b})$$

while the integrated matter density is given by

$$\int d^2 \, \mathbf{r} \, \sigma = \frac{4\pi}{\mu} \int_{-\infty}^{\infty} d\tau \, e^{\varphi(\tau)}. \qquad (6.71)$$

Since solutions of (6.70b) are periodic in τ, this diverges. A similar analysis can be carried out for (6.69).

B. Defining Representation

When matter is in the defining representation, *i.e.*, described by an N-component multiplet for $SU(N)$, there is no longer available the succinct geometrical formulation that we used in the adjoint representation, wherein matter and gauge field representations coincide.

A straightforward approach is to choose a gauge such that the matter density $\psi^\dagger T^a \psi$ is in the Cartan direction. This requires that ψ possess only one non-zero entry; without loss of generality we shall take it to be in the first position. This is the unitary gauge

$$\psi = \begin{pmatrix} u \\ 0 \\ \vdots \\ 0 \end{pmatrix}. \tag{6.72}$$

The definition

$$H^i \psi = \mu^i \psi \tag{6.73}$$

gives

$$\rho = i \sum_{i=1}^{r} \mu^i H^i |u|^2. \tag{6.74}$$

The simplest result emerges when the gauge potentials also involve only the Cartan matrices[21]

$$A_- = \sum_{i=1}^{r} A_i H^i. \tag{6.75}$$

Then the self-dual equation implies

$$\sum_{i=1}^{r} \mu^i A_i = -\partial_- \ln u \tag{6.76}$$

while the Chern-Simons equation requires

$$\partial_- A_i^* + \partial_+ A_i = \frac{2}{\kappa} \mu^i |u|^2. \tag{6.77}$$

Upon setting components of A_i perpendicular to μ^i to zero, and using (6.76), the above again becomes the Liouville equation for $\sigma = \sum_{i=1}^{r} \mu^i \mu^i |u|^2$

$$\nabla^2 \ln \sigma = -\frac{2}{\kappa} \sigma. \tag{6.78}$$

To survey the more intricate possibilities, we consider $SU(2)$ with the matter field multiplet in the unitary gauge,

$$\psi = \begin{pmatrix} u \\ 0 \end{pmatrix} \tag{6.79}$$

and a general form for the vector potential

$$A_- = iAT^3 + a_1 E^1 + a_{-1} E^{-1}. \tag{6.80}$$

The matter density is

$$\rho = -\frac{1}{2}|u|^2 T^3. \tag{6.81}$$

The self-dual equation gives,

$$A = -\partial_- \ln u^2 \tag{6.82}$$

and also require a_{-1} to vanish. The magnetic field has a component along $E^{\pm 1}$, which must vanish since the matter density is along T^3; this gives the equation

$$\partial_+ a_1 - A^* a_1 = 0. \tag{6.83}$$

With (6.82), a_1 is found in terms of an arbitrary analytic function annihilated by ∂_+

$$a_1 = \frac{F(x+iy)}{u^{*2}}. \tag{6.84}$$

The magnetic field along T^3 gives for the Chern-Simons equation,

$$\partial_- A^* + \partial_+ A + |a_1|^2 = \frac{1}{\kappa}|u|^2 \tag{6.85a}$$

from (6.82) and (6.84); we derive[21] for $\sigma = \frac{1}{4}|u|^2$

$$\nabla^2 \ln \sigma = -\frac{2}{\kappa}\sigma + \frac{|F|^2}{\sigma^2}. \tag{6.85b}$$

Thus the same equation arises within this general analysis of the defining $SU(2)$ representation that we previously encountered from a special *Ansatz* in the adjoint $SU(3)$ representation; *viz.* (6.68) with $\rho_1 = \rho_2$, and (6.69).

The above provides the general analysis of the $SU(2)$ problem in the defining representation. For higher groups, we saw that one still encounters the Liouville equation with the restricted *Ansatz* (6.75). Less restrictive *Ansätze* lead to non-linear equations of the type (6.85), but we do not pursue this any further.

VII. RELATIVISTIC ABELIAN MODEL

Another Abelian Chern-Simons theory, closer to the interests of particle physicists, makes use of relativistic dynamics for the matter degrees of freedom and supports self-dual solitons if the field potential is taken in a particular form. We consider the Lagrange density

$$\mathcal{L} = \frac{\kappa}{4}\epsilon^{\alpha\beta\gamma}A_\alpha F_{\beta\gamma} + (D_\mu\varphi)^*(D_\mu\varphi) - V(|\varphi|^2) \tag{7.1}$$

with a potential function[23,24]

$$V(|\varphi|^2) = \frac{1}{\kappa^2}|\varphi|^2(|\varphi|^2 - v^2)^2 \tag{7.2}$$

that exhibits two, zero-energy and degenerate minima: one at

$$|\varphi| = v \tag{7.3}$$

which breaks the $U(1)$ symmetry, and another, symmetry preserving one, at

$$\varphi = 0. \tag{7.4}$$

The form of V is dictated by our desire to arrive at a static, self-dual system. However, there is also a supersymmetry connection: the Lagrangian (7.1), (7.2) is the bosonic partner in an $N = 2$ supersymmetry of a fermionic Lagrangian[25]

$$\mathcal{L}_F = i\bar{\psi}\gamma^\mu D_\mu\psi + \frac{1}{\kappa}(3|\varphi|^2 - v^2)\bar{\psi}\psi. \tag{7.5}$$

The supersymmetry is preserved at both minima (7.3) and (7.4).

At the symmetric minimum, there are two bosonic excitations of mass

$$m = \frac{v^2}{\kappa} \tag{7.6}$$

associated with the charged scalar field; the fermions carry the same mass as is seen from (7.5), while the gauge field does not support any excitations. At the symmetry breaking minimum, the Higgs phenomenon sets in [in spite of the purely Chern-Simons kinetic term], the Higgs particle carries mass

$$m_H = 2\frac{v^2}{\kappa} \tag{7.7}$$

which is also the mass of the single degree of freedom now associated with the gauge field — for a total of two bosonic excitations as before; the fermion mass in the symmetry breaking case also is (7.7). The mass degeneracy in the two cases is of course a consequence of the supersymmetry, which remains unbroken.

Before proceeding with the analysis of the relativistic model, it is interesting to discuss its non-relativistic limit, which we consider in the symmetric sector upon writing φ as

$$\varphi = \frac{1}{\sqrt{2m}}e^{-imt}\psi + \frac{1}{\sqrt{2m}}e^{imt}\tilde{\psi}^* \,. \tag{7.8}$$

ψ becomes in the non-relativistic ($m \to \infty$) limit the annihilation operator for particles and $\tilde{\psi}^*$ the creation operator for anti-particles. Upon substituting (7.8) in (7.2), and dropping all oscillatory and sub-dominant terms at large m, we arrive at a model in which particles and anti-particles are separately conserved, and the particle, zero anti-particle sector coincides with our non-relativistic model [in which m is set to unity], where the non-linearity is precisely $1/\kappa$. Thus the previously considered self-dual $U(1)$ theory of (4.6) is the non-relativistic limit of the relativistic self-dual $U(1)$ theory (7.1), with a symmetric realization of the $U(1)$ gauge symmetry.

Static equations stationarize the static energy functional, which is the negative of the Lagrangian with all time-derivatives suppressed. Moreover, with the help of the Gauss law constraint

$$B = \boldsymbol{\nabla} \times \mathbf{A} = -\frac{1}{\kappa}\rho \,, \tag{7.9}$$

where the relativistic charge density for static fields reads,

$$\rho = i(\varphi^* D_0\varphi - \varphi D_0\varphi^*) = -2A^0|\varphi|^2 \,. \tag{7.10}$$

A^0 may be eliminated

$$A^0 = \frac{B}{2\kappa|\varphi|^2} \,. \tag{7.11}$$

Note that in contrast to Nielsen–Olesen vortices, here the magnetic field must vanish whenever φ does. With (7.11) the static energy functional is manifestly positive and becomes

$$E = \int d^2\mathbf{r}\left(\frac{\kappa^2}{4}\frac{B^2}{|\varphi|^2} + |\mathbf{D}\varphi|^2 + V\left(|\varphi|^2\right)\right) \tag{7.12}$$

which is further rewritten by using the identity

$$|\mathbf{D}\varphi|^2 = |(D_x \pm iD_y)\,\varphi|^2 \pm B|\varphi|^2 \pm \boldsymbol{\nabla} \times \mathbf{j} \qquad (7.13)$$

and dropping the last, total derivative when integrating (7.12) over all space

$$E = \int d^2\mathbf{r}\left\{|(D_x + iD_y)\,\varphi|^2 + \left|\,\frac{\kappa}{|\varphi|}B \pm \frac{1}{\kappa}|\varphi|\left(|\varphi|^2 - v^2\right)\,\right|^2\right\} \pm v^2\Phi. \qquad (7.14)$$

Here Φ is the total flux, which as always equals $-\frac{1}{\kappa}N$, by virtue of the Chern-Simons condition (7.9). Thus we conclude that for a fixed value of flux there is a lower bound on energy

$$E \geq v^2|\Phi|. \qquad (7.15)$$

Since static configurations that are stationary points of the energy are also stationary points of the action, the Euler-Lagrange equations of the theory will be satisfied by static configurations obeying the self-duality equation,

$$(D_x \pm iD_y)\varphi = 0\,, \qquad (7.16a)$$

$$\mathbf{D}\varphi = \mp i\,\mathbf{D} \times \varphi \qquad (7.16b)$$

and the Chern-Simons constraint,

$$eB = \pm\frac{m_H^2}{2}\frac{|\varphi|^2}{v^2}\left(1 - \frac{|\varphi^2|}{v^2}\right), \qquad (7.17)$$

where the upper [lower] sign corresponds to a positive [negative] value of Φ; these solutions achieve the lower bound in Eq. (7.15).

Equations (7.16) and (7.17) possess topologically stable vortex solutions for which $|\varphi| \to v$ at large distances and Φ is quantized.[24,25] But there also exist non-topological soliton solutions for which $\varphi \to 0$ asymptotically.[25,26] For these, the flux is not quantized but rather is an arbitrary parameter describing the solution. This parameter can be continuously varied; therefore it is evident, since the energy $E = v^2|\Phi|$, that these solutions are not stationary points of the energy. This is understood by recalling that the energy is stationary provided that the field variations vanish faster than $1/r$. Such variations are sufficient to establish the Euler–Lagrange equations, so the solutions of the self-duality equations are indeed solutions of the full field equations. The above-mentioned non-stationary variations have $\delta\mathbf{A} \sim 1/r$.

Since the relativistic theory is not conformally invariant, no argument has been constructed which would imply that [regular] static solutions are always self-dual, as in the non-relativistic case.

The energy of a non-topological soliton with a given charge $Q = N$ is

$$E = v^2 |\Phi| = \frac{v^2}{\kappa} |Q| = mQ, \qquad (7.18)$$

where m is the scalar mass (7.6) in the symmetric vacuum. Thus the energy per unit charge is identical to that of the elementary excitations in the symmetric phase. This indicates that the collective, non-topological excitations are just at the threshold of stability against emission of elementary particles. Consequently, stability does not impose an upper bound on the non-topological soliton charges. The topological solitons are of course stable for topological reasons; their flux is quantized.

Equation (7.16) implies for $\varphi = |\varphi| e^{i\omega}$ that

$$A^i = \partial_i \omega \mp \varepsilon^{ij} \partial_j \ln |\varphi|. \qquad (7.19)$$

When substituted into (7.17) this gives, away from zeroes of $|\varphi|$

$$\nabla^2 \ln |\varphi|^2 = -\frac{4v^2}{\kappa^2} |\varphi|^2 \left(1 - \frac{|\varphi|^2}{v^2}\right). \qquad (7.20)$$

Both topological $\left(|\varphi|^2 \xrightarrow[r \to \infty]{} v^2\right)$ and non-topological $\left(|\varphi|^2 \xrightarrow[r \to \infty]{} 0\right)$ solitons satisfy the same equation, which is the generalization to the relativistic context of the Liouville equation, the latter being regained in the non-relativistic limit, where $|\varphi|^2/v^2$ is dropped relative to 1.

Unfortunately (7.20) does not possess closed-form solutions. Only numerical integration has been performed[24] and a mathematical existence proof has been constructed.[27]

The nature of the solutions is as follows. [We discuss the rotationally symmetric case.]

For topological solitons/vortices in the broken symmetry phases, the constant large distance asymptote is approached exponentially and at the origin there is a zero, which can be multiple

$$|\varphi|^2 \xrightarrow[r \to 0]{} r^{2\mathcal{N}-2}, \qquad \mathcal{N} = 2, 3, \ldots \qquad \text{(topological)}. \qquad (7.21)$$

The integrality restriction on $\mathcal{N}$ comes from familiar arguments on the absence of singularities in the vector potential, together with a single valuedness requirement on φ. The flux is quantized

$$\Phi = \int d^2\mathbf{r}\, B = 2\pi(\mathcal{N} - 1)\,. \qquad (7.22)$$

The value of the integral is determined by the strength of the zero at the origin.

The non-topological solitons can possess the same zeroes as in (7.21) at the origin, but now there is also a solution which does not vanish at the origin, *i.e.*,

$$|\varphi|^2 \xrightarrow[r \to 0]{} r^{2\mathcal{N}-2}\,, \qquad \mathcal{N} = 1, 2, \ldots \qquad \text{(non-topological)}\,. \qquad (7.23)$$

[The $\mathcal{N} = 1$ solution is not available in the topological case.] The approach to the vanishing large-distance asymptote is power-law,

$$|\varphi|^2 \xrightarrow[r \to \infty]{} r^{-2(\alpha+1)} \qquad (7.24)$$

and one can prove that

$$\alpha \geq \mathcal{N}\,. \qquad (7.25)$$

The flux is no longer quantized,

$$\Phi = \int d^2\mathbf{r}\, B = 2\pi(\mathcal{N} + \alpha) \qquad (7.26)$$

with [quantized] contributions from the origin combining with [unquantized] contributions from infinity. In the non-relativistic limit, α approaches $\mathcal{N}$ and (7.26) coincides with the non-relativistic result. A useful point of view about the arbitrary-$\mathcal{N}$ solution is that the true non-topological soliton corresponds to $\mathcal{N} = 1$, while higher-$\mathcal{N}$ solutions describe $\mathcal{N} - 1$ vortices threading through the non-topological soliton.

Index theory may be used to count parameters and associate them with physical characteristics of the solitons. A picture consistent with the above description emerges.[24] In particular, the $\mathcal{N}$-vortex/topological soliton solution depends on $2\mathcal{N}$ parameters, identified as the locations of the vortices on the plane.

We record formulas for the angular momentum. From the energy momentum tensor

$$T_{\mu\nu} = (D_\mu\varphi)^* (D_\nu\varphi) + (D_\nu\varphi)(D_\mu\varphi)^* - g_{\mu\nu}\left[|D_\lambda\varphi|^2 - V(|\varphi|^2)\right] \qquad (7.27)$$

we infer that the angular momentum, which is constructed from the momentum density $\mathcal{P}^i = T^{0i}$ as,

$$J = \int d^2 \, \mathbf{r} \, \mathbf{r} \times \mathcal{P} \tag{7.28}$$

becomes after some manipulation

$$J = -\frac{\kappa}{4\pi} \Phi^2 + \kappa \alpha \Phi. \tag{7.29}$$

This formula holds for both the non-topological and topological solitons $[\alpha = -1]$.

One may consider the slow motion of the topological solitons/vortices and derive an effective Lagrangian from the underlying field theory. One finds a statistical interaction term with the value of the statistical factor in agreement with the generalized spin-statistic relation.[28]

VIII. CONCLUSIONS

Our understanding of the gauged, planar, non-linear Schrödinger equation is not as thorough as of the (1+1)-dimensional model: the only classical solutions that are completely cataloged [in the Abelian theory] are the static ones; but the complete integrability of the time-dependent equations remains an open question; in the quantum theory only partial results on the N-anyon problem are available — we have no Hans Bethe guessing the N-body solution as he did for the (1+1)-dimensional model.

The quantum meaning of the classical solution is of course well-known for the relativistic theory: upon quantization, the classical solutions give rise to quantum particle states.[1] In the non-relativistic model, quantization yields a spectrum of soliton states which may be expressed in terms of coherent states in the usual multi-particle Hilbert space.[29]

Quantum and thermal fluctuations have been computed in the relativistic theory, with the fermions included so that supersymmetry holds. The pleasing result is that renormalization does not change the shape of the classical potential; only v^2 in (7.2) acquires an uncalculable correction. The remaining corrections are calculable and small [in the perturbative regime]. Moreover, they do not remove the supersymmetry and both the $U(1)$ symmetric and asymmetric vacua persist. At high temperature only the $U(1)$ symmetry is realized; symmetry breaking configurations are not stable.[30]

Various further generalizations, which however retain self-duality, have been examined: adding the Maxwell term to both the non-relativistic and relativistic models so that the topologically massive equation (2.76) holds,[31] adding non-Abelian group structure to the relativistic theory,[32] using unconventional matter dynamics.[33] Such modifications complicate the dynamics significantly, explicit solution is never possible, but qualitative behavior is similar to what happens in the models that we have described.

Although the static, non-relativistic Abelian equations are completely integrated, a question remains. We have seen that this theory is the non-relativistic limit of the relativistic model in its $U(1)$ symmetric realization. Correspondingly, the Schrödinger theory solitons/vortices are the non-relativistic limits of the non-topological, relativistic solitons — this is also confirmed by the fact that the non-relativistic charge density vanishes at large distance, as it does for relativistic, non-topological solutions. Nevertheless, the non-relativistic solitons/vortices appear to possess topological characteristics, *eg.*, the flux is quantized. So one is left wondering where is the topology in the gauged, planar non-linear Schrödinger equation, and how does it arise as a non-relativistic limit is performed on non-topological excitations.

REFERENCES

1. For reviews see R. Jackiw, *Rev. Mod. Phys.* **49**, 681 (1977); R. Rajaraman, *Solitons and Instantons* (North-Holland, Amsterdam, 1982).
2. R. Jackiw and S. Templeton, *Phys. Rev. D* **23**, 2291 (1981); J. Schonfeld, *Nucl. Phys.* **B185**, 157 (1981); S. Deser, R. Jackiw and S. Templeton, *Phys. Rev. Lett.* **48**, 975 (1982), *Ann. Phys.* (NY) **140**, 372 (1982), (E) **185**, 406 (1988).
3. R. Jackiw and S.-Y. Pi, *Phys. Rev. Lett.* **64**, 2969 (1990), (C) **66**, 2682 (1991), *Phys. Rev. D* **42**, 3500 (1990).
4. C. Hagen, *Ann. Phys.* (NY) **157**, 342 (1984), *Phys. Rev. D* **31**, 848, 2135 (1985).
5. S. Takagi, *Prog. Theor. Phys.* (Kyoto) **84**, 1019 (1990), **85**, 463, 723 (1991), **86**, 783 (1991).
6. This conformal symmetry of non-relativistic dynamics was discussed in the early seventies, when conformal symmetry of relativistic theories was explored in connection with deep inelastic scattering, see "Introducing Scale Symmetry", reprinted in this volume on p. 217; C. Hagen, *Phys. Rev. D* **5**, 377 (1972); U. Niederer, *Helv. Phys. Acta* **45**, 802 (1972), **46**, 191 (1973), **47**, 119, 167 (1974), **51**, 220 (1978). In the context of Chern-Simons dynamics, the symmetry was identified in R. Jackiw, *Ann. Phys.* (NY) **201**, 83 (1990).
7. "Delta Function Potentials in Two- and Three-Dimensional Quantum Mechanics", reprinted in this volume on p. 35.
8. These issues are under further investigation by O. Bergman *Phys. Rev. D* **46**, 5474 (1992). In fact when the statistical (Chern-Simons) interaction is correlated

with the $|\psi^*\psi|^2$ (δ-function) interaction in the self-dual fashion (see below), infinities cancel and one is dealing with a conformally invariant quantum field theory in (2+1)-dimensional space-time; see G. Lozano, *Phys. Lett.* **B283**, 70 (1992); O. Bergman and G. Lozano, *Ann. Phys.* (NY) **229**, 416 (1993); D. Freedman, G. Lozano and N. Rius, *Phys. Rev.* D **49**, 1054 (1994); D. Bak and O. Bergman, *Phys. Rev.* D **51**, 1994 (1995).

9. V. de Alfaro, S. Fubini and G. Furlan, *Nuovo Cim.* **A34**, 569 (1976); R. Jackiw, *Ann. Phys.* (NY) **129**, 183 (1980) and last entry Ref. 6; Takagi, Ref. 5.

10. The idea of generating time-dependent solitons from static solitons by these transformations is due to Z. Ezawa, M. Hotta and A. Iwazaki, *Phys. Rev.* D **44**, 452 (1991).

11. M. Hotta, *Prog. Theor. Phys.* (Kyoto) **86**, 1289 (1991).

12. D. Freedman and A. Newell (unpublished).

13. S. Kim, K. Soh and J. Yee, *Phys. Rev.* D **42**, 4139 (1990).

14. P. Olesen, *Phys. Lett.* **B265**, 361 (1991).

15. A. Davey and K. Stewartson, *Proc. Roy. Soc. London* **A338**, 101 (1974); D. Anker and N. Freeman, *Proc. Roy. Soc. London* **A360**, 529 (1978).

16. R. Jackiw and S.-Y. Pi, *Phys. Rev.* D **44**, 2524 (1991) and Ref. 3.

17. Jackiw and Pi, Ref. 16.

18. R. Jackiw and S.-Y. Pi, *Phys. Rev. Lett.* **67**, 415 (1991) and Ref. 16.

19. Z. Ezawa, M. Hotta and A. Iwazaki, *Phys. Rev. Lett.* **67**, 441 (1991) and Ref. 10.

20. B. Grossman, *Phys. Rev. Lett.* **65**, 3230 (1990).

21. G. Dunne, R. Jackiw, S.-Y. Pi and C. Trugenberger, *Phys. Rev.* D **43**, 1332 (1991).

22. R. Ward, *Philos. Trans. Roy. Soc. London* **A315**, 451 (1985); N. Hitchin, *Proc. Math. Soc. London* **55**, 59 (1987); S. Donaldson, *Proc. Math. Soc. London* **55**, 127 (1987).

23. J. Hong, Y. Kim and P. Y. Pac, *Phys. Rev. Lett.* **64**, 2230 (1990).

24. R. Jackiw and E. Weinberg, *Phys. Rev. Lett.* **64**, 2234 (1990); R. Jackiw, K. Lee and E. Weinberg, *Phys. Rev.* D **42**, 3488 (1990).

25. C. Lee, K. Lee and E. Weinberg, *Phys. Lett.* **B243**, 105 (1990).

26. A. Khare, *Phys. Lett.* **B225**, 393 (1991).

27. R. Wang, *Comm. Math. Phys.* **137**, 587 (1991). Integrability of this equation and its generalizations when the spatial geometry is not flat have been examined by J. Schiff, *J. Math. Phys.* **32**, 753 (1991).

28. S. Kim and H. Min *Phys. Lett.* **B281**, 81 (1992); L. Hua and C. Chou, *Phys. Lett.* **B308**, 286 (1993); Q. Liu, *Phys. Lett.* **B321**; 219 (1994).

29. D. Kabat, *Phys. Lett.* **B281**, 265 (1992).

30. Y. İpekoğlu, M. Leblanc and M. T. Thomaz, *Ann. Phys.* (NY) **214**, 160 (1992).

31. a. Non-relativistic matter dynamics: G. Dunne and C. Trugenberger, *Phys. Rev. D* **43**, 1323 (1991).
    b. Relativistic matter dynamics: C. Lee, K. Lee and H. Min, *Phys. Lett.* **B252**, 79 (1990); C. Lee, H. Min and C. Rim, *Phys. Rev.* **D43**, 4100 (1991); J. Lee and S. Nam, *Phys. Lett.* **B261**, 437 (1991).
32. K. Lee, *Phys. Lett.* **B255**, 381 (1991); *Phys. Rev. Lett.* **66**, 553 (1991).
33. Lee and Nam, Ref. 31; Y. Yang, *Lett. in Math. Phys.* **23**, 179 (1991).

www.ingramcontent.com/pod-product-compliance
Lightning Source LLC
Chambersburg PA
CBHW070740220326
41598CB00026B/3715